Nucleoside Triphosphates *and their* Analogs

Chemistry, Biotechnology, *and* Biological Applications

Nucleoside Triphosphates *and their* Analogs

Chemistry, Biotechnology, *and* Biological Applications

Morteza Vaghefi

CRC Press
Taylor & Francis Group
Boca Raton London New York

CRC Press is an imprint of the
Taylor & Francis Group, an **informa** business
A TAYLOR & FRANCIS BOOK

CRC Press
Taylor & Francis Group
6000 Broken Sound Parkway NW, Suite 300
Boca Raton, FL 33487-2742

First issued in hardback 2019

ISBN-13: 978-1-57444-498-8 (hbk)
Library of Congress Card Number 2004065509

Library of Congress Cataloging-in-Publication Data

Nucleoside triphosphates and their analogs : chemistry, biotechnology, and biological applications / edited by Morteza M. Vaghefi.
p. cm.
Includes bibliographical references and index.
ISBN 1-57444-498-0 (alk. paper)
1. Nucleotides. 2. Nucleotides—Derivatives. 3. Adenosine triphosphate. I. Monir-Vaghefi, Seyed-Morteza. II. Title.

QP625.N89N838 2005
611'.01816—dc22 2004065509

Visit the Taylor & Francis Web site at
http://www.taylorandfrancis.com

and the CRC Press Web site at
http://www.crcpress.com

The Editor

Morteza M. Vaghefi, Ph.D., received his graduate training at the University of Utah and Brigham Young University in Utah, obtaining an M.S. degree (1980) in biochemistry and a Ph.D. (1985) in organic and medicinal chemistry under the guidance of Professor Ronald Robins. Dr. Vaghefi is an inventor of eight patents and has authored over 30 scientific publications. He is currently serving as the reviewer for the *Journal of Medicinal Chemistry and Nucleosides, Nucleotides and Nucleic Acid.*

From 1985 to 1989, Dr. Vaghefi served as head of the Department of Biomedicinal Chemistry at ICN Pharmaceuticals Inc., developing novel nucleoside and nucleotide analogs for various therapeutic indications. In 1989, he accepted a position at Genta Inc. in San Diego, California, serving as the head of oligonucleotides and nucleotide chemistry. In 1996, he cofounded TriLink Biotechnologies Inc. in San Diego.

From 1996 to 2004, Dr. Vaghefi served as cofounder and senior vice president and chief chemical officer of TriLink Biotechnologies Inc., which is involved in the research and synthesis of novel, labeled nucleotide, nucleoside, and oligonucleotides. Dr. Vaghefi currently serves as founder and president of a San Diego-based start-up company, ChemCyte Corporation. His current research interests include, labeled nucleotide triphosphate as a tool for diagnostics; oligonucleotide labeling technology; novel nucleoside drug target discovery; nucleic acids as targets for small molecule intervention; and fluorescent labeling.

Contributors

Wissam Abdel–Gawad
The Biophysics Program and Department of Biochemistry
The Ohio State University
Columbus, Ohio

Katsuyuki Aoki
Department of Materials Science
Toyohashi University of Technology, Tempaku-cho
Toyohashi 441-8580, Japan

Veeraiah Bodepudi
Department of Core Chemistry
Roche Molecular Diagnostics
Alameda, California

Erik De Clercq
Rega Institute for Medical Research
B-3000 Leuven, Belgium

Staffan Eriksson
Department of Molecular Biosciences, Section of Veterinary Medical Biochemistry
Swedish University of Agricultural Sciences
The Biomedical Center
Uppsala, Sweden

Joy Y. Feng
Gilead Sciences
Durham, North Carolina

Kevin A. Fiala
Department of Biochemistry
The Ohio State University
Columbus, Ohio

Ikuhide Fujisawa
Department of Materials Science
Toyohashi University of Technology, Tempaku-cho
Toyohashi 441-8580, Japan

Phillip A. Furman
Pharmasset, Inc.
Tucker, Georgia

Allison A. Johnson
Laboratory of Molecular Pharmacology, NCI, NIH
Bethesda, Maryland

Inna Koukhareva
Trilink Biotechnologies Inc.
San Diego, California

Alexandre Lebedev
Trilink Biotechnologies Inc.
San Diego, California

Glenn H. McGall
Affymetrix
Santa Clara, California

Zucai Suo
Department of Biochemistry
The Ohio State University
Columbus, Ohio

Seng-Lai Tan
Lilly Research Laboratories
Eli Lilly & Company
Indianapolis, Indiana

Morteza Vaghefi
ChemCyte Corporation
San Diego, California

Louise Wang
Institute of Cellular and Molecular Biology
University of Texas at Austin, Texas

Stephen G. Will
Department of Core Chemistry
Roche Molecular Diagnostics
Alameda, California

Introduction

The first nucleoside triphosphate (NTP), adenosine triphosphate (ATP), was discovered by Karl Lohmann in 1929. Some years later, its structure was clarified, and Alexander Todd synthesized ATP chemically in 1948. A major role of ATP is supplying the needed energy to synthesize the many thousands of macromolecules that a cell needs to exist. ATP is described as the "universal currency of energy" in the cell and functions as a carrier of energy in all living organisms from bacteria and fungi to plants and animals, including humans. The actual power source is the phosphate tail of ATP. Available energy is contained in the bonds between the phosphates and is released when they are hydrolyzed. Usually, only the outer phosphate is removed to yield energy and the nucleotide diphosphate. Although ATP is the main energy carrier in the cell, other energized nucleotides containing thymine, guanine, uracil, and cytosine are utilized. Nucleoside diphosphate kinase transfers energy contained in a nucleoside triphosphate (commonly ATP) to a nucleoside diphosphate, such as guanosine diphosphate (GDP), to form guanosine triphosphate (GTP).

Nucleoside triphosphates are utilized by a family of polymerases during the synthesis of nucleic acids on preexisting nucleic acid templates, assembling RNA from ribonucleotides, or DNA from deoxyribonucleotides. Polymerase chain reaction (PCR), a key technique in molecular genetics that permits the analysis of any short sequence of DNA (or RNA) in the test tube, relies on NTP. The PCR technology has greatly increased scientists' ability to study genetic material. Since its invention, PCR has revolutionized the way in which basic research and medical diagnosis are conducted.

Structurally, NTP consists of three main components: the nitrogenous base, the sugar, and a chain of three phosphate groups bound to ribose. Since the discovery of the first nucleotide and elucidation of its structure (ATP), scientists have synthesized various analogs of NTP modified at the base, sugar, or triphosphate chain. These compounds have served as valuable tools in the investigation of a variety of cellular processes for many years. Modification of the triphosphate chain has been mainly used to study enzymatic pathways, which results in hydrolysis or transfer of the phosphate from NTP to another molecule. Modification of sugar and base has served a number of different purposes, from pharmaceutical to diagnostic applications.

Structural analogs of nucleoside have diverse biological properties and therapeutic consequences. They comprise a diverse family of cytotoxic compounds, antiviral agents, and immunosuppressive molecules. Anticancer nucleoside analogs inhibit cellular DNA replication and repair, whereas antiviral nucleoside analogs inhibit replication of the viral genome. Nucleoside analogs are inactive prodrugs and require intracellular phosphorylation to their corresponding pharmacologically active triphosphate form.

A number of nucleosides are currently on the market (or may be marketed soon) for the treatment of various diseases. Presently, the backbone of effective antiretroviral therapy regimens for the treatment of HIV contains at least two nucleosides. Pyrimidine nucleoside analogs are essential components of hematological malignancy therapy and are also used in the treatment of solid tumors. Furthermore, immunosuppressive properties of these agents have been exploited for preventing rejection following transplant and for the treatment of nonmalignant disorders, including immunologic diseases and multiple sclerosis.

Due to their essential role in virtually all cellular processes, nucleotide analogs that facilitate tracking of global conformational changes in nucleic acids and nucleic acid complexes have been utilized extensively in cellular and molecular biology research. A variety of labeled nucleotides for molecular biology and cytogenetic applications, such as gene expression studies, mutation detection on arrays, microarrays, and *in situ* PCR and RT-PCR, have been developed and are commercially available.

Fluorescent labeled nucleotide analogs provide sensitive probes for studying the structure, dynamics, and interactions of nucleic acids. These analogs can be incorporated site specifically into oligonucleotides through standard automated synthetic procedures, allowing them to serve as sensitive probes of changes in the microenvironment that may result from variation in molecular interactions. A major advantage of fluorescent analogs is their similarity in chemical properties and molecular constitution to natural bases. In contrast to bulkier chromophores, they are incorporated into the oligonucleotides without introducing significant structural or chemical changes that might alter the measurement.

This book is the first dedicated to the field of nucleoside triphosphate and encompasses within its chapters the contributions of 18 scientists from academia as well as industry, each of whom has contributed to the field. The purpose of this text is to provide collective information on chemistry, physicochemical, and biological properties of natural and modified NTP, as well as their application in life sciences. In addition, we have included three chapters that review families of enzymes that depend on nucleotides for assembling DNA and RNA molecules and an appendix with supporting information.

Morteza M. Vaghefi

Contents

chapter one

Chemical synthesis of nucleoside 5′-triphosphate

Morteza M. Vaghefi
ChemCyte Corporation, San Diego, CA

Contents

1.1 Introduction

The first nucleoside 5′-triphosphate, adenosine triphosphate, was isolated from muscle and liver extracts in 1929.[1,2] The structure of adenosine triphosphate was a matter of dispute for years. Lohmann proposed a formula based on hydrolysis and behavior of the compound towards hydrolytic reagents. He noted that, by acid hydrolysis, it fragmented into 1 mol of adenine, 2 mol of phosphoric acid, and 1 mol of D-ribose-5-phosphate, which led him to formulate it as adenosine-5′-triphosphate.[3] His proposed structure was not universally accepted and it was argued that properties of isolated compound were incompatible with his proposed structure.[4] Finally, in 1945, after several years of additional biochemical studies and analysis, Masson and O'Farrell confirmed Lohmann's proposed structure.[5]

Because adenosine triphosphate was the only known organic derivative of triphosphoric acid, its chemical synthesis was of a matter of interest and under investigation for years. As might be expected, chemists believed that, as an anhydride of phosphoric acid, adenosine triphosphate was extremely unstable and readily degraded under chemical processing. Because adenosine was already synthesized, the major challenge in the chemical synthesis of adenosine triphosphate was the attachment of a triphosphate residue to the 5′-position of adenosine. In 1949, Baddiley et al. successfully synthesized adenosine 5′-triphosphate in a manner identical with adenosine triphosphate isolated from natural sources.[6]

The importance of nucleoside triphosphates (NTPs and dNTP) in biological systems and their involvement in a host of biochemical processes have led to the development of a number of methods for their synthesis. Some protocols work extremely well for a number of nucleosides and a few methods work moderately well for most nucleosides. No satisfactory protocol exists for preparing all nucleoside triphosphates. The field can benefit from some research innovations for preparation of these important molecules. For example, no procedures will allow for direct one-step triphosphorylation or synthesis of a diverse set of nucleoside triphosphates via a combinatorial or parallel synthesis.

This chapter reviews different approaches for the synthesis of natural and modified nucleoside triphosphates. It includes a historical aspect of synthetic evolution, which will be useful for evaluating the best strategy for the synthesis of nucleoside triphosphates. The synthesis of natural nucleoside triphosphates by enzymatic procedure is described in the next chapter. A number of alternative procedures for the synthesis of modified triphosphate moiety, including isotopic and radiolabeled, are also described in the following chapters.

1.2 Early synthetic efforts

Baddiley et al. reported the first chemical synthesis of adenosine 5′-triphosphate in 1949[6] (see Scheme 1.1). Tribenzyl ester of adenosine 5′-diphosphate was synthesized by coupling dibenzylchlorophosphonate with the silver salt of monobenzyl ester of adenosine 5′-monophosphate in acetic acid. Partial deprotection of tribenzyl ester of adenosine 5′-diphosphate at P^2 produced dibenzylpyrophosphate, which was coupled with another molecule of dibenzylchlorophosphonate yielding tetrabenyzyl-protected triphosphate analog. The benzyl protecting group was removed by catalytic hydrogenation.

This procedure, which was improved later that year by Michelson and Todd,[7] was very laborious and time consuming. It required initial synthesis of both components and careful protection of all the intermediates to limit the possibility of unwanted chemical reactions. Furthermore, purification of nucleoside triphosphate involved several precipitation steps under controlled conditions and then, finally, crystallization of the acridinium salt of adenosine 5′-triphosphate. The product was characterized using various

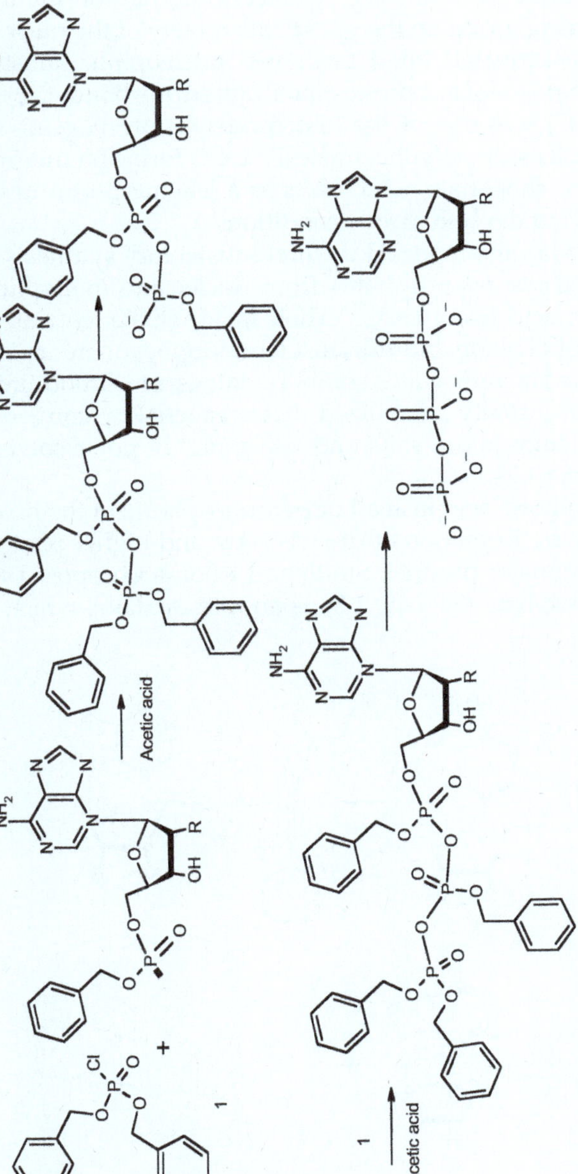

Scheme 1.1 First stepwise chemical synthesis of ATP by Baddiley and coworkers.

analytical tools available at the time, including melting points, X-ray crystallography, infrared spectroscopy, and absorption spectra; its biological activity was evaluated and compared with adenosine triphosphate isolated from natural sources.

Major efforts in the following years concentrated on the incorporation of an active leaving group on the phosphate moiety of the nucleoside monophosphate. The activated bond facilitated nucleophilic substitution with another phosphate to form a diphosphate anhydride bond. Dicyclohexylcarbodiimide (DCC) was one of the first condensation reagents used for the synthesis of nucleoside polyphosphates.[8,9] DCC forms an intermediate with nucleoside monophosphate, which acts as a leaving group in the presence of a nucleophile under anhydrous conditions.

Smith and Khorana reported the first satisfactory synthesis of ribo- and deoxyribonucleoside triphosphates from nucleoside monophosphate and pyrophosphoric acid using DCC.[10] Their major challenge was the low solubility of pyrophosphate and nucleoside monophosphate in inorganic solvents. They tried a few water-soluble analogs of carbodiimide without success. They eventually solubilized these reagents by converting them to various trialkylammonium salts and using highly polar solvents, such as pyridine (Scheme 1.2).

Using phosphoric acid instead of pyrophosphoric in the preceding reaction resulted in the formation of di-, tri-, tetra and higher phosphates, with triphosphate the major product. Smith and Khorana proposed formation of cyclictrimetaphosphate (Scheme 1.3) as an intermediate, which was hydro-

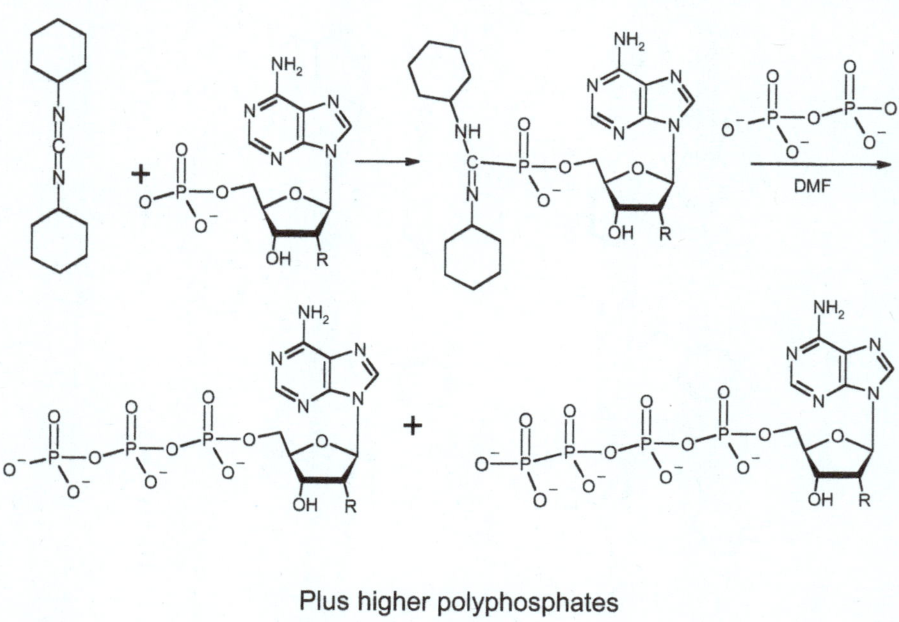

Scheme 1.2 Coupling tributylammonium pyrophosphate with adenosine monophosphate in pyridine.

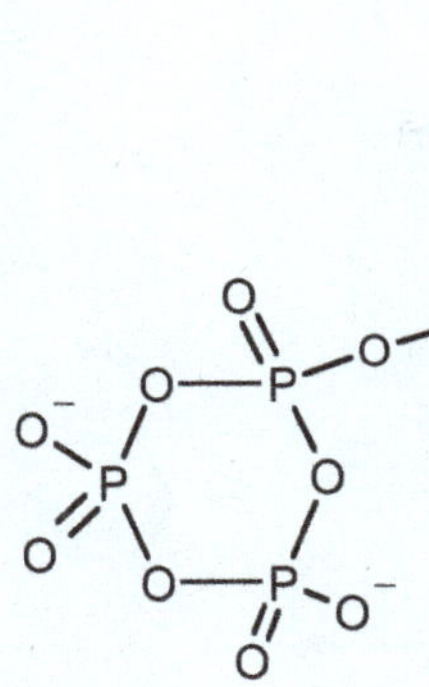

Scheme 1.3 Proposed structure of nucleoside cyclic trimetaphosphate intermediate.

lyzed to nucleoside triphosphate. This intermediate is much less reactive to further DCC activation and addition of more phosphates.

1.3 *Activation of phosphate*

1.3.1 *Phosphoramidates*

The development of a simple, general method for the activation of nucleoside monophosphate was the focus of many studies over a decade. Different alkylamines with various basicity and reactivity were studied for formation of a variety of nucleoside phosphoramidates. In 1961, Moffat and Khorana[11] introduced nucleoside phosphomorpholidate derivatives that opened the door for the synthesis of nucleoside triphosphate as well as diphosphates and a number of nucleoside coenzyme analogs (Scheme 1.4).

Nucleoside monophosphate was coupled with morpholine using DCC as a condensing reagent to form nucleoside phosphomorpholidate. Initially, pyridine was used as a solvent for coupling pyrophosphoric acid with nucleoside 5′-phosphomorpholidate. The reaction for the synthesis of triphosphate proceeded smoothly and resulted in a 57% conversion; however, when it was subjected to longer reaction times, it converted into a number of byproducts, including nucleoside diphosphate and dinucleotide-5′,5′-tetraphosphate. Moffatt improved this procedure using dimethylsulfoxide as the solvent and reported an isolated yield of 70 to 80% for different nucleotides.[12] The advantage of phosphomorpholidate nucleoside is its relative stability. This procedure is very reliable for forming P–O–P bond and is still practiced in laboratories.

Phosphoimidazolidate is another activated form of phosphate that was introduced in 1957 by Staab.[13] This activation was done by reacting phosphoric acid or organophosphate analogs with carbonyldiimidazole. Goldman and coworkers[14] were the first to report synthesis of adenosine diphosphate from adenosine monophosphate by imidazole activation (Scheme 1.5). This procedure was improved in the following years and applied to the

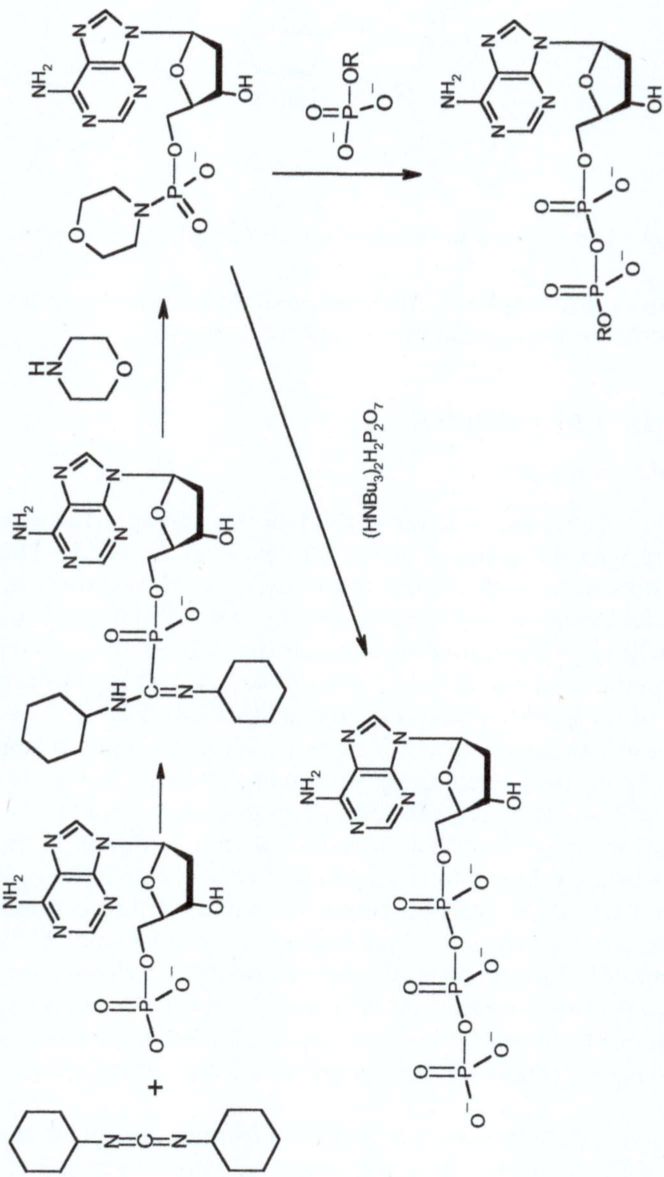

Scheme 1.4 Synthesis of deoxyadenosine 5′-phosphomorpholidate — a stable intermediate for the synthesis of nulcleoside polyphosphate analogs.

synthesis of a number of nucleoside di- and triphosphate analogs, as well as coenzymes and dinucleotide polyphosphates in high yields.[15–17]

The phosphoramidate couplings are routinely performed under anhydrous condition by several coevaporations with azeotropic solvents, such as anhydrous pyridine. Any trace of water dramatically influences the reaction yield. However, in the presence of a divalent metal ion such as Mn^{++} or Cd^{++}, it is possible to form a P–O–P bond through phosphoamidate intermediate in an aqueous condition in optimal yield.[18]

1.3.2 *Anhydride activation*

Michelson first reported the activation of phosphate by anhydride formation in 1964.[19] He used diphenylphosphoryl chloride in dimethylformamide (DMF) to activate nucleoside 5′-monophosphate under an anhydrous condition. The diphenylphosphoryl group then acted as a leaving group in a nucleophilic reaction with pyrophosphoric acid (Scheme 1.6). Although this method has no serious disadvantages, it has not been widely used.

Furusawa et al. reported an alternative anhydride-type activation procedure in 1976.[20] He used di-*tert*-butylphosphinothionyl bromide and nucleoside monophosphate to form nucleoside monophosphate-di-*n*-butylphosphinothioic anhydride. The mixed anhydride was stable even in the water and readily reacted with phosphoric acid or phosphomonoester in the presence of silver as a catalyst (Scheme 1.7).

1.3.3 *Phosphodiester activation*

Activation by diester has been mainly accomplished with compounds that interact through neighboring group coordination, such as 8-hydroxyquinoline.[21] Using triphenylphosphine/2,2′-dipyridyl diselenide as a coupling reagent, 8-quinolyl monophosphate reacted with an unprotected nucleoside and formed an 8-quinolyl nucleoside phosphodiester intermediate. The intermediate reacted with pyrophosphate through the addition of copper (II) as a catalyst (Scheme 1.8).

1.4 *Phosphorylation of nucleosides*

Early synthetic procedures of nucleoside triphosphate were mainly based on the natural nucleoside 5′-monophosphates. The discovery of the structure of natural nucleosides and nucleotides and their vital role in many critical cellular processes instigated chemists' interest in the design and synthesis of modified analogs and their potential biological applications. A wide variety of nucleosides modified at the base or sugar were synthesized. Because all previous procedures for the preparation of nucleoside 5′-triphosphates utilized a nucleoside 5′-monophosphate as a starting material, synthesis of nucleoside 5′-monophosphate for conversion to polyphosphate was an

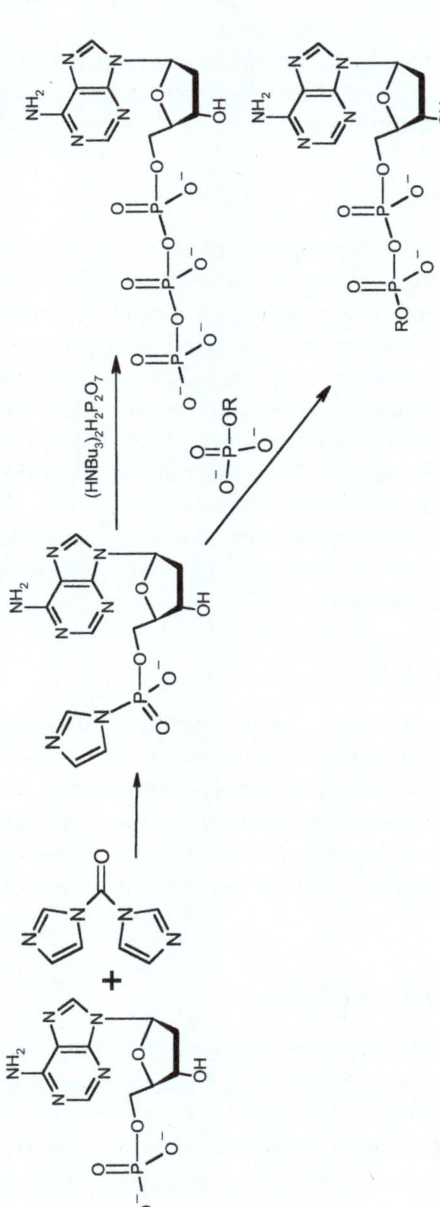

Scheme 1.5 Carbonyl diimidazole activation of nucleoside monophospahte for the synthesis of nucleoside polyphosphate analogs.

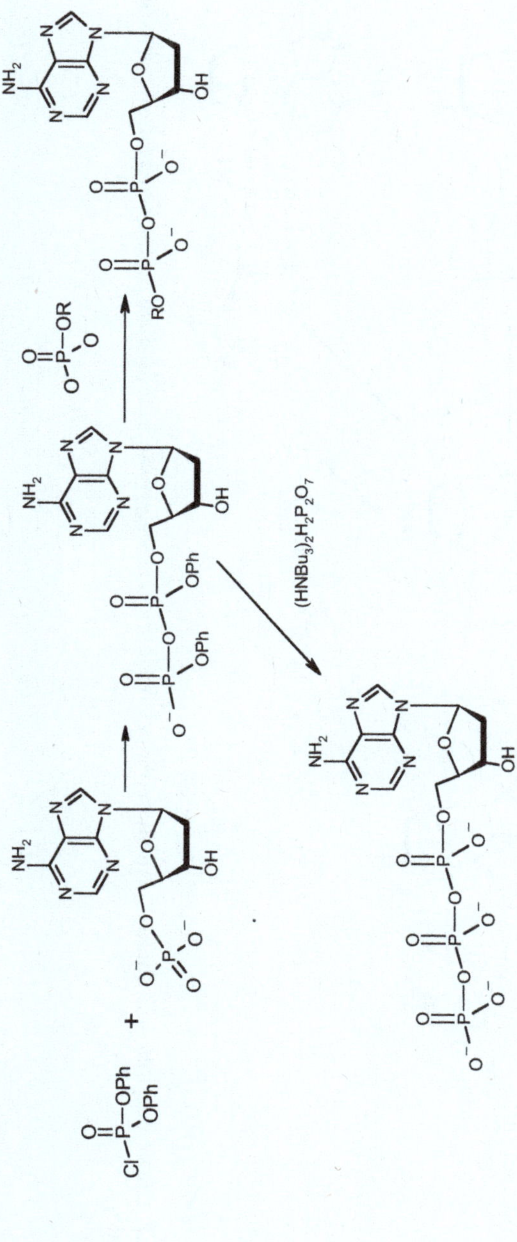

Scheme 1.6 Diphenylphosphate activation of nucleoside monophospahte — an intermediate for the synthesis of nucleoside polyphosphate.

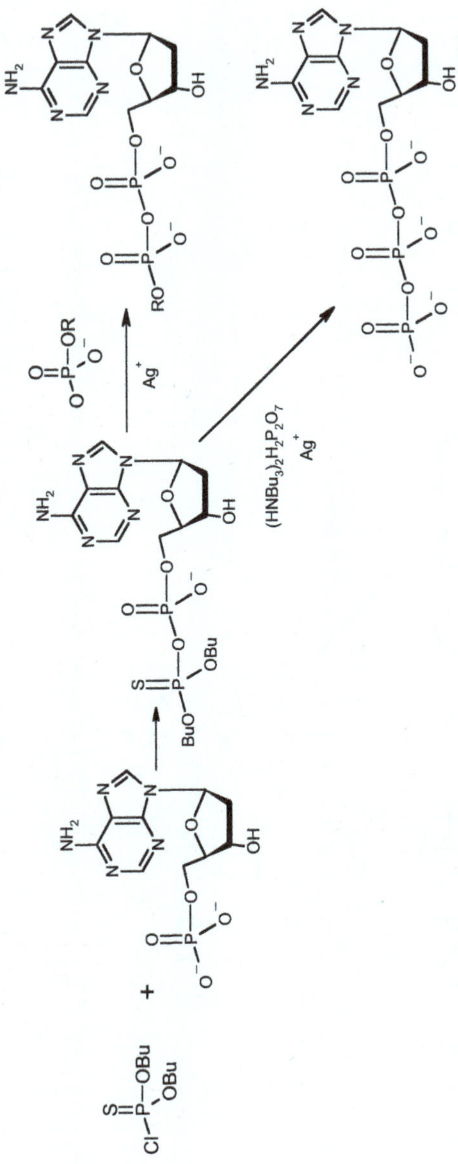

Scheme 1.7 Di-*tert*-butylphosphothionyl activation of nucleoside triphosphate — an intermediate for the synthesis of nucleoside polyphosphate analogs.

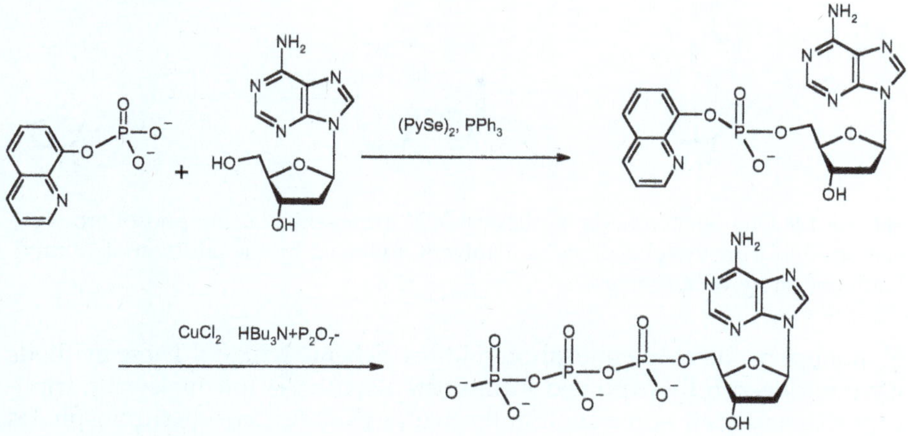

Scheme 1.8 8-Hydroxyquinoline activation of nucleoside 5′-monophosphate — an intermediate for the synthesis of nucleoside triphosphate.

important and challenging issue. Therefore, several methods for 5′-phosphorylation of nucleoside were developed.[22]

Phosphorylation of the hydroxyl groups requires the use of highly reactive phosphorus moieties. Early efforts were predominantly focused on using phosphorus compounds with P–Cl bonds. Because most nucleosides have more than one hydroxyl group, selective phosphorylation of the 5′-position imposed a serious challenge. Unselective phosphorylation resulted in a mixture of 2′-, 3′-, and 5′-phosphorylated compounds. Reagents, such as dichlorophosphoric acid and phosphoryl chloride, were routinely used for phosphorylation of the hydroxyl group. However, the breakthrough came when Yoshikawa et al.[23] discovered that using trimethylphosphate or other trialkylphosphates as a solvent accelerated the rate of reaction of nucleosides with phosphoryl chloride at the 5′ position.

Initially, they used 2′, 3′-protected nucleosides to perform 5′-phosphorylation in trialkylphosphate. The value of their innovation was materialized when they repeated the reaction with the corresponding unprotected nucleoside and observed selective phosphorylation of the 5′-hydroxy. Because this reaction performed best when no base was used, they concluded acidic media was necessary to suppress reaction at the 2′ and 3′ hydroxyls. However, it was later established that the reaction can be equally regioselective in a slightly basic medium and the relationship between regioselectivity of phosphorylation and pH remains unknown. Phosphorylation in trialkylphosphates is the primary method for the 5′-phosphorylation of nucleosides that do not require protection of the other functional groups (Scheme 1.9).

A number of alternative methods for 5′-phosphorylation of nucleosides have been described. For example, 5′-phosphate can be introduced by Mitsunobo reaction.[24,25] Other obvious reagents for synthesis of nucleoside

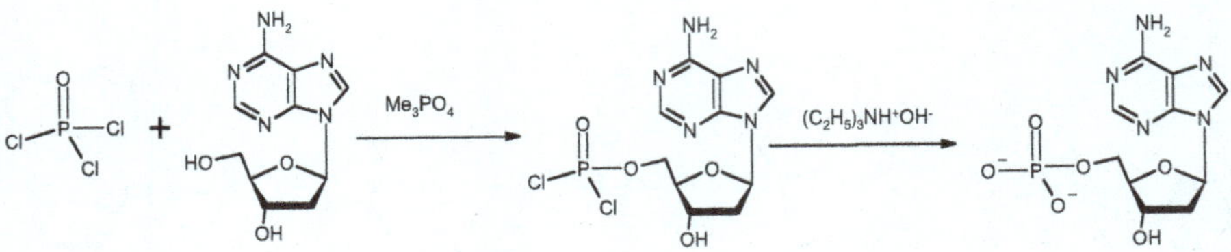

Scheme 1.9 One-pot synthesis of nucleoside 5′-triphosphate using phosphorous oxychloride in trimethylphosphate as a solvent, followed by the addition of tributylammonium pyrophosphate.

5′-monophosphates are phosphoramidites (Scheme 1.10).[26,27] These methods have not been fully exploited in multistep synthesis for nucleoside triphosphates and their application in the synthesis of nucleoside triphosphates is uncommon.

In recent years, several novel nucleoside phosphorylation procedures have been described. Uchiyama et al. reported an O-selective phosphorylation of nucleosides with an unprotected base.[28] They used metalo-organic bases (such as alkyllithium, potassium *tert*-butoxide, *tert*-butylmagnesium chloride, etc.) for hydroxyl activation. The activated 5′-hydroxyl was then reacted with phosphorochloridate or *p*-nitrophenylphosphate to form nucleoside 5′-phosphate in a yield exceeding 90% (Scheme 1.11).

Caputo et al.[29] reported a mild procedure for direct phosphorylation of acid-sensitive nucleosides. In this method, nucleosides are selectively phosphorylated at hydroxyl groups by a preactivated diphenylphosphate with a solid support-bond triphenylphosphine–I_2 complex in a yield exceeding 90% (Scheme 1.12).

Phosphorylation of alcohols using trialkylphosphite/iodone under mild conditions has been described by Stowell and Widlanski.[30] This method gives excellent yield and is compatible with a variety of functional groups, including hydroxyl of nucleosides. A typical reaction was run with triethylphosphite/Iodine and methanol in the presence of pyridine, which exclusively formed methyl-diethylphsophate. The triester was selectively hydrolyzed by bromotrimethylsilane (Scheme 1.13).

1.5 *Recent advances in nucleoside 5′-triphosphate synthesis*

The search for a simple, efficient method for the synthesis of nucleoside 5′-triphosphate has resulted in a number of specific routes. Yoshikawa's method of phosphorylation of unblocked nucleosides offers real advantages. The attractive feature of his method is that the resulting intermediate, nucleoside 5′-dichlorophosphate, directly reacts with ionized phosphate (or pyrophosphate) and forms a nucleoside 5′-triphosphate.

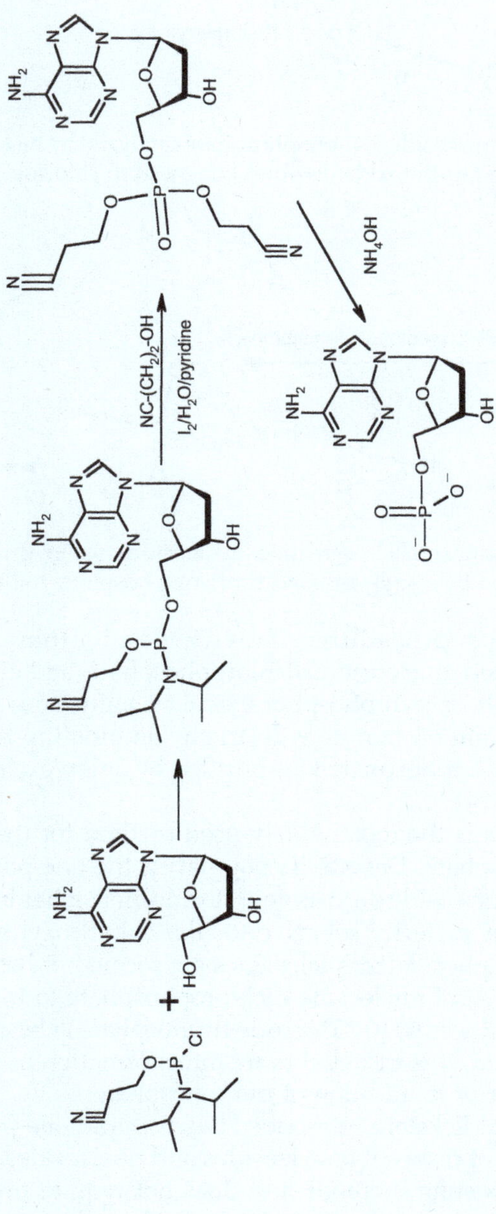

Scheme 1.10 Nucleoside 5′-phosphorylation using phosphorous P(III). Protection of other hydroxyl groups is necessary.

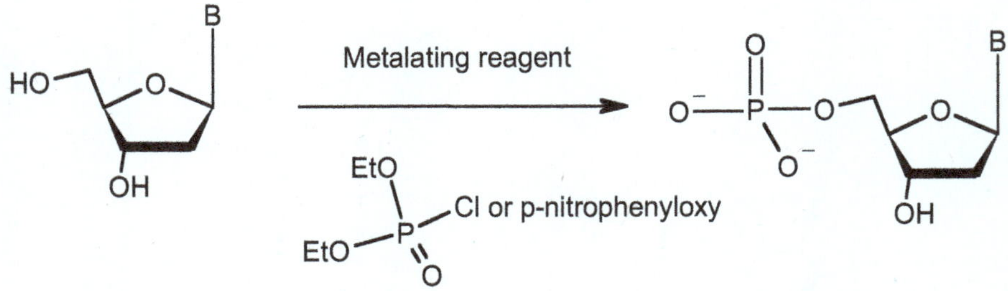

Scheme 1.11 Synthesis of nucleoside 5′-monophosphate catalyzed by metalating such as alkyllithium, potassium *tert*-butoxide, *tert*-butylmagnesium chloride, etc.

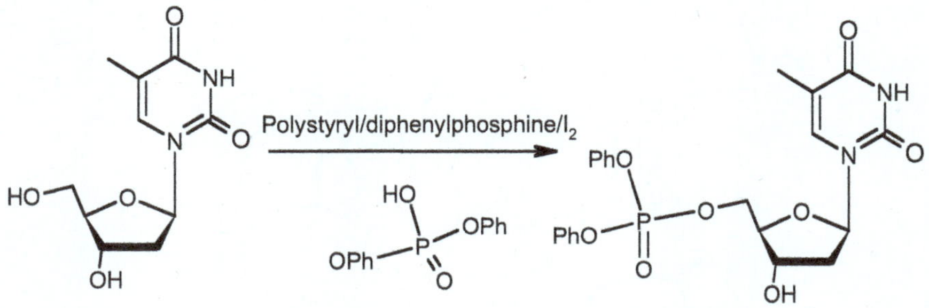

Scheme 1.12 Phosphorylation of acid-sensitive nucleosides using diphenylphosphates previously activated by a support bond triphenylphosphine iodine complex.

Using Yoshikawa's procedure, Ludwig[31] and others working independently[32] prepared nucleoside dichlorophosphate and then added tributhylammonium salt of pyrophosphoric acid in anhydrous DMF. The reaction was quenched after 1 minute with triethylammonium bicarbonate buffer and nucleoside 5′-triphosphate was purified by anion exchange chromatography (Scheme 1.14).

Ludwig's procedure is the most widely used method for the synthesis of nucleoside 5′-triphosphate. Despite its popularity, the one-pot synthesis is not applicable to all nucleoside derivatives.[33] In addition, it has been shown that the first step is not perfectly selective for the 5′-hydroxyl group, and traces of 2′- and 3′-phosphorylated nucleosides are formed. It has also been noted that the conversion of nucleoside dichlorophosphate to triphosphate is not complete and produces 60 to 70% crude triphosphate at best. To obtain the highest possible yield, it is essential to monitor formation of nucleoside 5′-dichlorophosphate prior to addition of pyrophosphate.

In 1989, Ludwig and Eckstein[34] described a new, rapid one-pot reaction for the phosphitylation of deoxyribonucleosides and nucleosides. It has certain advantages over existing methods and does not require protection of nucleobases. They reacted sugar-protected nucleosides with 2-chloro-4H-1,3,2-benzodioxaphosphrin-4-one, which was subsequently reacted with pyrophosphate to form a cyclic intermediate, P^2,P^3-dioxo-P^1-5′-

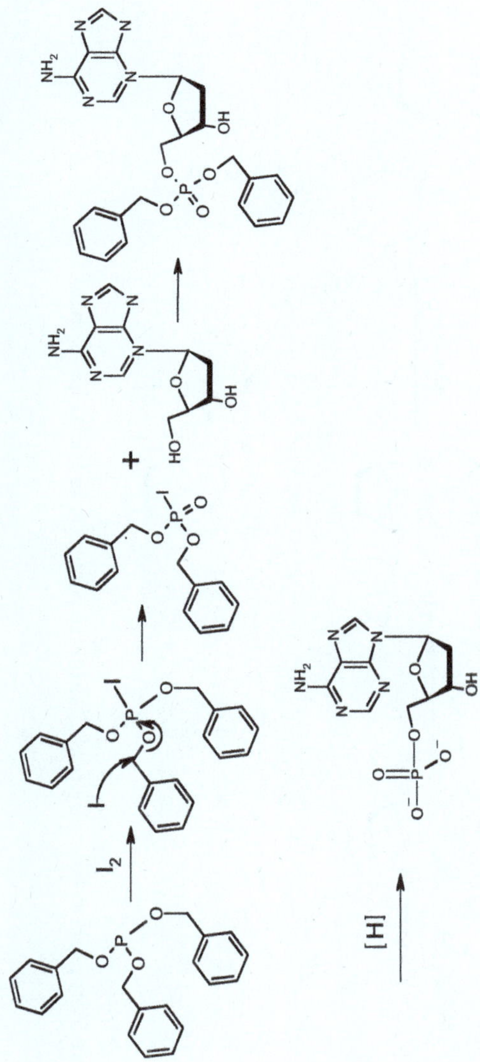

Scheme 1.13 Phosphorylation of 5′-alcohols using trialkylphosphite/iodone.

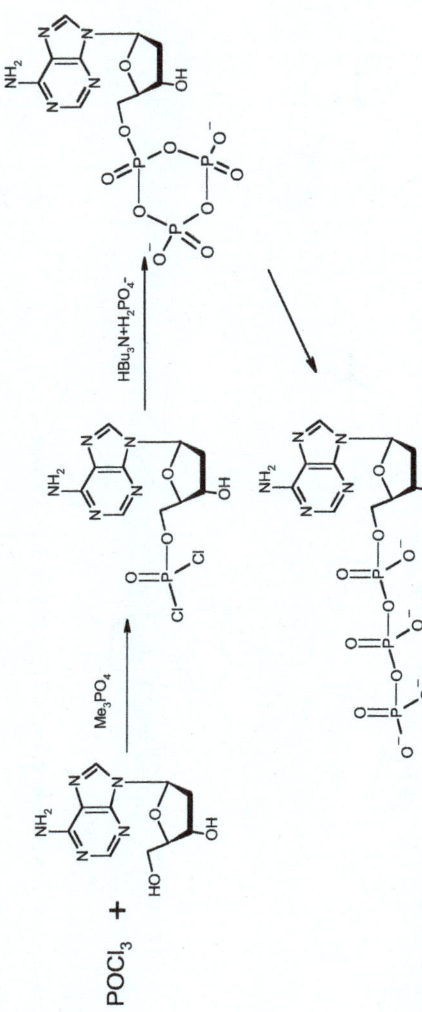

Scheme 1.14 Selective 5′-phosphorylation by phosphorous oxychloride in trialkylphosphate.

nucleosidylcyclotriphosphite, in a double displacement process. Oxidation of the cyclic intermediate with iodine/water gave nucleoside triphosphate, produced nucleoside-α-thiotriphosphate with sulfur, and gave α-P-boranotriphosphate with aminoborane complex.[35] The reaction is performed in anhydrous pyridine, which neutralizes the HCl byproduct. Because a phosphitylating reagent does not discriminate between primary and secondary alcohol, protection of the 2′- and 3′-hydroxyl is necessary. Ludwig and Eckstein's method is best suited for 2′-3′-dideoxynucleosides (Scheme 1.15).

More recently, Bogachev[36] developed a new one-pot procedure using trifluoro acetic anhydride (TFAA) as an activating agent for the conversion of nucleoside monophosphate to nucleoside triphosphate in the presence of a tertiary amine. The advantages of this procedure are largely due to the use of TFAA, which is highly reactive in acylating phosphate, hydroxyl, and amino group of deoxynucleosidemonophosphate, and can be easily eliminated from the reaction mixture. This one-pot nucleoside triphosphate synthesis is carried out in three main steps and the reaction time does not exceed 2 to 5 minutes at each step.

Bogachev reacted nucleoside monophosphates with TFAA in an aprotic solvent in the presence of *N*-methyl imidazole. TFAA not only protected potential reactive sites such as hydroxyls and amines, but also formed a mixed anhydride with phosphate. Subsequent treatment with a nucleophilic catalyst such as *N*-methylimidazole and inorganic pyrophosphate generated a nucleoside triphosphate in 89 to 92% yields (Scheme 1.16).

1.6 Conclusion

Since the discovery of nucleoside triphosphates, ample efforts have been directed at designing reliable procedures for their syntheses. There is no universal method for preparing all nucleoside triphosphates. Currently, Yoshigawa's $POCl_3$ phosphorylation method; Ludwig's "one pot"; Ludwig–Eckstein's "three-step" triphosphorylation procedure; and Bogachev's method of preparing triphosphate from monophosphate are the most widely used methods for syntheses of nucleoside triphosphates.

Advances in biochemical research have demonstrated the unique importance of nucleoside triphosphates in numerous biochemical and pharmacological processes. Modified nucleoside triphosphates have been investigated in searches for potential diagnostic and therapeutic agents. Nucleoside triphosphates are utilized in *in-vitro* transcription methods to prepare aptamers, polymerase chain reactions, and DNA sequencing, and as inhibitors of polymerases as therapeutic agents. All biological applications require high-quality nucleoside triphosphates. The preferred methods for preparation of nucleoside triphosphates are those that do not require multistep protection and deprotection of the functional groups and provide high-quality materials in high yields at a reasonable cost. Currently, most of the commercially available nucleoside triphosphates are prepared synthetically, with unprotected nucle-

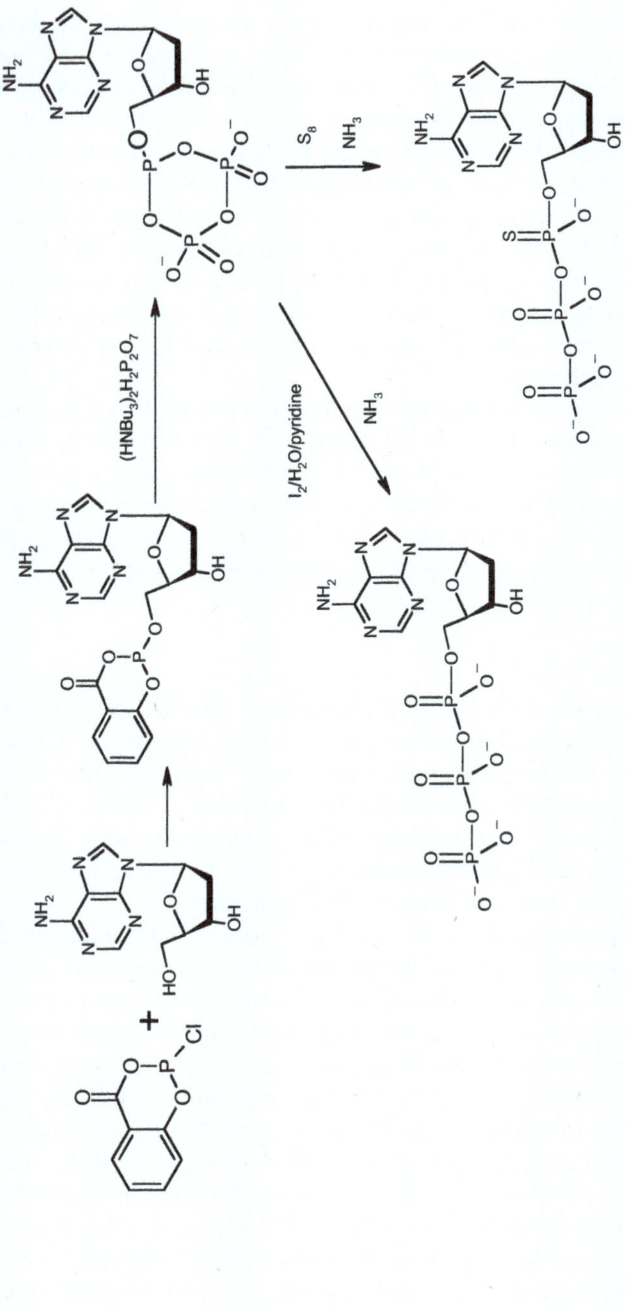

Scheme 1.15 One-pot nucleoside triphosphate synthesis using 2-chloro-4H-1,3,2-benzodioxaphosphrin-4-one as phosphitylating reagent.

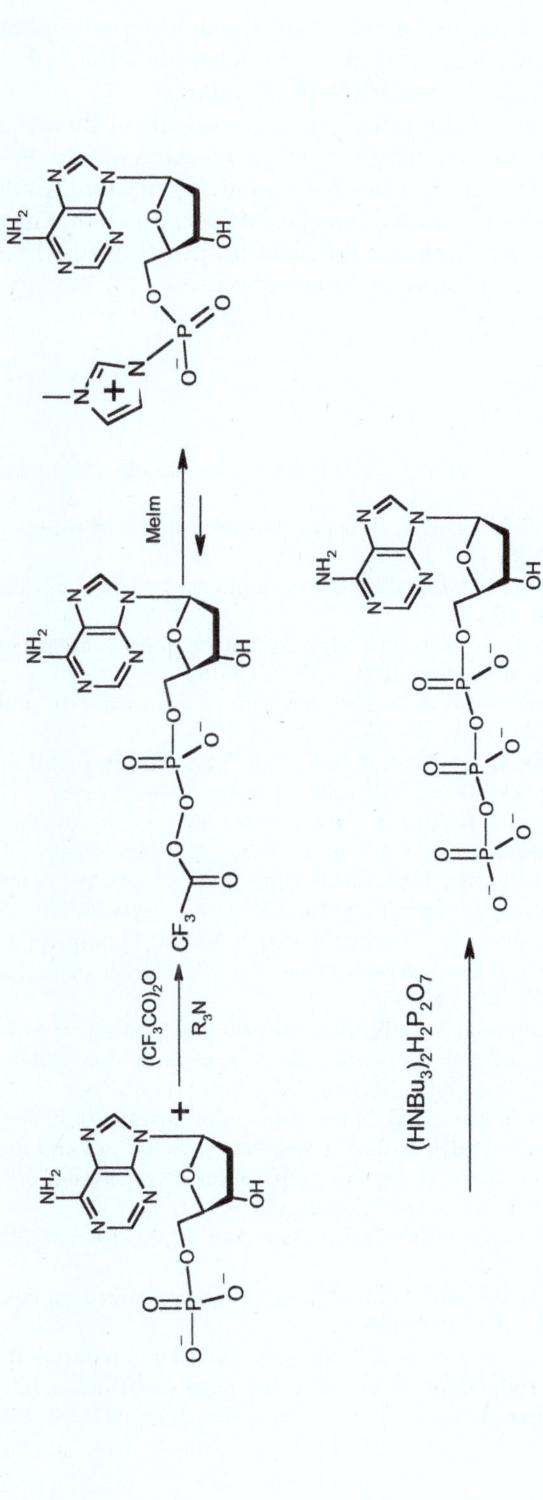

Scheme 1.16 Activation of monophosphates with trifluoroacetic anhydride in an aprotic solvent in the presence of a tertiary amine. The activated phosphate couples with pyrophosphate in anhydrous condition.

oside and phosphoryl chloride, or enzymatically in a stepwise phosphorylation. Either method produces 50 to 80% of nucleoside triphosphates in the reaction mixture and requires extensive purification.

Taking an overview of the information presented in this chapter, it is important to hint at areas wherein further research might lead to the development of superior approaches for preparing these molecules. These include (1) direct, one-step triphosphorylation procedures that improve the yield of the synthesis and minimize the need for purification; (2) improved solid-phase synthesis amenable to automation; and (3) improved enzymatic procedures.

References

1. Lohmann, K. The pyrophosphate fraction in the muscle. *Naturwiss*, 1929, 17, 624–625.
2. Fiske, C.H. and Subbarow, Y. Phosphorus compounds of muscle and liver. *Science*, 1929, 70, 381–382.
3. Lohmann, K. Preparation of adenylpyrophophoric acid from muscle. *Biochem. Z.*, 1931, 233, 460–466.
4. Satoh, T. Hydrolysis of adenosine triphosphate by phosphomonoesterase and pyrophosphatase. *J. Biochem. Jpn.*, 1935, 2, 19–36.
5. Gulland, J.M. and Walsh, E.O. The structure of adenosine di- and tri-phosphate. *J. Chem. Soc.*, 1945, 169–172.
6. Baddiley, J., Michelson, A.M., and Todd, A.R. Nucleosides. Part II. A synthesis of adenosine triphospahte. *J. Chem. Soc. London*, 1949, 582–586.
7. Michelson, A.M. and Todd, A.R. Nucleosides. Part IV. A novel synthesis of adenosine triphospahte. *J. Chem. Soc. London*, 1949, 2487–2490.
8. Hall, R.H. and Khorana, H.G. Nucleoside polyphosphates. A synthesis of uridine-5′-di- and –triphosphate. *J. Am. Chem. Soc.*, 1954,76, 5056–5060.
9. Potter, R.L., Schlesinger, S., Buettner–Janusch, V., and Thompson, L. Isolation of pyrimidine deoxyribonucleotides from the acid soluble extract of thymus. *J. Biol. Chem.*, 1957, 226, 381–386.
10. Smith, M. and Khorana, H.G. Nucleoside polyphosphates. VI. An improved and general method for the synthesis of ribo- and deoxyribonucleoside 5′-triphosphate. *J. Am. Chem. Soc.*, 1958, 80, 1141–1145.
11. Moffatt, J.G. and Khorana, H.G. Nucleoside polyphosphates. X. The synthesis and some reactions of nucleoside-5′-phosphomorpholidates and related compounds. Improved method for the preparation of nucleoside-5′ polyphosphates. *J. Am. Chem. Soc.*, 1961, 83, 649–658.
12. Moffatt, J.G. A general synthesis of nucleoside-5′ triphosphates. *Can. J. Chem.*, 1964, 42, 599–604.
13. Staab, H.A. Reactive *N*-carboxylic acid esters and *N*-carboxamides of imidazole and tetrazole. *Ann* 1957, 609, 75–82.
14. Goldman, L., Marsico, J.W., and Anderson, G.W. The preparation of adenosine-5′ imidazol-1-ylphosphonate and its reactions with nucleophiles. A novel synthesis of nucleoside coenzymes. *J. Am. Chem. Soc.*, 1960, 82, 2969–2970.

15. Cramer, F. and Neunhoeffer, H. Reaktionen von Adenosin-5′-phosphorsäure-imidazolid-eine neue Synthese von Adenosindiphosphat und Flavin-adenin-dinuleotid. *Chem. Ber.*, 1962, 95, 1665–1669.
16. Hoard, D.E. and Ott, D.G. Conversion of mono- and oligodeoxyribonucleosides to 5′-triphosphates. *J. Am. Chem. Soc.*, 1965, 87, 1785–1788.
17. Kozarich, J.W., Cinault, A.C., and Hecht, S.M. Ribonucleoside phosphate via phosphorimidazolidate intermediates. Synthesis of pseudoadenosine 5′-triphosphate. *Biochemistry*, 1973, 12, 4458–4463.
18. Shimazu, M., Shinazuka, K., and Sawai, H. Facile synthesis of nucleotides containing polyphosphates by Mn(II) and Cd(II) ino-catalyzed pyrophosphate bond formation in aqueous solution. *Tetrahedron Lett.*, 1990, 31, 235–238.
19. Michelson, A.M. Synthesis of nucleoside anhydrides by anion exchange. *Biochim. Biophys. Acta*, 1964, 91, 1–13.
20. Furusawa, K., Sekine, M., and Hata, T. Studies on pyrophosphates. Part III. A new method for the synthesis of nucleotide coenzymes by means of di-n-butylphosphinothionyl bromide. *J. Chem. Soc. Perkin*, 1976, 1, 1711–1716.
21. Takaku H, Konishi T, and Hata T. 8-Quinolyl nucleoside 5′-phosphates as a useful intermediate for the synthesis of nucleoside 5′-di and 5′-triphosphates. *Chem. Lett.*, 1977, 655–658.
22. Slotin, L.A.,Current methods of phosphorylation of biological molecules. *Synthesis*, 1977, 737–752.
23. Yoshikawa, M., Kato, T., and Takenishi, T. A novel method for phosphorylation of nucleosides to 5′-nucleotides. *Tetrahedron Lett.*, 1967, 50, 5065–5068.
24. Mitsunobo, O. The use of diethyl azodicarboxylate and triphenylphosphine in synthesis and transformation of natural products. *Synthesis*, 1981, 1–28.
25. Campbell, D.A. The synthesis of phosphonate esters, an extension of the Mitsunobu reaction. *J. Org. Chem.*, 1992 57, 6331–6335.
26. Coe, D., Flitsch, S.L., Hilpert, H., Liebster, M., Roberts, S.M., and Turner, N.J. Synthesis of some mimics of nucleoside triphosphates. *Chem. Ind.*, 1989, 21, 724–728.
27. Perich, J.W. and Johns, R.W. Di-tert-butyl N,N-diethylphosphoramidite. A new phosphitylating agent for the efficient phosphorylation of alcohols. *Synthesis*, 1988, 142–144.
28. Uchiyama, N., Aso, Y., Noyori, I., and Hayakawa, Y. O-selective phosphorylation of nucleosides without N-protection. *J. Org. Chem.*, 1993, 58, 373–379.
29. Caputo, R, Guaraga, A., Pedatella, S., and Palumbo, G., Mild and regiospecific phosphorylation of nucleosides. *Synlett*,1997, 8, 917–918.
30. Stowell, J.K. and Widlanski, T.S. A new method for the phosphorylation of alcohols and phenols. *Tetrahedron Lett.*, 1995, 36, 1825–1826.
31. Ludwig, J. A new route to nucleoside-5′-triphosphates. *Acta Biochim. Biophys. Acad. Sci. Hung.*, 1981, 16, 131–133.
32. Ruth, J.L. and Cheng, Y.C. Nucleoside analogues with clinical potential in antivirus chemotherapy. The effect of several thymidine and 2′-deoxycytidine analogue 5′-triphosphates on purified human (alpha, beta) and herpes simplex virus (types 1, 2) DNA polymerases. *Mol. Pharmacol.*, 1981, 20, 415–422.
33. Burgess, K. and Cook, D. Synthesis of nucleoside triphosphate. *Chem. Rev.*, 2000, 100, 2047–2059.

34. Ludwig, J. and Eckstein, F. Rapid and efficient synthesis of nucleoside 5′-(1-thiotriphosphates), 5′-triphosphates and 2′,3′-cyclicphosphorothioates using 2-chloro-4h-1,3,2-benzodiaxaphosphorin-4-one. *J. Org. Chem.*, 1989, 54, 631–635.
35. He, K., Hassan, A., Krzyzanowska, B., and Shaw, B.R. Synthesis and separation of diastereomers of ribonucleoside 5′-(α-p-borano)triphosphate. *J. Org. Chem.*, 1998, 63, 5769–5773.
36. Bogachev, V.S. Synthesis of deoxynucleoside 5′-triphosphates using trifluoroacetic anhydride as an activation reagent. *Russ. J. Bioorg. Chem.*, 1996, 22, 599–604.

chapter two

Enzymatic synthesis of nucleoside triphosphates

Staffan Eriksson
Department of Molecular Biosciences, Section of Veterinary Medical Biochemistry, Swedish University of Agricultural Sciences, Uppsala, Sweden

Contents

2.1 Introduction

This chapter attempts to survey the enzymatic methods utilized to synthesize natural and modified ribo- and deoxyribonucleosides and their analogs. Burgess and Cook have previously provided an excellent review partly covering this topic.[1]

2.2 Nucleoside phosphorylating enzymes

Chemical phosphorylation of nucleosides in a specific position is often relatively complex because other functional groups competing for the phosphate must be reversibly protected. Therefore, in the early days of nucleotide chemistry, enzymes were already used to carry out the 5′-phosphorylation of natural or modified nucleosides.

2.2.1 Nucleoside phosphotransferases

A group of phosphotransferases were discovered by Chargaff and co-workers and used for this reaction with phenylphosphate as phosphate donor and crude enzyme preparations from malt, mammalian tissues, or, preferentially, from carrot and wheat germ (Figure 2.1).[2–8] Crude extracts from *Serratia marcescens* or wheat shoots could easily be prepared.[6,8,9] These phosphotransferases are very reliable as general 5′-phosphorylating agents with 0.1 to 0.6 *M* *p*-nitro-phenyl phosphate in buffers at pH 4.0 or 5.0 at 37°C for 1 to 7 days; they give varying but often high yields of many different nucleosides, such as; 2′-O-methylcytidine, 3′-O-methylcytidine, 6-methyluridine,

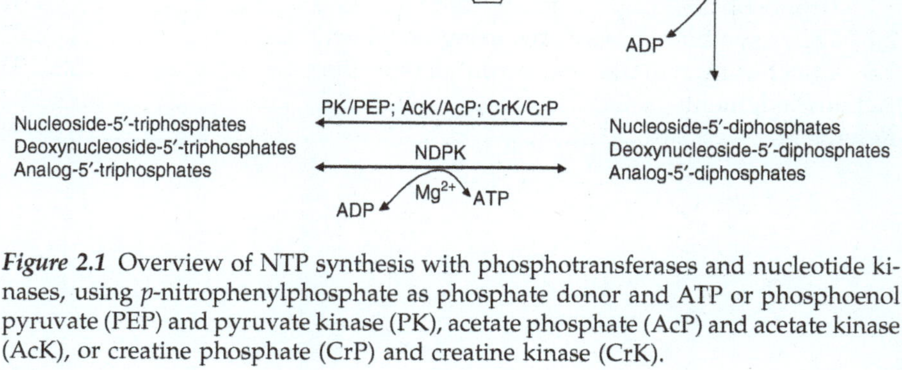

Figure 2.1 Overview of NTP synthesis with phosphotransferases and nucleotide kinases, using *p*-nitrophenylphosphate as phosphate donor and ATP or phosphoenol pyruvate (PEP) and pyruvate kinase (PK), acetate phosphate (AcP) and acetate kinase (AcK), or creatine phosphate (CrP) and creatine kinase (CrK).

formycine, 2′-deoxy-5-fluorouridine, acyclovir, and 5-aminoimidazol-D-ribofuranosyl.[6,8,9]

There are two problems with these types of enzyme preparations; first, several of these enzymes (even in purified form) could also phosphorylate O-2′ and O-3′ in addition to O-5′ and, second, contaminating activities, e.g., phosphorylases, nucleotidases and deaminases, were also present. However, enzyme purification procedures could eliminate most of these contaminating activities[4,5] and whole cell extracts from the bacteria *Erwinia herbicola* were reported to be naturally devoid of many of these interfering enzyme reactions and could be used with low molar excess of phosphate donors.[10] Overall, these phosphorylation reactions have been and still are valuable for production of radioactive or small amounts of nucleotides and nucleotide analogs of a large variety with yields up to 30%.[4–10] However, their role in present day synthetic procedures is limited and, apparently, no pure recombinant enzymes of this type are available.

2.2.2 *Nucleoside and deoxyribonucleoside kinases*

These types of enzymes carry out the 5′-phosphorylation of deoxy- and ribonucleosides and related analogs using a nucleoside triphosphate (usually ATP) as phosphate donor in the presence of Mg^{2+} ions (Figure 2.2). A drawback for these types of enzymes in synthetic reactions is that they usually have a rather narrow specificity and thus are highly selective for certain types of nucleosides. The introduction of adeosine kinase (E.C. 2.7.1.20) as a synthetic tool has been pioneered by Whitesides and colleagues[11] and adenosine has been converted to AMP using polyacrylamide cross linked commercially available adenosine kinase (Figure 2.2).

Usually, the reactions were carried out in a mixture also containing adenylate kinase and acetate kinase with ATP, ADP, AMP, Mg^{2+} acetate, and

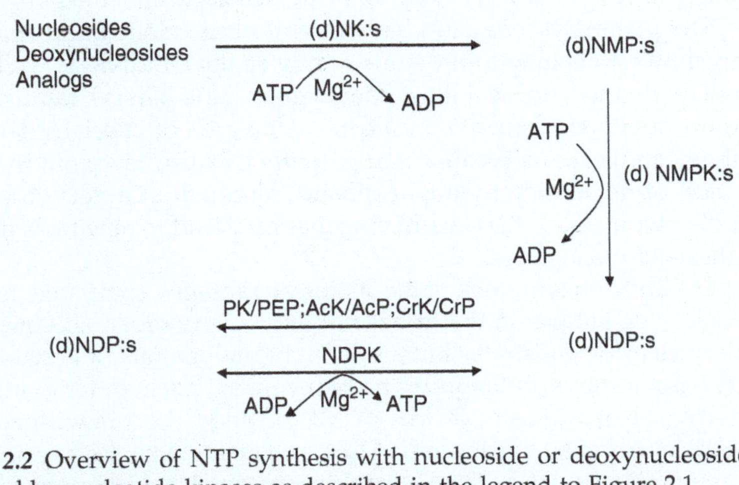

Figure 2.2 Overview of NTP synthesis with nucleoside or deoxynucleoside kinase followed by nucleotide kinases as described in the legend to Figure 2.1.

the reducing agent dithiotreitol (DTT); the overall yield of the final product ATP was 98%.[11] A similar approach based on crude preparations of yeast adenosine kinase and adenylate kinase for the continues synthesis of ATP has been described.[12] The rationale for the coupled reactions will be described in the following sections. Several crude preparations or partially purified preparations have been used, e.g., uridine–cytidine kinase to phosphorylate 5-azacytidine to 5-aza-CMP in small scale (at pH 6.6 to stabilize the analog) with about 12% yields.[13]

Adenosine (AK) regulates the extracellular adenosine and intracellular adenylate concentrations and inhibition of the enzyme elevate adenosine to levels that activate nearby adenosine receptors and produce a wide variety of physiological activities. The structure of the human adenosine kinase has been determined.[14] The overall structure is similar to the structure of ribokinase from *Escherichia coli*, which belongs to the family of carbohydrate kinases.

Two molecules of adenosine were present in the initial AK crystals with one adenosine molecule located in a site that matches the ribose site in ribokinase and probably represents the substrate-binding site. The second adenosine site probably represents the ATP site. An Mg^{2+} ion-binding site is observed in a trough between the two adenosine sites. The structure of the active site is consistent with the observed substrate specificity. The specificity of AK is relatively restricted to adenosine, but the enzyme can accept deoxyadenosine as well as certain analogs (see below) with much lower, but still significant, activity.[15] Uridine–cytidine kinase, which belongs to the same enzyme family, has been used for synthetic purposes,[13] but only to a very limited extent.

Some recent synthetic applications of the broadly specific deoxynucleoside kinases from *Drosophila melanogaster* (DmdNK) and plants, e.g., tomato-dNK, will be presented here (Figure 2.2). DmdNK has recently been shown to have considerable potential as a more general and efficient phosphorylating agent for a large variety of nucleosides and nucleoside analogs.[16,17] The plant dNK enzymes have recently been cloned and characterized and shown to belong to the same family as the DmdNK as well as the mammalian deoxycytidine kinase/deoxyguanosine kinase family.[19] The dNK from tomato (*Lycopericum escultentum*) may be of special interest for biosynthetic applications because it apparently can also carry out the phosphorylation of nucleoside monophosphates, although at a very low rate.[18] Further development of this class of enzymes may lead to new tools of value for synthesis of nucleotides.

The DmdNK enzyme has some major advantages compared to other deoxynucleoside kinases in that it has not only a very broad acceptance for different nucleoside substrates but also a high catalytic rate; it is stable even at higher temperatures and requires no reducing agents in order to maintain full activity. The synthesis of dCMP, dGMP, or dAMP occurs with approximately 80% yields in 60 h using ATP, Mg^{2+}, and creatine kinase as phosphate donor regenerating system. The Drosophila enzyme has been used for commercial production of dideoxynucleotide analogs.[17]

The structure of DmdNK was recently determined[19] and the results demonstrated that it belongs to the nucleotide/deoxynucleoside kinase family, which also includes HSV1-TK. These enzyme structures consist of a dimer of two 31-kDa subunits, each with an alpha/beta-structure made of 9 to 10 helices and five beta-sheets with an active site cleft in which a nucleoside-binding domain and the ATP-binding loop are key features. This latter loop is formed by a strand-turn-helix that forms a large anion hole that coordinates the phosphate group of the phosphate donors.

This type of ATP-binding structure is found in all nucleoside/nucleotide kinases studied so far. On top of the active site, a helix containing basic amino acids forms the s c lid region. The length and size of the lid region varies in different nucleoside/nucleotide kinases, but there is always this kind of function because it allows shielding of the active site in order to limit the access for water molecules that otherwise will compete with the 5′-OH group of the nucleosides and result in the transfer reaction. The capacity of DmdNK to phosphorylate many different nucleoside and nucleoside analogs at high rates has been described as well as many important structure–function relationships.[16–19]

Another member of this enzyme family that has been in focus for pharmacological investigations for two decades is herpes virus 1 thymidine kinase. The enzyme is a dimer of two 44-kDa subunits; its specificity towards nucleosides is very broad and many acyclic analogs such as acyclovir and ganciclovir are efficiently phosphorylated.[7,8,20] This is the underlying mechanism for the antiviral selectivity of most antiherpes drugs. The literature on the structure and activity of this enzyme in complex with many analogs[20] is extensive.

The herpes virus type 1 enzyme belongs to a different enzyme family than the adenosine kinases do, but shows high structural similarity with the insect and mammalian deoxycytidine/deoxyguanosine kinases as described earlier.[19] Herpes TK enzymes have recently received much attention in gene therapy applications.[21] This is again due to its broad substrate specificity, which includes L nucleosides and many other antiviral analogs not phosphorylated by the cellular deoxynucleoside kinases. Therefore, treatment with such analogs can be made very selective with minimal toxic side effect to nontransfected cells. Nevertheless, so far these viral enzymes have not been extensively used for synthetic approaches.

2.3 Nucleoside monophosphate kinases

This enzyme family contains many members; usually there is one type of monophosphate kinase for each nucleoside with regard to the bases, but not the sugar moiety — i.e., adenylate, guanylate, uridine/cytidine, and thymidylate kinases (AMPK, GMPK, UMP-CMPK, and dTMPK), respectively. This family also has several isoenzymes, e.g., mammalian cells have at least five different AMPK genes. These are differently expressed in tissues and show somewhat varying substrate specificities and, to a certain extent, also structures.[22] However, the basic reaction catalyzed is shown in Figure 2.1 and Figure 2.2.

2.3.1 *Adenylate kinase*

Purified adenylate kinase (EC 2.7.4.3) (myokinase, AMPK) from mammalian muscle tissues has been commercially available for many years and is the enzyme that has been used most frequently in synthetic procedures. A number of detailed studies have been performed by Whitesides and colleagues,[11,23,24] but only a few applications can be summarized here.

As mentioned earlier, adenylate kinase was combined with adenosine kinase in the highly effective generation of ATP from adenosine; the same enzyme could also carry out phosphorylation of CMP and ribavirin monophosphate (RMP), but with lower efficiency. In case of AMP phosphorylation, acetyl phosphate was the donor and acetate kinase (EC 2.7.2.1) was also needed. For CTP and RTP generation, phosphoenol pyruvate (PEP) was the donor and pyruvate kinase (EC 2.7.1.40) was added as a catalyst to produce the final phosphorylation reaction. This latter combination has involved the most commonly used reactants in the generation of nucleotides, as will be discussed later.

The overall yield for the synthesis of CTP and RTP was 74 and 93%, respectively, but the levels of AMPK were high and long reaction times were necessary. Apparently, AMPK has a relatively broad specificity, but requires the presence of reducing agents such as mercaptoethanol for optimal activity.[23,24] Protocols for the phosphorylation of ATP analogs (e.g., formycin A- and tubericidine-monophosphate) with AMPK followed by further phosphorylation by PEP and pyruvate kinase has been described with yields around 70%.[25] A recent optimization study in which the production of dATP from dAMP was determined with AMPK and PK with different Mg^{2+} and K^+ concentrations, etc. led to very high yields.[26] The most favorable conditions were a ratio of AMPK to PK of 1:2; a KH_2PO_4 buffer with pH 7.5 to 8.0; and a K^+ of 500 m*M* and Mg^{2+} concentrations of 30 m*M*.

2.3.2 *Nucleoside monophosphate kinase preparations*

Several crude or partially purified preparations of nucleoside monophosphate kinases from *E. coli*[27,28] and yeast[7,29] have been used in the synthesis of nucleotides. Imazawa and Eckstein isolated and partially purified three different monophosphate kinases from calf thymus and could get high-yield production of AraCMP, 2′-deoxy-2′-azido-CMP, dCMP, 2-deoxy-2′-azido UMP and 2′,3′-dideoxy-3′-amino-dAMP with these enzyme preparations.[29] In the previously mentioned synthesis of 5-Aza-CTP, a partially purified preparation of UMP-CMPK was included.[13]

The structures of many nucleoside monophosphate kinases are known and overall are similar to the ones described for nucleoside kinases mentioned earlier.[19,22,30] However, the subunit structure as well as the length of the lid region varies considerably among the enzyme family members. Schultz et al. have conducted elegant studies in which the large conformational rearrangements associated with the phosphorylation reactions are well characterized and displayed.[30] However, despite the considerable knowledge of the structure–function relationships in this enzyme family, the development of new biotechnologically favorable variants of nucleoside monophosphate kinase has so far not been reported. Especially, new enzymes with broader specificity, suitable as general reactants for many different nucleoside analogs, are much needed tools in synthetic work.

2.4 *Nucleoside diphosphate phosphorylating enzymes*

The formation of nucleoside triphosphates from nucleoside diphosphates can be catalyzed by several different enzyme systems of two general types (Figure 2.1 and Figure 2.2):

- Key enzymes in energy metabolism responsible for phosphorylation of ADP such as phosphoglycerate kinase (EC 2.7.2.3, PgK); pyruvate kinase (EC 2.7.1.40, PK); acetate kinase (EC 2.7.4.3, AcK); creatine kinase (EC 2.7.3.2, CrK) utilizing 1,3-disphosphoeglycerate, phosphoenol private (PEP); acetyl phosphate (AcP); and creatine phosphate (CrP)
- Nucleoside diphosphate kinase (EC 2.7.4.6), which catalyzes the reversible phosphorylation of diphosphates with nucleoside triphosphates as donors

The first category of enzymes clearly has the capacity to phosphorylate many different nucleoside diphosphates in addition to the natural substrates; a few examples are included later. However, there are also well established procedures based on nucleoside diphosphate kinases (e.g., reference 11 and reference 29) and the previously mentioned enzymes are commercially available. In several of the earlier studies with crude enzyme preparations, they contain sufficient amounts of these types of enzymes to guarantee the transformation of diphosphate to triphosphate, as long as an excess of ATP was present.

2.4.1 *Pyruvate kinase, acetate kinase, and creatine kinase*

The combination of PK and PEP can be regarded as the golden standard for obtaining highly efficient conversion of nucleoside diphosphates (including a large variety of analogs) to triphosphates. However, a high concentration of enzyme and PEP (up to 200 m*M*[26]) is required to drive the reaction; the costs involved are a limiting factor. Whitesides and colleagues have often preferred to utilize the less expensive AcP and AcK, preferentially from *B. stearothermophilus*, because this enzyme is stable, is not susceptible to auto-oxidation, and can be coupled to polyacrylamide beads retaining full activity. However, AcP is less stable than PEP and, for situations with long reaction times, the latter is superior, e.g., in RTP or formycin A-triphosphate synthesis.[23,25] The effective preparation of dATP using PEP and PK has been described earlier.[26] Other examples with this enzyme system are the synthesis of purine and pyrimidine alpha,beta-imido-triphosphate analogues,[31] demonstrating that natural phosphodiester bonds in the diphosphates are not necessary for PK.

CrK and CrP are also standard reagents in many phosphorylation reactions and the acceptance of L-nucleotides by CrK has been utilized in a recent protocol.[32] Relatively long reaction times were necessary in this case. However, in a two-step reaction, first with AMPK, followed by incubations with CrK/CrP, a quantitative conversion of beta-L2′, 3′ddAMP to the corresponding triphosphate was obtained.

2.4.2 *Nucleoside diphosphate kinase*

Highly purified nucleoside diphosphate kinase (EC 2.7.4.6 NDPK) preparations are also available commercially and have been successfully applied in several phosphorylation reactions. However, this type of reaction is reversible and it is difficult to get high-yield conversions. The reaction proceeds via a histidine–phosphate intermediate; this amino acid is relatively exposed on the surface of the enzyme.[33] The reaction is a typical example of a ping-pong mechanism and shows relatively little selectivity for different sugar or base compositions of the substrates.

The catalytic rate of this reaction is very high — several orders of magnitude higher that those of the nucleoside kinases. However, some important analogs are very inefficiently phosphorylated by NDPK, e.g., 3′-analogs and L-nucleosides.[34] In fact, it was recently shown that several pharmacologically important L-nucleosides apparently are phosphorylated by PgK and not by NDPK *in vivo*.[35] The unpredictable enatiomeric specificities of these enzymes are also illustrated by the fact that both enatiomers of carbovir (the carbocyclic analog of 2′, 3′dideoxyguanosine) can be phosphorylated by PK, PgK, and CrK; however, GMPK can only use the (+) enatiomer and NDPK preferred the (–) enatiomer of this analog.[36]

It has also been shown that NDPK can phosphorylate 8-hydroxy-2′-dGDP and 2-hydroxy-2′-dADP with relatively high yields and could be used to prepare radioactively labeled nucleotides.[37] Several different isoenzymes of the NDPK family are found in animal cells and are expressed differentially in tissues. However, they all apparently have similar enzyme kinetic properties,[33,34] although some of these isoenzymes may have other physiological roles unrelated to their capacity to catalyze phosphorylation reactions.

It is important to note that nucleoside monophosphate kinases (e.g., UMP-CMPK) can catalyze the formation of nucleoside triphosphate in a reverse reaction (Figure 2.3). This reaction has recently been shown to have different substrate requirements compared to the standard reaction in which diphosphates are formed so that the antiviral analog beta L-2′,3′-dideoxy-3′-thiacytidine (L-3TCMP) could be converted to diphos-

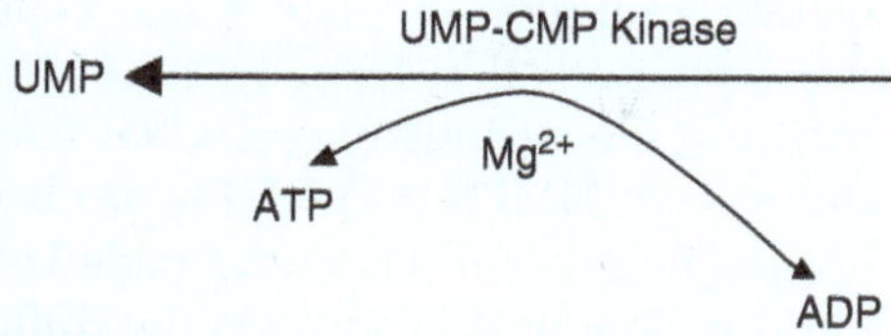

Figure 2.3 The reversibility of the NMPK reaction as exemplified by the forward and reverse reactions catalyzed by UMP-CMP kinase.

phate by human UMP-CMP kinase. However, it could not react in the reverse reaction and add a third phosphate.[38]

It is likely that, by screening other organisms or by setting up *in-vitro* mutagenesis and selection procedures, new types of nucleoside phosphorylating enzymes can be isolated with new and preferable properties regarding a broaden substrate specificity, high catalytic rates, and capacity to give quantitative conversion to triphosphate end products.

2.5 *Combined phosphorylation systems*

Mixtures of nucleoside triphosphates have been produced starting with RNA as source.[39] The nucleic acid was first degraded to RNA oligomers by nuclease P1 (EC 3.1.4) followed by a second step in which the oligonucleotides were converted to nucleoside diphosphates by treatment with polynucleotide phosphorylase (EC 2.7.7.8, PNPas) immobilized on polyacrylamid gels. This step is a reversible reaction and favors oligonucleotides; however, sufficient production of the diphosphates occurred and they were phosphorylated to triphosphates with PK and PEP.

The limiting step in this procedure was reported to be the costs and efficiency of the PNPase step, but the overall yields were: ATP: 24%; UTP: 28%; GTP: 30%; and CTP: 18%, all obtained as a mixture. In a later modification of this approach, complete degradation of RNA to the four ribonuceoside monophosphates was done, followed by stepwise phosphorylation using purified (and immobilized) preparations of AMPK, GMPK, and UMP/CMPK from yeast.[40] The final phosphorylation step was achieved with AcK and AcP as outlined previously. This procedure gave a mixture of all four ribonucleotides with the following yields: ATP: 90%; GTP: 90%; CTP: 60%; and UTP: 40%.

Large quantities of nuclease P1 were necessary due to its instability; it was not possible to immobilize this enzyme so that it could be reused. The monophosphate kinases also had limited stability and the reducing agent DTT had to be added. The reaction was carried out in two steps: first, the AMP was converted to ATP (for 2 days) and then the two other kinases were added in order to synthesize UTP, GTP, and CTP. Long reaction times (total of 9 days with continuous additions of several components, including pH control) were needed to get high yields. The main rate-limiting factor was reported to be due to substrate inhibition of UMP/CMP kinase. A general description of this type of coupled reaction for generation of ATP and the methods to prepare and measure the various reagents has been reported.[41]

A combination of bovine spleen NMPK and NDPK has been applied to prepare all the natural 2′-deoxyribonucleotides with yields between 25 and 75%.[42] A general problem in this type of procedure is the difficulty in separating the final end products from the nucleoside triphosphates added as phosphate donors. In the latter case, they utilized UTP instead of ATP as phosphate donor when synthesizing dGTP, leading to better separations during the final HPLC step.[42] Recently, a combination of synthetic and enzymatic procedures, using NDPK and PK and PEP for the final phosphorylation in the synthesis of azole carboxamide ribo- and deoxyribonucleoside triphosphates, was described.[43] No ADP was added in this reaction but a twofold excess of PEP compared to diphosphate analogs was present to ensure complete conversion to triphosphate. An efficient purification step for the analog triphosphates based on boronate affinity gels was also presented.[43]

2.6 Deoxyribonucleotide triphosphate synthesis from ribonucleoside triphosphates

Current gene technology and therapeutic applications have an increasing need for deoxyribonucleotides. Traditionally, 2′-deoxyribonucleosides and deoxyribonucleoside monophosphates are produced with herring and salmon sperm DNA as source. An alternative approach, based on RNA or, more specifically, ribonucleoside triphosphates, as starting material has been described.[44] The key enzyme in this case was pure recombinant ribonucleotide reductase from *Lactobacillus leichmanni* (Figure 2.4).[45]

This enzyme was very stable (with a half-life of 21 to 23 days) and the allosteric effectors needed for its full activity with ATP, CTP, GTP, ITP and UTP (at 10 m*M*-concentrations) could be substituted by the addition of 1 *M* sodium acetate or dipotassium phosphate. DTT was the preferred reducing agent; however, other reducing agents such as 1,4 dithioerythriol or 1,3-propanedithol could be used but had to be added repeatedly. Adenosine cobalamin was also needed (at 0.1 m*M*) and was apparently stable during these

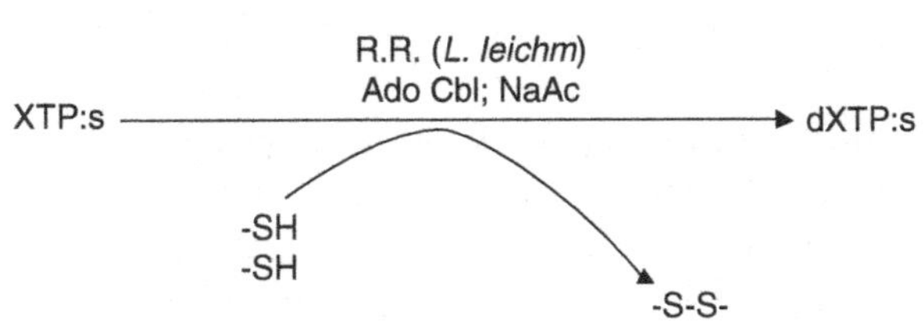

Figure 2.4 Synthesis of deoxyribonucleoside triphosphates with ribonucleotide reductase (RR) from *Lactobacillus lechmanni.* The reaction requires a dithiol as hydrogen donor and adenosine cobalamin (Ado Cbl) and high concentrations of sodium acetate.

reaction conditions — i.e., pH 8.2 and 25°C for 6 h. The overall yields were for dITP, 96%; dATP, 97%; dGTP, 92%; and dUTP, 91%. These results indicate that *L. leichmanni* ribonucleotide reductase could be a suitable tool for the production of dNTPs; however, a sufficient source of inexpensive and suitable XTPs has so far not been reported.

2.7 *Other synthetic procedures using enzymes*

Adenosine deaminase has a relatively high degree of tolerance to substitutions in the carbohydrate moiety of adenosine analogs; 2′- and 3′-deoxyadenosine serve as substrates. The preparation of radioactive inosine, deoxyinosine, and 2′-3′-dideoxyinosine by deamination is well established.[46] However, 2′,3′-didehydro carbocyclic, e.g., the enatiomer of carbovir, could also be resolved from a racemic mixture of the 2,6-diaminopurin analog only producing the (–) G containing nucleoside.[36] This analog was then phosphorylated using the crude phosphotransferase from *Serrata marcescens*, followed by treatment with GMPK and PK, CrK, or PgK as described earlier. A set of other carbocyclic analogs — aristeromycin, dideoxyaristeromycin, and cycladine — was deaminated at a slow but significant rate, provided enough adenosine deaminase was added to the solutions.[47] Considerable batch-to-batch variation in the activity of the enzyme preparations was observed and, although no yields were presented, the procedure was reported to be synthetically useful.

A somewhat unusual phosphorylation procedure was studied for (–) carbovir with human placenta 5′-nucleotidase (the CNII enzyme[46,47]). The reaction was carried out with 5 m*M* IMP as phosphate donor, 20 m*M* Mg^{2+}, and ATP as activator; the efficiency of phosphorylation of (–) carbovir was high — about 50% of that of inosine — and no phosphorylation of the (+) enatiomer was observed. This reverse reaction of the CNII enzyme has been demonstrated to be of *in-vivo* significance for activation of antiviral nucleoside analogs, e.g, ddI[48]; however, so far it has not been further tested in synthetic procedures.

Another method to obtain enatiomeric pure intermediates in the synthesis of carbocycylic nucleoside analogs was based on a lipase from *Pseudomonas cepacia*.[49] A racemic *trans*-2-(hydroxymethyl) cyclopentanol and the corresponding silyloxy alcohol could be resolved using the lipase in a *trans*-estrification reaction with vinyl acetate. Only one of the enatiomers could thus be acetylated and then the two forms could be separated and used for chemical synthesis of the two pure carbocyclic nucleosides. Surprisingly, both enantiomers of the carbocyclic adenosine analog triphosphates were efficient inhibitors of terminal deoxynucleotidiyl transferase.[49]

2.8 *Concluding remarks and future perspectives*

In this chapter, several examples of synthetic procedures for obtaining nucleoside triphosphates based on enzyme technology have been presented.

Clearly, several enzymes from carbohydrate metabolism and nucleoside/nucleotide metabolism have been utilized successfully for this purpose for many decades. Therefore, enzymes have an established role in this field that will likely increase considerably in the near future — partly because of increasing demand on industry for more environmentally friendly production methods, but also because of improved methods to isolate and design enzymes optimized for their specific production tasks.

The genomes and proteoms of many new organisms are rapidly disclosed; there are many new opportunities to use their great variety in life styles to identify enzymes, which can perform useful reactions. Furthermore, *in vitro* selection and enzyme design can now be accomplished with a series of new and powerful methods that can lead to isolation of enzymes tailor-made for specific reactions and conditions. Relatively little has so far been done in this respect with the enzymes in nucleotide synthesis; thus the potential for future biotechnological initiatives in this area is great.

Acknowledgments

The author wishes to thank M. Vaghefi for valuable suggestions and discussions in connection with the preparation of this chapter. S. Eriksson's work on enzymes in nucleotide metabolism is supported by grants from the Swedish Research Council and the Swedish Foundation for Strategic Research.

References

1. Burgess, K. and Cook, D. Synthesis of nucleoside triphosphates. *Chem. Rev.*, 2000, 100, 2047–2060.
2. Brawerman, G. and Chargaff, E. On the synthesis of nucleotides by nucleoside phosphotransferases. *Biochim. Biophys. Acta*, 1954, 15, 549–559.
3. Tunis, M. and Chargaff, E. Studies on the nucleoside phosphotransferase of carrot III. On the synthetic ability of the transfer enzymes. *Biochim. Biophys. Acta*, 1960, 40, 206–210.
4. Strider, W., Harvey, C., and Nussbaum, L. Enzymatic and chemical synthesis of 1-beta-D-arabinosyl-5-fluoropyrimidine 5′-phosphates. A comparative study. *J. Med. Chem.*, 1968, 11, 524–527.
5. Harvey, C.L., Clericuzio, E.M., and Nussbaum, A. Small-scale preparation of 5′-nucleotides and analogs by carrot phosphotransferase. *Anal. Biochem.*, 1970, 36, 413–421.
6. Gizievicz, J. and Shugar, D. Enzymatic phosphorylation of nucleosides to the 5′-phosphates. In: *Nucleic Acid Chemistry*, Townsend, L.B. and Tipson, R.S., Eds, Plenum Press, New York, 1979, 955–961.
7. Preussel, B., Fritz, M., and Langen, P. An enzymatic procedure for the synthesis of ^{3}H-thymidine-5′-triphosphate. *J. Labl. Comp. Radiopharmaceut.*, 1977, 14, 617–624.

8. Fyfe, J.A., Keller, P.M., Furman, P.A., Miller, R.L., and Elion, G.B. Thymidine kinase from herpes simplex virus phosphorylates the new antiviral compound, 9-(2-hydroxyethoxymethyl)guanine. *J. Biol. Chem.*, 1978, 253, 8721–7.
9. Humble, R.W., Mackenzie, G., and Shaw, G. Phosphorylation of some 5-aminoimidazole nucleosides to the 5′-phosphates using a phosphotransferase from wheat. *Nucleoside Nucleotides*, 1984, 3, 363–367.
10. Zinchenko, A.I., Barai, V.N., Zalashko, L.M., Pupeiko, N.E., Pirkota, T.I., Sivets, G.G., and Mikhailopoulo, I.A. Enzymatic synthesis of nucleoside 5′-mono- and triphosphates. *FEBS Lett.*, 1990, 260, 254–256.
11. Pollack, A., Baughn, R.L., Adalsteinsson, O., and Whitesides, G.M. Large-scale enzyme-catalysed synthesis of ATP from adenosine and acetyl phosphate. Regeneration of ATP from AMP. *J. Am. Chem. Soc.*, 1978, 100, 302–306.
12. Asada, M., Nakanishi, K., Matsuno, R., Kariya, Y., Kimura, A., and Kamikubo, T. Continuous ATP regeneration utilizing glycolysis and kinase systems of yeast. *Agric. Biol. Chem.*, 1978, 42, 1533–1538.
13. Lee, T.T. and Momparler, R.L. Enzymatic synthesis of 5-azacytidine 5′-triphosphate from 5-azacytidine. *Anal. Biochem.*, 1976, 71, 60–67.
14. Mathews, II, Erion, M.D., and Ealick, S.E. Structure of human adenosine kinase at 1.5. A resolution. *Biochemistry*, 1998, 37, 15607–15620.
15. Gomtsyan, A., Didomenico, S., Lee, C.H., Matulenko, M.A., Kim, K., Kowaluk, E.A., Wismer, C.T., Mikusa, J., Yu, H., Kohlhaas, K., Jarvis, M.F., and Bhagwat, S.S. Design, synthesis, and structure — activity relationship of 6-alkynylpyrimidines as potent adenosine kinase inhibitors. *J. Med. Chem.*, 2002, 45, 3639–3648.
16. Munch–Petersen, B., Piskur, J., and Sondergaard, L. Four deoxynucleoside kinase activities from *Drosophila melanogaster* are contained within a single monomeric enzyme, a new multifunctional deoxynucleoside kinase. *J. Biol. Chem.*, 1998, 273, 3926–3931.
17. Ihlenfeldt, H.G., Munch–Petersen, B., Piskur, J., and Soendergard, L. Deoxynukleosidekinase aus Insektenzellen zur Nukleoside-monophosphatsynthese. 1999. Patent DE 199114644.
18. Knecht, W., Munch–Petersen, B., and Piskur, J. Plant thymidine kinases and their use. 2003. Patent WO 03/100045A1.
19. Eriksson, S., Munch–Petersen, B., Johansson, K., and Eklund, H. Structure and function of cellular deoxynucleoside kinases. *Cell Mol. Life Sci.*, 2002, 59, 1327–1346.
20. Brown, D.G., Visse, R., Sandhu, G., Davies, A., Rizkallah, P.J., Melitz, C., Summers, W.C., and Sanderson, M.R. Crystal structures of the thymidine kinase from herpes simplex virus type-1 in complex with deoxythymidine and ganciclovir. *Nat. Struct. Biol.*, 1995, 2, 876–881.
21. Fillat, C., Carrio, M., Cascante, A., and Sangro, B. Suicide gene therapy mediated by the Herpes Simplex virus thymidine kinase gene/ganciclovir system: fifteen years of application. *Curr. Gene Ther.*, 2003, 3, 13–26.
22. Van Rompay, A.R., Johansson, M., and Karlsson, A. Substrate specificity and phosphorylation of antiviral and anticancer nucleoside analogues by human deoxyribonucleoside kinases and ribonucleoside kinases. *Pharmacol. Ther.*, 2003, 100, 119–139.

23. Kim, M.-J. and Whitesides, G.M. Enzyme-catalysed synthesis of nucleoside triphosphates from nucleoside monophosphates. ATP from AMP and ribavirin 5′-triphosphate from ribavirin 5′-monophosphate. *Appl. Biochem. Biotech.*, 1987, 16, 95–108.
24. Simon, E.S., Bednarski, M.D., and Whitesides, G.M. Generation of cytidine 5′-triphosphate using adenylate kinase. *Tetrahedron Lett.*, 1988, 29, 1123–1126.
25. Schobert, B. Enzymatic synthesis of ATP analogs and their purification by reverse-phase high-performance liquid chromatography. *Anal. Biochem.*, 1995, 226, 288–292.
26. Bao, J., Imamura, K., Sakiyama, T. and Nakanisshi, K. Enzymatic synthesis of deoxyadenosine triphosphate from deoxyadenosine monophosphate catalysed by a binary enzyme system of adenylate kinase and pyruvate kinase. *Kag Kog Nen Ken Hap Koe Yos*, 2002, 67, 1023.
27. Canellakis, E.S., Gottesman, M.E., and Kammen, H.O. A method for the synthesis of ribonucleoside and deoxyribonucleoside-5`-triphosphates. *Biochim. Biophys. Acta*, 1960, 39, 82–87.
28. Symon, R.H. Synthesis of alpha-^{32}p-ribo-and deoxyribonucleoside-5′-triphosphates. *Methods Enzymol.*, 1974, 29, 103–115.
29. Imazawa, M. and Eckstein, F. Synthesis of sugar modified nucleoside 5′-triphosphates with partially purified nucleotide kinases from calf thymus. *Biochim. Biophys. Acta*, 1979, 570, 284–290.
30. Abele, U. and Schulz, G.E. High-resolution structures of adenylate kinase from yeast ligated with inhibitor Ap5A showing the pathway of phosphoryl transfer. *Protein Sci.*, 1995, 4, 1262–1271.
31. Li, A., Muscate, A., and Kenyon, G.L. Synthesis, characterization and inhibitory activities of nucleoside alpha/beta-imido triphosphate analogues on human immunodeficiency virus-1 reverse transcriptase. *Bioorg. Chem.*, 1996, 24, 251–261.
32. Shoshani, I., Boudou, V., Pierra, C., Gosselin, G., and Johnson, R.A. Enzymatic synthesis of unlabeled and beta-^{32}p-labeled-beta-L-2′, 3′-dideoxyadenosine-5′-triphosphate as a potent inhibitor of adenylyl cyclases and its use as reversible binding ligand. *J. Biol. Chem.*, 1999, 274, 34735–34741.
33. Chen, Y., Gallois–Montbrun, S., Schneider, B., Veron, M., and Morera, S., Deville–Bonne, D., Janin, J. Nucleotide binding to nucleoside diphosphate kinases: x-ray structure of human NDPK-A in complex with ADP and comparison to protein kinases. *J. Mol. Biol.*, 2003, 332, 915–926.
34. Bourdais, J., Biondi, R., Sarfati, S., Guerreiro, C., Lascu, I., Janin, J., and Veron, M. Cellular phosphorylation of anti-HIV nucleosides. Role of nucleoside diphosphate kinase. *J. Biol. Chem.*, 1996, 271, 7887–7890.
35. Krishnan, P., Gullen, E.A., Lam, W., Dutschman, G.E., Grill, S.P., and Cheng, Y.C. Novel role of 3-phosphoglycerate kinase, a glycolytic enzyme, in the activation of *L*-nucleoside analogs, a new class of anticancer and antiviral agents. *J. Biol. Chem.*, 2003, 278, 36726–36732.
36. Miller, W.H., Daluge, S.M., Garvey, E.P., Hopkins, S., Reardon, J.E., Boyd, F.L., and Miller, R.L. Phosphorylation of carbovir enantiomers by cellular enzymes determines the stereoselectivity of antiviral activity. *J. Biol. Chem.*, 1992, 267, 21220–21224.
37. Kamiya, H. and Kasai, H. Preparation of 8-hydroxy-dGTP and 2-hydroxy-dATP by a phopshate transfer reaction by nucleoside diphosphate kinase. *Nucleoside Nucleotides*, 1999, 18, 307–310.

38. Pasti, C., Gallois–Montbrun, S., Munier–Lehmann, H., Veron, M., Gilles, A.M., and Deville–Bonne, D. Reaction of human UMP-CMP kinase with natural and analog substrates. *Eur. J. Biochem.*, 2003, 270, 1784–1790.
39. Wong, C.-H., Haynie, S.L., and Whitesides, G.M. Preparation a mixture of nucleosides triphosphates from yeast RNA: use in enzymatic synthesis requiring nucleoside triphosphate regeneration and conversion to nucleoside diphosphate sugars. *J. Am. Chem. Soc.*, 1983, 105, 115–117.
40. Haynie, S.L. and Whitesides, G.M. Preparation of a mixture of nucleoside triphosphates suitable for use in synthesis of nucleotide phosphate sugars from ribonucleic acid using nuclease P1, a mixture of nucleoside monophosphokinases and acetate kinase. *Appl. Biochem. Biotechnol.*, 1990, 23, 205–220.
41. Crans, D.C., Kazlauskas, R.J., Hirschbein, B.L., Wong, C.H., Abril, O., and Whitesides, G.M. Enzymatic regeneration of adenosine 5′-triphosphate: acetyl phosphate, phosphoenolpyruvate, methoxycarbonyl phosphate, dihydroxyacetone phosphate, 5-phospho-alpha-D-ribosyl pyrophosphate, uridine-5′-diphosphoglucose. *Methods Enzymol.*, 1987, 136, 263–280.
42. Pal, S. and Nair, V. Enzymatic synthesis of 2′-deoxynucleoside diphosphates and triphosphates. *Biotech. Lett.*, 1998, 20, 1149–1151.
43. Wu, W., Bergstrom, D.E., and Davidsson, J.V. A combination chemical and enzymatic approach for the preparation of azole carboxamide nucleoside triphosphate. *J. Org. Chem.*, 2003, 68, 3860–3865.
44. Brunella, A. and Ghishalba, O. Recombninant *Lactobacillus leichmanni* ribonucleosidetriphosphate reductase as biocatalyst in the preparative synthesis of 2′-deoxyribonucleoside-5′-triphosphates. *J. Mol. Catal. B*, 2000, 10, 215–222.
45. Brunella, A., Abrantes, M., and Ghisalba, O. Preparative 2′-reduction of ATP catalyzed by ribonucleotide reductase purified by liquid–liquid extraction. *Biosci. Biotechnol. Biochem.*, 2000, 64, 1836–1841.
46. Johnson, M.A. and Fridland, A. Phosphorylation of 2′,3′-dideoxyinosine by cytosolic 5′-nucleotidase of human lymphoid cells. *Mol. Pharmacol.*, 1989, 36, 291–295.
47. Gala, D. and Schumacher, D.P. Deamination of carbocyclic adenosine analogs with Adenosine deaminase and its synthetic utility. *Synthesis Lett.*, 1992, 61–63.
48. Bianchi, V. and Spychala, J. Mammalian 5′-nucleotidases. *J. Biol. Chem.*, 2003, 278, 46195–46198.
49. Theil, F., Ballschuh, S., Flatau, S., von Janta–Lipinsky, M., and Matthes, E. Chemo-enzymatic synthesis of a new type of enansiomerically pure carbocyclic nucleoside analogues with strong inhibitory effects on terminal deoxynucleotidyl transferase. *Bioorg. Med. Chem.*, 1998, 6, 701–706.

chapter three

Synthesis and properties of NTP analogs with modified triphosphate chains

Inna Koukhareva*
Trilink Biotechnologies Inc., San Diego, CA

Morteza Vaghefi
ChemCyte Corporation, San Diego, CA

Alexandre Lebedev
Trilink Biotechnologies Inc., San Diego, CA

Contents

* Corresponding author.

3.1 Introduction

Nucleoside 5′-triphosphates (NTP) are known as fundamental components of all living organisms. They act as substrates for many NTP-binding proteins and as regulators or mediators of the enzymatic activity. ATP and GTP are the primary energy resources for the cell and all of the NTPs are important participants in a large variety of cellular processes.

The NTP analogs and derivatives with modifications in the triphosphate chain have been used widely in studies of the structure and mechanism of NTP-binding protein action. The effects of these modifications have been explored in order to determine the stereochemical and electronic requirements for binding of NTPs to the different proteins. Modifications of the triphosphate fragment have led to the development of NTP analogs that serve as stable crystallographic and spectroscopic probes of the proteins. Triphosphate chain modifications become absolutely necessary when stabilization of a specific iso-form of the protein is required to grow homogeneous crystals or to obtain a single state of protein.

Studies using different DNA or RNA polymerases to investigate incorporation of NTP analogs with modified triphosphate fragments into the growing polynucleotide chain or inhibition of the enzymatic activity of these polymerases produce valuable information for understanding the mechanisms of enzyme actions. These findings can also contribute to the development of potential new drugs for the treatment of viral diseases.

Another important function of the NTPs is the participation in transfer of the phosphoryl group to different substrates. Nonhydrolysable analogs of NTPs (mostly ATP) have been valuable tools for biochemical and structural studies in which hydrolysis of the P–O–P linkages and/or transfer of the phosphate group occur. Nonhydrolysable and stable NTP analogs are also important for carrying special reporter or ligand groups such as biotin and fluorescien required in the investigation of cell membrane penetration, diffusion, and distribution of these compounds inside the cell.

In this chapter the synthetic approaches for the preparation of NTP analogs and derivatives containing different types of modification of the triphosphate chain are discussed. Substitutions for bridging oxygen atoms and nonbridging oxygen and hydrogen atoms are among the most common ways of modifying the triphosphate chain. Many examples of NTP derivatives exist in which the triphosphate chain, as well as the nucleoside fragment (base and/or sugar), is modified. Finally, the isotopic substitutions in the triphosphate chain, including oxygen and/or phosphorus atoms, represent another important class of NTP analogs.

A general structure for NTP analogs with modifications in the triphosphate chain is shown here. Throughout this chapter these designated symbols are used to identify the exact location of each modification.

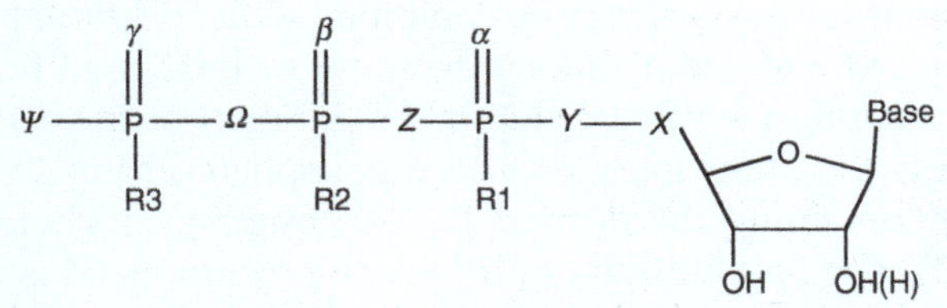

In many instances, common synthetic schemes, reagents, or techniques are used for the preparation of different derivatives containing a specific modification of the triphosphate chain. Most of the commonly used synthetic schemes are presented in this chapter. However, in many examples, the preparation of a specific NTP derivative requires a unique chemical route. Those specific approaches can be found in the cited original literature.

Several excellent reviews have been published on the preparation and biological applications of the NTP analogs with modified triphosphate chains.[1–5] These reviews contain valuable information and data analysis relevant to a majority of today's research needs. Therefore, our efforts were concentrated primarily on expanding the existing database to include the NTP analogs synthesized and studied since those reviews were written.

In this chapter NTP analogs and their derivatives are classified and tabulated according to the position of modification in the triphosphate chain. The relevant biological applications (mostly enzymatic) of these modified NTPs, if known, are indicated in the tables and summarized and discussed in Section 3.8. A large variety of phosphorus NMR data for NTP analogs was analyzed and tabulated in a general form. These data may be helpful for identification of the relevant compounds and for a structural analysis of newly synthesized NTP analogs with modification in the triphosphate fragment.

3.2 *Bridged modifications*

3.2.1 *Substitutions in P^1–O–CH_2(5′) fragment (structure I type analogs)*

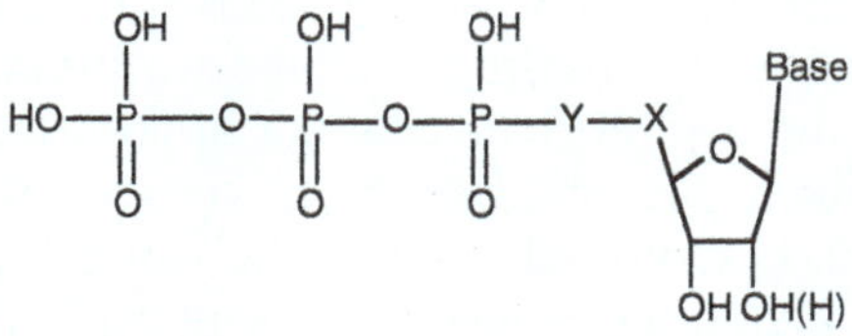

In many examples of NTP analogs, the original P^1–O–CH_2(5′) fragment, or its equivalent, has been modified (Structure I). One common type of modification represents the substitution for an oxygen atom (Y in the P^1–Y–CH_2(5′) fragment) with S, NH, or CR_2 (R is H, alkyl, aryl group and their derivatives). The substitutions for oxygen may be combined with additional substitutions for CH_2 group (X) and can also include sugar or base modifications (Table 3.1). The key common synthetic step in preparation of the Structure I type NTP analogs is the attachment of a pyrophosphate group to a nucleotide precursor (1) containing the desired P^1–Y–X fragment (Scheme 3.1, routes 3.1A and 3.1B). The preparation of nucleotide precursors (1) is not reviewed in this chapter, but information can be found in the relevant cited references.

Generally, there are two possibilities for the attachment of the pyrophosphate moiety to the precursor (1), as shown in route 3.1A. One approach relies on the activation of the nucleotide precursor (1), and the other involves the activation of the pyrophosphate precursor (2), as shown in route 3.1B.

Nucleotide precursors (1) are commonly activated by preparing N-morpholidate, N-imidazolidate, N-pyridinium or another derivative (1a) with a high-energy P–N or anhydride bond (route 3.1A). P-activation can be achieved through the use of different types of coupling reagents (1,1-carbonyldiimidazole; carbodiimides; arylsulfonylchlorides; triphenylphosphine/α,α-dipyridyl disulfide; and others) — usually in the presence of a weak organic base (imidazole, triazole, or pyridine). 1,1-Carbonyldiimidazole is the most common reagent that efficiently converts phosphomonoesters (or their analogs) to their respective phosphoryl imidazolidate derivatives. Effective P-activation may also be achieved through a formation of a mixed anhydride intermediate of phosphomonoester and some hindered aromatic carboxylic acid derivatives, such as mesitylene carbonyl chloride. The activated nucleoside precursor (1a) then reacts with the trialkylammonium salt of pyrophosphoric acid in anhydrous aprotonic media. In some cases, the

Table 3.1 Nucleoside 5′-Triphosphate Analogs with Bridged Substitutions in P^1–O–CH_2(5′) Fragment

X	Y	Sugar type or its substitute	Base	Synthetic route[a]	Biological applications: Enzyme	Biological applications: Substrate properties	Ref.
O	CH_2	But-2-enyl	A, G T, C	3.1A	HIV-1	Substrate (CT)[b]	15
					AMV RT	Substrate (CT)	
					DNA pol (various)	Inactive	
O	CH_2	Cyclopentene type sugar (D isomer)	A	3.1A	HIV RT	Substrate (CT)	16, 17
					AMV RT	Substrate (CT)	
					TDT	Poor substrate (CT)	
O	CH_2	Cyclopentene type sugar (L isomer)	A	3.1A	HIV RT	Substrate (CT)	16
					AMV RT	Substrate (CT)	
					DNA pol α	Inactive	
					DNA pol β	Poor substrate (CT)	
					TDT	Substrate (CT)	
O	CH_2	Cyclopentene type sugar (L-isomer)	G	3.1A	HIV RT	Substrate (CT)	16
					AMV RT	Substrate (CT)	
					DNA pol α	Inactive	
					DNA pol β	Poor substrate (CT)	
					TDT	Substrate (CT)	
O	CH_2	Cyclopentene type sugar (D & L isomers)	G	3.1A	HIV RT	Inhibition (probably CT)	18
O	CH_2	Cyclopentane type sugar (D isomer)	T	3.1A	HIV RT	Inhibition (probably CT)	19
O	CH_2	CH2-CH2-	A	3.1A	None	None	20
O	CH_2	Cyclopentene type sugar (L isomer)	G	3.1A	HIV RT	Inhibition (probably CT)	21
O	CH_2O	Rib	A, G,	NS[c]	Uridine kinase	Substrate	22

Table 3.1 Nucleoside 5′-Triphosphate Analogs with Bridged Substitutions in P[1]–O–CH_2(5′) Fragment (Continued)

X	Y	Sugar type or its substitute	Base	Synthetic route[a]	Biological applications: Enzyme	Biological applications: Substrate properties	Ref.
CH=	=CH	Rib	A	NS	Methionine adenosyl transferase	Inhibitor	23
CH_2	CH_2	3′-azido d-Rib	T	3.1A	HIV RT	Inactive	24
					PDE	Substrate	
					Alkaline phosphatase	Substrate	
CH_2	CH_2	Rib	A	3.1A	Adenosyltransferase	Inhibitor	23, 25, 26
CHR R = (L)SCH2CH2 -CH(NH2)CO2H (two isomers)	CH_2	Rib	A	3.1A	Adenosyltransferase	Weak and moderate inhibitors (R and S isomers)	27, 28
CHR R = (L)SCH2CH2 -CH(NH2)CO2H (two isomers)	CH_2	Rib	A	3.1A	Adenosyltransferase	Inhibitor	25, 29
CH_2	CH(CN)	Rib	A		AMP kinase	Substrate (one isomer)	30
CH_2	CH_2	Rib	A	3.1A	None	None	31
CH_2CH_2	NH	Rib	A	3.1C	Adenosyltransferase	Inhibitor	27

CH2CH(R) R = (L)SCH2CH2 -CH(NH2)CO2H (two isomers)	NH	Rib	A	3.1C	Adenosyltransferase	Inhibitor (R isomer) Weak inhibitor (S isomer)	27, 29
CH_2	NH	d-Rib	T	3.1C	DNA pol I	Substrate	32
CH_2	NH	Rib	A	3.1C	Hexokinase	Substrate	33
					Kreatine kinase	Substrate	34
CH_2	NH	d-Rib	T	3.1C	DNA pol I	Substrate	35
CH_2	NH	d-Rib	A, C G, T U, I	3.1C	Klenow DNA pol	Substrate	36
CH_2	S	Rib	U, A	3.1A, 3.1D	RNA pol (T7, *E. coli*)	Inactive or weak inhibitor	37
					Hexokinase	Substrate	38
					Alkaline phosphatase	Substrate	
CH_2	S	d-Rib	T	3.1A	DNA pol 1 (*E.coli*)	Inactive	39
CH_2	None	d-Rib	T	3.1A	DNA pol I	Inactive	40
CH_2	None	Rib	U	3.1A	PME	Substrate	41
					PDE	Substrate	

[a] For route see Scheme 3.1.

[b] CT = chain terminating substrate.

[c] NS = not specified or commercial product.

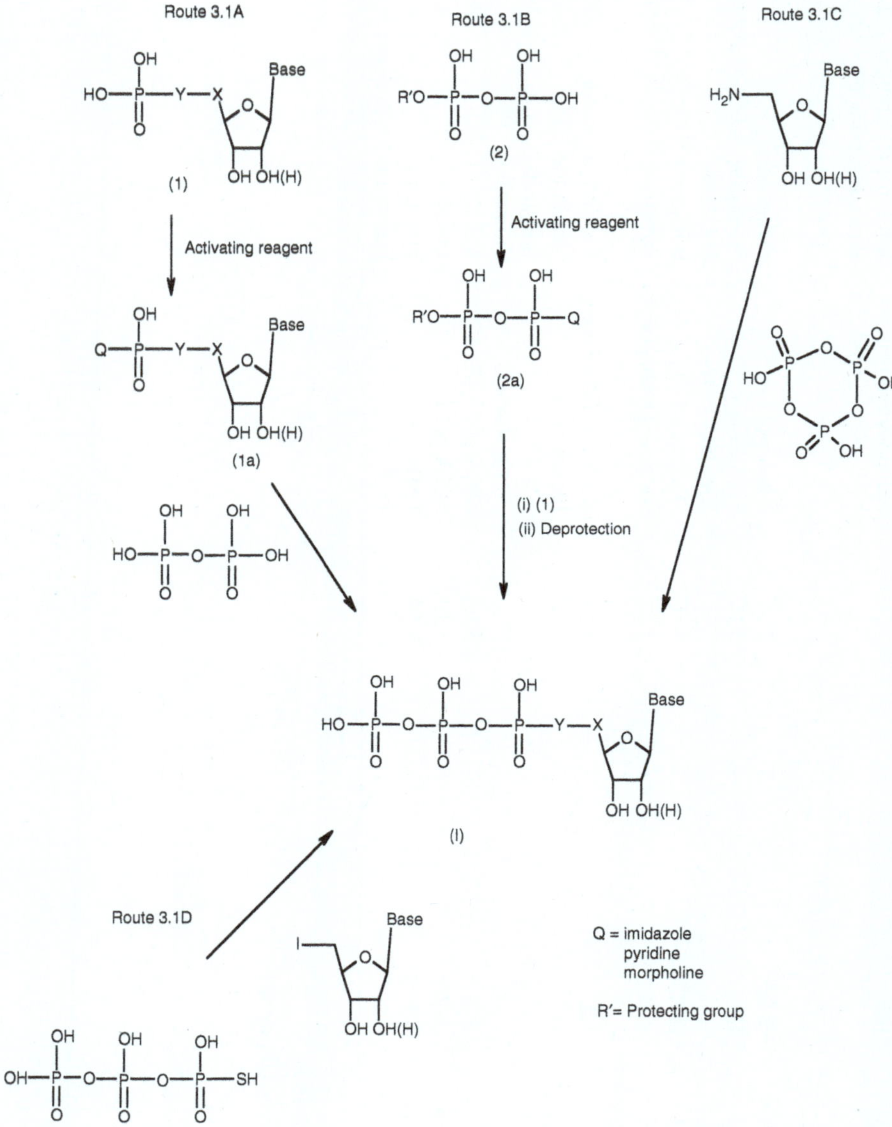

Scheme 3.1 Approaches to the synthesis of NTP analogs with modification(s) in P^1-O-CH_2(5′) fragment.

coupling reaction can be performed in a buffered aqueous solution catalyzed by a transition metal ion.[6,7]

Activation of the pyrophosphate moiety in a form of pyrophosphate monoester derivative (2a) may be achieved through a reaction with coupling reagents, usually 1,1-carbonyldiimidazole, leading to the formation of an intermediate with high-energy P–N bond (route 3.1B). Deprotection of the terminal phosphate group and/or nucleoside moiety may be required after completion of the coupling reaction. The authors are not aware of any reports utilizing route 3.1B for the synthesis of Structure I type NTP analogs. However, this route was widely used for the synthesis of m^{7}GpppG and its derivatives,[8–14] as discussed in Section 3.3.3.

Structure I type compounds with a P–NH–CH_2(5′) fragment were prepared in a very specific way. Inorganic cyclic trimetaphosphate is reacted with 5′-deoxy-5′-aminonucleoside at room temperature to give the corresponding NTP analog, as shown in route 3.1C.

3.2.2 *Substitutions for oxygen in P^2–O–P^1 fragment (structure II type analogs)*

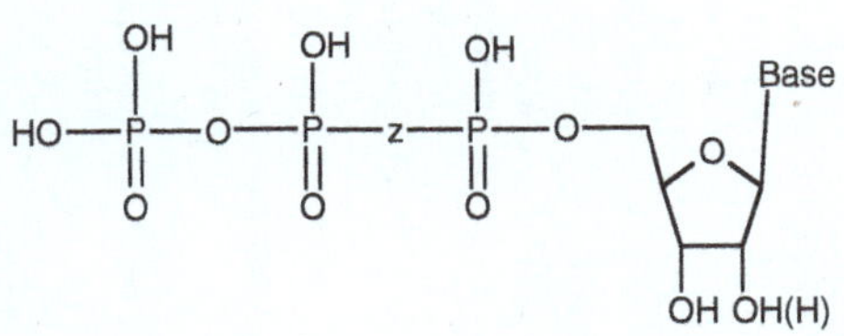

Synthesis of P^2–Z–P^1-substituted NTP analogs (Table 3.2), where Z = NH, CH_2, CH(Hal), $C(Hal)_2$ instead of oxygen atom, typically involves preparing a nucleoside 5′-diphosphate (NDP) analog (3) by coupling of nucleoside (4) and pyrophosphate analog (5) containing a P^2–Z–P^1 fragment as illustrated in Scheme 3.2. This reaction may be achieved by an *in situ* reaction of the P^2–Z–P^1 derivative and 5′-hydroxyl group of nucleoside in the presence of a coupling reagent (route 3.2A) or by activating the 5′-hydroxyl group of the protected nucleoside followed by coupling with P^2–Z–P^1 derivative (6) (route 3.2B). An alternative approach, route 3.2C, has been reported for the preparation of an NDP analog (3) with Z = NH. Nucleoside is treated with trichloro[(dichlorophosphoryl)imino]phosphorane followed by alkaline hydrolysis of the intermediate.

The synthesis of an NTP analog (II) from the NDP precursor (3) is completed by attaching the terminal phosphate group. The phosphomonoester derivative (7), such as 2-cyanoethylphosphate, is activated (usually as imidazolidate or morpholidate) and coupled with the NDP precursor (3) to yield the NTP analog (II) with a modified triphosphate chain (route 3.2E). Removal of the protecting group(s) may require an additional step after synthesis of the modified triphosphate chain is completed. An alternative

Table 3.2 Nucleoside 5′-Triphosphate Analogs with Bridged Substitutions in P^1–O–P^2 Fragment

Z	Sugar	Base	Synthetic route	Biological application: Enzyme or cell lines	Biological application: Substrate properties	Ref.
NH	Rib	A	3.2B + 3.2E	T7 pol	Substrate	42
				Creatine kinase	Substrate	
				Hexokinase	Substrate	
NH	Rib	A	3.2B + 3.2E	Adenosylmethionine synthesase	Substrate	43
NH	d-Rib or d-Rib-3′-azido	A, G, C, U(5-I), T	3.2C + 3.2D, 3.2C + 3.2E	HIV RT	Inhibitors	44
NH	d-Rib	T	3.2C + 3.2E	HIV RT	Inhibitor	45
				DNA pol	Inactive	
NH	d-Rib-3′-azido	T	3.2C + 3.2E	HIV RT	Inhibitor	46
				DNA pol	Inactive	
NH	2′,3′-dd-Rib	T	3.2C + 3.2E	HIV RT	Inhibitor	45
CH_2	Rib	A	3.2A + 3.2D	None	None	46
CH_2	Rib	A	NS	Adenosyltransferase	Weak inhibitor	47
CH_2	Rib	A	NS	Creatine kinase	Poor substrate	48
CH_2	Rib	A	NS	Adenylate cyclase	Inhibitor	49
CH_2	Rib	A	NS	ATP-diphosphohydrolase	Inhibitor	50
CH_2	Rib	G	NS	Cell lines	Interferes with cell functions	51
CH_2	dRib-3′-azido-	T	3.2A + 3.2D	Tumor cell lines	Inactive	52
				MSV virus	Inactive	
				HIV virus	Inactive	
				MSV and HIV RTs	Inactive	
				PDE I	Stable	
				PDE II	Stable	
CH_2	d-Rib	A, T	3.2B + 3.2D	None	None	53
CH_2	Rib	A	NS	ATP-diphosphohydrolase	Inhibitor	50
CHF	d-Rib	A, T	3.2B + 3.2D	None	None	53
CF_2	d-Rib	A, T	3.2B + 3.2D	None	None	53
CF_2	Rib	N^2-(4-n-butyl-phenyl-dG	3.2B + 3.2D	GTP-binding protein Ha-*ras*-p21	Inactive	54
CHBr	Rib	A	3.2B + 3.2D	None	None	55

Note: NS = not specified or commercial product; for route see Scheme 3.2.

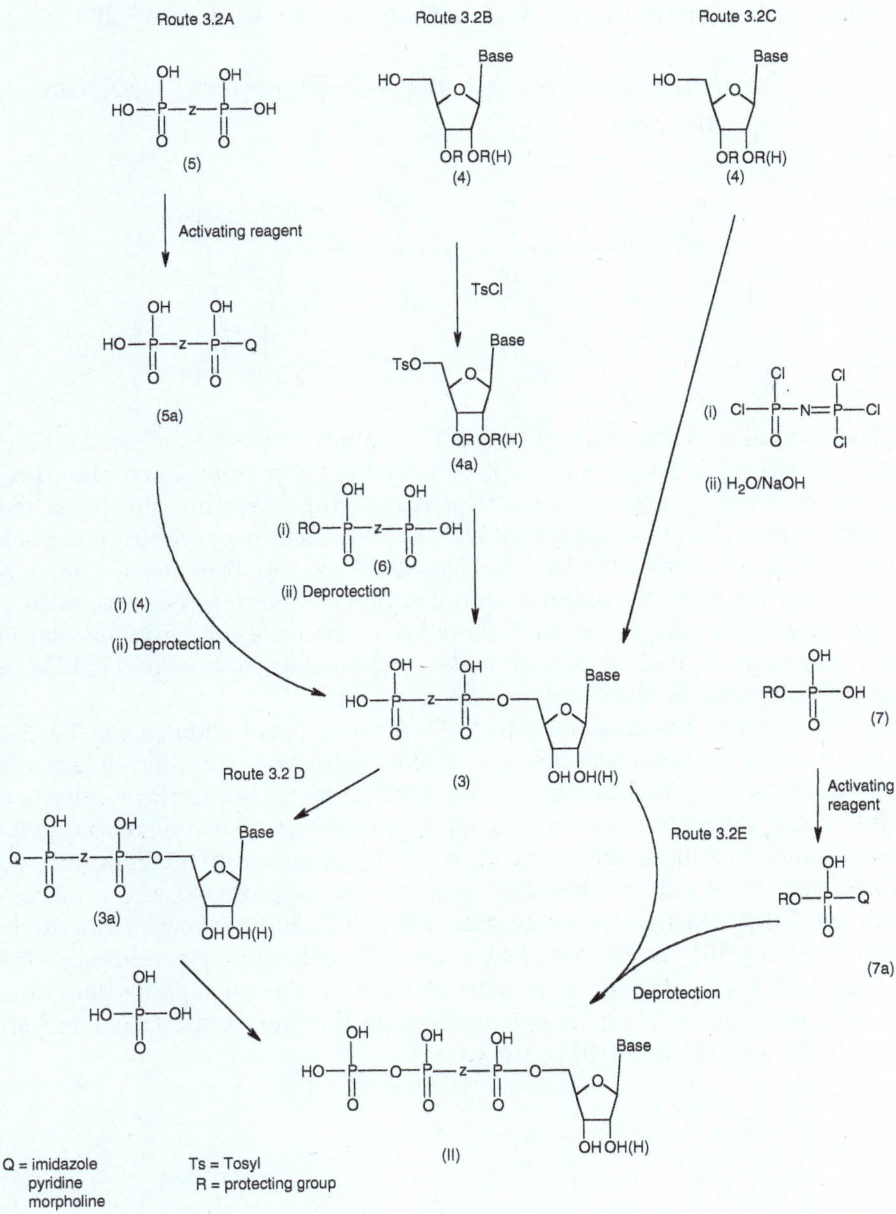

Scheme 3.2 Approaches to the synthesis of P[2]–Z–P[1]-substituted NTP analogs.

approach converts the NDP precursor (3) to imidazolidate (3a) with 1,1′-carbonyldiimidazole treatment followed by coupling with trialkylammonium salt of orthophosphoric acid to give desired analog (II) (route 3.2D).

3.2.3 *Substitutions for oxygen in P^3–O–P^2 fragment (structure III type analogs)*

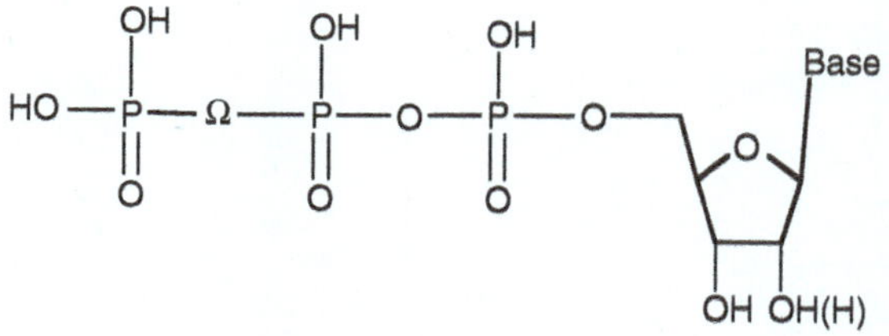

The syntheses of Structure III type NTP analogs (Table 3.3), where Ω = NH, CH_2, CH(Hal), $C(Hal)_2$, etc. in place of the oxygen atom, have often been accomplished by reacting an activated nucleoside 5′-monophosphate (8a) with a trialkylammonium salt of a P^3–Ω–P^2 analog of pyrophosphoric acid (5) (Scheme 3.3, route 3.3A). Nucleoside 5′-monophosphates (8) may be activated through the reaction with coupling/activating reagents, such as diphenylchlorophosphate; 1,1-carbonyldiimidazole; 2,4,6-mesitylene carbonyl chloride; carbodiimides; arylsulfonylchlorides; triphenylphosphine/α; α-dipyridyl disulfide; as well as others.

Moderately reactive nucleoside 5′-phosphomorpholidates can be prepared from mononucleotides via N,N-dicyclohexylcarbodiimide-assisted coupling as described in the original publication reporting the synthesis of these compounds.[56] Even a nucleoside 5′-phosphoramidate will react directly with methylenediphosphonic acid to yield a structure (III) compound. An activated nucleoside 5′-phosphomonoester can be prepared as an intermediate (10a) by reacting the nucleoside with $POCl_3$ and then converting to the NTP analog (III), as illustrated by route 3.3C. Alternatively, (although this route is not very popular), an activated P–Ω–P pyrophosphate derivative can be prepared and reacted with nucleoside 5′-monophosphate (route 3.3B) to obtain the structure (III) compound.

Table 3.3 Nucleoside 5′-Triphosphate Analogs with Bridged Substitutions in P^2–O–P^3 Fragment

W	Sugar	Base	Synthetic route	Biological applications		Ref.
				Enzyme/protein	Substrate properties	
NH	Rib	A	NS	Glycerol kinase	Inhibitor	57
NH	Rib	A	NS	None	None	58
NH	Rib	A	NS	Protein kinase A	Inhibitor	59
NH	Rib	A	NS	ATP phosphohydrolase	Inhibitor	60
NH	Rib	A	NS	ATP diphosphohydrolase	Inhibitor	50
NH	Rib	A	NS	N-acetyl-L-glutomate kinase	Forms crystals with enzyme	61
NH	Rib	G	NS	Adenylate cyclase	Effector	62
NH	Rib	G	NS	GTP binding protein	Effector	63
NH	Rib	A	3.3A	Myosin	Binds	64
NH	Rib	A	3.3A	None	None	65
NH	Rib	G	3.3A	None	None	65
NH	Rib	G	3.3A	None	None	66
NH	Rib	A	NS	Adenosyltransferase	Inhibitor	47
NH	Rib	G	NS	GTP-binding protein (Ras)	Binds, substrate	67
NH	Rib	A	3.3A	SVPD	Substrate	68
				Alkaline phosphatase	Substrate	
				Hexokinase	Inactive	
				Myokinase	Inactive	
NH	d-Rib-3′-azido	T	3.3C	HIV RT	Substrate (CT)[a]	45
NH	Rib-2′(3′)-BODIPY(FL)	G	NS	G-binding protein	Binds	69

Table 3.3 Nucleoside 5′-Triphosphate Analogs with Bridged Substitutions in P^2–O–P^3 Fragment (Continued)

W	Sugar	Base	Synthetic route	Biological applications: Enzyme/protein	Biological applications: Substrate properties	Ref.
CH_2	d-Rib-3′-azido	T	3.3C	Tumor cell lines	Selective inhibitor	52
				MSV virus	Inhibitor	
				HIV virus	Poor inhibitor	
				MSV and HIV RTs	Inactive	
				PDE I	Substrate	
				PDE II	Stable	
CH_2	Rib	A	NS	Glycerol kinase	Inhibitor	57
CH_2	Rib	A	NS	Polynucleotide phosphorylase	Inhibitor	70
CH_2	Rib	A	NS	Adenosyltransferase	Inactive	47
CH_2	Rib	A	NS	ATP diphosphohydrolase	Inhibitor	50
CH_2	Rib	A	NS	ATP phosphohydrolase	Inhibitor	50, 60
CH_2	Rib	G	NS	GTP-binding protein (Ras)	Binds, substrate	67
CH_2	d-Rib	G	NS	DNA pol III	Substrate	54
CH_2	Rib	A	NS	None	None	58
CHF	Rib	A, G	3.3A	None	None	58
CF_2	Rib	A	3.3A	Glycerol kinase	Inhibitor	57
CF_2	Rib	A, G	3.3A	None	None	58
CF_2	Rib	N^2-(4-n-butyl-phenyl)-G	3.3A	GTP binding protein (Ha-*ras*-p21)	Inactive	54
CF_2	d-Rib	G	3.3A	DNA pol III	Substrate	54
				DNA pol I	Poor substrate	
				DNA pol α	Inactive	
CF_2	d-Rib	N^2-(4-n-butyl-phenyl)-G	3.3A	DNA pol α	Inhibitor	54
CF_2	Rib	A	3.3A	Glycerol kinase	Strong inhibitor	57

CCl_2	Rib	A, G	3.3A	None	None	58
CHBr	Rib	A	3.3A	None	None	55
CBr_2	d-Rib	T	3.3A	AMV RT	Substrate	71
CBr_2	d-Rib	U(5-linker-azido)	3.3A	AMV RT	Substrate	71
CBr_2	d-Rib	U(5-linker-amino)	3.3A	AMV RT	Inactive	71
CBr_2	d-Rib	U(5-linker-biotin)	3.3A	AMV RT	Inactive	71
CBr_2	d-Rib	U(5-linker-fluorescein	3.3A	AMV RT	Inactive	71
CBr_2	d-Rib	U(5-linker-rhodamine	3.3A	AMV RT	Inactive	71
O-O	Rib	A	3.3A	Lucferase	Poor substrate	72
				NAD^+ pyrophosphorylase	Substrate	
				Alkaline phosphatase	Substrate	
				SVPD	Substrate	
				PRPP synthesase	Inhibitor	
				Various kinases	Inhibitor	
O-O	Rib	A	3.3A	None	None	58
O-O	Rib	A	3.3A	NAD^+ pyrophosphatase	Substrate	73
				Various kinases	Inhibitor/Inactive	
				PRPP synthesase	Inhibitor	
				RNA pol	Inactive	
O-O	d-Rib	A	3.3A	DNA pol	Inactive	73
O-O	d-Rib	G, T, C, U	3.3A	None	None	73
O-O	Rib	G, U, C, A	3.3A	None	None	73
-C ≡ C-	Rib	A	3.3A	None	None	58
None	Rib	A	3.3A	None	None	74, 75
None	Rib	G	3.3A	Protein synthesis in free cell system	Inhibitor	76

[a] CT = chain terminating substrate.

Note: NS = not specified or commercial product; for route see Scheme 3.3.

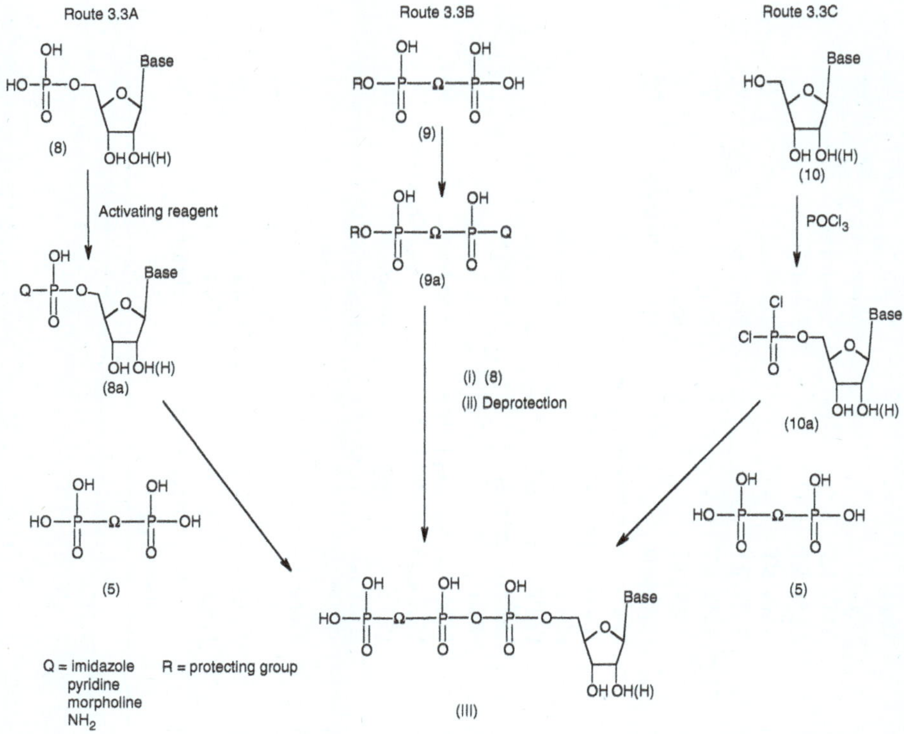

Scheme 3.3 Approaches to the synthesis of P²–Ω–P³-substituted NTP analogs.

3.3 Nonbridged modifications

3.3.1 Substitutions for OH group at P¹ (structure IV type analogs)

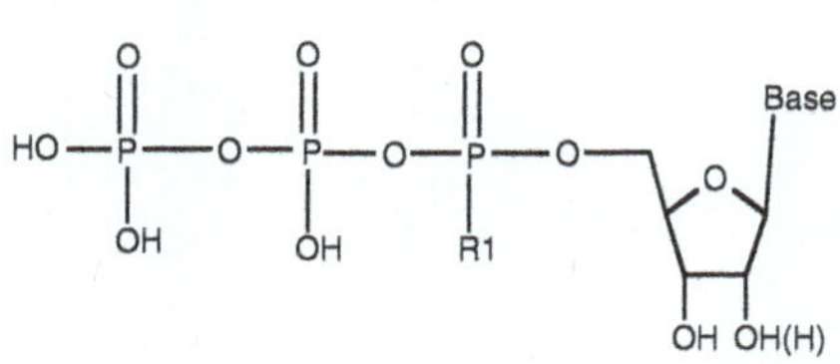

Syntheses of Structure IV type analogs (Table 3.4), where R1 = SH, BH_3^- replaces the usual OH group, have been performed most successfully through P(III) chemistry (Scheme 3.4, route 3.4A) using the Ludwig–Eckstein procedure.[80] The first step is the reaction of a protected nucleoside (4) with phosphitylating reagent (11) to form the intermediate nucleoside 5′-(4H-1,3,2-benzodioxaphosphorin-4-one) (12). The intermediate (12) is then coupled with pyrophosphate (tributylammonium salt) to form P²,P³-dioxa-P¹-nucleosidyl cyclic triphosphite (13). Finally, a sulfurization or boronation of the triphosphite (13), followed by hydrolysis of trimeta-

Table 3.4 Nucleoside 5′-Triphosphate Analogs with Nonbridged Substitution at P^1 Fragment

R1	Sugar	Base	Synthetic route	Biological application		Ref.
				Enzyme/protein	Substrate properties	
SH	Rib-2′-SH	A, U, C, G	3.4B	RNA pol (T7)	Substrate	77
SH	Rib	A, U, C, G	3.4B	None	None	78, 79
SH	Rib	A, U, C, G	3.4A	None	None	80
SH	Rib	A	3.4A	None	None	81
SH	Rib	A	NS	Mevalonate kinase	Substrate	82
				Phosphoglycerate kinase	Substrate	83
SH	d-Rib	A, T, C, G	3.4A or 3.4B	None	None	79, 80
SH	d-Rib	A, T, C, G	3.4B	None	None	84
SH	d-Rib-3′-F	T	3.4B	None	None	84
SH	d-Rib-3′-azido	T	3.4B	None	None	84
SH	d-Rib-3′-amino	T	3.4B	None	None	84
SH	Rib-2′-OMe	A, U, C, G	3.4A	None	None	25
SH	Rib	A	NS	ATPase	Substrate (both diastereomers)	85
SH	Rib	A	NS	Creatine kinase	Substrate (both diastereomers)	86
SH	Rib	A	3.4E	Acetate kinase	Substrate (both diastereomers)	87
SH	Rib	A	NS	Adenylate cyclase	Substrate (Rp diasereomer)	88, 89
SH	dRib	A, G, C, T	NS	DNA pol (Taq)	Substrate (Sp diastereomer)	90
SH	dRib	A	NS	DNA pol (T4)	Substrate (Sp diastereomer)	91
				DNA pol I	Substrate (Sp diastereomer)	92
SH	dRib	A	NS	DNA pol (T4)	Substrate (Sp diastereomer)	93
SH	dRib	T	NS	AMV RT	Substrate (Sp diastereomer)	94
SH	dRib	A, T, C, G	NS	DNA pol (Taq)	Substrate (Sp diastereomer)	90
SH	Rib	G	3.4B[a] or 3.4E	Acetate kinase	Substrate (one diastereomer)	95
				RNA pol	Substrate (one diastereomer)	
				Succinyl-CoA synthesase	Substrate (both diastereomers)	

Table 3.4 Nucleoside 5′-Triphosphate Analogs with Nonbridged Substitution at P[1] Fragment (Continued)

R1	Sugar	Base	Synthetic route	Biological application: Enzyme/protein	Biological application: Substrate properties	Ref.
SH	Rib	G	NS	Transducin, GTPase activity	Substrate/Inhibitor (Sp diastereomer)	96
					Inhibitor (Rp diastereomer)	
SH	Rib	A	3.4B	Hexokinase	Substrate	97
SH	Rib	A	NS	Mevalonate kinase	Substrate	82
SH	Rib	A	NS	Hexokinase	Substrate (one diastereomer)	98
SH	Rib	A	NS	Adenylate kinase	Substrate	99
SH	Rib	A	NS	Glycerate kinase	Substrate (Sp diastereomer)	83
SH	Rib	A	NS	ATPase	Substrate (both diastereomers)	100
SH	Rib	A	3.4D	tTRNA synthesase (Met)	Substrate (one isomer)	101
Me	d-Rib	T	3.4C	HIV RT	Substrate	102
				AMV RT	Substrate	
				DNA pol (I, β)	Substrate	
				DNA pol (α, ε)	Inhibitor	
				TdT	Substrate	
Me	d-Rib-3′-F	T	3.4C	HIV RT	Substrate (CT)[b]	103
				AMV RT	Substrate (CT)	
				DNA pol (I, β)	Substrate (CT)	
				DNA pol (α, ε)	Inhibitor	
				TdT	Substrate	
Me	d-Rib	T	3.4C	TdT	Substrate	103

Ph	d-Rib	T	3.4C	HIV RT	Substrate	17
				AMV RT	Substrate	
				DNA pol (I, β)	Substrate	
				TdT	Substrate	
Ph	d-Rib-3′-F	T	3.4C	HIV RT	Substrate (CT)	17
				AMV RT	Substrate (CT)	
				DNA pol (I, β)	Substrate (CT)	
				TdT	Substrate	
$C_{10}H_{21}$	d-Rib	T	3.4C	HIV RT	Substrate	17
				AMV RT	Substrate	
				DNA pol (I, β)	Substrate	
				TdT	Substrate	
BH_3	Rib	A, U, G, C	3.4A	None	None	104
BH_3	d-Rib	A, C, T, G, U	3.4A	None	None	105
BH_3	d-Rib	C(5-R) R = Me, Et, Br, I	3.4A	DNA pol (T7)	Substrate (one diastereomer)	106
BH_3	d-Rib	G	3.4A	DNA pol (T7)	Substrate	107
NH_2	Rib	A	3.4D	None	None	108
NH_2	d-Rib	T	3.4D	None	None	109

[a] Modified route 3.4B.

[b] CT = chain terminating substrate.

Note: NS = not specified or commercial product; for route see Scheme 3.4.

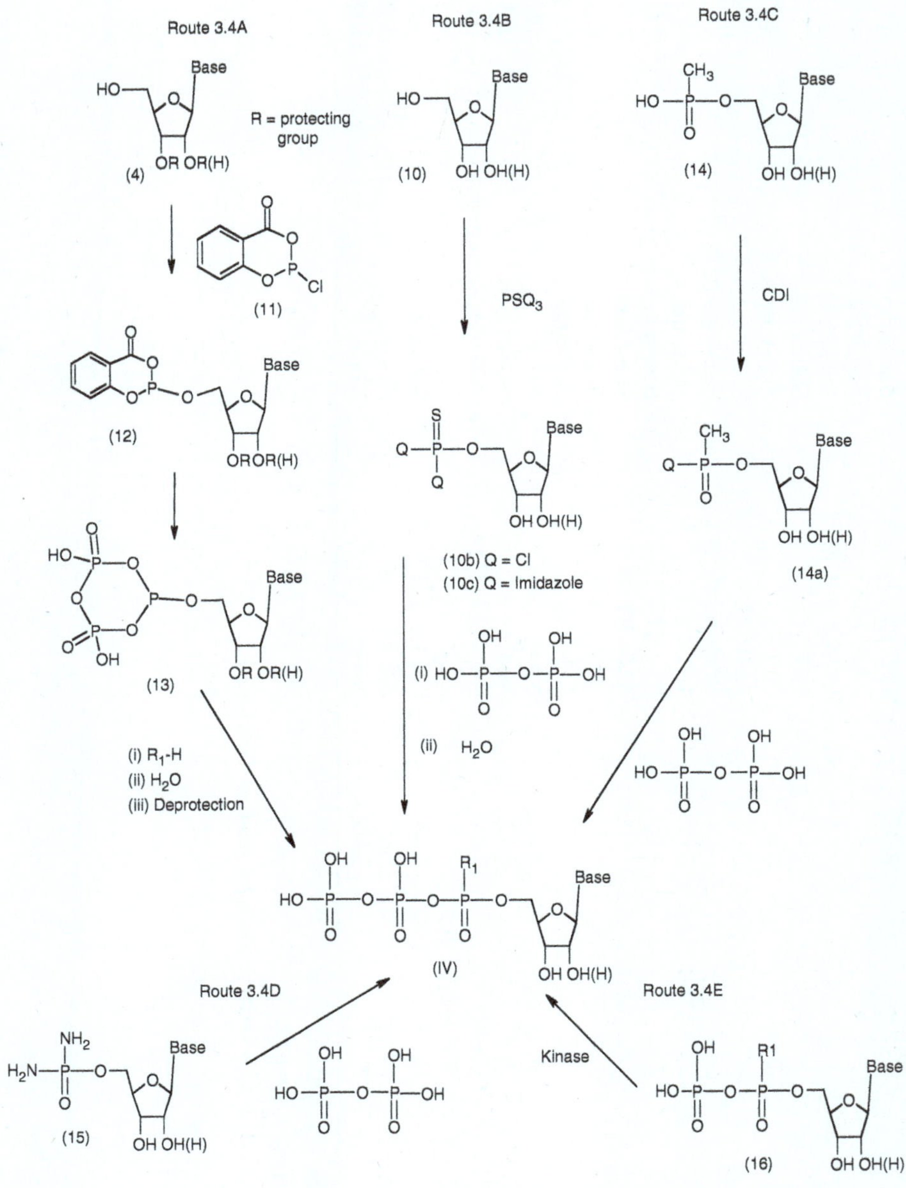

Scheme 3.4. Approaches to the synthesis of nonbridged NTP analogs with substitution for hydroxyl group at P[1].

phosphate cycle and removal of the protecting groups, results in the formation of the desired compound.

Another approach to the synthesis of P[1]-thio derivatives of NTPs relies on P(V) chemistry using $PSCl_3$ (or tris-imidazolyl-1-phosphinesulfide) that forms an activated precursor intermediate (10b or 10c), which then couples with pyrophosphate (route 3.4B). This method is similar to the routine synthesis of nonmodified NTP using the phosphoryl chloride method.

Synthesis of the P[1]-CH_3 analog of thymidine 5′-triphosphate is approached somewhat differently (route 3.4C). Nucleoside 5′-methylphosphonate (14) is first synthesized by DCC assisted coupling of nucleoside and methylphosphonic acid. The resulting compound (14) is then activated with 1,1-carbonyldiimidazole to obtain the phosphoroimidazolidate derivative (14a) and coupled with pyrophosphosphate to give a final product (IV). Syntheses of P[1]-amido, P[1]-O-thymidine 5′-triphosphate and P[1]-amido, P[1]-O-adenosine 5′-triphosphate are performed in a different manner. This is accomplished through a reaction of tributylammonium pyrophosphate with nucleoside 5′-phosphorodiamidate (15) in DMF, as can be seen in route 2.1D.

Finally, an enzymatic synthesis for certain P[1]-thio NTP analogs, route 2.1E, has been described. Some nucleoside 5′-(P[1]-thio)diphosphate analogs (16) are substrates for particular kinases that are obtained from a variety of sources. The kinase converts these NDPs to their corresponding NTP analogs. This approach is convenient and has been used widely for the preparation of different P[1]-thio NTP analogs at small scales.

3.3.2 *Substitutions for OH group at P[2] (structure V type analogs)*

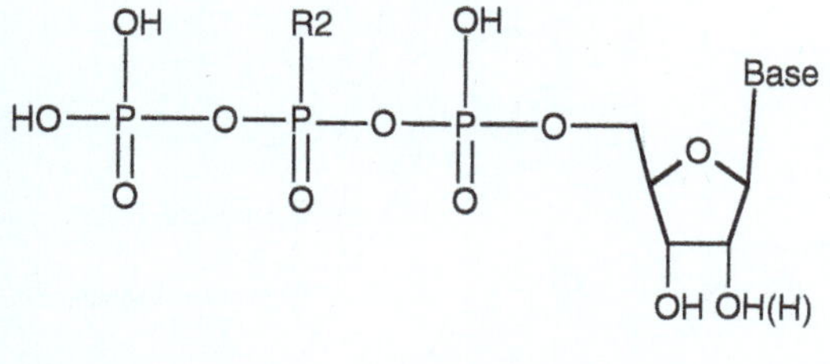

Substitution for OH group at P[2] position of an NTP has not been well documented. These compounds are not stable; if they form, they easily degrade or rearrange to more stable isomers. However, Eckstein and coworkers developed an enzymatic method to prepare P[2]-thio NTPs, which to our knowledge is the only route to these compounds (Table 3.5; Scheme 3.5). The method involves conversion of a chemically or enzymatically synthesized P[2]-thio NDP (17) to its NTP analog. Because of the enzymes specificity this approach allows the preparation of individual diastereomers of P[2]-thiophosphate analogs of NTPs.

Table 3.5 Nucleoside 5′-Triphosphate Analogs with Nonbridged Substitution at P^2 Fragment

R2	Sugar	Base	Biological application: Enzyme	Biological application: Substrate properties	Ref.
SH	Rib	G	Acetate kinase	Substrate (one diastereomer)	95
			RNA pol	Substrate (one diastereomer)	
			Succinyl-CoA synthesase	Substrate (both diastereomers)	
SH	Rib	G	Transducin, GTPase activity	Inactive (both diastereomers)	96
SH	Rib	A	ATPase	Substrate (Sp >> Rp, metal dependent)	85
SH	Rib	A	ATPase	Substrate (both diastereomers, metal dependent)	100
SH	Rib	A	Creatine kinase	Substrate (both diastereomers, metal dependent)	86
SH	Rib	A	Hexokinase	Substrate (both diastereomers, metal dependent)	98
SH	Rib	A	Acetate kinase	Substrate (both diastereomers, metal dependent)	87
SH	dRib	T	DNA polymerase (T4)	Substrate (Sp diastereomer)	91
SH	dRib	A	DNA polymerase I	Inactive	92

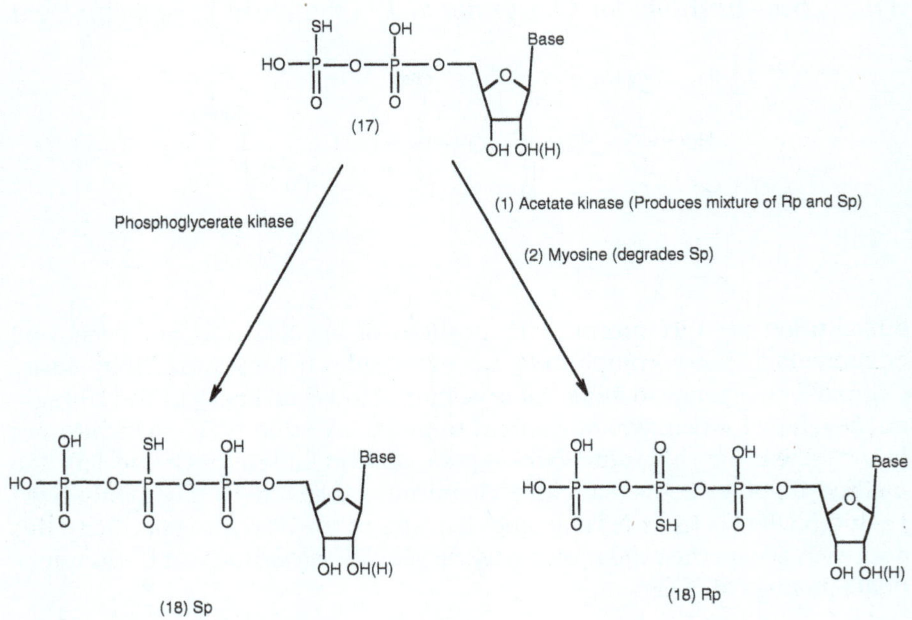

Scheme 3.5 Enzymatic synthesis of P^2 -thio NTPs.

3.3.3 Substitutions for OH group at P^3 (structure VI type analogs)

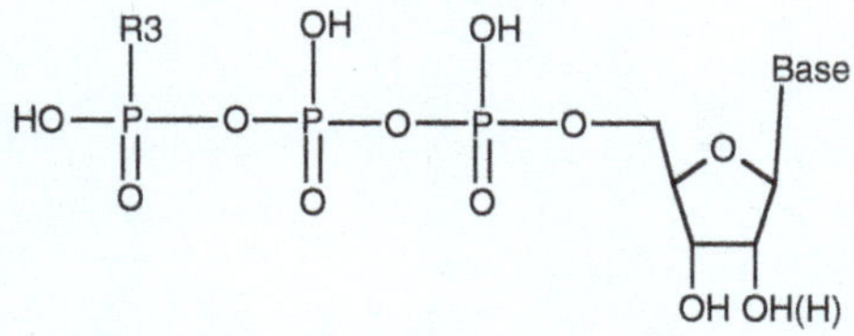

The family of NTP analogs with substitution of an OH group at the P^3 position is represented by several natural compounds and by a large number of chemically synthesized analogs (Table 3.6).

Several approaches have been developed for the preparation of Structure VI type analogs. The first approach relies on the activation of the substituted monophosphate derivative (19) and subsequent reaction with a respective NDP (Scheme 3.6, route 3.6A). Methods used to activate phosphate monoesters or any P-substituted derivative (19) are very similar to those described in Sections 3.2.1 and 3.2.3 (route 3.1A and 3.3A). The substituted monophosphate compound (19) is converted to N-morpholidate, N-imidazolidate, N-pyridinium, or another derivative with a high-energy P–N or anhydride bond. The activated derivative (19a) reacts with the trialkylammonium salt of the NDP (20) in anhydrous aprotonic media to yield the final substituted NTP analog.

A variant of this synthetic scheme includes activating the substituted pyrophosphate precursor (21) (route 3.6B), which becomes the R3–P^3–O–P^2 fragment of the final NTP analog after reacting with the nucleoside 5′-monophosphate. Activation of the pyrophosphate derivative (21), usually in a form of monoester or other P-substituted derivative, can be achieved by formation of the N-imidazolidate pyrophosphate precursor (21a) similar to that shown in routes 3.1B (Scheme 3.1) and 3.3B (Scheme 3.3).

The choice between routes 3.6A and 3.6B relies on the availability of the respective precursors and the convenience of the purification procedure(s). In reality, route 3.6A is used in the majority of syntheses, probably due to the ready availability of NDPs and P-substituted analogs of phosphoric acid. On the other hand, route 3.6B is used extensively in the synthesis of m^7GpppG and its analogs (5′-CAP structure of the eukaryotic messenger RNAs[8–13]). Interestingly, the coupling step of route 3.6B leading to a final Structure VI type NTP can be facilitated in the presence of some divalent metal ions (such as Zn^{2+}, Mn^{2+}). This reaction may be performed in aqueous and nonaqueous media by using an appropriate choice of solvents or aqueous buffers.[6,7,14,110]

An efficient method for preparation of Structure VI type compounds was introduced in the early 1970s when nucleoside 5′-trimetaphosphate was prepared (see Section 3.5) as a key intermediate in the synthesis of P^3-phosphoramidate derivatives of the NTP analogs. This procedure is very convenient and reproducible and produces the desired P^3-phosphoramidate deriv-

Table 3.6 Nucleoside 5′-Triphosphate Analogs with Substitution at P^3 Fragment

R3	Sugar	Base	Synthetic route	Biological application: Enzyme	Biological application: Substrate properties	Ref.
CH_3	d-Rib	T	3.6A	HIV RT	Substrate	111
				AMV RT	Substrate	
				TdT	Substrate	
				DNA pol α	Poor substrate	
				DNA pol β	Substrate	
CH_3	d-Rib-3′-azido	T	3.6A	HIV RT	Subst. (CT)[b]	111
				AMV RT	Substrate (CT)	
				TdT	Inhibitor	
				DNA pol α	Inhibitor	
$N_3CH_2CH_2$ -	d-Rib	T	3.6A	AMV RT	Substrate	71
				DNA pol α	Poor substrate	
$NH_2CH_2CH_2$ -	d-Rib	T	3.6A	AMV RT	Substrate	71
				DNA pol α	Poor substrate	
DNP-$NHCH_2CH_2$ -	d-Rib	T	3.6A	AMV RT	Substrate	71
				DNA pol α	Poor substrate	
DNP-NH-$(CH_2)_5$ C(O)CH_2CH_2-	d-Rib	T	3.6A	AMV RT	Substrate	71
				DNA pol α	Poor substrate	
	d-Rib	U(5-alkyl-azido)	3.6A	AMV RT	Substrate	71
Ph	d-Rib	U(5-alkyl-amino)	3.6A	AMV RT	Substrate	71
Ph	Cyclopentene type sugar (L-isomer)	A	3.6A	AMV RT	Substrate (CT)	16
				DNA pol α	Inactive	
				DNA pol β	Inactive	
				TDT	Poor substrate (CT)	

Ph	d-Rib	T	3.6A	HIV RT	Substrate	111
				AMV RT	Substrate	
				TdT	Substrate	
				DNA pol α,	Poor substrate	
				DNA pol β	Inactive	
Ph	d-Rib-3′-azido	T	3.6A	HIV RT	Substrate	111
				AMV RT	Substrate	
				TdT	Inhibitor	
				DNA pol α	Inhibitor	
H_2N	Rib	A	3.6C[a]	None	None	112
H_2N	Rib	A	3.6C	Adenosyl transferase	Substrate	47
H_2N	Rib	G	3.6C	GTP-binding protein (Ras)	Binds, substrate	67
Ph-NH-	d-Rib	T	3.6C[a]	HIV RT	Substrate	111
				AMV RT	Substrate	
				TdT	Substrate	
				DNA pol α	Substrate	
				DNA pol β	Substrate	
Ph-NH-	d-Rib-3′-azido	T	3.6C[a]	HIV RT	Substrate	111
				AMV RT	Substrate	
				TdT	Inhibitor	
				DNA pol α	Inhibitor	
$ClCH_2CH_2(Me)NC_6H_4CH_2NH$-	Rib	A	3.6C	None	None	113, 114
$[ClCH_2CH_2]_2NC_6H_4CH_2NH$-	Rib	A	3.6C	None	None	114
4-N_3-Ph-NH-	Rib	G	3.6C	RNA pol	Substrate	115
4-N_3-Ph-NH-	Rib	A	3.6C	Phenylalanyl-tRNA synthease	Inhibitor	116 117
Ph-NH-	Rib	A	3.6C	None	None	113

Table 3.6 Nucleoside 5′-Triphosphate Analogs with Substitution at P^3 Fragment (Continued)

				Biological application		
R3	Sugar	Base	Synthetic route	Enzyme	Substrate properties	Ref.
Ph-NH-	Rib	G	3.6C	Tubulin	Inactive	118
$NH_2CH_2CH_2NH$-	Rib	A	3.6C	None	None	113
(1-Naphthalene-5-sulfonate)-NH-	d-Rib	A, C, T, G	3.6C	SVPDE	Substrate	119
(1-Naphthalene-5-sulfonate)-NH-	Rib	A, T, C, G	3.6C	SVPDE	Substrate	119
(1-Naphthalene-5-sulfonate)-NH-	Rib	A, U	3.6C	RNA pol (yeast and *E. coli*)	Substrate	119–123
(1-Naphthalene-5-sulfonate)-NH-	Rib	G	NS	Tubulin	Binds, inactive	124
(1-Naphthalene-5-sulfonate)-NH-	d-Rib	T, G	3.6C	AMV RT	Substrate	119
Ph-NH-NH-	Rib	A	3.6C	None		113
N_3	Rib	A	3.6C	SVPD	Substrate	125
				Alkaline phosphatase	Inactive	
N_3	Rib	G	3.6C	Ribosome-EF-GTPase	Binds, substrate	125
H	Rib	A	3.6A	None	None	126
MeO-	Rib	A	3.6C[a]	DNA pol	Substrate	127
MeO-	Rib	A	3.6C	None	None	113
MeO-	Rib	G	3.6A	None	None	128

MeO-	Rib	A	3.6C	Ribosome-dependent EF-G GTPase	Inactive	129
MeO-	Rib	G	3.6C		Inhibitor	129
EtO-	Rib	G	3.6C		Inhibitor	129
$NH_2CH_2CH_2O$-	Rib	G	3.6A		Inhibitor	129
$CH_3C(O)NHCH_2CH_2O$-	Rib	A	3.6A		Inactive	129
$CH_3C(O)NHCH_2CH_2O$-	Rib	G	3.6A		Inhibitor	129
PrO-	Rib	G	3.6C		Inhibitor	129
BuO-	Rib	G	3.6C		Inhibitor	129
PhO-	Rib	G	3.6C		Inhibitor	129
PhO-	d-Rib	T	3.6A	HIV RT	Substrate	111
				AMV RT	Substrate	
				TdT	Substrate	
				DNA pol α	Poor substrate	
				DNA pol β	Inactive	
PhO-	d-Rib-3′-azido	T	3.6A	HIV RT	Substrate	111
				AMV RT	Substrate	
				TdT	Inhibitor	
				DNA pol α	Inhibitor	
O_2N-C_6H_4O-	Rib	A	3.6C	None	None	113
Adenosine-5′-O-	Rib	A	3.6A	None	None	130
Adenosine-5′-O-	Rib	A	3.6D	Fhit-binding protein	Substrate	131
7-Methylguanosine 5′-O-	Rib Rib-3′-OMe Rib-3′-deoxy	G, A	3.6B	none	None	8-13, 110
7-Methylguanosine 5′-O-	Rib-3′-OMe Rib-3′-deoxy	G	3.6B	RNA pol (SP6)	Primer Primer	110
7-Methylguanosine 5′-O-	Rib-3′-OMe	G	3.6A	RNA pol (SP6)	Primer	132

Table 3.6 Nucleoside 5′-Triphosphate Analogs with Substitution at P[3] Fragment (Continued)

				Biological application		
R3	Sugar	Base	Synthetic route	Enzyme	Substrate properties	Ref.
7-Methylguanosine 5′-O-	Rib	G	NS	RNA pol (SP6, T3, T7)	Primer	110, 133, 134, 135
BODIPY(FL)-S-	Rib	G	3.6D	G-binding protein	Binds	69
				Fhit-binding protein	Substrate	131
BODIPY(515)-S-	Rib	G	3.6D	G-binding protein	Binds	69
BODIPY(FL)-S-	Rib	A	3.6D	Fhit-binding protein	Substrate	131
BODIPY(TR)-S-	Rib	G	3.6D	G-binding proteins	Binds	69
-S-acetylamidoethyl-(1-aminonaphthalene-5-sulfonate)	Rib	G	3.6C	Tubuline	Binds, Inactive	118,124
SH	Rib	G	3.6E	Tubulin	Supports assembly	118
SH	Rib	A	NS	Tubulin	Inactive	118
SH	Rib	G	NS	Transducin, GTPase activity	Inhibitor	96
SH	d-Rib	A	3.6E	DNA pol II	Substrate	99
SH	Rib	A, G	NS	RNA pol	Substrate	136, 137
SH	Rib	A, G, I (S^6), I (Cl^6)	3.6E	RNA pol	Substrate	99
SH	Rib	A	NS	Mevalonate kinase	Substrate	82
				Phosphoglycerate kinase	Substrate	83

SH	Rib	A	3.6E	RNA Pol	Substrate	99
				ATPase	Substrate	
				tRNA synthesase	Substrate	
				SVPDE	Substrate	
				PNP	Substrate	
SH	Rib	U	3.6E	RNA pol	Substrate	99
				NDP kinase	Substrate	
SH	Rib	G	3.6E	NDP kinase	Substrate	99
SH	Rib	A, G, U	3.6A	Alkaline phosphatase	Inhibitor	138
SH	Rib	G	NS	Ribisome-dependent EF-G GTPase	Inhibitor	129
SH	Rib	G	NS	GTP-binding protein (Ras)	Binds, substrate	67
F	Rib	A	3.6A	Hexokinase Myokinase Myosin	Inhibitor Inhibitor	139
F	Rib	A	3.6A	DNA pol	Substrate	127
F	Rib	G	3.6C[a]	GTP-binding protein (Ras)	Binds, substrate	67
F	Rib	G	3.6A	Ribisome-dependent EF-G GTPase	Inhibitor	129

[a] Modified 3.6C route.

[b] CT = chain terminating substrate.

Note: NS = not specified or commercial product; for route see Scheme 3.6.

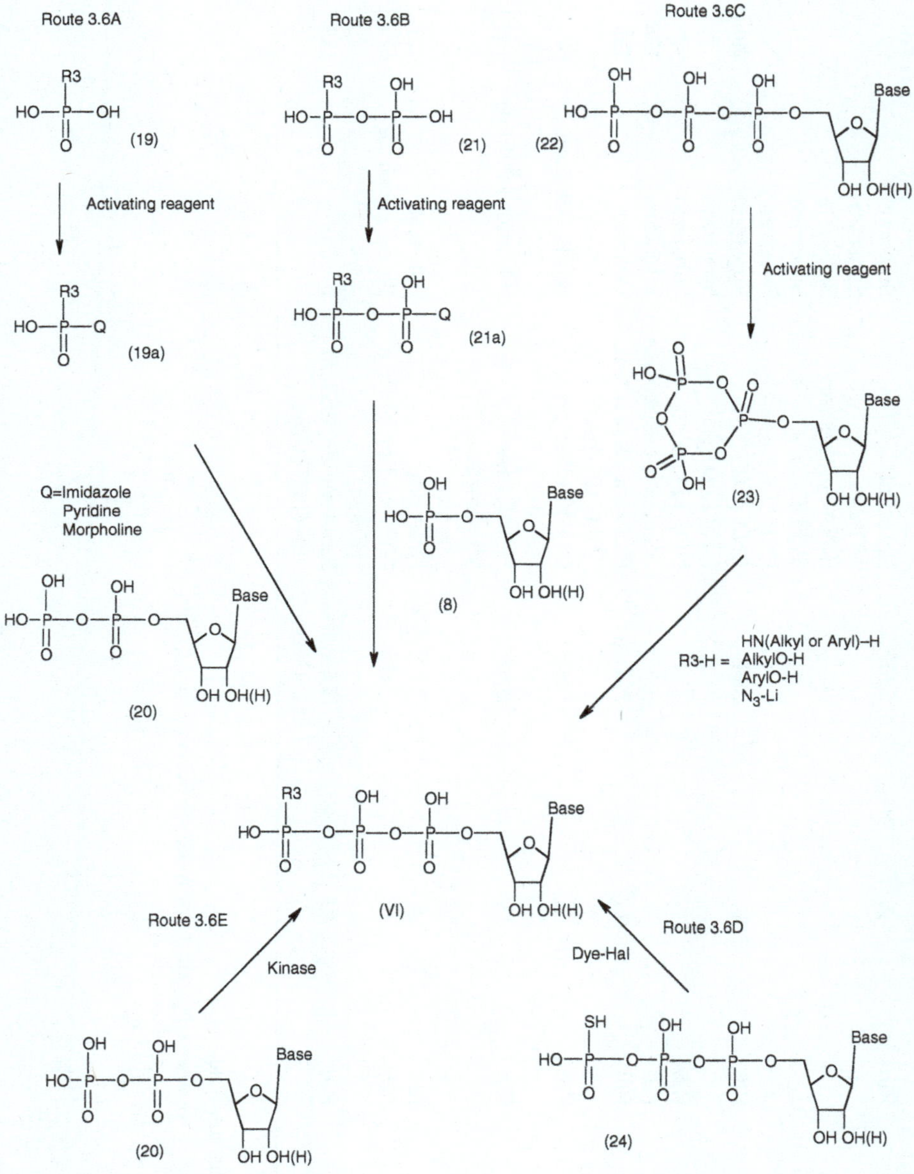

Scheme 3.6 Approaches to the synthesis of nonbridged P³-substituted NTP analogs.

ative in high yield (route 3.6C). The P^3-esters of NTPs can be prepared in a similar fashion, but less efficiently.

Route 3.6D was used specifically to prepare a number of fluorescent derivatives of P^3-thioesters of NTPs. It was shown that the alkylation of P^3–thio–NTP derivatives with thiol-reactive dyes (Dye-Hal) can achieve high yields when performed in aqueous media. Finally, the enzymatic preparation of P^3-thio NTPs has also been illustrated in route 3.6E.

3.4 Multiple substitutions in triphosphate chains

3.4.1 Bridged substitutions (structure VII type analogs)

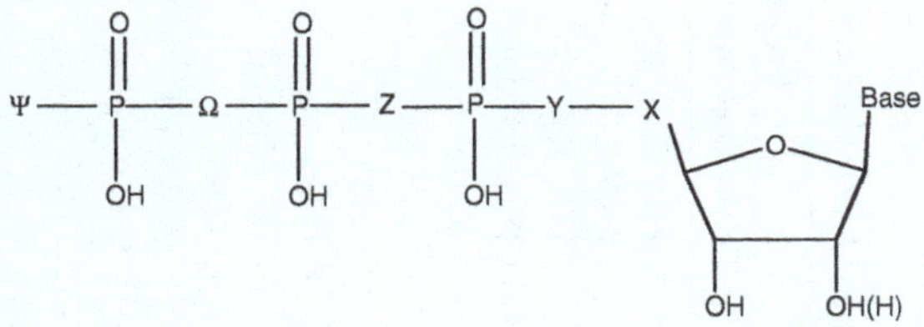

The strategy for the synthesis of NTPs with multiple bridged substitutions (and sometimes in combination with nonbridged substitutions) in the triphosphate chain relies on coupling two premodified fragments of the final NTP analog being prepared. Most of the synthetic routes discussed in previous sections are applicable in one way or another to the synthesis of Structure VII type NTP analogs. Choosing the specific fragments to be coupled depends on several factors, including the availability and cost of the precursors, the method of activating and coupling the fragments and the difficulty of the isolating and purifying the final compound. Specific information on the synthesis of Structure VII type compounds can be found in the literature cited in Table 3.7.

Table 3.7 Nucleoside 5′-Triphosphate Analogs with Multiple Bridged (and Nonbridged) Substitutions in Triphosphate Chain

X	Y	Z	Ω	Ψ (R3)	Sugar or its substitute	Base	Biological application		Ref.
							Enzyme/media	Substrate properties	
O	CH_2	O	CH_2	OH	Cyclopentene type sugar (L-isomer)	G	HIV-RT	Inactive	21
O	CH_2	O	CHF	OH	Cyclopentene type sugar (L-isomer)	G	HIV-RT	Weak inhibitor	21
O	CH_2	O	CF_2	OH	Cyclopentene type sugar (L-isomer)		HIV-RT	Inhibitor	21
							Blood serum	Enhanced stability	
CH_2	O	CH_2	CH_2	OH	d-Rib	T	None	None	140
CH_2	O	NH	NH	OH	d-Rib	T	HIV RT	Inhibitor	45
O	CH_2	O	O	Ph	Cyclopentene type sugar (L-isomer)	A	AMV RT	Poor substrate (CT)[a]	16
							DNA pol α	Inactive	
							DNA pol β	Inactive	
							TDT	Poor substrate (CT)	
O	CH_2	O	CH_2	Me	Cyclopentene type sugar (L-isomer)	A	AMV RT	Poor substrate (CT)	16
							DNA pol α	Inactive	
							DNA pol β	Inactive	
							TDT	Poor substrate (CT)	
O	CH_2	O	CF_2	Me	Cyclopentene type sugar (L-isomer)	G	HIV RT	Substrate (CT)	16
							AMV RT	Poor substrate (CT)	
O	CH_2	O	CF_2	Ph	Cyclopentene type sugar (L-isomer)	G	HIV RT	Substrate (CT)	16
							AMV RT	Poor substrate (CT)	
							TDT	Poor substrate (CT)	

O	CH_2	O	CF_2	OH	Cyclopentene type sugar (L-isomer)	A	HIV RT	Substrate (CT)	16
							AMV RT	Substrate (CT)	
							DNA pol α	Inactive	
							DNA pol β	Inactive	
							TDT	Substrate (CT	
O	CH_2	O	CF_2	OH	Cyclopentene type sugar (L-isomer)	G	HIV RT	Substrate (CT)	16
							AMV RT	Substrate (CT)	
							DNA pol α	Inactive	
							DNA pol β	Inactive	
							TDT	Substrate (CT)	
O	CH_2	O	CBr_2	OH	Cyclopentene type sugar (L-isomer)	A	HIV RT	Inactive	16
							AMV RT	Poor substrate (CT)	
							DNA pol α	Inactive	
							DNA pol β	Inactive	
							TDT	Substrate (CT)	
O	CH_2	O	CBr_2	OH	Cyclopentene type sugar (L-isomer)	G	HIV RT	Poor substrate (CT)	16
							AMV RT	Poor substrate (CT)	
							DNA pol α	Inactive	
							DNA pol β	Inactive	
							TDT	Substrate (CT)	

[a] CT = chain terminating substrate.

3.4.2 Nonbridged substitutions (structure VIII type analogs)

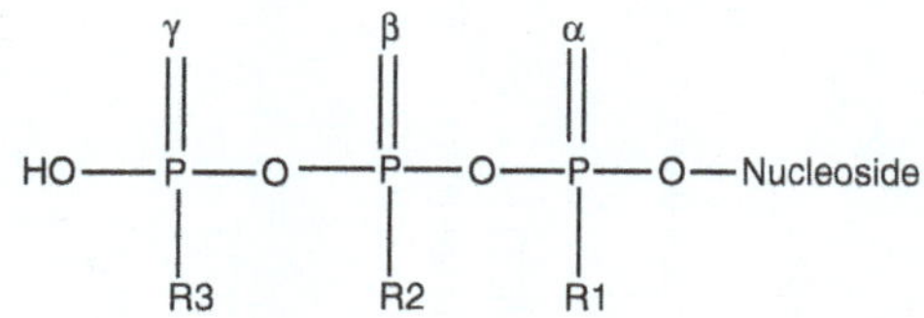

Another family of NTP analogs with multiple substitutions in triphosphate fragment is represented by analogs with nonbridged substitutions, such as P^1,P^1-, P^1,P^2-, and P^1,P^3-dithiotriphosphate derivatives of NTPs (Table 3.8; Scheme 3.7). These compounds can be prepared using a modified version of the Ludwig–Eckstein procedure through P(III) chemistry by forming a nucleoside 5′-(4H-1,3,2-benzodioxaphosphorin-4-one) intermediate (12). Once this intermediate is obtained, two routes (3.7A and 3.7B) lead to the NTP dithiotriphosphate derivatives (27), (28), and (29).

Route 3.7A incorporates the first sulfur atom after reaction of the intermediate with pyrophosphate. The second sulfur is added during a reaction with lithium sulfide that opens the ring of the compound (25). Route 3.7B incorporates two sulfur atoms in the cyclic trimetaphosphate derivative (26), which then undergoes a hydrolitical ring opening. In both cases, the opening of the trimetaphosphate ring leads to the formation of two isomeric dithiotriphosphate derivatives (Scheme 3.7).

Another approach based on P(III) chemistry was used to prepare P^3-[(7-methylguanosine-5′-O-), P^1-(adenosine-5′-O-), P^1-thio-triphosphate and P^1,P^3-bis-O-adenosine-5′, P^1-thio-triphosphate derivatives of NTP (route is not shown; Table 3.8). Although low yields of these compounds were obtained, they were successfully isolated by HPLC and characterized by proton and phosphorus NMR.

Synthesis of nucleoside 5′-O-(P^1-borano, P^1-thio)triphosphates was also accomplished using a modified Ludwig–Eckstein approach through boronation of the P^2,P^3-dioxa-P^1-nucleosidyl cyclic triphosphite intermediate (13). The reaction to open the ring is performed with lithium sulfide following ammonia deprotection to yield the final product (Scheme 3.7, route 3.7A).

The P^1, P^2, P^3-(trinucleoside-5′-O-)tripolyphosphates (33) may be considered as examples of Structure VIII type analogs. These compounds are readily available as the intermediates formed in the reaction of nucleoside 5′-phosphomonoesters with various coupling reagents (such as N,N′-dicyclohexylcarbodiimide, arenesulfonylchlorides, arenesulfonyltriazolides, triphenylphosphine/α,α-dipyridyl disulfide and others) in DMF or pyridine solution (Scheme 3.8). The P^1, P^2, P^3-(trinucleoside-5′-O-) tripolyphosphate (33) is unstable in the aqueous media (especially in the presence of pyridine, imidazole, or other nitrogen heterocyclic compounds), but can be isolated under the appropriate conditions. The hydrolysis of P^1, P^2, P^3-(trinucleoside-5′-O-)tripoly-phosphate (33) leads to a formation of nucleoside

Table 3.8 Nucleoside 5′-Triphosphate Analogs with Multiple Nonbridged Substitutions in Triphosphate Chain

R1	R2	R3	a	b	g	Nucleoside	Route	Biological application: Enzyme	Biological application: Substrate properties	Ref.
SH	SH	OH	O	O	O	3′-Azidothymidine-5′-	3.7B	None	None	141
SH	OH	SH	O	O	O	3′-Azidothymidine-5′-	3.7B	SVPDE	Substrate	141
								Alkaline phosphatase	Substrate	
								HIV RT	Inhibitor	
SH	OH	2-Cyanoethyl	O	O	S	3′-Azidothymidine-5′-	3.7B	None	None	141
SH	OH	SH	O	O	O	Thymidine-5′-	3.7B	SVPDE	Substrate	141
								Alkaline phosphatase	Substrate	
SH	OH	OH	S	O	O	Thymidine-5′-	3.7A	DNA pol	Inactive	141
SH	OH	OH	S	O	O	5′-Guanosine	3.7A	None	None	141
BH_3	OH	OH	S	O	O	Thymidine-5′-	3.7A	DNA pol	Poor substrate (one isomer)	142
BH_3	OH	OH	S	O	O	Adenosine-5′-	3.7A	None	None	142
SH	O	Adenosine 5′-	O	O	O	Adenosine-5′-	3.7A[a]	None	None	81
SH	O	7-Methyl-guanosine 5′-	O	O	O	Adenosine-5′-	3.7A	None	None	81
BnzO-	BnzO-	BnzO- (ψ = BnzO-)	O	O	O	2′,3′-Isopropyledene-uridine-5′-	3.8C	None	None	143
BnzO-	BnzO-	BnzO-	O	O	O	3′-O-benzylthymidine-5′-	3.8C[b]	None	None	144
OH	dT(Ac)	dT(Ac)	O	O	O	3′-O-acetylthymidine-5′-[dT(Ac)]	3.8C	None	None	145 146
OH	ibudG(iBu)	ibu-dG-(iBu)	O	O	O	N[6], 3′-O-diisobutyryl-2′-deoxy-guanosine-5′-[ibudG(iBu)]	3.8C	None	None	147

[a]3.7A = modified Route 3.7A (different version of P(III) chemistry), see Scheme 3.7.

[b]3.8C = modified Route 3.8C (reaction of nucleoside 5′-O-benzyl-chlorophosphate with di- or tribenzylpyrophosphate), see Scheme 3.8.

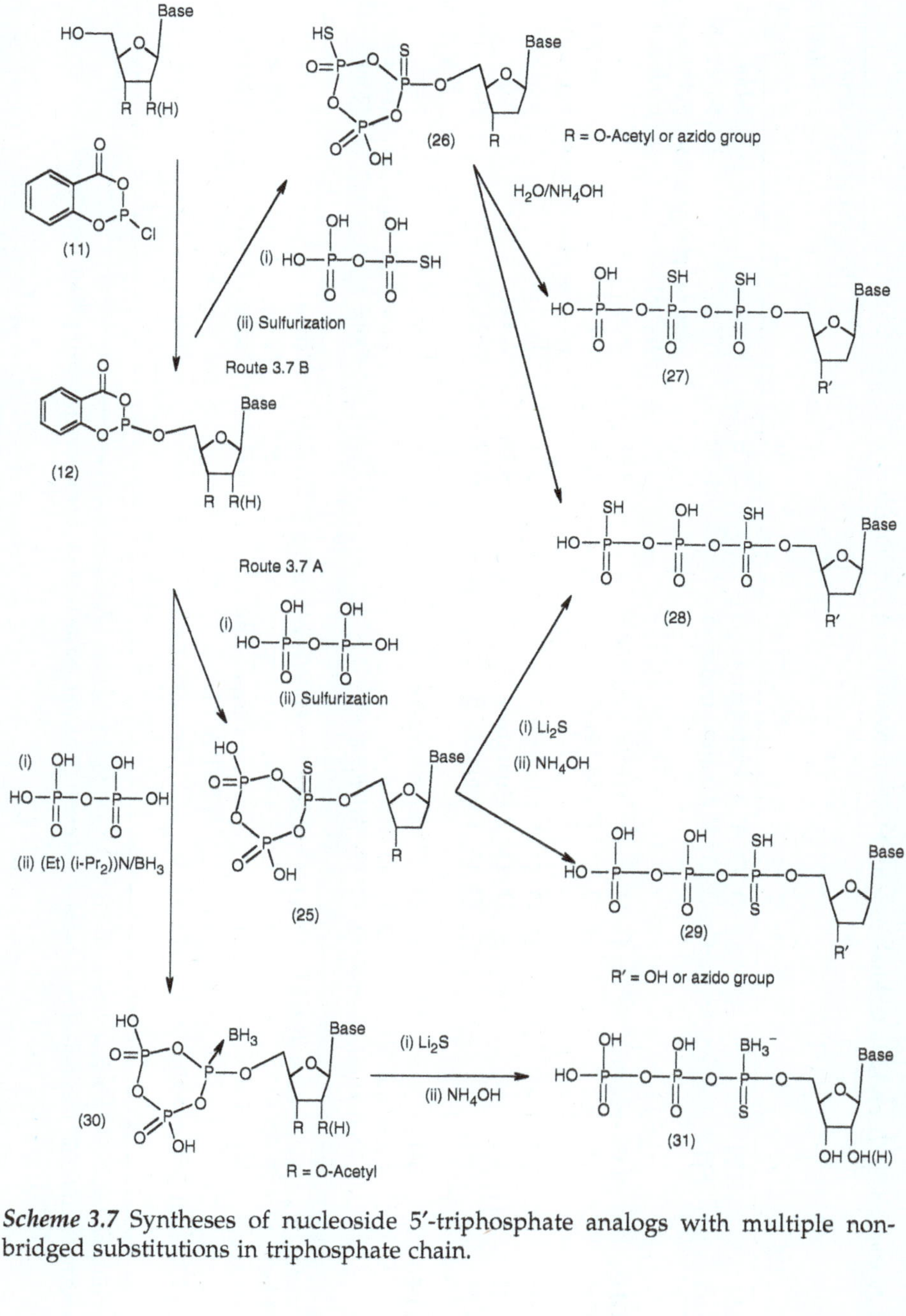

Scheme 3.7 Syntheses of nucleoside 5′-triphosphate analogs with multiple non-bridged substitutions in triphosphate chain.

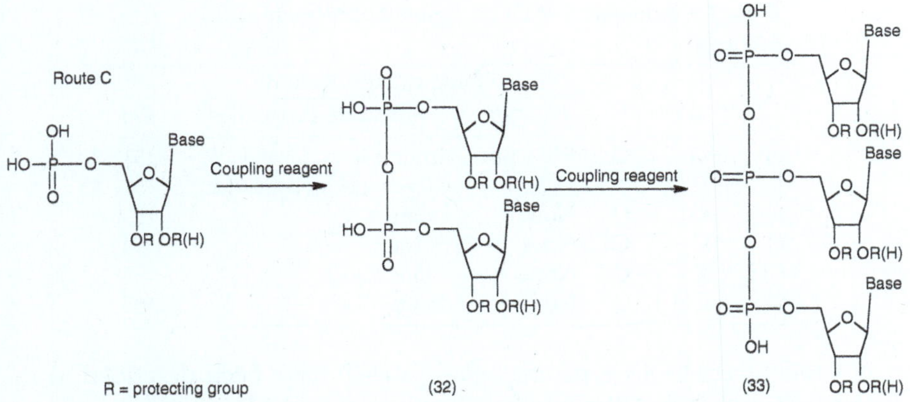

Scheme 3.8 Preparation of P^1-, P^2-, P^3-substituted tripolyphosphates.

5′-monophosphate and its symmetrical P^1, P^2-(dinucleoside-5′-O-)pyrophosphate (32). Tetra- and trisubstituted benzyl esters of NTPs were obtained in the 1950s as intermediates in the syntheses of standard NTPs (routes are not shown; see Table 3.8).

3.5 Nucleoside cyclic 5′-trimetaphosphate (structure IX type analogs)

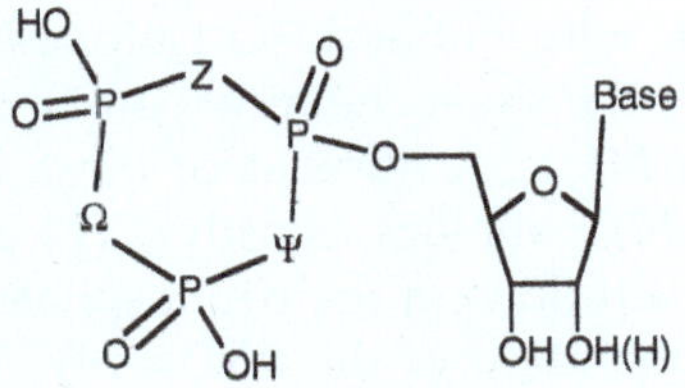

Nucleoside 5′-cyclic trimetaphosphate (NTMP, compound (23), Scheme 3.6) was first proposed as an intermediate in the synthesis of NTPs.[148] This hypothesis was proven in the works of Khorana[149] and Hecht.[150] NTMP can be routinely prepared by the treatment of trialkylammonium salt of nucleoside 5′-triphosphates with coupling reagents such as N,N′-dicyclohexylcarbodiimide, arylsulfonylchlorides, arylsulfonyltriazolides, triphenylphosphine/α,α-dipyridyl disulfide, etc. in DMF, DMSO, DMSO–methanol, or in pyridine. Although not stable in a solution containing pyridine, this compound is quite stable in DMSO and DMF. NTMP slowly hydrolyzes in water, but it can be isolated with fast chromatography (ion exchange, reverse phase) and can be used for chemical or enzymatic studies. NTMP can also be prepared from NTP and water-soluble carbodiimides in aqueous media as an intermediate for preparation of P^3-modified NTPs.

Table 3.9 Adenosine 5′-Cyclic Trimetaphosphate and its Analogs

			Biological application		
Z	W	Y	Enzyme	Substrate properties	Ref.
O	O	O	RNA pol	Inhibitor	151
				(enzyme modification)	152
O	O	O	None	None	153
O	O	O	None	None	154
CH_2	O	O	None	None	155
O	CH_2	O	None	None	155

A few Structure IX type analogs of NTMP have been described in the literature (Table 3.9). Although these analogs were not isolated as pure compounds, their structures were verified by ^{31}P NMR spectroscopy. Among the known analogs of NTMP, in some the bridging oxygen atom is substituted with a methylene group and in others sulfur or borane groups replace nonbridging oxygen atoms (Scheme 3.7).

3.6 Isotopic substitutions in triphosphate chain of NTP

3.6.1 Oxygen isotopic substitutions

Substitution of the natural and most commonly occurring isotope, ^{16}O, in NTP triphosphate chain with ^{17}O and ^{18}O isotopes has become one of the tools in investigations of the stereochemical course of enzymatic and chemical reactions involving NTPs. A number of these investigations are done with phosphorothioate NTP analogs (mostly ATP) in combination with ^{17}O or/and ^{18}O isotope substitutions in the triphosphate fragment (Table 3.10).

A stereospecific preparation of the ATP with ^{17}O or ^{18}O isotopes at P^1 position can be achieved through enzymatic phosphorylation of P^1-[^{17}O or ^{18}O]-ADP (36). The latter compound is prepared (Scheme 3.9) by the reaction of a single diastereomer of P^1-thio ADP (34) with bromine or N-bromosuccinimide in the presence of $H_2{}^{17}O$ or $H_2{}^{18}O$.[3,4,156] The reaction in acidic conditions presumably proceeds through the intermediate (35) with activated P–SX function, which hydrolyses with inversion of configuration at P^1 position to isotopically labeled ADP (36). It should be noted that, when the preceding reaction takes place at neutral pH or when the reaction of compound (34) with cyanogene bromide occurs at any pH, it leads to a distribution of the oxygen isotope between P^1 and P^2 positions. It is believed that this is caused by a formation of cyclo-diphosphate intermediate.[157–159] The isotopomers of the ATP with ^{16}O, ^{17}O, and ^{18}O isotopes at P^1 position were prepared by enzymatic ptrophosphorylation (adenylate kinase and pyruvate kinase) of Sp-[^{16}O,^{17}O,^{18}O]AMP.[160]

Table 3.10 Examples of Various [^{17}O and ^{18}O] NTP Analogs and Their Use in Studies of Stereochemistry of Enzymatic Reactions

Position of the isotope or sulfur							Enzymatic reaction		
α or R1	β or R2	Ω	γ, ψ, or R3	Sugar	Base	Synthetic route	Enzyme	Stereochemical course	Ref.
^{17}O, ^{18}O				Rib	A	E-PP[a]	SVPDE	Retention	160
		^{18}O	^{18}O, S	Rib	A	3.11A	Adenylate kinase	Inversion	165
			^{17}O, ^{18}O	Rib	A	3.11C	Yeast hexokinase	Inversion	167
			^{17}O, ^{18}O	Rib	A	3.11C	Polynucleotide kinase	Inversion	168
			^{18}O, S	Rib	A	3.11A[b]	T4 polynucleotide kinase	Inversion	169
			^{17}O, ^{18}O	Rib	A	3.11B	Creatine kinase	Inversion	170
			^{17}O, ^{18}O	Rib	A	3.11B	Acetate kinase	Inversion	166
							Hexokinase puruvate	Inversion	
							kinase	Inversion	
							Glycerol kinase	Inversion	
			^{18}O	Rib	A	3.11B	Glycerol kinase	Inversion	171
	^{18}O, S	^{18}O		Rib	A	3.10	Nucleoside diphosphate kinase	Retention	163
			^{17}O, ^{18}O, S	Rib	A	3.11A	ATPase		
							Mitochondrial	Inversion	172
							Myosin	Inversion	173 174
							Sarcoplasmic reticulum	Retention	
^{18}O, ^{18}O				dRib	A	E-PP[a]	T4 DNA polymerase (Sp-diastereomer)	Inversion	93
			^{17}O, ^{18}O, S	Rib	G	3.11A	GTPase	Inversion	175 176
^{18}O				Rib	G	3.9	Guanylate cyclase	Inversion	177

[a] Enzymatic pyrophosphorylation of isotopically labeled adenosine (Sp)AMP.

[b] Modified route; see routes in Schemes 3.9–3.11.

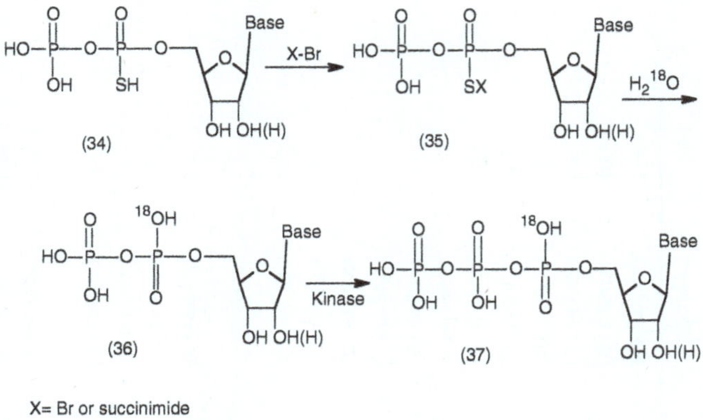

Scheme 3.9 Synthesis of NTP with oxygen's isotopes labeling at P¹ position.

Oxygen isotope labeling at P^2 position of NTP has been reported according to Scheme 3.10. First, adenosine 5′-[$^{18}O_2$]-phosphorothoate (38) was prepared by the reaction of nucleoside with thiophosphoryl chloride followed by hydrolysis with $H_2{}^{18}O$.[161,162] Compound (38) was then activated with diphenylchlorophosphate and subsequently coupled with 2′,3′-methoxymethylidene AMP. The reaction produced a mixture of two diastereomers of P^1,P^2-dinucleoside-(5′,5′)-diphosphate (39). After separation by anion-exchange chromatography, the individual Rp and Sp diastereomers were oxidized at unblocked ribose by sodium periodate treatment. The 2′,3′-ribose protecting group was removed under mild acidic conditions, and oxidized nucleoside residue was removed by β-elimination reaction and producing the individual diastereomers of P^2-thio-[^{18}O]ADP (40).[161] Phosphorylation of P^2-thio-[^{18}O]ADP (40) to P^2-thio-[^{18}O]ATP (41) was accomplished using acetyl phosphate and acetate kinase.[161,163,164] A similar scheme (with $POCl_3$ instead of $PSCl_3$) can be used to prepare P^2-[^{18}O]ATP.

The P^3-thio ATP analog labeled with ^{18}O simultaneously at P^3 position and bridged (P^2–O–P^3) position was prepared according to Scheme 3.11, route A. A nucleoside is first reacted with thiophosphoryl chloride followed by hydrolysis with $H_2{}^{18}O$ to give nucleoside 5′-[$^{18}O_2$]phosphorothioate (42). The latter is converted enzymatically with kinase to P^1-thio-[$^{18}O_2$]ADP (43) and coupled with 2′,3′-methoxymethylidene AMP activated by diphenylchlorophosphate. The resulting P^1,P^3-dinucleotide-(5′,5′)-triphosphate (44) is finally converted to P^3-thio-[$^{18}O_2$]ATP (45) by oxidation with sodium periodate followed by ribose deprotection and β-elimination.[165]

NTP analogs labeled with a combination of ^{17}O and ^{18}O isotopes at the P^3 position were prepared according to Scheme 3.11, route B. Ephedrine was reacted with ^{17}O -phosphoryl chloride (obtained from PCl_5 and $H_2{}^{17}O$) to form a cyclic chlorophosphoramidate (46). The major diastereomer of compound (46) was isolated and hydrolyzed with $Li^{18}OH/H_2{}^{18}O$ to form cyclic phosphoramidate (47) and then coupled with ADP to obtain P^3-ephe-

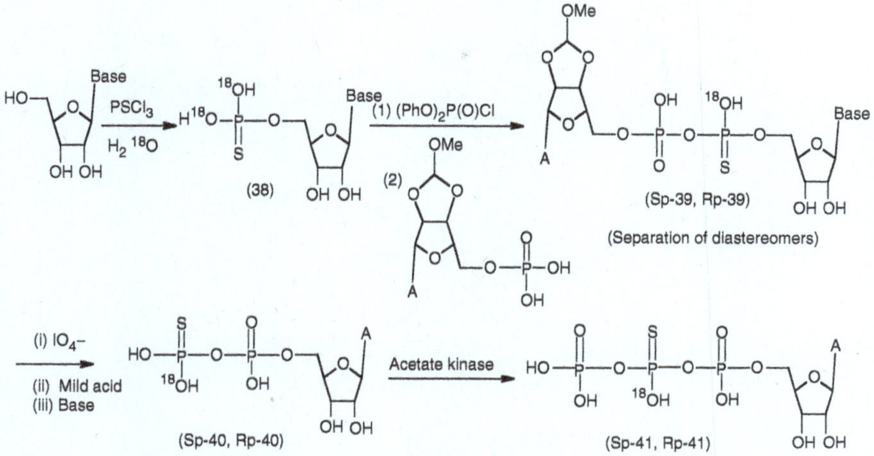

Scheme 3.10 Synthesis of ATP analog labeled with oxygen isotope at P^2 position.

drine-ATP (48). Catalytic reduction was used to remove ephedrine residue in the final ^{17}O and ^{18}O labeled product (49).[166]

The alternative route to the NTP analog labeled with combination of ^{17}O and ^{18}O isotopes at P^3 position is shown in Scheme 3.11, route C. An individual isomer of [^{18}O]-1,2-diphenyl-1,2-ethylene glycol was treated with $^{17}OPCl_3$, followed by reaction with ADP. A final product, Sp-[^{16}O,^{17}O,^{18}O]ATP (49), was isolated after reduction of the intermediate (50) on Pd/C.[167]

3.6.2 *Phosphorus isotopic substitutions*

NTPs labeled with phosphorus radioisotopes have been used for decades in biological research. While ^{32}P and ^{33}P labels at P^1 position are used for incorporation into DNA or RNA, NTPs with radioisotopes at the P^3 position are used for enzymatic transfer of the radiolabel to another biological molecule. These NTPs have been synthesized by enzymatic procedures and by chemical methods and they are commercially available. The reliable approaches for preparation of NTPs radiolabeled at P^1, P^2, or P^3 positions were developed in 1950s and1960s (Wehrli et al.[178] and references therein) and improved later.

Labeling at the P^3 position (compound 51; Scheme 3.12) can be achieved by equilibration of the corresponding NTP and ^{32}P-labeled orthophosphoric acid in an exchange reaction catalyzed by 3-phosphoglycerate kinase (route A).[87] Chemical synthesis can be performed by activation of nucleoside 5′-diphosphate and coupling with ^{32}P-labeled orthophosphoric acid, as illustrated.[179,180]

The NTPs with radiolabel at the P^2 position were prepared according to Scheme 3.13.[178] The ^{32}P orthophosphate was reacted with p-nitrobenzyl phosphomorpholidate to give p-nitrophenyl P^2-[^{32}P]diphosphate (52). The latter

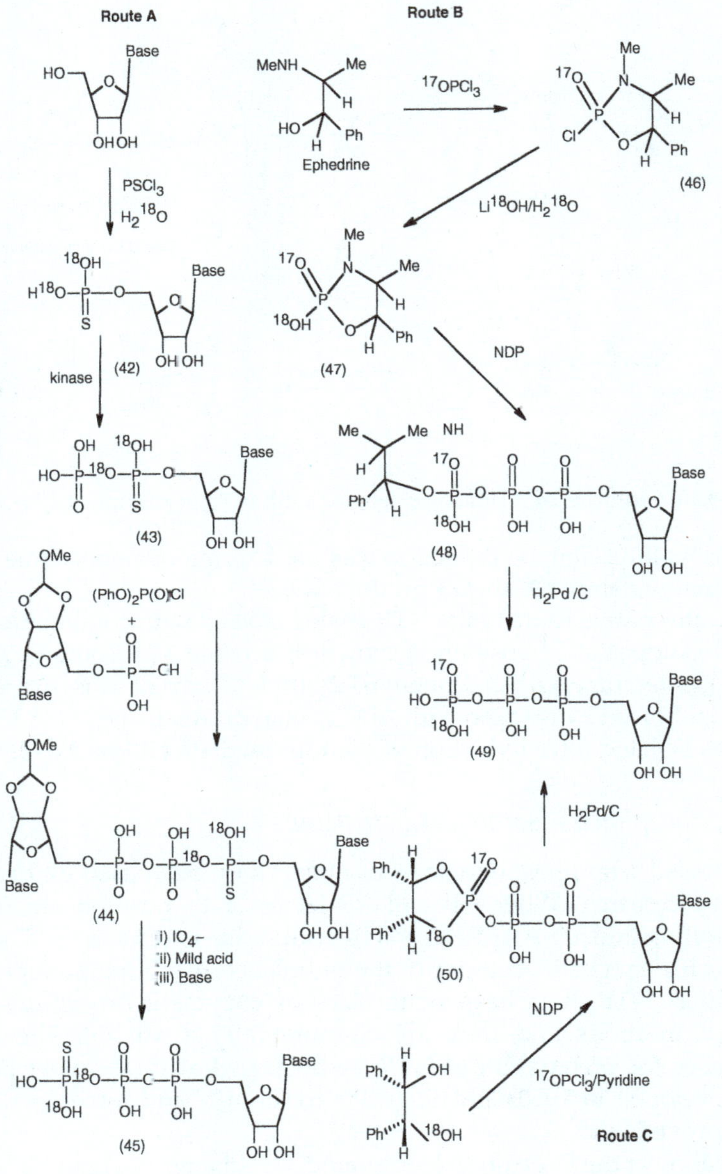

Scheme 3.11 Synthesis of NTP analogs labeled with oxygen isotopes at P^3 and P^2–O–P^3 positions.

compound was coupled to morpholidate of AMP and P^3-[p-nitrophenyl], P^2-[^{32}P]ATP intermediate (53) was hydrogenated to P^2-[^{32}P]ATP (54).

The NTPs labeled at P^1 position can be prepared from ^{32}P-labeled orthophosphoric acid on a preparative scale enzymatically (Scheme 3.14). The sequential steps of the method involve the 5′[^{32}P]-phosphorylation of nucleoside 3′-monophosphate with T4 polynucleotide kinase and P^3-[^{32}P]ATP to

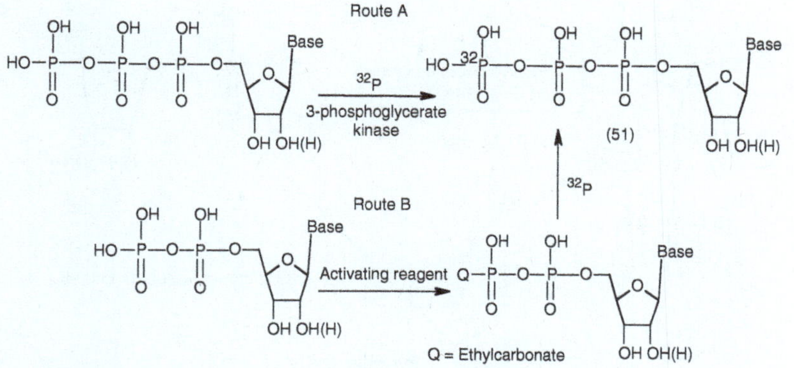

Scheme 3.12 Synthesis of P[3]-radiolabeled NTP.

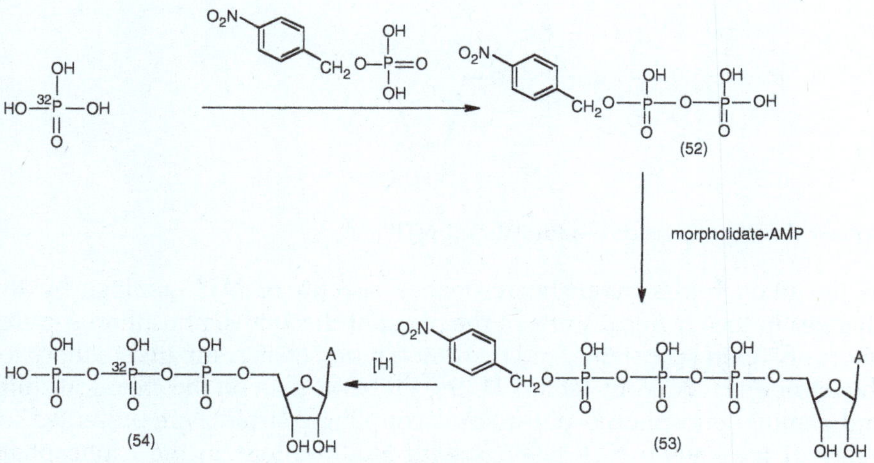

Scheme 3.13 Synthesis of P[2]-radiolabeled ATP.

yield nucleoside 3′, 5′[^{32}P]-diphosphate (55). The hexokinase is used to degrade any remaining P[3]-[^{32}P]ATP. The nucleoside 3′,5′[^{32}P]-diphosphate (55) is treated with P1 nuclease to remove the 3′-phosphate and produce the nucleoside 5′-[^{32}P]monophosphate (56). The latter is phosphorylated to the nucleoside 5′-P[1]-[^{32}P]triphosphate (57) with nucleoside monophosphate kinase and pyruvate kinase. More than 99.9% of detectable radioactivity was found at the P[1] position.[181]

3.7 ^{31}P NMR data for NTP analogs with modified triphosphate chains

Modification of the triphosphate chain in an NTP changes the electronic and structural properties of the triphosphate fragment, especially around the modified phosphate groups. In many instances, these changes are reflected

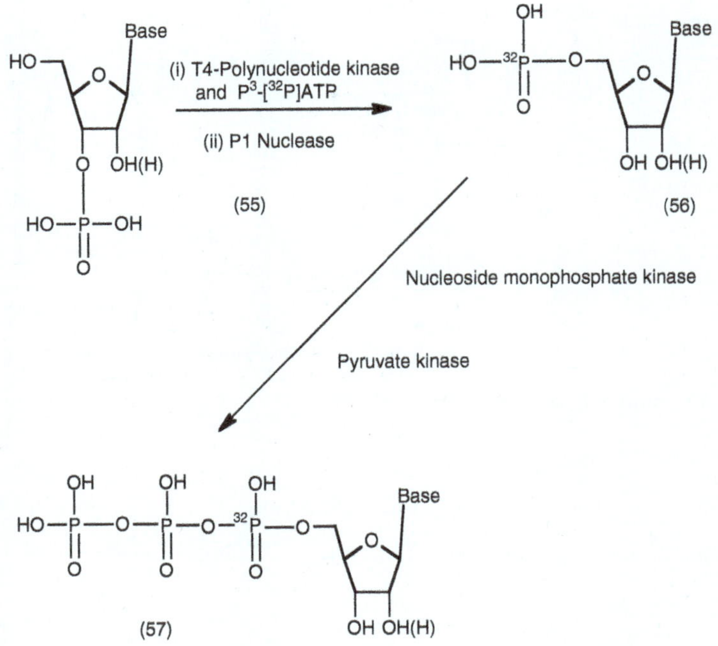

Scheme 3.14 Synthesis of P¹-radiolabeled NTP.

in the phosphorus magnetic resonance spectra of NTP analogs by the changes in the chemical shifts of the phosphorus atoms, phosphorus–phosphorus coupling constants, and relaxation parameters compared to those for the unmodified NTP. In Table 3.11, the ^{31}P NMR data on the chemical shifts and geminal phosphorus–phosphorus coupling constants are presented for many of the known NTP analogs with modifications in the triphosphate fragment. Instead of presenting the original NMR data, all the information was generalized in the form of differences of chemical shift and coupling constant values ($\Delta\delta$, ΔJ) compared to those for the unmodified NTP.

Data in Table 3.11 were generated based on the average values of chemical shifts and P–O–P coupling constants (absolute values) of the phosphorus resonances for nonmodified NTPs.[182] These values are: –10 ppm (P^1), –21 ppm (P^2), and –6 ppm (P^3) for chemical shifts and 20 Hz (J_{12} and J_{23}) for coupling constants. All positive $\Delta\delta P$ values in Table 3.11 signify a downfield shift compared to chemical shifts of the respective phosphorus atoms in the parent NTP molecule (and vice versa for the negative $\Delta\delta P$ values). The positive value for the coupling constants signifies that the absolute value of the respective coupling constant in the modified NTP is larger than that in the parent NTP molecule (and vice versa for negative values).

Prior to evaluation of the NMR data presented in Table 3.11, a number of factors should be taken into consideration. The nature of the ^{31}P NMR chemical shifts and coupling constants is very sensitive to conditions of the media (solvent composition; pH; ionic strength; nature of the counter ions,

Table 3.11 ^{31}P Nuclear Magnetic Resonance Data of NTPs with Triphosphate Chain Modification

Compound	X	Y	Z	Ω	R1	R2 or β	R3 or ψ or γ	ΔδP¹ ppm	ΔδP² ppm	ΔδP³ ppm	ΔJ12 Hz	ΔJ23 Hz	Ref.
I.1		NH						+9	0	0	NR	NR	36
I.2		CH_2						+38	–1	0	NR	NR	15
I.3	O	CH_2						+20	–1	–4	+7	0	15, 18, 19
I.4		S						+19	–1	–1	+7	0	37
II.1			NH					+12	+10	+1	–14	+1	44, 45
II.2			CH_2					+29	+27	0	–12	+4	53
II.3			CHF					+22	+21	+1	–5	+5	53
II.4			CHBr					+18	+19	–7	–16	+5	55
II.5			CF_2					+14	+15	–4	NR	+8	54
III.1				NH				–1	+14	+7	+1	–14	29, 58
III.2				CH_2				0	+35	+18	+7	–12	58
III.3				CHF				0	+23	+15	+8	–6	58
III.4				CHBr				–4	+21	+16	+5	–15	55
III.5				CF_2				–1	+15	+10	+11	NR	58
III.6				CBr_2				0	+22	+14	+7	–5	71
III.7				CCl_2				0	+23	+15	+8	–2	58
III.8				OO				–2	+10	+12	+1	–7	72
III.9				CC				–1	0	–4	+2	–8	58
IV.1					NH_2			+12	–2	–1	–2	+1	109
IV.2					CH_3			+35	–2	–3	+2	+1	102
IV.3					SH			+53	–1	–1	?	0	97
IV.4					BH_3			+95	0	–1	–1	–1	104, 105, 106
V.1						SH		–1	+51	0	–14	+7	187, 97
VI.1							NH_2	–1	–1	+9	0	–1	67
VI.2							NHAlkyl	–2	0	+7	+2	+2	113, 114
VI.3							NHAryl	–1	–2	0	–1	+1	113, 114, 122
VI.4							NHNHAryl	–1	–1	+6	NR	NR	113, 114
VI.5							Alkyl	–1	–1	+23	0	+3	71

Table 3.11 [31]P Nuclear Magnetic Resonance Data of NTPs with Triphosphate Chain Modification (Continued)

Compound	X	Y	Z	Ω	R1	R2 or β	R3 or ψ or γ	ΔδP[1] ppm	ΔδP[2] ppm	ΔδP[3] ppm	ΔJ12 Hz	ΔJ23 Hz	Ref.
VI.6							Aryl	−1	−1	+17	+5	−1	71
VI.7							H	−1	0	+6	−1	−1	126
VI.8							OAlkyl	−1	−1	+1	−2	−2	128, 129
VI.9							OAryl	−2	−2	−6	NR	NR	129
VI.10							SH	0	−1	+41	0	+9	97, 188
VI.11							SAlkyl	+1	0	+17	NR	NR	131
VI.12							N_3	−2	−2	−3	−1	+1	125
VI.13							F	−1	−2	−8	0	−2	67, 129
VII.1			NH	NH				+11	+25	+9	−13	−16	45
VII.2			CH_2	CH_2				+28	+49	+21	NR	NR	140
VII.3				CH_2			CH_3	−2	+28	+48	+6	−20	71
VII.4	O	CH_2		CH_2			CH_3	+19	+30	+46	NR	NR	16
VII.5	O	CH_2		CH_2				+20	+37	+16	NR	NR	21
VII.6	O	CH_2		CHF				+20	+23	+15	+9	NR	21
VII.7	O	CH_2		CF_2				+21	+19	+11	+12	+45	16, 21
VII.8	O	CH_2		CF_2			CH_3	+21	+17	+41	+12	NR	16
VII.9	O	CH_2		CF_2			Ph	+21	+16	+30	+13	+29	16
VII.10	O	CH_2					Ph	+19	+1	+17	+6	+4	16
VIII.1					SH	SH		+53	+49	−1	+15	+8	141
VIII.2					SH		SH	+53	+45	+40	+8	+10	141
VIII.3					SH		OAlkyl (R3) & S (γ)	+53	−2	+53	+7	+7	141
VIII.4					SH		OAlkyl	+55	−2	0	+5	0	81
VIII.5					SH and S			+110	−2	−1	+14	0	141
VIII.6					BH_3 and S			+163	−1	−4	+15	−2	142
VIII.7						OAlkyl	OAlkyl	−2	−1	−2	−2	−2	146, 147

especially metal ions; concentration of the sample; temperature). The chemical shift values also depend on the way in which a reference signal was set up during the recording of NMR spectrum (internal or external standard). In most publications, the conditions for ^{31}P NMR measurements are not specified; therefore, all data are cautiously rounded to whole parts per million and hertz units. Differences in the chemical shifts of less than 2 ppm were not considered a reliable indication of the effect of a substitute. (It should be noted, however, that if all the precautions and corrections are made properly, one can consider the changes of 1 ppm or less to be significant enough to indicate a structural and/or electronic change in the vicinity of that phosphorus atom.) A similar statement is applicable to changes of 5 Hz or less for the coupling constants. Whenever possible, the averaged values (Δδ, ΔJ) for similar compounds were presented in Table 3.11.

In addition, we should emphasize that the chemical shift of the phosphorus in the terminal phosphate residue is sensitive to its ionization state at pH range within 4 to 9. The difference in the ^{31}P NMR chemical shifts of phosphate monoester groups for P^3-OH and P^3-O^- +H^+ states may be as large as 4 ppm. Therefore, a different approach has been chosen for calculating the $\Delta\delta P^3$ values for P^3-substituted and P^3-nonsubstituted NTP analogs. For derivatives containing a P^3-OH function capable of establishing a deprotonation ↔ protonation equilibrium at pH range from 4 to 9, the reference chemical shift (for parent NTP) was chosen to be –10 ppm. For P^3-substituted derivatives (which have no such equilibrium at pH range 4 to 9), the reference signal was chosen to be –6 ppm.

For the sake of simplicity, we did not discriminate positions "α" and "R1," positions "β" and "R2," and positions "γ," "ψ," and "R3." Therefore, the exact structural representation of the substitute may not match that in Table 3.11 and in previous tables.

3.7.1 Modifications of triphosphate fragment at a single position: chemical shifts

When bridged or nonbridged oxygen (or a hydroxyl group) is substituted for another atom (or group), the electronic and structural changes in the triphosphate fragment lead to pronounced changes in the ^{31}P NMR chemical shifts.[182] Analysis of the chemical shift data for NTP analogs with a single modification, presented in Table 3.11, leads to several observations.

The extent to which a substitute affects the ^{31}P chemical shift is specifically determined by the nature of the atom directly attached to the phosphorus atom. Thus, with a nitrogen atom (NH) substituted for oxygen (or NH_2 for OH), the chemical shift changes are in the range of +9 to +14 ppm for P^1 and P^2 atoms, but are smaller for P^3 atom (from +6 to +9 ppm). The exception to this rule is found for compound VI.3 (Table 3.11); here the $\Delta\delta P^3$ value is "0," indicating the apparent absence of the effect. In reality, two events (first, -NH_2 substitution for -OH and, second, -NHPh substitution for NH_2) work in opposite directions to cancel each other. Note that aromatic group attached to the phosphorus atom through oxygen and nitrogen causes an upfield shift of the ^{31}P signal (–6 to –9 ppm) compared to alkyl-NH or alkyl-O P^3-substituted NTP analogs.

Carbon substitution (CH_2) for oxygen causes a change of the chemical shift from +27 to +38 ppm for P^1 and P^2 atoms, and a smaller change from +18 to +23 ppm for P^3 atom. The CH(Hal) and C(Hal)$_2$ substitutions for oxygen lead to smaller shift changes (+15 to +23 ppm for P^1 and P^2 atoms; +10 to +15 ppm for P^3 atom). Borane NTP analog modified at P^1 position (compound IV.4) shows the extreme shift of +95 ppm compared to the parent NTP. Generally, the ^{31}P NMR signals shift toward a lower field in the order of P–O >P–N > P–C > P–B analogs; this correlates with the electron withdrawal effect of those atoms (or groups).

When sulfur substitutes for oxygen in the triphosphate fragment, the value of the ^{31}P NMR chemical shift is very sensitive to its position. Therefore, the NTP phosphorothioate analogs are discussed here as a separate group of NTP analogs. The ^{31}P NMR chemical shift for phosphorothioate esters is known that depends on the degree of S = P double bonding.[182,183] In the extreme case of S = P(OR)$_3$ type compounds (R = alkyl), the ^{31}P chemical shift is observed at approximately 70 ppm (compare with –2 ppm for O = P(OR)$_3$). For compounds from the second group (with less S = P double bonding), such as S = P(OR)$_2$(OH) ↔ O = P(OR)$_2$(SH), the chemical shift is observed at 50 to 55 ppm. The third group of phosphorothioate monoesters includes compounds S = P(OR)(OH)$_2$ ↔ O = P(OR)(SH)(OH) with a chemical shift at 40 to 45 ppm. Finally, for phosphate thioester compounds O = P(OR)$_2$(SR) with minimal or no S = P bonding, the chemical shift is observed at 20 ppm (fourth group).

Previously mentioned groups two, three, and four are represented among the thio NTP analogs in Table 3.11. The changes in the ^{31}P NMR chemical shifts for those P^1-, P^2-, and P^3-substituted thio NTP analogs are in agreement with the discussed rules (compounds I.4; IV.3; V.1; VI.10; and VI.11).

The last observation is that a substitute has little or no effect on the ^{31}P chemical shift ($\Delta\delta \sim 0$ ppm) unless it is attached directly to the phosphorus atom; see:

- $\Delta\delta P^1$ for compounds III.1 to III.9; and VI.1 to VI.13
- $\Delta\delta P^2$ for compounds I.1 to I.4; IV.1 to IV.4; and V.1 to VI.13
- $\Delta\delta P^3$ for compounds I.1 to I.4; II.1 to II.5; and IV.1 to IV.4

The exceptions may be explained by taking into account the ionization state of P^3 terminal phosphate group (compounds I.3, II.4, and II.5 show significant value for $\Delta\delta P^3$ from –4 to –7 ppm).

3.7.2 *Modifications of triphosphate fragment at multiple positions: chemical shifts*

Information on ^{31}P NMR chemical shifts for NTP analogs with multiple modifications in the triphosphate fragment is limited. However, analysis of the available data suggests that all the simple rules mentioned earlier for singly modified NTP analogs are applicable to the double- or triple-substituted NTP analogs. The chemical shift effects of the substitutes are additive.

As an example, for P^1 resonance of double-substituted compound VIII.5 (R1 = SH; α = S), the value $\Delta\delta P^1$ is +110 ppm. This is almost double the effect observed for monosubstituted P^1-thio NTP analog IV.3, for which $\Delta\delta P^1$ is +53 ppm. For compound VIII.1 (Z = NH; Ω = NH), the resonance for P^2 is shifted downfield and the value $\Delta\delta P^2$ equals +25 ppm. This change in the chemical shift is equal to the sum (12 + 14 = 26 ppm) of the effects observed for P^2 resonance in compounds 1.2.1 ($\Delta\delta P$ is +12 ppm) and III.1 ($\Delta\delta P$ is +14 ppm).

Examination of other double- and triple-modified NTP analogs will further illustrate the additive properties of the chemical shifts caused by individual modifications, although these tendencies are not always strictly quantitative. Thus, for compound VIII.2 (Z = CH_2; Ω = CH_2) the resonance for P^2 is shifted downfield for +49 ppm, which is close (but not equal) to the sum (27 + 35 = 62 ppm) of effects observed for P^2 signal in compounds II.2 ($\Delta\delta P$ is +27 ppm) and III.2 ($\Delta\delta P$ is +35 ppm).

3.7.3 ^{31}P NMR of the NTPs with oxygen isotopic substitutions

The sensitivity of ^{31}P NMR chemical shifts to the presence of different oxygen isotopes ^{16}O, ^{17}O, and ^{18}O was demonstrated in a series of papers published in the 1970s and1980s. The presence of ^{18}O attached to the phosphorus atom results in an upfield shift of the ^{31}P resonance; the magnitude of the shift depends on the bond order between oxygen and phosphorus atoms.[184–186] Therefore, in ^{31}P NMR isotopomer species containing ^{18}O atoms are seen as separate ^{31}P signals shifted from signals of isotopomers that contain only ^{16}O atoms. In other words, ^{31}P NMR can follow the process of substituting one oxygen isotope with another.

In many instances when NTPs participate in enzymatic and chemical reactions, the result is hydrolysis of existing P–O–P linkages or formation of new P–O–P linkages. Therefore, when individual isotopomers are used, a stereochemical course of chemical and enzymatic reactions can be followed through the identification of ^{16}O- and ^{18}O-containing species in ^{31}P NMR spectra. It should be noted that, due to its quadrupole moment, the phosphorus resonances for ^{17}O-containing species are not normally observed in ^{31}P NMR spectra.[184,187]

3.7.4 Phosphorus–phosphorus coupling constants in NTP analogs

Phosphorus–phosphorus coupling constants significantly change with substitutions in the triphosphate chain of NTP analogs. The nature of the effect of the substitute on geminal coupling constants is complicated and involves contributions from several factors. Generally, it reflects the changes in electronic effects of the substitute and also structural changes in P–Z–P or P–Ω–P fragments.

A few tendencies are observed for the changes in coupling constants. For instance, substitutions of NH or CH_2 for oxygen in P–O–P fragments lead to smaller J_{12} and J_{23} values (for P–NH–P or P–CH_2–P) by 12 to 16 Hz, compared to J_{12} and J_{23} values in the parent NTPs. On the other hand, CH(Hal) and $C(Hal)_2$ substitutes cause an increase in J_{12} and J_{23} values. A very large change in coupling constants — increases by 30 to 45 Hz — was observed when CF_2 group substitutes for oxygen in P–Z–P or P–Ω–P fragments. The effect of sulfur on ΔJ_{12} or ΔJ_{23} varies from –14 to +15 Hz, depending on the position of the substitution. It is noteworthy that variations in solvents, nature, and concentration of the counter ions and salt, as well as the temperature, may all lead to large variations in the coupling constant values for the same compound.

Although the information in Chapter 3.7 is not comprehensive, it might be helpful in the identification and structural analysis of known and newly synthesized NTP analogs with modified triphosphate fragments. A summary of data is presented in Table 3.11.

3.8 *Approaches to biological applications of NTPs with modified triphosphate chains*

Various aspects of biological applications of the NTPs are described in detail in several other chapters of this book. NTP analogs with modified triphosphate fragments are used in many similar applications to those of other types of modified and nonmodified NTPs. In this chapter, we briefly discuss only the general issues related to the biological applications of this particular class of NTPs.

The nucleoside 5′-triphosphate analogs with modified triphosphate chain are primarily designed to serve as substitutes for natural NTPs in biological studies in which the interaction or chemical transformation of the triphosphate chain is a critical element of the biochemical process. The presence of different atoms (or groups) instead of natural oxygen in a modified triphosphate chain alters the electronic and structural properties of the NTP molecule and, therefore, possibly how the modified NTP interacts with a particular enzyme. The modification may or may not change the ability of NTP to be recognized by the enzyme (or other biological molecule) and to bind the enzyme at active center. Even if binding does occur, the modification may affect the ability of the enzyme to perform a catalytic act and allow the NTP to vacate the active center for the next incoming substrate. Therefore, NTP analogs with modified triphosphate chain can be substrates, effectors, or inhibitors for enzymes, or they can be inactive with those enzymes.

The NTP analog acts much as a natural "substrate" when a modification does not interfere with normal enzymatic functions and a conversion of the NTP analog into the product takes place. If the NTP analog stimulates the activity of the enzyme toward the natural substrate, it is acting as an "effector." Alternatively, if the modified NTP analog interacts with the enzyme in such a way that it reduces (or even prevents) the enzyme's ability to perform its natural biochemical function, the NTP analog acts as an "inhibitor." If an enzyme does not recognize an NTP analog and the analog does not interfere with its normal functions, it is considered "inactive" toward the enzymatic reaction. The extent of the substrate, effector, or inhibitor properties is varied for different NTP analogs and may be compared to that for the natural NTP substrates or known effectors or inhibitors. These properties are qualitatively indicated in Table 3.12.

Some specific modifications in the triphosphate chain may cause a phosphorus atom (P^1, P^2, or P^3) to become a chiral center, and therefore two (or more) diastereomers of NTP analog may be obtained. In many instances, pure diastereomers can be prepared by separation of their mixtures or by performing stereospecific chemical or enzymatic synthesis. Due to a natural stereospecificity observed in most enzymatic reactions, one should expect the behavior of the diastereomers of NTP analogs to be quite different in the biological systems. The individual diastereomers of NTP analogs are regarded as one of the most important molecular tools in the investigation of mechanisms and stereochemistry of enzymatic reactions involving NTPs.

The determining role of any substitution in the triphosphate chain of an NTP in preserving, changing, or losing the substrate properties, or in acquiring inhibitory properties, is obviously related to the specific mechanism of the enzyme action. When NTP becomes a part of the enzyme-assisted chemical conversion process, one would like to know the particular structural elements of the NTP and enzyme directly involved in the chemical transformation and the elements critical (or noncritical) to support that transformation. There are examples when the same NTP analog may be a substrate for one enzyme, but completely inactive or be an inhibitor for another one. The choice of which NTP modification to use should be relevant to the goal of a specific study. Choosing an appropriate NTP analog among hundreds of possibilities makes it likely that an adequate answer to a specific question related to the biological (biochemical) system can be found.

In addition to being a substrate, an effector, or an inhibitor, the NTP analog may perform some other functions, as described in the following paragraphs.

First, the modified NTP may carry a reactive chemical function (such as different kinds of activated phosphate residues, nitrogen mustard derivatives, photoreactive groups, platinum complexes, etc.) attached to the triphosphate chain. Different groups of the macromolecule may be modified after interacting with those NTPs in the complexes. This process is called "affinity modification" of biopolymers, and it is one of the key techniques used in structural elucidation of biomolecules and, especially, in enzymology studies.

Second, the modified NTP may carry a reporter chemical function attached to the triphosphate chain. This feature allows detection of the exact location of that NTP molecule in a biological system or it allows monitoring changes in the microenvironment near the location of the NTP-binding site. Examples of those groups are fluorescent dyes; spin labels; metal ions; biotin

Table 3.12 Evaluation of Properties of NTPs with Triphosphate Chain Modification(s) in Selected Enzymatic Systems

Structure type	DNA pol	RNA pol	RT	TdT	ATase	Kinase	PME	PDE	AP	tRNA synth	ATP & GTPase	AC
I												
X = CH_2, Y = O	PS, IA		S, I	S, PS		S						
X = NH, Y = CH_2, CHR	S				MI, WI							
X = CHR, Y = CHR (R = H, Alkyl)			IA		WI, I	S, IA		S	S			
X = S, Y = CH_2	IA	WI, IA				S			S			
X = None, Y = CH_2	IA						S	S				
II												
Z = NH	IA	S (?)	I		WI	S						
Z = CH_2						PS		IA	IA			
III												
Ω = CH_2	S, IA		IA		IA	I		IA, S				
Ω = CF_2	S, IA, I					SI						
Ω = CBr_2			S, IA									
Ω = O-O	IA	IA				I, IA		S	S			
Ω = None										I		
Ω = NH			S		I	I, IA		S	S		I	E
IV												
R1 = SH	S	S	S	S		S				S	S	S
R1 = Me	S, I		S	S								
R1 = BH_3	S											
V												
R2 = SH	S, IA					S					S	
VI												
R3 = NH_2	IA				S							
R3 = NH-aryl, arylazido		S	S, I	S, I				S		I		
R3 = N_3	S							S	IA		I	

R3 = O-alkyl, aryl			S	S, I					
R3 = SH, S-alkyl	S	S			S	S	I	S	I
R3 = F	S								I
R3 = alkyl, aryl	S, PS		S	S, I					
VII									
Y = CH_2, Z = O, Ω = O, ψ = Ph	IA	PS		PS					
Y = CH_2, Z = O, Ω = CH_2, ψ = Me	IA		PS	PS					
Y = CH_2, Z = O, Ω = $C(Hal)_2$, ψ = Ph or Me			S, PS	PS					
Y = CH_2, Z = O, Ω = $C(Hal)_2$ ψ = OH	IA		S, PS	S					
Y = CH_2O, Z = O, Ω = CH_2, ψ = OH			IA						
Y = CH_2O, Z = O, Ω = CHF, ψ = OH			WI						
Y = O, Z = CH_2, Ω = CF_2, ψ = OH			I						
Y = CH_2, Z = NH, Ω = NH, ψ = OH			I						
VIII									
α = SH, β = SH, γ = OH, R1 = O, ψ = OH			I			S	S		
α = SH, β = OH, γ = OH, R1 = S, ψ = OH	IA								
α = BH_3, β = OH, γ = OH, R1 = S, ψ = OH	PS								

Notes: AC = adenylate cyclase; AP = alkaline phosphatase; ATase = adenosyltransferase; ATPase = adenosine 5′-triphosphatase; DNA pol = DNA polymerase; E = effector; GTPase = guanosine 5′-triphosphatase; I = inhibitor; IA = inactive; PS = poor substrate; RNA pol = RNA polymerase; RT = reverse transcriptase; S = substrate; SI = strong inhibitor; TdT = terminal deoxynucleotidyl transferase; tRNA synth = tRNA synthesase; WI = weak inhibitor; PDE = phosphodiesterase; PME = phosphomonoesterase.

and related compounds; and radioactive and nonradioactive isotopes (see below). Also, some molecules attached to the triphosphate chain of the NTP may serve as a vehicle delivering the NTP analog to a specific location across the cellular membrane.

Third, the triphosphate fragment of the NTPs may be labeled with isotopes of the oxygen (^{17}O and ^{18}O) or phosphorus (^{32}P or ^{33}P). The isotopically modified NTP is normally an adequate substrate when compared to the parent NTP (with a natural distribution of atom isotopes). Using physical methods of investigation such as ^{31}P NMR, mass-spectrometry, and radiochemical detectors, one may follow the steps of the biochemical process and the fate of the labeled NTP in the biological system. This is an important approach in understanding the various mechanisms by which NTPs participate in biochemical processes. NTP analogs containing specific oxygen isotopes have been used (often in combination with sulfur in phosphorothioate fragment) for investigation of the stereochemistry of the enzymatic and chemical reactions. NTPs with radioisotopes of phosphorus have been used for decades as the most reliable approach to investigate biological processes such as replication, transcription, translation, and other *in vitro* and *in vivo* systems. The isotope labeling approaches have been valuable in a number of pharmacokinetic studies for drug discovery.

The multiple modifications of the triphosphate chain (in combination with modifications of the sugar and the heterocyclic base) may result in new and sometimes unexpected properties for NTP analogs.

In Table 3.12, the compounds are classified according to the type of structure discussed previously. The purpose of this table is to give a preliminary overview of all types of NTP analogs and the types of enzymatic systems tested with those NTP analogs. The substrate, effector, or inhibitory properties of the NTP analogs are indicated qualitatively and relative to the substrate properties of a natural NTP or compared to known inhibitors of the enzyme (when this information is available).

When analyzing the data presented in Table 3.12, one should keep in mind two issues. First, the enzymes from the same class, but from different sources (for example, DNA polymerases) are different in the way in which they recognize substrates and perform the catalytic act. Second, in some instances the data show a cumulative effect of modifications of triphosphate chain and nucleoside fragments. Therefore, the properties of the NTP analogs demonstrate apparent contradictions in respect to the same class of enzymes. For example, the NTP analog could be a substrate or inhibitor, or could be inactive with those enzymes. One should refer to the tables in the previous sections for more detailed information on the NTP analog, enzyme source, and cited literature.

The general rule for an NTP analog to be a substrate is that it possess the same cleavable linkage (for example P^1–O–P^2) present in the natural substrate. On the other hand, to be a potent inhibitor, the specific cleavable linkage must be modified (for example, substitute P^1–CH_2–P^2 fragment for P^1–O–P^2) enough that the linkage cannot be cleaved or is cleaved at a much slower rate than the natural substrate. One should always keep in mind, however, that these rules have many exceptions.

References

1. Yount, R.G. ATP analogs. *Adv. Enzymol.*, 1975, 43, 1–56.
2. Scheit, K.-H. *Nucleotide Analogs.* J. Wiley & Sons, New York, 1980, 96–141.
3. Eckstein, F. Phosphorithioate analogues of nucleotides — tools for the investigation of biochemical processes. *Angew Chem.* (Int. Ed.), 1983, 22, 423–439.
4. Eckstein, F. Nucleoside phosphorothioates. *Ann. Rev. Biochem.*, 1985, 54, 367–402.
5. Burgess, K. and Cook D. Syntheses of nucleoside triphosphates. *Chem. Rev.*, 2000, 100, 2047–2059.
6. Sawai, H., Wakai, H., and Shimazu, M. Facile synthesis of CAP portion of messenger RNA by Mn (II) ion-catalyzed pyrophosphate formation in aqueous solution. *Tetrahedron Lett.*, 1991, 32, 6905–6906.
7. Chlebicka, L., Wieczorek, Z., Stolarski, R., Stepinski, J., Darzynkiewicz, E., and Sugar, D. Synthesis and properties of mRNA 5′-cap analogues with 7-methylguanosine replaced by benzimidazole or 3-methylbenzimidazole. *Nucleoside Nucleotides,* 1995, 14, 771–775.
8. Nakagawa, I., Konya, S., Ohtani, S., and Hata, T. A "capping" agent: P^1-S-Phehyl P^2-7-methylguanosine-5′ pyrophosphorothioate. *Synthesis,* 1980, 556–557.
9. Kamimura, T., Osaki, Y., Sekine, M., and Hata, T. An effective method for the synthesis of the cap structure of eukaryotic messenger ribonucleic acid. *Tetrahedron Lett.*, 1984, 25, 2683–2686.
10. Kohno, K., Nishiyama, S., Kamimura, T., Sekine, M., Hata, T., Kumagai, I., and Miura, K. Chemical synthesis of capped RNA fragments and their ability to complex with eukaryotic ribosomes. *Nucl. Acids Res. Simp. Series,* 1985, 16, 233–236.
11. Sekine, M., Nishiyama, S., Kamimura, T., Osaki, Y., and Hata, T. Chemical synthesis of capped oligoribonucleotides, $m^7G^{5'}$pppAUG and $m^7G^{5'}$pppAUGACC. *Bull. Chem. Soc. Jpn.*, 1985, 58, 850–860.
12. Fukuoka, K., Suda, F., Suzuki, R., Ishikawa, M., Takaku, H., and Hata, T. Large-scale synthesis of the cap part in messenger RNA using a new type of bifunctional phosphorylating reagent. *Nucleosides Nucleotides,* 1994, 13, 1557–1567.
13. Fukuoka, K., Suda, F., Suzuki, R., Takaku, H., Ishikawa, M., and Hata, T. One-pot reaction for the synthesis of $m^7G^{5'}$pppG and $m^7G^{5'}$pppA by using an activatable bifunctional phosphorylating reagent. *Tetrahedron Lett.*, 1994, 35, 1063–1066.
14. Sawai, H., Wakai, H., and Nakamura–Ozaki, A. Synthesis and reactions of nucleoside 5′-diphosphate imidazolide. A non-enzymatic capping agent for 5′-monophosphorylated oligoribonucleotides in aqueous solution. *J. Org. Chem.*, 1999, 64, 5836–5840.
15. Shirokova, E.A., Tarussova, N.B., Shipitsin, A.V., Semizarov, D.G., and Krayevsky, A.A. Novel acyclic nucleotides and nucleoside 5′-triphosphates imitating 2′,3′-dideoxy-2′,3′-didehydronucleotides: synthesis and biological properties. *J. Med. Chem.*, 1994, 37, 3739–3748.

16. Shipitsin, A.V., Victorova, L.S., Shirokova, E.A., Dyatkina, N.B., Goryunova, L.E., Beabealashvilli, R.Sh., Hamilton, C.J., Roberts, S.M., and Krayevsky, A. New modified nucleoside 5′-triphosphates: syntheses, properties towards DNA polymerases, stability in blood serum and antiviral activity. *J. Chem. Soc. Perkin Trans.*, I, 1999, 1039–1050.
17. Dyatkina, N.B., Arzumanov, A.A., Victorova, L.S., Kukhanova, M.K., and Krayevsky, A.A. Synthesis and substrate properties of thymidine 5′-triphosphate analogs with large hydrophobic substituent groups at α-P atom. *Nucleosides Nucleotides*, 1995, 14, 91–103.
18. Merlo, V., Roberts, S.M., Storer, R., and Bethell, R.C. Synthesis and biological activity of the diphosphorylphosphonate derivatives of (+)- and (–)-cis-9-(4′-hedroxycyclopent-2′-enyl)guanine. *J. Chem. Soc. Perkin Trans.*, I 1994, 1477–1481.
19. Coe, D.M., Roberts, S.M., and Storer, R. The potential of carbocyclic nucleosides for the treatment of AIDS: synthesis of some diphosphorylphosphonates possessing potent activity against HIV-coded reverse transcriptase. *J. Chem. Soc. Perkin Trans.*, I. 1992, 2695–2704.
20. Holy, A., and Rosenberg, I. Synthesis of 9-(2-phosphonylnethoxyethyl)adenine and related compounds. *Collect. Czech. Chem. Commun.*, 1987, 52, 2801–2809.
21. Hamilton, C.J., Roberts, S.M., and Shipitsin, A. Synthesis of a potent inhibitor of HIV transcriptase. *Chem. Commun.*, 1998, 10, 1087–1088.
22. Vesely, J., Rosenberg, I., and Holý, A. Donor activity of 5′-phosphonylmethyl analogues of ATP and GTP in the phosphorylation of uridine catalyzed by uridine kinase from mouse leukemic cells. *Collect. Czech. Chem. Commun.*, 1983, 48, 1783–1787.
23. Kappler, F., Hai, T.T., and Hampton, A. Use of a vinyl phosphonate analog of ATP as a rotationally constrained probe of the C5′-O5′ torsion angle in ATP complexed to methionine adenosyltransferase. *Bioorg. Chem.*, 1985, 13, 289–295.
24. Freeman, G.A., Rideout, J.L., Miller, W.H., and Reardon, J.E. 3′-Azido-3′,5′-dideoxythymidine-5′-methylphosphonic acid diphosphate: synthesis and HIV-1 reverse transcriptase inhibition. *J. Med. Chem.*, 1992, 35, 3192–3196.
25. Gaur, R.K., Sproat, B.S., and Krupp, G. Novel solid phase synthesis of 2′-O-methylribonucleoside 5′-triphosphates and their α-thio analogs. *Tetrahedron Lett.*, 1992, 33, 3301–3304.
26. Kappler, F., Hai, T.T., Abo, M., and Hampton, A. Species- or isozyme-selective enzyme inhibitors. 8. Synthesis of disubstituted two-substrate condensation products as inhibitors of rat adenylate kinases. *J. Med. Chem.*, 1982, 25, 1179–1184.
27. Vrudhula, V.M., Kappler, F., and Hampton, A. Isozyme-specific enzyme inhibitors. 13. S-[5′(R)-[′N-triphosphoamino)methyl]adenosyl]-L-homocysteine, a potent inhibitor of rat methionine adenosyltransferases. *J. Med. Chem.*, 1987, 30, 888–894.
28. Kappler, F., Hai, T.T., Cotter, R.J., Hyver, K.J., and Hampton, A. Isozyme-specific enzyme inhibitors. 11. (L-homocysteine)-ATP S-C5′ covalent adducts as inhibitors of rat methionine adenosyltransferases. *J. Med. Chem.*, 1986, 29, 1030–1038.

29. Kappler, F., Vrudhula, V.M., and Hampton, A. Isozyme-specific enzyme inhibitors. 14. 5′(R)-C-[(L-Homocystein-S-yl)methyl]adenosine 5′-(β,γ-imidotriphosphate), a potent inhibitor of rat methionine adenosyltransferases. *J. Med. Chem.*, 1987, 30, 1599–1603.
30. Hampton, A., Sasaki, T., and Paul, B. Synthesis of 6′-cyano-6′-deoxyhomoadenosine-6′-phosphonic acid and its phosphoryl and pyrophosphoryl anhydrides and studies of their interactions with adenine nucleotide utilizing enzymes. *J. Am. Chem. Soc.*, 1973, 95, 4404–4414.
31. Matulis–Adamic, J., Haeberli, P., and Usman, N. Synthesis of 5′-deoxy-5′-difluoromethyl phosphonate nucleotide analogs. *J. Org. Chem.*, 1995, 60, 2563–2569.
32. Letsinger, R.L., Wilkes, J.S., and Dumas, L.B. Incorporation of 5′-amino-5′-deoxythymidine 5′-phosphate polynucleotides by use DNA polymerase I and phiX174 DNA template. *Biochemistry*, 1976, 15, 2810–2816.
33. Wilkes, J.S., Hapke, B., and Letsinger, R.L. 5′-Amino analog of adenosine diphosphate. *Biochem. Biophys. Res. Commun.*, 1973, 53, 917–922.
34. Trowbridge, D.B., Yamamoto, D.M., and Kenyon, G.L. Ring openings of trimetaphosphoric acid and its bismethylene analog. Syntheses of adenosine 5′-bis(dihydroxyphosphinylmethyl) phosphinate and 5′-amino-5′-deoxyadenosine 5′-triphosphate. *J. Am. Chem. Soc.*, 1972, 94, 3816–3824.
35. Letsinger, R.L., Wilkes, J.S., and Dumas, L.B. Nucleotide chemistry. XVII. Enzymic synthesis of polydeoxyribonucleotides possessing internucleotide phosphoramidate bonds. *J. Am. Chem. Soc.*, 1972, 94, 292–293.
36. Wolfe, J.L., Kawate, T., Belenky, A., and Stanton, V., Jr. Synthesis and polymerase incorporation of 5′-amino-2′,5′-dideoxy-5′-N-triphosphate nucleotides. *Nucl. Acids Res.*, 2002, 30, 3739–3747.
37. Patel, K.B., and Eckstein, F. 5′-Deoxy-5′-thioribonucleoside-5′-triphosphates. *Tetrahedron Lett.*, 1997, 38, 1021–1024.
38. Stutz, A., and Scheit, K.H. Properties of ATP and UTP analogues with P–S–C-5′ bonds. *Eur. J. Biochem.*, 1975, 50, 343–349.
39. Scheit, K.H., and Stutz, A. Properties of dATP-analog with phosphorus-sulfur-carbon-5′ bond. *J. Carbohydrate Nucleoside Nucleotide*, 1974, 1, 485–490.
40. Rammler, D.H., Yengoyan, L., Paul, A.V., and Bax, P.C. Nucleoside phosphonic acids. II. The synthesis of 5′-deoxythymidine 5′-phosphonic acid and its pyrophosphate derivatives. *Biochemistry*, 1967, 6, 1828–1837.
41. Yengoyan, L., and Rammler, D.H. Nucleoside phosphonic acids. I. The synthesis of 5′-deoxyuridine 5′-phosphonic acids and derivatives. *Biochemistry*, 1966, 5, 3629–3638.
42. Ma, Q., Babbitt, P.C., and Kenyon, G.L. Adenosine 5′-[α,β]-imido-triphosphate, a substrate for T7 RNA polymerase and rabbit muscle creatine kinase. *J. Am. Chem. Soc.*, 1988, 110, 4060–4061.
43. Ma, Q., Kenyon, G.L., and Markham, G.D. Specificity of S-adenosylmethionine synthetase for ATP analogs mono- and disubstituted in bridging positions of the polyphosphate chain. *Biochemistry*, 1990, 29, 1412–1416.
44. Li, R., Muscate, A., and Kenyon, G.L. Synthesis, characterization and inhibitory activities of nucleoside α,β-imido triphosphate analogues on human immunodeficiency virus-1 reverse transcriptase. *Bioorg. Chem.*, 1996, 24, 251–261.

45. Ma, Q.-F., Bathurst, I.C., Barr, P.J., and Kenyon, G.L. New thymidine triphosphate analog inhibitors of human immunodeficiency virus-1 reverse transcriptase. *J. Med. Chem.*, 1992, 35, 1938–1941.
46. Myers, T.C., Nakamura, K., and Flesher, J.W. Phosphonic acid analogs of nucleoside phosphates. I. The synthesis of 5-adenylyl methylenediphosphonate, a phosphonic acid analog of ATP. *J. Am. Chem. Soc.*, 1963, 85, 3292–3295.
47. Kappler, F., Hai, T.T., and Hampton, A. Isozyme-specific enzyme inhibitors. 10. Adenosine 5′-triphosphate derivatives as substrates or inhibitors of methionine adenosyltransferases of rat normal and hepatoma tissues. *J. Med. Chem.*, 1986, 29, 318–322.
48. Milner–White, E.J., and Rycroft, D.S. The alpha beta-methylene analogues of ADP and ATP act as substrates for creatine kinase. ΔG_0 for this reaction and for the hydrolysis of the alpha beta-methylene analogue of ATP. *Eur. J. Biochem.*, 1983, 133, 169–172.
49. Krug, F., Parikh, I., Illiano, G., and Cuatrecasas, P. α,β-Methylene-adenosine 5′-triphosphate: a competitive inhibitors of adenylate cyclase in fat and liver cell membranes. *J. Biol. Chem.*, 1973, 248, 1203–1206.
50. Picher, M., Sevigny, J., D'Orleans–Juste, P., and Beaudoin, A.R. Hydrolysis of P_2-purinoceptor agonists by a purified ectonucleotidase from the bovine aorta, the ATP-diphosphohydrolase. *Biochem. Pharmacol.*, 1996, 51, 1453–1460.
51. Wehland, J., and Sandoval, I. Cells injected with guanosine 5′-[α,β-methylene]triphosphate, an α,β-nonhydrolyzable analog of GTP, show anomalous patterns of tubulin polymerization affecting cell translocation, intracellular movement, and the organization of Golgi elements. *Proc. Natl. Acad. Sci. USA*, 1983, 80, 1938–1941.
52. Balzarini, J., Herdewijn, P., Pauwels, R., Broder, S., and De Clercq, E. α,β- and β,γ-methylene 5′-phosphonate derivatives of 3′-azido-2′,3′-dideoxythymidine 5′-triphosphate. Correlation between affinity for reverse transcriptase, susceptibility to hydrolysis by phosphodiesterases and anti-retrovirus activity. *Biochem. Pharmacol.*, 1988, 37, 2395–2403.
53. Blackburn, G.M., and Langston, S.P. Novel P1,P2-substituted phosphonate analogues of 2′-deoxyadenosine and 2′-deoxythymidine 5′-triphosphates. *Tetrahedron Lett.*, 1991, 32, 6425–6428.
54. Arabshahi, L., Khan, N.N., Butler, M., Noonan, T., Brown, N.C., and Wright, G.E. (Difluoromethylene)phosphates of guanine nucleosides as probes of DNA polymerases and G proteins. *Biochemistry*, 1990, 29, 6820–6826.
55. Tarussova, N.B., Osipova, T.I., Biriukov, A.I., Pokrovskaya, M.J., Meshkov, C.V., and Gnuchev, N.V. Phosphonate and halophospahonate analogs of P1,P4-bis(5′-adenosyl)tetraphosphonate and ATP. *Nucl. Acid Res. Symp. Series*, 1984, 14, 287–288.
56. Moffatt, J.G., and Khorana, H.G. Nucleoside polyphosphates. X. The synthesis and some reactions of nucleoside 5′-phosphomorpholidates and related compounds. Improved methods for the preparations of the nucleoside 5′-polyphosphates. *J. Am. Chem. Soc.*, 1961, 83, 649–659.
57. Bystrom, C.E., Pettigrew, D.W., Remington, J., and Branchaud, B.P. ATP analogs with non-transferable groups in the γ position as inhibitors of glycerol kinase. *Bioorg. Med. Chem. Lett.*, 1997, 7, 2613–2616.
58. Blackburn, C.M., Kent, D.E., and Kolkmann, F. The synthesis and metal binding characteristics of novel isopolar phosphonate analogues of nucleotides. *J. Chem. Soc. Perkin Trans.*, I. 1984, 1119–1125.

59. Wang, Z.W., and Kotlikoff, M.I. Activation of KCa channels in airway smooth muscle cells by endogenous protein kinase A. *Am. J. Physiol.*, 1996, 271, L100–L105.
60. Tobin, T., Akera, K., Lee, C.Y., and Brody, T.M. Ouabain binding to (Na^+ + K^+)-ATPase: Effects of nucleotide analogues and ethacrynic acid. *Biochem. Biophys. Acta*, 1974, 354, 102–117.
61. Gil–Ortiz, F., Fita, I., Ramon–Maiques, S., Marina, A., and Rubio, V. A crystallographic glimpse of a nucleotide triphosphate (AMPPNP) bound to a protein surface: external and internal AMPPNP molecules in crystalline N-acetyl-L-glutamate kinase. *Acta Crystallogr. D Biol. Crystallogr.*, 2002, 58, 1892–1895.
62. Paz, M.M., Ramos, M., Ramirez, G., and Souza, D. Differential effects of guanine nucleotides on kainic acid binding and on adenylate cyclase activity in chick optic tectum. *FEBS Lett.*, 1994, 355, 205–208.
63. Miller, L.D., Petrozzino, J.J., and Conner, J.A. G protein-coupled receptors mediate a fast excitatory postsynaptic current in CA3 pyramidal neurons in hippocampal slices. *J. Neurosci.*, 1995, 15, 8320–8330.
64. Schliselfeld, L.H. Binding of adenylyl imidodiphosphate, an analog of adenosine triphosphate, to myosin and heavy meromysin. *J. Biol. Chem.*, 1974, 249, 4985–4989.
65. Yount, R.G. Adenylylimidodiphosphate and guanylylimidodiphosphate. O'Malley, B., and Hardman, J., Eds., *Methods Enzymol.*, 1974, 38C, 420–427.
66. Eckstein, F., Kettler, M., and Parmeggiani, A. Guanylylimidodiphosphate and its interaction with amino acid polymerization factor. *Biochem. Biophys. Res. Commun.*, 1971, 45, 1151–1158.
67. Stumber, M., Herrman, H.C., Wohlgemuth, S., Kalbitzer, H.R., Jahn, W., and Geyer, M. Synthesis, characterization and application of two nucleoside triphosphate analogues, GTPγNH2 and GTPγF. *Eur. J. Biochem.*, 2002, 269, 3270–3278.
68. Yount, R.G., Babcock, D., Ballantyne, W., and Ojala, D. Adenylyl imidodiphosphate, an adenosine triphosphate analog containing a P–N–P linkage. *Biochemistry*, 1971, 10, 2484–2489.
69. McEwen, D.P., Gee, K.R., Kang, H.C., and Neubig, R.R. Fluorescent BODIPY-GTP analogs: real-time measurement of nucleotide binding to G proteins. *Anal. Biochemistry*, 2001, 291, 109–117.
70. Simon, L.N., and Myers, T.C. The effect of some phosphonic acid analogues of adenosine 5′-diphosphate on polynucleotide phosphorylase. *Biochem. Biophys. Acta*, 1961, 51, 178–180.
71. Alexandrova, L.A., Skoblov, A.Yu., Jasko, M.V., Victorova, L.S., and Krayevsky, A.A. 2′-Deoxynucleoside 5′-triphosphates modified at α-, β- and γ-phosphates as substrates for DNA polymerases. *Nucl. Acids Res.*, 1998, 26, 778–786.
72. Leonard, N.J., and Rosendahl, M.S. Synthesis and biological activity of the peroxy-ATP analogue, adenylyl 5′-peroxydiphosphate. *Nucl. Acids Res. Symp. Series*, 1981, 9, 141–144.
73. Gibson, K., and Leonard, N.J. Nucleoside 5′-(beta,gamma-peroxytriphosphates). *Biochemistry*, 1984, 23, 78–85.
74. Remy, P., Dirheimer, G., and Ebel, J.P. Analogues de nucléosides polyphosphates: I. Synthèse de l'adénosine 5′-phosphohypophosphate. *Biochim. Biophys. Acta.*, 1967, 136, 99–107.

75. Remy, P., Dirheimer, G., and Ebel, J.P. Analogs of nucleoside polyphosphates. Synthesis of an analog of ATP containing a hypophosphoric bond. *Acta Biochim. Pol.*, 1966, 13, 337–342.
76. Remy, P., Engel, M.L., Dirheimer, G., Ebel, J.P., and Revel, M. Analogs of nucleoside polyphosphates II. Guanosine-5′-phosphohypophosphate: preparation and effects on protein synthesis *in vitro*. *J. Mol. Biol.*, 1970, 48, 173–176.
77. Schwans, J.P., Cortez, C.N., Olvera, J.M., and Piccirilli, J.A. 2′-Mercaptonucleotide interference reveals regions of close packing within folded RNA molecules. *J. Am. Chem. Soc.*, 2003, 125, 10012–10018.
78. Arabshahi, A., and Frey, P.A. A simplified procedure for synthesizing nucleoside 1-thiotriphosphates: dATPαS, dGTPαS, UTPαS and dTTPαS. *Biochem. Biophys. Res. Commun.*, 1994, 204, 150–155.
79. Goody, R.S., and Isakov, M. Simple synthesis and separation of the diastereomers of α-thio analogs of ribo- and deoxyribo-, di-, and triphosphates. *Tetrahedron Lett.*, 1986, 27, 3599–3602.
80. Ludwig, J., and Eckstein, F. Rapid and efficient synthesis of nucleoside 5′-0-(1-thiotriphosphates), 5′-triphosphates and 2′,3′-cyclophosphorothioates using 2-chloro-4H-1,3,2-benzodioxaphosphorin-4-one. *J. Org. Chem.*, 1989, 54, 631–635.
81. Puri, N., Hunsch, S., Sund, C., Ugi, I., and Chattopadhyaya, J. The synthesis and reactivity of new 2-(N,N-diisopropylamino)-3-methylsulfonyl-1,3,2-benzoxazaphospholes. The utility of the 5-chloro analogue in the one-pot synthesis of oligothiophosphates: [Ap$_s$ppA, Ap$_s$pppA, ppp5′A2′p$_s$5′A, m^{7}Gp$_s$ppA, Ap$_s$pppp, Ap$_s$pp]. *Tetrahedron*, 1995, 51, 2991–3014.
82. Lee, C.S., and O'Sullivan, W.J. Phosphorothioate analogues of adenosine triphosphate as substrates of the mevalonate kinase reaction. *Biochim. Biophys. Acta*, 1984, 787, 131–137.
83. Jaffe, E.K., Nick, J., and Cohn, M. Reactivity and metal-dependent stereospecificity of the phosphorothioate analogs of ADP and ATP and reactivity of Cr(III)ATP in the 3-phosphoglycerate kinase reaction. Structure of the metal nucleotide substrates. *J. Biol. Chem.*, 1982, 257, 7650–7656.
84. Dyatkina, N.B., and Arzumanov, A.A. Terminating substrates of DNA polymerases: synthesis and functional study. *Nucl. Acids Res. Simp. Series*, 1987, 18, 117–120.
85. Connolly, B.A., and Eckstein, F. Structures of the mono- and divalent metal nucleotide complexes in the myosin ATPase. *J. Biol. Chem.*, 1981, 256, 9450–9456.
86. Burgers, P.M.J., and Eckstein, F. Structure of metal x nucleotide complex in the creatine kinase reaction. A study with diastereomeric phosphorothioate analogs of adenosine di- and triphosphate. *J. Biol. Chem.*, 1980, 255, 8229–8233.
87. Romaniuk, P.J., and Eckstein, F. Structure of the metal–nucleotide complex in the acetate kinase reaction. A study with γ-^{32}P-labeled phosphorothioate analogs of ATP. *J. Biol. Chem.*, 1981, 256, 7322–7328.
88. Gerlt, J.A., Coderre, J.A., and Wolin, M.S. Mechanism of the adenylate cyclase reaction. Stereochemistry of the reaction catalyzed by the enzyme from *Brevibacterium liquefaciens*. *J. Biol. Chem.*, 1980, 255, 331–334.
89. Eckstein, F., Romaniuk, P.J., Heideman, W., and Storm, D.R. Stereochemistry of the mammalian adenylate cyclase reaction. *J. Biol. Chem.*, 1981, 256, 9118–9120.

90. Nakamaye, K.L., Gish, G., Eckstein, F., and Vosberg, H.-P. Direct sequencing of polymerase chain reaction amplified DNA fragments through the incorporation of deoxynucleoside α-thiotriphosphates. *Nucl. Acids Res.*, 1988, 16, 9947–9959.
91. Romaniuk, P.J., and Eckstein, F. A study of the mechanism of T4 DNA polymerase with diastereomeric phosphorothioate analogues of deoxyadenosine triphosphate. *J. Biol. Chem.*, 1982, 257, 7684–7688.
92. Burgers, P.M.J., and Eckstein, F. A study of the mechanism of DNA polymerase I from *Escherichia coli* with diastereomeric phosphorothioate analogs of deoxyadenosine triphosphate. *J. Biol. Chem.*, 1979, 254, 6889–6893.
93. Gupta, A., DeBrosse, C., and Benkovic, S.J. Template-prime-dependent turnover of (Sp)-dATP α S by T4 DNA polymerase. The stereochemistry of the associated 3′ goes to 5′-exonuclease. *J. Biol. Chem.*, 1982, 257, 7689–7692.
94. Bartlett, P.A., and Eckstein, F. Stereochemical course of polymerization catalyzed by avian myeloblastosis virus reverse transcriptase. *J. Biol. Chem.*, 1982, 257, 8879–8884.
95. Connolly, B.A., Romaniuk, P.J., and Eckstein, F. Synthesis and characterization of diastereomers of guanosine 5′-O-(1-thiotriphosphate) and guanosine 5′-O-(2-thiotriphosphate). *Biochemistry*, 1982, 21, 1983–1989.
96. Yamanaka, G., Eckstein, F., and Stryer, L. Stereochemistry of the guanyl nucleotide binding site of transducin probed by phosphorothioate analogs of GTP and GDP. *Biochemistry*, 1985, 24, 8094–8101.
97. Jaffe, E.K., and Cohn, M. Diastereomers of the nucleoside phosphorothioates as probes of the structure of the metal nucleotide substrates and of the nucleotide binding site of yeast hexokinase. *J. Biol. Chem.*, 1979, 254, 10839–10845.
98. Darby, M.K., and Trayer, I.P. Metal-nucleotide structure at the active sites of the mammalian hexokinases. *Eur. J. Biochem.*, 1983, 129, 555–560.
99. Goody, R.S., Eckstein, F., and Schirmer, R.H. The enzymatic synthesis of thiophosphate analogs of nucleotides. *Biochim. Biophys. Acta*, 1972, 276, 155–161.
100. Senter, P.D., Eckstein, F., and Kagawa, Y. Substrate metal-adenosine 5′-triphosphate chelate structure and stereochemical course of reaction catalyzed by the adenosinetriphosphatase from the thermophilic bacterium PS3. *Biochemistry*, 1983, 22, 5514–5518.
101. Smith, L.T., and Cohn, M. Reactions of thio analogs of adenosine 5′-triphosphate catalyzed by methionyl-t-RNA synthetase from *Escherichia coli* and metal dependence of stereospecificity. *Biochemistry*, 1982, 21, 1530–1534.
102. Victorova, L.S., Dyatkina, N.B., Mozzherin, D.J., Atrazhev, A.M., Krayevsky, A.A., and Kukhanova, M.K. Formation of phosphonester bonds catalyzed by DNA polymerase. *Nucl. Acids Res.*, 1992, 20, 783–789.
103. Higuchi, H., Endo, T., and Kaji, A. Enzymic synthesis of oligonucleotides containing methylphosphonate internucleotide linkages. *Biochemistry*, 1990, 29, 8747–8753.
104. He, K., Hasan, A., Krzyzanowska, B., and Shaw, B.R. Synthesis and separation of diastereomers of ribonucleoside 5′-(α-borano)triphosphates. *J. Org. Chem.*, 1998, 63, 5769–5773.
105. Krzyzanowska, B.K., He, K., Hasan, A., and Shaw, B.R. A convenient synthesis of 2′-deoxyribonucleoside 5′-(α-P-borano)triphosphates. *Tetrahedron*, 1998, 54, 5119–5128.

106. He, K., Porter, K.W., Hasan, A., Briley, J.D., and Shaw, B.R. Synthesis of 5-substituted 2′-deoxycytidine 5′-(α-P- borano)triphosphates, their incorporation into DNA and effects on exonuclease. *Nucl. Acids Res.*, 1999, 27, 1788–1794.
107. Li, H., Porter, K., Huang, F., and Shaw, B.R. Boron-containing oligodeoxyribonucleotide 14-mer duplexes: enzymatic synthesis and melting studies. *Nucl. Acids Res.*, 1995, 23, 4495–4501.
108. Simoncsits, A., and Tomasz, J. A new type of nucleoside 5′-triphosphate analogue: P^1- (nucleoside 5′-) P^1-amino-triphosphates. *Tetrahedron Lett.*, 1976, 17, 3995–3998.
109. Tomasz, J. P^1-(Thymidine 5′-)P^1-amino-triphosphate: synthesis, hydrolysis and reaction with cytidine in alkali. *Nucleosides Nucleotides*, 1983, 2, 63–79.
110. Stepinski, J., Waddell, C., Stolarski, R., Darzynkiewicz, E., and Roads, R.E. Synthesis and properties of mRNAs containing the novel "anti-reverse" cap analogs 7-methyl(3′-O-methyl)GpppG and 7-methyl(3′-deoxy)GpppG. RNA. 2001, 7, 1486–1495.
111. Arzumanov, A.A., Semizarov, D.G., Victorova, L.S., Dyatkina, N.B., and Krayevsky, A.A. γ-Phosphate-substituted 2′-deoxynucleoside 5′-triphosphates as substrates for DNA polymerases. *J. Biol. Chem.*, 1996, 271, 24389–24394.
112. Nosova, V.V., Sokolova, N.I., and Shabarova, Z.A. Modification of the terminal phosphate groups in nucleoside di- and triphosphates. *Bioorg. Khim.*, (Russian) 1975, 1, 1130–1133.
113. Knorre, D.G., Kurbatov, V.A., and Samukov, V.V. General method for the synthesis of ATP gamma-derivatives. *FEBS Lett.*, 1976, 70, 105–108.
114. Zarytova, V.F., Knorre, D.G., Kurbatov, V.A., Lebedev, A.V., and Shishkin, G.V. Synthesis of alkylating derivatives of ATP γ-phosphoramidates. *Bioorg. Khim.* (Russian), 1975, 1, 793–799.
115. Sverdlov, E.D., Tsarev, S.A., and Kuznetsova, N.F. Derivatives of guanosine triphosphate-photoreactive substrates of *Escherichia coli* RNA polymerase. *FEBS Lett.*, 1980, 112, 296–298.
116. Badashkeyeva, A.G., Gall, T.S., Efimova, E.V., Knorre, D.G., Lebedev, A.V., and Mysina, S.D. Reactive derivative of adenosine-5′-triphosphate formed by irradiation of ATP γ-p-azidoanilide. *FEBS Lett.*, 1983, 155, 263–266.
117. Ankilova, V.N., Knorre, D.G., Kravchenko, V.V., Lavrik, O.I., and Nevinsky, G.A. Investigation of the phenylalanyl-tRNA synthetase modification with γ-(p-azidoanilide)-ATP. *FEBS Lett.*, 1975, 60, 172–175.
118. Kirsch, M., and Yarbrough, L.R. Assembly of tubulin with nucleotide analogs. *J. Biol. Chem.*, 1981, 256, 106–111.
119. Pollack, S.E., and Auld, D.S. Flurometric assays for nuclotidyl transferases and endonucleases. *Fed. Proc.*, 1980, 39, 2242.
120. Pollack, S.E., and Auld, D.S. Fluorescent nucleotide triphosphate substrates for snake venome phosphodiesterase. *Anal. Biochem.*, 1982, 127, 81–88.
121. Yarbrough, L.R., Schlageck, J.G., and Boughnam, M. Synthesis and properties of fluorescent nucleotide substrates for DNA-dependent RNA polymerases. *J. Biol. Chem.*, 1979, 254, 12069–12073.
122. Yarbrough, L.R., Schlageck, J.G., and Boughnam, M. Stacking interactions in fluorescent nucleotide analogs containing 1-aminonaphthalene-5-sulfonate at the phosphoryl terminus. *J. Biol. Chem.*, 1980, 255, 9907–9911.

123. Schlageck, J.G., and Block, J.L. Spectroscopic techniques for study of phosphodiester bond formation by *Escherichia coli* RNA polymerase. *J. Biol. Chem.*, 1979, 254, 12074–12077.
124. Yarbrough, L.R., and Kirsch, M. Binding of fluorescent analogs of GTP to the exchangeable nucleotide binding site of tubulin. *J. Biol. Chem.*, 1981, 256, 112–117.
125. Chladek, S., Quiggle, K., Chinali, G., Kobut, J., III, and Ofengand, J. Synthesis and properties of nucleoside 5′-phosphoazidates derived from guanosine and adenosine nucleotides: effect on elongation factors G and Tu dependent reactions. *Biochemistry*, 1977, 16, 4312–4319.
126. Baba, Y., Sumiyama, T., Tsuhako, M., and Yoza, N. Phosphonylation of biomolecules with inorganic diphosphonate. III. Formation of an H-phosphonate analogue of adenosine 5′-triphosphate by phosphonylation of adenosine 5′-diphosphate. *Bull. Chem. Soc. Jpn.*, 1989, 62, 1587–1592.
127. Armstrong, V., and Eckstein, F. Interaction of substrate analogues with *Escherichia coli* DNA-dependent RNA polymerase. *Eur. J. Biochem.*, 1976, 70, 33–38.
128. Kadokura, M., Wada, T., Urashima, C., and Sekine, M. Efficient synthesis of γ-methyl-capped guanosine 5′-triphosphate as a 5′-terminal unique structure of U6 RNA via a new triphosphate bond formation involving activation of methyl phosphorimidazolidate using $ZnCl_2$ as a catalyst in DMF under anhydrous conditions. *Tetrahedron Lett.*, 1997, 38, 8359–8362.
129. Eckstein, F., Bruns, W., and Parmeggiani, A. Synthesis of guanosine 5′-di and triphosphate derivatives with modified terminal phosphates: effect on ribosome-elongation factor G-dependent reactions. *Biochemistry*, 1975, 14, 5225–5232.
130. Shimazu, M., Shinozuka, K., and Sawai, H. Facile synthesis of nucleotides containing polyphosphates by Mn(II) and Cd(II) ion-catalyzed pyrophosphate bond formation in aqueous solution. *Tetrahedron Lett.*, 1990, 31, 235–238.
131. Draganescu, A., Hodawadekar, S.C., Gee, K.R., and Brenner, C. Fhit-nucleotide specificity probed with novel fluorescent and fluorogenic substrates. *J. Biol. Chem.*, 2000, 275, 4555–4560.
132. Peng, Z.-H., Sharma, V., Singleton, S.F., and Gershon, P.D. Synthesis and application of a chain-terminating dinucleotide mRNA cap analog. *Organic Lett.*, 2002, 4, 161–164.
133. Contreras, R., Cheroutre, H., Degrave, W., and Fiers, W. Simple, efficient *in vitro* synthesis of capped RNA useful for direct expression of cloned eukaryotic genes. *Nucl. Acids Res.*, 1982, 10, 6353–6362.
134. Konarska, M.M., Padgett, R.A., and Sharp, P.A. Recognition of cap structure in splicing *in vitro* of mRNA precursors. *Cell*, 1984, 38, 731–736.
135. Yisraeli, J.K., and Melton, D.A. Synthesis of long capped transcripts *in vitro* by SP6 and T7 RNA polymerases. *Methods Enzymol.*, 1989, 180, 42–50.
136. Reeve, A.E., Smith, M.M., Pigiet, V., and Huang, R.C. Incorporation of purine nucleoside 5′-[γ-S]triphosphates as affinity probes for initiation of RNA synthesis *in vitro*. *Biochemistry*, 1977, 16, 4464–4469.
137. Smith, M.M., Reeve, A.E., and Huang, R.C. Analysis of RNA initiated in isolated mouse myeloma nuclei using purine nucleoside 5′[γ-S]triphosphates as affinity probes. *Cell*, 1978, 15, 615–626.
138. Goody, R.S., and Eckstein, F. Thiophosphate analogs of nucleoside di- and triphosphates. *J. Am. Chem. Soc.*, 1971, 93, 6252–6257.

139. Haley, B., and Yount, R.G. γ-Fluoroadenosine triphosphate. Synthesis, properties, and interaction with myosin and heavy meromyosin. *Biochemistry*, 1972, 11, 2863–2871.
140. Stock, J.A. Synthesis of phosphonate analogs of thymidine di- and triphosphate from 5′-O-toluenesulfonylthymidine. *J. Org. Chem.*, 1979, 44, 3997–4000.
141. Ludwig, J., and Eckstein, F. Synthesis of nucleoside 5′-O-(1,3-dithiotriphosphates) and 5′-O-(1,1-dithiotriphosphates). *J. Org. Chem.*, 1991, 56, 1777–1783.
142. Lin, J., Porter, K.W., and Shaw, B.R. Synthesis and properties of novel triphosphate analogues: Ribonucleoside and deoxyribonucleoside (α-P-borano, α-P-thio)triphosphates. *Nucleosides Nucleotides Nucl. Acids*, 2001, 20, 1019–1023.
143. Kenner, G.W., Todd, A.R., Webb, R.F., and Weymouth, F.J. Nucleotides. Part XXVIII. Synthesis of uridine 5′-triphosphate. *J. Chem. Soc.*, 1954, 2288–2293.
144. Griffin, B.E., and Todd, A. Nucleotides. Part XLIV. Thymidine 5′-pyrophosphate and triphosphate. *J. Chem. Soc.*, 1958, 1389–1393.
145. Weimann, G., and Khorana, H.G. Studies on polynucleotides. XVII. On the mechanism of internucleotide bond synthesis by the carbodiimide method. *J. Am. Chem. Soc.*, 1962, 84, 4329–4341.
146. Knorre, D.G., Lebedev, A.V., Levina, A.S., Rezvukhin, A.I., and Zarytova, V.F. Active monomeric nucleotide intermediate in the oligonucleotide synthesis. *Tetrahedron*, 1974, 30, 3073–3079.
147. Zarytova, V.F., Knorre, D.G., Lebedev, A.V., Levina, A.S., and Rezvukhin, A.I. Mechanism of the interaction of nucleotides with triisopropylbenzenesulphonyl chloride. *Izv. Sib. Otd. Akad. Nauk. USSR, Ser Khim Nauk* (in Russian) 1974, 1, 85–96.
148. Michelson, A.M., and Todd, A.R. Nucleotides. Part IV. A novel synthesis of adenosine triphosphate. *J. Chem. Soc.*, 1949, 2487–2490.
149. Smith, M., and Khorana, H.G. Nucleoside polyphosphates. VI. An improved and general method for synthesis of ribo- and deoxyribonucleoside 5′-triphosphates. *J. Am. Chem. Soc.*, 1958, 80, 1141–1145.
150. Kozarich, J.W., Chinault, A.C., and Hecht, S.M. Ribonucleoside phosphates via phosphorimidazolidate intermediates. Synthesis of pseudoadenosine 5′-triphosphate. *Biochemistry*, 1973, 12, 4458–4463.
151. Smirnov, Yu.V., Lipkin, V.M., and Ovchinnikov, Yu. A. Affinity modification of the binding site of initiation substrate of *E. coli* DNA dependent RNA polymerase. *Bioorgan. Khim.* (Russian), 1981, 7, 1113–1116.
152. Grachev, M.A., and Mustaev, A.A. Cyclic adenosine 5′-trimetaphosphate phosphorylates a histidine residue nearby the initiating substrate binding site of *E. coli* DNA dependent RNA polymerase. *FEBS Lett.*, 1982, 137, 89–94.
153. Glonek, T., Kleps, R.A., and Myers, T.C. Cyclization of the phosphate side chain of adenosine triphosphate: Formation of monoadenosine 5′-trimetaphosphate. *Science*, 1974, 185, 352–355.
154. Babkina, G.T., Zarytova, V.F., and Knorre, D.G. Preparation of γ-amides of nucleoside 5′-triphosphates in aqueous solution with the use of water soluble carbodiimide. *Bioorgan. Khim.* (in Russian), 1975, 1, 611–615.
155. Glonek, T., Kleps, R.A., Van Wazer, J.R., and Myers, T.C. Cyclization of the phosphorus chain of the methylene-bridged analogs of adenosine triphosphate. *Bioinorg. Chem.*, 1976, 6, 295–304.
156. Lowe, G., Tansley, G., and Cullis, P.M. Synthesis of adenosine 5′[(R)α-^{17}O]triphosphate. *J. Chem. Soc. Chem. Commun.*, 1982, 595–598.

157. Sammons, R.D., Ho, H-.T., and Frey, P.A. Evidence implicating cyclo-diphosphates as intermediates in reactions of nucleoside phosphorothioates with cyanogen bromide. *J. Am. Chem. Soc.*, 1982, 104, 5841–5842.
158. Iyengar, R., Ho, H-.T., Sammons, R.D., and Frey, P.A. Cyclo-diphosphates as intermediates in reactions of nucleoside phosphorothioates with cyanogen bromide. *J. Am. Chem. Soc.*, 1984, 106, 6038–6049.
159. Lowe, G., Sproat, B.S., Tansley, G., and Cullis, P.M. A stereochemical and positional isotope exchange study of the mechanism of activation of isoleucine by isoleucyl-tRNA synthetase from *Escherichia coli. Biochemistry,* 1983, 22, 1229–1236.
160. Jarvest, R.L., and Lowe, G. The stereochemical course of hydrolysis catalysed by snake venom 5′-nucleotide phosphodiesterase. *Biochem J.*, 1981, 199, 447–451.
161. Richard, J.P., and Frey, P.A. Synthesis of chiral [^{18}O]phosphorothioate analogs of adenine nucleotides. *J. Am. Chem. Soc.*, 1982, 104, 3476–3481.
162. Brody, R.S., and Frey, P.A. Unambiguous determination of the stereochemistry of nucleotidyl transfer catalyzed by DNA polymerase I from *Escherichia coli. Biochemistry,* 1981, 20, 1245–1252.
163. Sheu, K.R., Richard, J.P., and Frey, P.A. Stereochemical courses of nucleotidyltransferase and phosphotransferase action. Uridine diphosphate glucose pyrophosphorylase, galactose-1-phosphate uridylyltransferase, adenylate kinase, and nucleoside diphosphate kinase. *Biochemistry,* 1979, 18, 5548–5556.
164. Eckstein, F., and Goody, R.S. Synthesis and properties of diastereoisomers of adenosine 5′-(O-1-thiotriphosphate) and adenosine 5′-(O-2-thiotriphosphate). *Biochemistry,* 1976, 15, 1685–1691.
165. Richard, J.P., and Frey, P.A. Stereochemical course of thiophosphoryl group transfer catalyzed by adenylate kinase. *J. Am. Chem. Soc.*, 1978, 100, 7757–7758.
166. Blättler, W.A., and Knowles, J.R. Stereochemical course of phosphokinases. The use of adenosine [γ-(S)-^{16}O,^{17}O,^{18}O]triphosphate and the mechanistic consequences for the reactions catalyzed by glycerol kinase, hexokinase, pyruvate kinase, and acetate kinase. *Biochemistry,* 1979, 18, 3927–3933.
167. Lowe, G., and Potter, B.V. The stereochemical course of yeast hexokinase-catalysed phosphoryl transfer by using adenosine 5′[γ(S)-^{16}O,^{17}O,^{18}O]triphosphate as substrate. *Biochem. J.*, 1981, 199, 227–233.
168. Jarvest, R.L., and Lowe, G. The stereochemical course of phosphoryl transfer catalysed by polynucleotide kinase (bacteriophage-T4-infected *Escherichia coli* B). *Biochem. J.*, 1981, 199, 273–276.
169. Bryant, F.R., Benkovic, S.J., Sammons, D., and Frey, P.A. The stereochemical course of thiophosphoryl group transfer catalyzed by T4 polynucleotide kinase. *J. Biol. Chem.*, 1981, 256, 5965–5966.
170. Hansen, D.E., and Knowles, J.R. The stereochemical course of the reaction catalyzed by creatine kinase. *J. Biol. Chem.,* 1981, 256, 5967–5969.
171. Pliura, D.H., Schomburg, D., Richard, J.P., Frey, P.A., and Knowels, J.R. Stereochemical course of a phosphokinase using a chiral [^{18}O]phosphorothioate. Comparison with the transfer of a chiral [^{16}O,^{17}O,^{18}O] phosphoryl group. *Biochemistry,* 1980, 19, 325–329.
172. Webb, M.R., and Trentham, D.R. The stereochemical course of phosphoric residue transfer during the myosin ATPase reaction. *J. Biol. Chem.,* 1980, 255, 8629–8632.
173. Webb, M.R., Grubmeyer, C., Penefsky, H.S., and Trentham, D.R. The stereochemical course of phosphoric residue transfer catalyzed by beef heart mitochondrial ATPase. *J. Biol. Chem.*, 1980, 255, 11637–11639.

174. Webb, M.R., and Trentham, D.R. The stereochemical course of phosphoric residue transfer catalyzed by sarcoplasmic reticulum ATPase. *J. Biol. Chem.*, 1981, 256, 4884–4887.
175. Eccleston, J.F., and Webb, M.R. Characterization of the GTPase reaction of elongation factor Tu. Determination of the stereochemical course in the presence of antibiotic X5108. *J. Biol. Chem.*, 1982, 257, 5046–5049.
176. Webb, M.R., and Eccleston, J.F. The stereochemical course of the ribosome-dependent GTPase reaction of elongation factor G from *Escherichia coli*. *J. Biol. Chem.*, 1981, 256, 7734–7737.
177. Senter, P.D., Eckstein, F., Mülsch, A., and Böhme, E. The stereochemical course of the reaction catalyzed by soluble bovine lung guanylate cyclase. *J. Biol. Chem.*, 1983, 258, 6741–6745.
178. Wehrli, W.E., Verheyden, D.L.M., and Moffat, J.G. Dismutation reactions of nucleoside polyphosphates II. Specific chemical synthesis of α-, β-, and γ-^{32}P-nucleoside 5′-tripolyphosphates. *J. Am. Chem. Soc.*, 1965, 87, 2265–2276.
179. Hecht, S.E., and Kozarich, J.W. A chemical syntheses of adenosine 5′-γ[^{32}P]triphosphate. *Biochim. Biophys. Acta*, 1973, 331, 307–309.
180. Koziolkiewicz, W., Pankowski, J., and Janecka, A. A method for the rapid preparation of adenosine 5′-γ-[^{32}P]triphosphate by chemical synthesis. *Prep. Biochem.*, 1978, 8, 471–478.
181. Martin, B.R., and Voorheis, H.P. A simple enzymic method for the synthesis of adenosine 5′-[α-^{32}P]triphosphate on a preparative scale. *Biochem. J.*, 1977, 161, 555–559.
182. Crutchfield, M.M., Dungan, C.H., Letcher, J.H., Mark, V., and Van Wazer, J.R. P^{31} Nuclear magnetic resonance, in *Topics in Phosphorus Chemistry*, Grayson, M., and Griffith, E.J., Eds. Interscience Publishers, New York. 1967, 5, 227–447.
183. Lebedev, A.V., and Rezvukhin, A.I. Tendencies of ^{31}P chemical shifts changes in NMR spectra of nucleotide derivatives. *Nucl. Acids Res.*, 1984, 12, 5547–5566.
184. Lowe, G., Potter, B.V.L., Sproat, B.S., and Hull, W.E. The effect of ^{17}O and the magnitude of the ^{18}O-isotope shift in ^{31}P nuclear magnetic resonance spectroscopy. *J. Chem. Soc. Chem. Commun.*, 1979, 17, 733–735.
185. Cohn, M., and Hu, A. Isotopic ^{18}O shifts in ^{31}P NMR of adenine nucleotides synthesized with ^{18}O in various positions. *J. Am. Chem. Soc.*, 1980, 102, 913–916.
186. Lowe, G., and Sproat, B.S. ^{18}O-Isotope shift on the ^{31}P nuclear magnetic resonance of adenosine-5′-phosphate and inorganic phosphate. *J. Chem. Soc. Chem. Commun.*, 1978, 565–566.
187. Tsai, M.D. Use of phosphorus-31 nuclear magnetic resonance to distinguish bridge and nonbridge oxygens of oxygen-17-enriched nucleoside triphosphates. Stereochemistry of acetate activation by acetyl coenzyme A synthetase. *Biochemistry*, 1979, 18, 1468–1472.
188. Jaffe, E.K., and Cohn, M. Divalent cation-dependent stereospecificity of adenosine 5′-O-(2-thiotriphosphate) in the hexokinase and pyruvate kinase reactions. The absolute stereochemistry of the diastereoisomers of adenosine 5′-O-(2-thiotriphosphate). *J. Biol. Chem.*, 1978, 253, 4823–4825.

chapter four

Chemical synthesis of oligonucleotide triphosphates

Morteza M. Vaghefi
ChemCyte Corporation, San Diego, CA

Contents

4.1 Introduction

The nonenzymatic replication of oligonucleotides has been extensively studied for many years because it models the type of reaction thought to have been important in the origin of life, particularly those that invoke the RNA world.[1,2] Much of this work has focused on template-directed polymerization. For the most part, these studies have employed activated mononucleotides in the form of triphosphate as substrate,[1,3] but a few have used short activated oligonucleotides.[4]

Szosotak's group pioneered the study of a nonenzymatic template-directed ligation reaction of triphosphate-activated oligonucleotide. Gao and Orgel introduced an alternate ligation reaction of tetraphosphate-activated oligonucleotide to make a longer oligonucleotide and suggested that this type of ligation is more likely to have occurred under prebiotic conditions. They prepared oligonucleotide 5′-polyphosphates from oligonucleotide 5′-phosphates and sodium trimetaphosphate. Trimetaphosphate has been considered as a polyphosphorylating reagent in the context of prebiotic chemistry.[5]

Oligonucleotides containing a 5′-phosphate are finding widespread applications in biochemistry and molecular biology: e.g., as linkers and adapters; in cloning and gene construction; in capping of RNA; and in the ligase chain reaction. In addition, they may serve as a source for preparation of γ-derivatives of oligonucleotide 5′-triphosphates.[6] This chapter deals with currently employed methods for the synthesis of oligonucleotide-5′-triphosphate.

4.2 Synthesis of oligonucleotide 5′-triphosphate

Oligonucleotide 5′-polyphosphates and their modified analogs have become important tools in modern biological sciences. Their syntheses are of particular significance and have been carried out by various methods. The conventional procedures for the synthesis of nucleoside 5′-triphosphate are described in Chapter 1. These methods require anhydrous conditions and are not suited for phosphorylation of oligonucleotides, which are insoluble in organic solvents and contain multiple functional groups that limit the activity of 5′-hydroxyl moiety.

4.2.1 Synthesis of oligonucleotide 5′-polyphosphate in aqueous media

Synthesis of oligonucleotide 5′-triphosphate requires preparation of the desired oligonucleotide (DNA synthesizer), 5′-phosphorylation (synthetic or enzymatic), and activation of 5′-phosphate to a reactive intermediate, such as phosphoimidazolide or phosphomorpholidate, followed by displacement of the leaving group by pyrophosphate. The activation and conversion to triphosphate are done in an aqueous solution in the presence of metal ions.

Sawai et al. reported pyrophosphate (P–O–P) bond formation in aqueous solutions in 1981.[7] They observed Pb(II) ion-catalyzed dimerization of adenosine 5′-phosphoimidazolide (ImpA) and formation of *bis*-adenosine-5′,5′-pyrophosphate in an aqueous solution among other products. ImpA is an activated nucleotide that readily hydrolyzes in water to form adenosine 5′-monophosphate. A divalent metal ion promotes formation of oligoadenylate, as well as pyrophosphate bond.

In 1990, Shimazu and coworkers reported Mn(II) and Cd(II) catalyzed pyrophosphate bond formation using activated phosphoroimidazolide.[8] They improved the procedure later and produced oligonucleotide-polyphosphate.[9] The phosphorylating reagent was phosphorotrimidazolide (PIM) or phosphorotribenzimidazolide (PBIM), which was prepared from phosphorous oxychloride and imidazole or benzimidazole in tetrahydrofuran. A typical phosphorylation reaction was carried out according to Scheme 4.1.

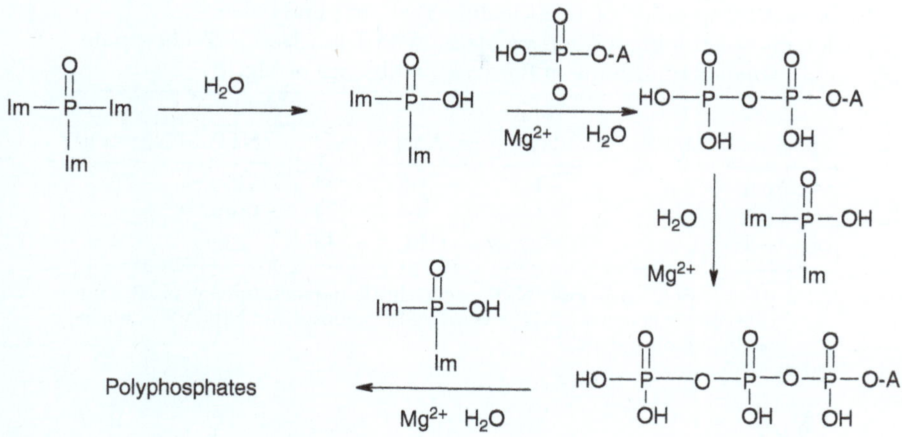

Scheme 4.1 Formation of a pyrophosphate bond in an aqueous media with phosphoimidazolide and ademosine 5′-monophosphate in the presence of Mg(II). The number of pyrophosphate bond formations is random and increases by the length of the reaction.

In a typical ion-mediated pyrophosphorylation, the reaction mixture contained 10 m*M* adenosine monophosphate; 0.5 *M* phosphoimidazole; 0.5 *M* $MgCl_2$; and 0.2 *M* *N*-ethylmorpholine buffer (pH 7.0). Reaction was carried out at 50°C for 8 h. The Mg(II) was removed from the reaction mixture by treatment with a chelater (Versenol) and products were analyzed by HPLC. Similar reactions were performed using Mn(II) in place of Mg(II) or phosphobenzimidazole, instead of phosphoimidazole, at 50°C, as well as at room temperature. In general, adenosine di- and triphosphates were the major products in the reaction mixtures. The reaction performed at room temperature required a longer time to complete and produced more adenosine triphosphate with Mg(II) and more adenosine diphosphate in the presence of Mn(II). Only traces of adenosine di- or triphosphate were formed in the absence of divalent ions. A longer reaction time (72 h) at 50°C resulted in hydrolysis of the product into monophosphate.

The reaction was repeated with trinucleotide 5′-phosphate, containing 2′-5′-phosphodiester linkages. The reaction was carried out with and without Mg(II) for different lengths of time. The results of this study are summarized in Table 4.1.

Mg(II) is necessary for polyphosphate formation. In the absence of Mg(II), hydrolysis of phosphoimidazole takes place, which diminishes the reactivity of the phosphate group. More recently, Gao and Orgel[4] reported the synthesis of oligonucleotide polyphosphates using synthetic oligonucleotide monophosphorylated at 5′ position by T4 polynucleotide kinase. The oligonucleotide 5′-phosphate coupled with sodium trimetaphosphate in the presence of Mg(II).[4] The reaction produced oligonucleotides containing 4, 7, and 10 phosphates at the 5′-end (Scheme 4.2).

Table 4.1 Formation of Trinucleotide 5′-di-, tri-, and Other Polyphosphates in an Aqueous Media with Trinucleotide 5′-Phosphate and Phosphoimidazolide in Presence or Absence of Mg(II)

pNuc	$MgCl_2$	Time (days)	Yield NMP	NDP	NTP	N(poly)P
pApApA	+	3	10	34	26	21
pApApA	–	3	93	4	Trace	
pApApA	+	7	14	48	29	8

Notes: pNuc = poly nucleotide; NMP = nucleoside monophosphate; NDP = nucleoside diphosphate; NTP = nucleoside triphosphate; N(poly)P = nucleoside polyphosphate.

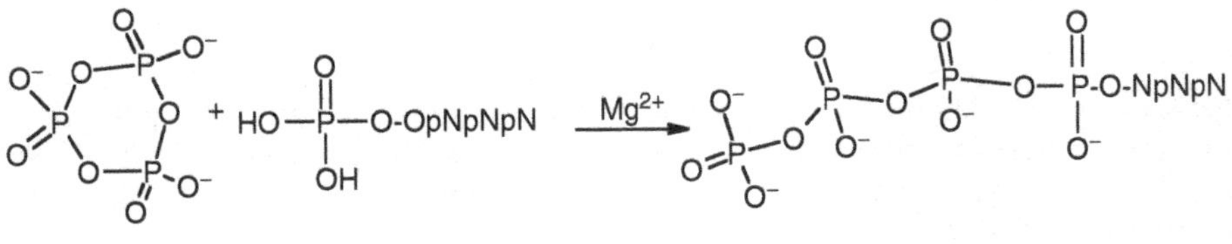

Scheme 4.2 Aqueous synthesis of oligonucleotide polyphosphate by the addition of trimetaphosphate to the oligonucleotide 5′-monophosphate in the presence of Mg(II).

4.2.2 *Synthesis of oligonucleotide 5′-polyphosphate in anhydrous media*

The synthetic procedure for oligonucleotide 5′-triphosphate must be designed with great care owing to low organic solubility and a high degree of complexity of oligonucleotide molecules. The conventional methods for preparation of nucleoside 5′-triphosphate are not applicable to pure oligonucleotides. To overcome these difficulties, scientists have applied the basics of solid support chemistry and improved solid phase methods for phosphorylation of 5′-oligonucleotide. The anhydrous conditions required for phosphorylation chemistry are applicable to solid support-bound oligonucleotides.

A few examples for the synthesis of oligonucleotide 5′-triphosphates can be found in the literature.[10–12] Gaur[14] adapted Ludwig and Eckstein's chemical phosphorylation method[13] to the solid support and developed a novel procedure for the synthesis of nucleoside triphosphate on CPG (controlled pore glass). Later on, Brownlee et al. described a solid phase method for the addition of 5′-diphosphate to synthetic oligodeoxyribonucleotides.[11] They used the method of Ludwig and Eckstein because the reaction conditions were mild and nucleobase protection was not necessary. Trivalent phosphorus chemistry is the most promising way for the preparation of oligonucleotide 5′-triphosphates of any length, sequence, and nature (DNA, RNA, or their analogs) on the solid support. A synthetic procedure allowing conversion of oligodeoxy-nucleotides to their corresponding 5′-triphosphate derivatives has been developed[15] (Scheme 4.3).

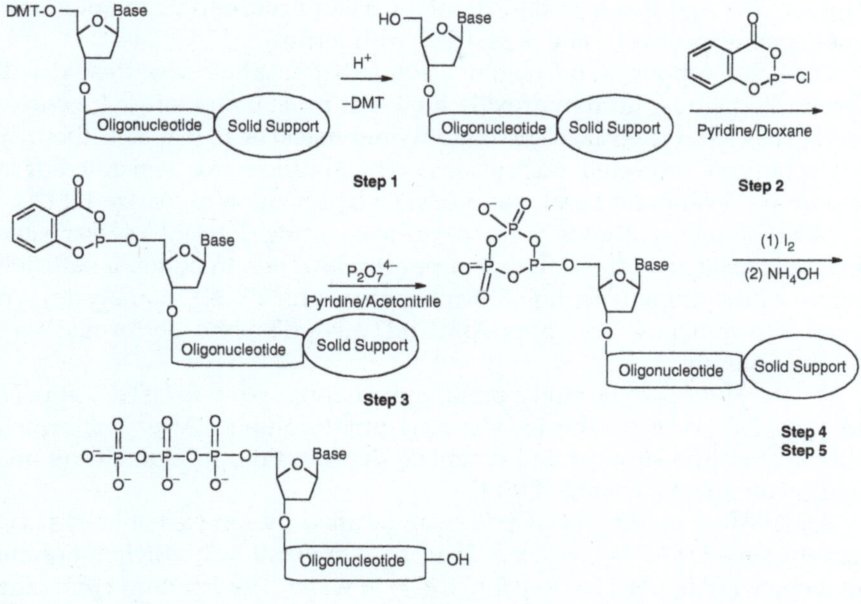

Scheme 3

Scheme 4.3 Solid phase synthesis of oligonucleotide 5′-triphosphate. First, the oligonucleotides were synthesized on the solid support using standard phosphoramidite protocols. Next, the DMT protection group was removed (step 1) and the 5′-OH was phosphitylated using 2-chloro-4H-1,3,2-benzodioxa-phosphorin-4-one (step 2), followed by reaction with pyrophosphate (step 3) and iodine oxidation (step 4). After subsequent removal from the support and complete deprotection (step 5), the product was analyzed and purified by HPLC.

4.2.2.1 Syntheses and purification

Synthesis of oligonucleotides was performed on a PerSeptive Biosystems Expedite Model 8909 synthesizer using a standard phosphoramidite protocol. At the end of the synthesis, the dimethoxytrityl protecting group was removed and the column was washed with acetonitrile and dried with argon flow.

For 5′-phosphorylation, the column was washed with 20 mL of dry dioxane (2 × 10 mL) and was treated with 1 mL of freshly prepared phosphitylating solution (1 *M* solution of 2-chloro-4H-1, 3, 2-benzodioxa-phosphorin-4-one) in dioxane:pyridine (4:1). After 2 min, another 1mL of the phosphitylating solution was pushed through the column and the reaction was allowed to proceed for 5 min. The previous step was repeated and the reaction continued for another 10 min. The column was washed with 10 mL of dioxane and 10 mL of acetonitrile, respectively. Synthesis was continued by adding 10 mL of 0.5 *M* tributylammonium pyrophosphate in anhydrous acetonitrile:pyridine (1:1) in portions of 0.5 mL with slow flow over 20 min. The column was washed with 10 mL of acetonitrile and treated with 10 mL of oxidizer solution for 5 min. Then, the column was washed once with

10 mL of 70% pyridine in water, 20 mL of acetonitrile, and 20 mL of dichloromethane, respectively, and was dried with airflow.

The solid support bond oligonucleotide triphosphate was treated with concentrated ammonium hydroxide for 1 h at room temperature (phenoxyacetyl protected nucleosides and oligothymidylates) or 15 h at 55°C (benzoyl and isobutyryl protected nucleosides). The solution was separated from support and evaporated and was analyzed by an anion exchange HPLC.

Most of the syntheses were performed using 1 μmol of respective deoxynucleosides on the polymer support. However, to obtain a sufficient amount of the oligonucleotide 5′-triphosphates for ^{31}P NMR analysis, syntheses of pppdCpdC and pppdApdCpdTpdGpdT were performed on a 15-μmol column.

For the HPLC profile study, pppT, pppdA, pppTpT, pppTpTpT, ppp$(T)_9$, and ppp$(T)_{15}$ were synthesized on a 1-μmol column. After subsequent removal from the support and complete deprotection, products were analyzed by an anion-exchange HPLC.

For NMR analysis, pppdCpdC was purified by ion-exchange chromatography on a DEAE Sephadex A-25 using a gradient of triethylammonium bicarbonate (TAB, pH 7.5) from 0 to 0.8 *M* in water. The fractions containing pppdCdpC were combined and TEAB was removed by repetitive coevaporation with methanol under reduced pressure. The products were repurified by HPLC on a Pharmacia Resource Q anion exchange column (1 × 5 cm), using a gradient of LiCl from 0 to 1 *M* in water. Fractions containing pppdCpdC were combined and evaporated. The product was precipitated in and washed with acetone twice and dried under a high vacuum; 5.3 mg of pppdCdpC was recovered and a ^{31}P NMR spectrum was recorded in the deuterium oxide.

The pentamer-TP, pppdApdCpdTpdGpdT, was purified by anion-exchange chromatography on a DEAE Sephadex and the addition of a gradient of LiCl from 0 to 1.5 *M* in water. Fractions containing pppdApdCpdTpdGpdT were combined, evaporated, and precipitated in acetone and washed with acetone twice and dried under high vacuum to obtain 20.5 optical units (at 260 nm). A ^{31}P NMR spectrum of pppdApdCpdTpdGpdT was recorded in the deuterium oxide.

4.2.2.2 Analysis

The HPLC profiles for several of the reaction mixtures are presented in Figure 4.1A and 4.1B. Typically, a major HPLC peak with a longer retention time than the corresponding 5′-OH precursor oligonucleotide was present that was the respective 5′-triphosphate derivative. For the pppdT and pppdA, the major peaks were identified by comparison with authentic samples. For compounds pppdTpdT, pppdTpdTpdT, pppdApdCpdTpdGpdT, and pppdCpdC, the authentic samples were unavailable. However, the 15-μmol syntheses of pppdCpdC and pppdApdCpdTpdGpdT provided sufficient amounts of material to record ^{31}P NMR spectra (Figure 4.2).

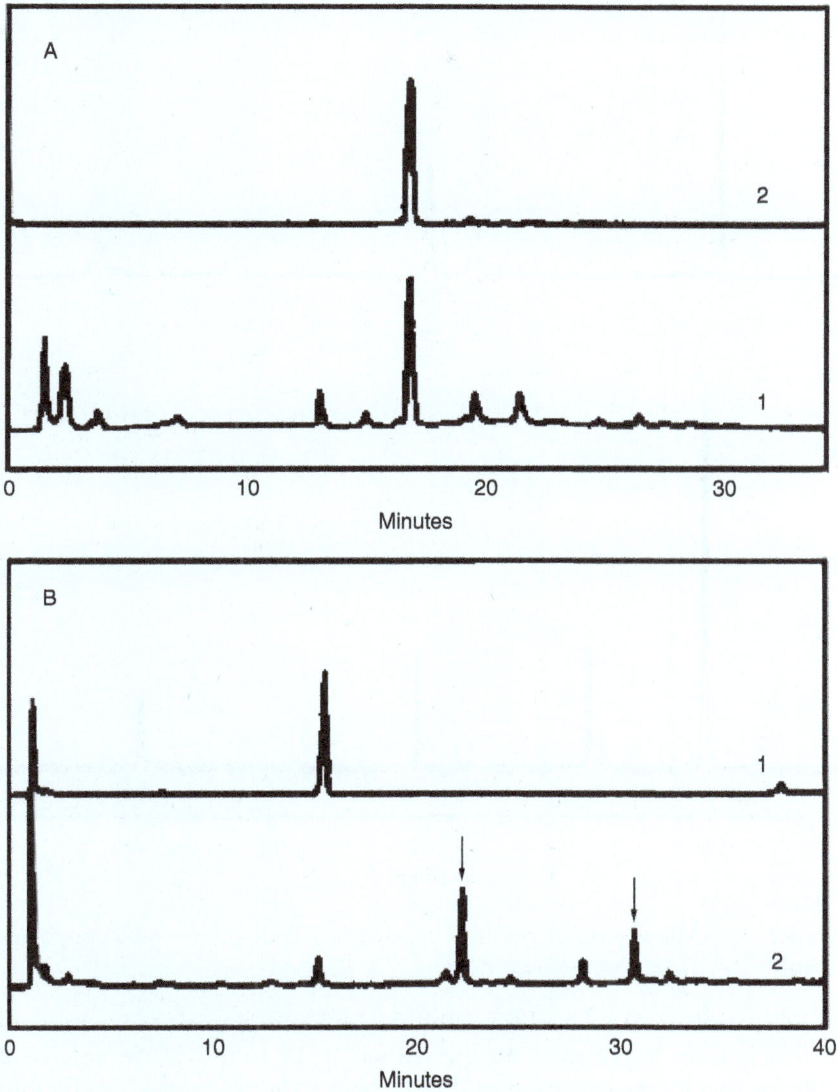

Figure 4.1 Anion exchange chromatogram: A: dimer-triphosphate synthesized on the solid support. Crude triphosphate reaction mixture after ammonia deprotection (1). Purified triphosphate on DEAE sephadex (2). B: Pentamer–triphosphate synthesis. Crude pentamer prior to phosphorylation (1). Crude triphosphate reaction mixture after amonia deprotection (2).

The signals representing tripolyphosphate chains and phosphodiester bonds were clearly seen in both NMR spectra. In addition, a respective molecular ion was detected in mass spectrum of pppdCpdC. All the data obtained were in agreement with the proposed structures for 5′-triphosphate derivatives.

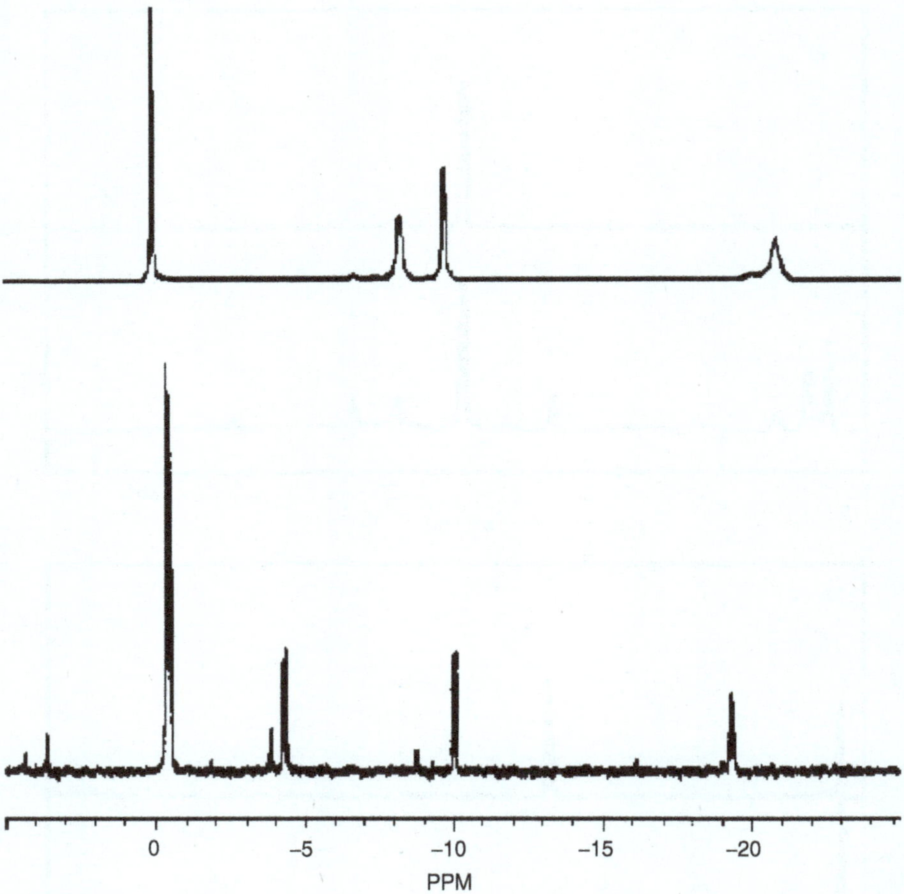

Figure 4.2 ^{31}P NMR spectra of purified pppdApdCpTpdGpT (top) and pppdCpdC (bottom).

The overall isolated yields of the desired compounds were moderate and varied from experiment to experiment (~15 to 30%). We believe the reason for this inconsistency was due to poor experimental reproducibility of the manual syringe technique used in our synthetic procedure.

We evaluated the preceding procedure for the synthesis of longer oligonucleotide 5′-triphosphates, such as $(T)_9$ and $(T)_{15}$. The HPLC analyses of both reaction mixtures showed the presence of oligonucleotide derivatives possessing several additional ionized groups compared to 9-mer or 15-mer precursors. The longer retention times were in agreement with the presence of oligomers possessing 5′-polyphosphate chain(s).

References

1. Joyce, G.F. and Orgel, L.E. Prospects of understanding the origin of the RNA world. In: Gesteland, R.F. and Atkins, J.F., Eds. In: *The RNA World*, Cold Spring Harbor Laboratory Press, Cold Spring Harbor, New York, 1993, 1–25.
2. Orgel, L.E. RNA catalysis and the origin of life. *J. Theor. Biol.*, 1986, 123, 127–149.
3. Joyce, G.F. Nonenzymatic template-directed synthesis of informational macromolecules. *Cold Spring Harbor Symp. Quant. Biol.*, 1987, 52, 41–51.
4. Gao, K. and Orgel, L.E. Polyphosphorylation and nonenzymatic template-directed ligation of oligonucleotides, origins of life and evolution of the biosphere, 2000, 30, 45–51.
5. Lohmann, R. Formation of nucleoside 5′-polyphosphates from nucleotides and trimetaphosphate. *J. Mol. Evol.*, 1975, 6, 237–252.
6. Grachev, M.A., Knorre, D.G., and Lavrik, O.I. Biology reviews. *Sov. Sci. Rev.*, Sculachev, V.P., Ed., Section D. Switzerland: Harwood Academic Publishers. 1981, 2, 107–148.
7. Sawai, H., Shibata, T., and Ohno, M. Preparation of oligoadenylate with 2′-5′ linkage using Pb^{2+} ion catalyst. *Tetrahedron*, 1981, 37, 481–485.
8. Shimazu, M., Shinozuka, K., and Sawai, H. Facile synthesis of nucleotides containing polyphosphates by Mn(II) and Cd(II) ion-catalyzed pyrophosphate bond formation in aqueous solution. *Tetrahedron Lett.*, 1990, 31, 235–238.
9. Sawai, H., Inaba, Y., Hirano, A., Wakai, H., and Shimazu, M. Mg(II) Ion-mediated conversion of mono- and oligonucleotides to 5′-polyphosphates in aqueous solution. *Tettrahedron Lett.*, 1993, 34, 4801–4804.
10. Tomasz, J., Simoncsits, A., Kajtar, M., Krug, R.M., and Shatkin, A.J. Chemical synthesis of 5′-pyrophosphate and triphosphate derivatives of 3′-5′ ApA, ApG, GpA, and GpG. CD study of the effect of 5′-phosphate groups on the conformation of 3′-5′ GpG. *Nucleic Acids Res.*, 1978, 5, 2945–2957.
11. Brownlee, G.G., Fodor, E., Pritlove, D.C., Gould, K.G., and Dalluge, J.J. Solid phase synthesis of 5′-diphosphorylated oligoribonucleotides and their conversion to capped m7Gppp-oligoribonucleotides for use as primers for influenza A virus RNA polymerase *in vitro*. *Nucleic Acids Res.*, 1995, 23, 2641–2647.
12. Hartog, J.A.J. den, Wijnands, R.A. and van Boom, J.H. Chemical synthesis of an effective inhibitor of protein synthesis in euharyitic cells: pppA2′,5′A2′,5′A and some analogs. *Nucleic Acids Res. Symp.* Series No. 7, 1980, 157–166.
13. Ludwig, J. and Eckstein, F. Rapid and efficient synthesis of nucleoside 5′-0-(1-thiotriphosphates), 5′-triphosphates and 2′,3′-cyclophosphorothioates using 2-chloro-4H-1,3,2-benzodioxaphosphorin-4-one. *J. Org. Chem.*, 1989, 54, 631–635.
14. Gaur, R.K., Sproat, B.S., and Krupp, G. Novel solid phase synthesis of 2′-O-methylribonucleoside 5′-triphosphates and their α-thio analogues. *Tetrahedron Lett.*, 1992, 33, 3301–3304.
15. Lebedev, A.V., Koukhareva, I.I., Beck, T., and Vaghefi, M.M. Preparation of oligonucleotide 5′-triphosphates using solid support approach, *Nucleosides Nucleotides Nucleic Acids*, 2001, 20, 1403–1409.

chapter five

Solid state structures of nucleoside triphosphate metal complexes

Katsuyuki Aoki*
Department of Materials Science, Toyohashi University of Technology, Tempaku-cho, Toyohashi, Japan

Ikuhide Fujisawa
Department of Materials Science, Toyohashi University of Technology, Tempaku-cho, Toyohashi, Japan

Contents

* Corresponding author.

5.1 Introduction

Nucleic acids and nucleotides including nucleoside 5′-triphosphates (NTPs) generally occur as complexes coordinated with metal ions because metal ions like Mg^{2+}, Ca^{2+}, Na^{+}, and K^{+} are present in the body in millimolar concentrations.[1] It has now been well documented[1–6] that these complexes are of importance for the biochemical action of nucleic acids, nucleotides, and their derivatives. A large body of data has been accumulated, including structures formed,[1,7–13] which are prerequisites for understanding of a variety of metal participation phenomena in nucleic acid processes. In this field, X-ray investigations have played a major part in providing well defined information on structural aspects, including metal binding to nucleic acid derivatives; however, care must be taken in relating solid state structures to solution structures.

The objective here is to describe structural aspects of NTPs in the solid state. NTPs as metal complexes, especially metal complexes of adenosine 5′-triphosphate (ATP), may function as substrates in many biochemical reactions involving the phosphoryl and nucleotidyl transfers or hydrolysis, source of metabolic energy, cofactor, and/or modulator of enzymatic activity.[1,14] Although a large number of crystal structures of nucleotides,[15] most of which are metal complexes,[13] are now available, the number of crystal structures of NTPs reported to date is very small, mainly due to the difficulty in preparing suitable single crystals. Moreover, they are all metal complexes of ATP[16–24] and thus the present chapter deals with structural aspects of ATP–metal complexes.

Structures of ATP–metal complexes were once reviewed by Cini[10] and, although this chapter relies heavily on that review, an additional emphasis is on deriving any general conclusions from solid state studies of ATP–metal complexes. NTP–metal complexes in protein crystal structures are not dealt with here and the reader should refer to the literature[25] describing some ATP–metal complexes in protein crystal structures. Chemical structure of ATP^{4-} in its dominating *anti* conformation is shown in Figure 5.1. The definition of the conformational terms is according to the literature.[26]

5.2 Structures of alkali and alkaline earth–metal ion complexes

Table 5.1 lists crystal data of a total of 10 ATP–metal complexes; their coordination data are summarized in Table 5.2 and Table 5.3. Conformational parameters of ATP molecules in ATP–metal complexes are listed in Table 5.4.

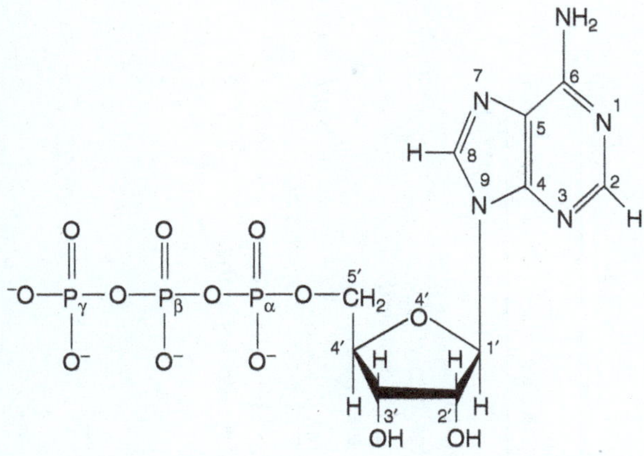

Figure 5.1 Chemical structure of adenosine 5′-triphosphate (5′-ATP^{4-}) in its dominating *anti*conformation, where the six-membered ring of the purine base is in a position opposite to the ribose ring.

5.2.1 Sodium complexes of adenosine 5′-triphosphate

The first crystal structure of an ATP complex, $Na_2{\cdot}H_2ATP{\cdot}3H_2O$,[16–18] was reported by Kennard and coworkers in 1970. Four crystallographically independent Na^+ ions and two H_2ATP^{2-} molecules exist in the asymmetric unit, where the adenine base is protonated at N1 and the triphosphate group is protonated at a γ-phosphate oxygen for each ATP molecule. The sodium ion Na1 (or Na2) binds simultaneously to the base moiety and the phosphate group of the *same* molecule, thus forming an intramolecular N7–Na–O (γ-phosphate of ATP) chelation as shown in Figure 5.2 and Table 5.2. In addition, the Na1 (or Na2) binds to the phosphate group of the neighboring ATP molecule through α-, β-, and γ-phosphate oxygens, thus forming a metal ion-bridged dimeric structure, at which a pseudo twofold symmetry exits at the midpoint of the Na1–Na2 vector. The third sodium ion, Na3, is octahedrally coordinated by an oxygen atom of the β-phosphate from each of two ATP molecules and by four water molecules (not shown in Figure 5.2; see Table 5.3). The coordination around the fourth sodium ion, Na4, is from one phosphate oxygen and four ribose oxygen atoms (not shown in Figure 5.2; see Table 5.3).

Many polymorphic crystals form, depending on different states of hydration,[16] and this trihydrate form transforms into the dihydrate form, $Na_2{\cdot}H_2ATP{\cdot}2H_2O$, and its crystal structure was determined.[19] The structural feature involving the coordination mode in the Na1- and Na2-bridged ATP dimer unit is common to that in the trihydrate form. A major difference exists around Na3, which is coordinated by a phosphate oxygen atom from each of two ATP molecules (one from the α-phosphate and the other from the β-phosphate), a hydroxyl oxygen of ribose, and two water molecules (Table 5.3). ATP molecules in the tri- and dihydrate forms adopt largely the

Table 5.1 Crystal Data of ATP–Metal Complexes[a]

Complex	Space group	Cell constants [Å], [°]	Z^b V^c [Å^3]	R-factor[d]; additional remarks	Ref.
$Na_2H_2ATP\cdot 3H_2O$	$P2_12_12_1$	$a = 30.451(38)$ $b = 20.881(29)$ $c = 7.075(10)$	8 4498(5)	0.123 0.120: restrained refinement (Ref. 18) Needles from 50% dioxane/water. Many polymorphic crystals depending on different states of hydration. Two molecules in the asymmetric unit.	16, 17, 18
$Na_2H_2ATP\cdot 2H_2O$	$P2_12_12_1$	$a = 27.572(5)$ $b = 21.066(3)$ $c = 7.0854(9)$	8 4116(1)	0.096 Crystals from 1-propanol/water. Trihydrate form transforms into this form. Two molecules in the asymmetric unit.	19
$[Mg(HATP)_2][Mg(H_2O)_6](Hdpa)_2\cdot 12H_2O$	$C222_1$	$a = 22.734(3)$ $b = 10.233(3)$ $c = 30.997(4)$	4 7211	0.111 Crystals from water (pH 5.0)	20, 21
$[Ca(HATP)_2][Ca(H_2O)_6](Hdpa)_2\cdot 9H_2O$	$C222_1$	$a = 22.965(3)$ $b = 10.154(3)$ $c = 32.390(4)$	4 7553	0.124 Crystals from water (pH 5.0)	20, 21
$[Cu(H_2ATP)(phen)]_2\cdot 7H_2O$	$P2_1$	$a = 11.807(3)$ $b = 24.824(5)$ $c = 10.693(2)$ $\beta = 94.98(3)$	2 3122	0.069: anisotropic temperature factors for Cu, P, and O atoms only Blue-green crystals from water (adjusted to pH 2.8)	22

$[Zn(H_2ATP)(bpy)]_2 \cdot 4H_2O$	$P2_1$	$a = 11.105(3)$ $b = 25.223(7)$ $c = 10.539(3)$ $\beta = 91.34(4)$	2 2951.2	0.098: anisotropic temperature factors for 2Zn atoms only Small transparent plates from 50% ethanol/water (adjusted to pH 4 with H_2SO_4)	23
$[Mn(HATP)_2]$ $[Mn(H_2O)_6]$ $(Hdpa)_2 \cdot 12H_2O$	$C222_1$	$a = 10.234(3)$ $b = 22.699(3)$ $c = 31.351(4)$	4 7283	0.072: hydrogen positions determined Crystals from water (pH 4.5). Both Mn atoms on a twofold axis.	20
$[Co(HATP)_2]$ $[Co(H_2O)_6]$ $(Hdpa)_2 \cdot 9H_2O$	$C222_1$	$a = 10.212(3)$ $b = 22.710(3)$ $c = 31.017(4)$	4 7193	0.111 Crystals from water (pH 4.5)	20
$[Zn(HATP)_2]$ $[Mg(H_2O)_6]_{0.72}$ $[Zn(H_2O)_6]_{0.28}$ $(Hdpa)_2 \cdot 12H_2O$	$C222_1$	$a = 22.666(3)$ $b = 10.131(2)$ $c = 30.893(6)$	4 7094(1)	0.090 Crystals from water (pH 4.5)	24
$[Cd(HATP)_2]$ $[Ca(H_2O)_6]$ $(Hdpa)_2 \cdot 9H_2O$	$C222_1$	$a = 22.846(3)$ $b = 10.252(2)$ $c = 31.914(6)$	4 7475(1)	0.101 Crystals from water (pH 4.5)	24

[a] Abbreviations: Hdpa = protonated 2,2′-dipyridylamine; phen = 2,10-phenanthroline; bpy = 2,2′-bipyridyl.

[b] Z = number of molecules in the unit cell.

[c] V = volume of the unit cell.

[d] $R = \Sigma||F_o| - |F_c||/\Sigma|F_o|$.

Table 5.2 Coordination Data of ATP–Metal Complexes[a]

Complex	Coordination sites or modes	Additional remarks; chirality of α, β, γ-P chelation[b]	Ref.
$Na_2 \cdot H_2ATP \cdot 3H_2O$	N7(B)-M-O(γ-P), O(α, β, γ-P),	Intramolecular M-binding Tridentate chelation: Δ-*endo* and Λ-*endo*	16, 17, 18
	O2′(S)-M-O3′(S)	Intramolecuar M-binding	
$Na_2 \cdot H_2ATP \cdot 2H_2O$	N7(B)-M-O(γ-P), O(α, β, γ-P), O3′(S)	Intramolecular M-binding Tridentate chelation: Δ-*endo* and Λ-*endo*	19
$[Mg(HATP)_2]^{4-}$	O(α, β, γ-P)	Tridentate chelation: Λ-*exo*	20, 21
$[Ca(HATP)_2]^{4-}$	O(α, β, γ-P)	Tridentate chelation: Λ-*exo*	20, 21
$[Cu(H_2ATP)(phen)]_2$	O(α, β, γ-P)	Tridentate chelation: Δ-*exo* and Λ-*exo*	22
$[Zn(H_2ATP)(bpy)]_2$	O(α, β, γ-P)	Tridentate chelation: Δ-*exo* and Λ-*exo*	23
$[Mn(HATP)_2]^{4-}$	O(α, β, γ-P)	Tridentate chelation: Λ-*exo*	20
$[Co(HATP)_2]^{4-}$	O(α, β, γ-P)	Tridentate chelation: Λ-*exo*	20
$[Zn(HATP)_2]^{4-}$	O(α, β, γ-P)	Tridentate chelation: Λ-*exo*	24
$[Cd(HATP)_2]^{4-}$	O(α, β, γ-P)	Tridentate chelation: Λ-*exo*	24

[a] Abbreviations: Hdpa = protonated 2,2′-dipyridylamine; phen = 2,10-phenanthroline; bpy = 2,2′-bipyridyl; B = base; S = sugar; and P = phosphate.

[b] Definition of chirality of coordination is from Saenger, W. In: Saenger, W., Ed. *Principles of Nucleic Acid Structure*. New York, Berlin: Springer–Verlag, 1984, 201–219.

usual conformations. That is, the adenine base is oriented *anti* with respect to sugar, where the six-membered ring moiety of adenine is away from the ribose ring; sugar puckering mode is C2′-*endo* or C2′-*endo*-C3′-*exo*, which is close to C2′-*endo* puckering; and the torsion angle about the C4′–C5′ bond is *gauche–gauche*, where O5′ resides over the ribose ring (Table 5.4).

The triphosphate chains are folded into right- and left-handed helical forms for Mol. 1 and Mol. 2, respectively, and the adenine base is located over the six-membered chelate ring formed by β- and γ-phosphate for Mol. 1 and Mol. 2. That is, adenosine is placed *endo* with respect to the chelate; the configuration of ATP is described as Δ-*endo* for Mol. 1 and Λ-*endo* for Mol. 2 (Figure 5.2 and Table 5.2; for the definition of Δ, Λ, *endo*, and *exo* configurations, see Saenger[1]). Dehydration from the trihydrate-form to the dihydrate-form causes the conformational change in one (Mol. 2 in Figure 5.2) of the two ATP molecules from C2′-*endo*-C3′-*exo* and *gauche-gauche* in the trihydrate-form to uncommon C4′-*endo* and *trans-gauche* in the dihydrate-form (Table 5.4).

The formation of the intramolecular N7–metal–O (γ-phosphate) chelation is reminiscent of the earlier proposed structure for the $[Mg(ATP)]^{2-}$ complex by Szent–Györgi.[27] The metal chelation by the base and α-phosphate group on the *same* nucleotide is also of interest, but has not been proven in the solid state, although its existence has been demonstrated in solution.[28] It is of interest to note that the formation of a hydrogen bond (a broken line

Table 5.3 Metal–Ligand Distances in ATP–Metal Complexes[a]

Complex	Geometry about metal (M)	Coordination sites (L)	M–L Distance [Å]	Coordination sites (L)	M–L Distance [Å]	Ref.
$Na_2 \cdot H_2ATP3H_2O$	Distorted	Na1–N7(B)	2.90(5)	Na1–O(γ–P)	2.65(5)	16, 17, 18
	octahedral	–O(α–P)	2.59(5)	–O(γ–P)	2.53(5)	
		–O(β–P)	2.42(5)	–O(γ–P)	2.47(5)	
	Distorted	Na2–N7(B)	2.69(5)	Na2–O(γ–P)	2.54(5)	
	octahedral	–O(α–P)	2.45(5)	–O(γ–P)	2.32(5)	
		–O(β–P)	2.48(5)	–O(γ–P)	2.32(5)	
	Distorted	Na3–O(β–P)	2.32(5)	Na3–O(W)	2.23(5)	
	octahedral	–O(β–P)	2.32(5)	–O(W)	2.72(5)	
		–O(W)	2.56(5)	–O(W)	2.82(5)	
	Distorted	Na4–O(α–P)	2.72(5)	Na4–O2′(S)	2.80(5)	
	square-	–O2′(S)	2.84(5)	–O4′(S)	2.80(5)	
	pyramidal	–O3′(S)	2.69(5)			
$Na_2 \cdot H_2ATP2H_2O$	octahedral	Na1–N7(B)	2.71(2)	Na1–O(γ–P)	2.79(2)	19
		–O(α–P)	2.47(2)	–O(γ–P)	2.42(2)	
		–O(β–P)	2.41(2)	–O(γ–P)	2.34(2)	
	Octahedral	Na2–N7(B)	2.51(2)	Na2–O(γ–P)	2.89(2)	
		–O(α–P)	2.31(2)	–O(γ–P)	2.38(2)	
		–O(β–P)	2.50(2)	–O(γ–P)	2.39(2)	
	Distorted	Na3–O3′(S)	2.46(2)	Na3–O(W)	2.38(2)	
	trigonal-	–O(α–P)	2.32(2)	–O(W)	2.51(5)	
	bipyramidal	–O(β–P)	2.34(2)			
	Distorted	Na4–O(α–P)	2.35(2)	Na4–O(W)	2.42(3)	
	trigonal-	–O(β–P)	2.37(2)	–O(W)	2.47(5)	
	bipyramidal	–O(W)	2.33(2)			
$[Mg(HATP)_2]^{4-}$	Octahedral	Mg–O(α–P)	2.10(2)	Mg–O(α–P)	2.10(2)	20, 21
		–O(β–P)	2.08(2)	–O(γ–P)	2.08(2)	
		–O(β–P)	2.01(2)	–O(γ–P)	2.01(2)	
$[Ca(HATP)_2]^{4-}$	Octahedral	Ca–O(α–P)	2.28(3)	Ca–O(α–P)	2.28(3)	20, 21
		–O(β–P)	2.30(2)	–O(γ–P)	2.30(2)	

Table 5.3 Metal–Ligand Distances in ATP–Metal Complexes[a] (Continued)

Complex	Geometry about metal (M)	Coordination sites (L)	M–L Distance [Å]	Coordination sites (L)	M–L Distance [Å]	Ref.
		–O(γ–P)	2.25(2)	–O(γ–P)	2.25(2)	
[Cu(H$_2$ATP)(phen)]$_2$	Distorted octahedral	Cu1–O(α–P)	2.878(9)	Cu1–N(phen)	1.989(10)	22
		–O(β–P)	1.942(9)	–N(phen)	2.013(10)	
		–O(γ–P)	1.925(8)	–O(γ–P)	2.284(8)	
	Distorted octahedral	Cu2–O(α–P)	2.273(9)	Cu2–N(phen)	2.050(10)	
		–O(β–P)	1.977(9)	–N(phen)	1.993(10)	
		–O(γ–P)	1.919(8)	–O(γ–P)	2.273(9)	
[Zn(H$_2$ATP)(bpy)]$_2$	Distorted octahedral	Zn1–O(α–P)	2.71(4)	Zn1–N(bpy)	2.13(3)	22
		–O(β–P)	2.09(3)	–N(bpy)	2.14(4)	
		–O(γ–P)	2.02(3)	–O(γ–P)	2.00(3)	
	Distorted octahedral	Zn2–O(α–P)	2.39(3)	Zn2–N(bpy)	2.09(4)	
		–O(β–P)	2.05(4)	–N(bpy)	2.20(4)	
		–O(γ–P)	2.00(3)	–O(γ–P)	2.01(4)	
[Mn(HATP)$_2$]$^{4-}$	Octahedral	Mn–O(α–P)	2.205(6)	Mg–O(α–P)	2.205(6)	20, 21
		–O(β–P)	2.156(4)	–O(β–P)	2.156(4)	
		–O(γ–P)	2.144(5)	–O(γ–P)	2.144(5)	
[Co(HATP)$_2$]$^{4-}$	Octahedral	Co–O(α–P)	—[b]	Co–O(α–P)	—[b]	20
		–O(β–P)	—	–O(β–P)	—	
		–O(γ–P)	—	–O(γ–P)	—	
[Zn(HATP)$_2$]$^{4-}$	Octahedral	Zn–O(α–P)	—[b]	Zn–O(α–P)	—[b]	24
		–O(β–P)	—	–O(β–P)	—	
		–O(γ–P)	—	–O(γ–P)	—	
[Cd(HATP)$_2$]$^{4-}$	Octahedral	Cd–O(α–P)	—[b]	Cd–O(α–P)	—[b]	24
		–O(β–P)	—	–O(β–P)	—	
		–O(γ–P)	—	–O(γ–P)	—	

[a] Abbreviations: Hdpa = protonated 2,2′-dipyridylamine; phen = 2,10-phenanthroline; bpy = 2,2′-bipyridyl; M = metal ion, L = ligand; B = base; S = sugar; P = phosphate; and W = water ligand.

[b] Not reported.

Table 5.4 Conformational Parameters of ATP Molecules in ATP–Metal Complexes[a]

Complex	Conformation about C1′–N9 bond[b]	Sugar puckering[b]	Conformation about C4′–C5′ bond[b]	P_α–P_β–P_γ angle [°]	Ref.
$Na_2{\cdot}H_2ATP{\cdot}3H_2O$	*anti*	C3′-*endo*	*gauche–gauche*	98.3	16, 17,
	anti	C2′-*endo*-C3′-*exo*	*gauche–gauche*	91.2	18
$Na_2{\cdot}H_2ATP{\cdot}2H_2O$	*anti*	C3′-*endo*	*gauche–gauche*	100.6	19
	anti	C4′-*endo*	*trans–gauche*	103.3	
$[Mg(HATP)_2]^{4-}$	*anti*	C2′-*endo*	*gauche–gauche*	85.4(7)	20, 21
$[Ca(HATP)_2]^{4-}$	*anti*	C2′-*endo*	*gauche–gauche*	87.5(9)	20, 21
$[Cu(H_2ATP)(phen)]_2$	*anti*	C3′-*endo*	*gauche–gauche*	93.5	22
	anti	C3′-*endo*	*gauche–gauche*	88.8	
$[Zn(H_2ATP)(bpy)]_2$	*anti*	C3′-*endo*	*gauche–gauche*	94(2)	23
	anti	C3′-*endo*	*gauche–gauche*	87(2)	
$[Mn(HATP)_2]^{4-}$	*anti*	C2′-*endo*-C3′-*exo*	*gauche–gauche*	85.9(2)	20
$[Co(HATP)_2]^{4-}$	*anti*	C2′-*endo*-C3′-*exo*	*gauche–gauche*	84.8(8)	20
$[Zn(HATP)_2]^{4-}$	*anti*	—[c]	*gauche–gauche*	85(1)	24
$[Cd(HATP)_2]^{4-}$	*anti*	—[c]	*gauche–gauche*	86(1)	24

[a] Abbreviations: Hdpa = protonated 2,2′-dipyridylamine; phen = 2,10-phenanthroline; bpy = 2,2′-bipyridyl.

[b] For definition of these conformational terms, see Saenger, W. In: Saenger, W., Ed. *Principles of Nucleic Acid Structure*. New York, Berlin: Springer–Verlag, 1984, 9–28.

[c] Not reported.

in Figure 5.2) between the amino N6 of the base and a phosphate oxygen from the β-phosphate of the neighboring ATP may be a factor affecting the metal bonding to N7 of the adenine base. This suggests that, when a metal ion is bound to hydrogen-bonding acceptor ligands, the metal ion interacts preferentially with adenine at N7 due to an interligand hydrogen bond with the amino N6.[13]

This is also the case for $Li{\cdot}NAD{\cdot}2H_2O$,[29] where the metal ion binds to N7 of the adenine base and an interligand hydrogen bond is formed from the amino N6 to a phosphate oxygen (of a different molecule) that ligates to the same metal ion. On the other hand, when a metal ion is bound to hydrogen-bonding donor ligands like water and amines, it prefers guanine N7, possibly due to the formation of an interligand hydrogen bond with the carbonyl O6.[13]

5.2.2 *Magnesium and calcium complexes of adenosine 5′-triphosphate*

A Mg^{2+} complex, $[Mg(HATP)_2]{\cdot}[Mg(H_2O)_6]{\cdot}(Hdpa)_2{\cdot}12H_2O$[20,21] and a Ca^{2+} complex, $[Ca(HATP)_2]{\cdot}[Ca(H_2O)_6]{\cdot}(Hdpa)_2{\cdot}9H_2O$[20,21] in which Hdpa is proto-

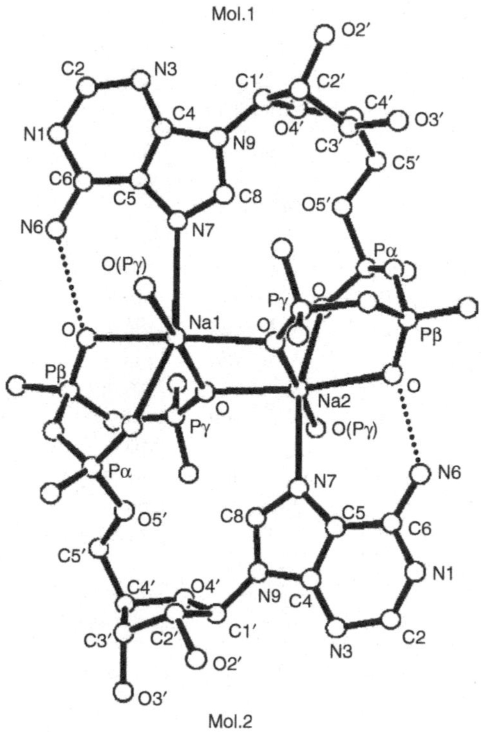

Figure 5.2 Dimeric structure of the $Na_2{\cdot}H_2ATP{\cdot}3H_2O$ complex (Kennard, O. et al., *Proc. R. Soc. Lond. A*, 1971, 325, 401–436), showing coordination of Na1 and Na2 ions with the nucleotide molecules Mol. 1 and Mol. 2. Note the intramolecular N7–Na-O(γ-phosphate) chelation in addition to the α,β,γ-tridentate chelation to the triphosphate group. The same structure is observed in the $Na_2{\cdot}H_2ATP{\cdot}2H_2O$ complex (Sugawara, Y. et al., *J. Am. Chem. Soc.*, 1991, 113, 5440–5445). Broken lines denote interligand hydrogen bonds.

nated 2,2′-dipyridylamine (Figure 5.3) have been reported. These complexes are isostructural to each other and the Mg^{2+} complex is described here.

Figure 5.4 shows the structure of the $[Mg(HATP)_2]^{4-}$ structural unit in which the metal ion is located on a twofold axis. The ion is octahedrally bonded to two symmetry-related $HATP^{3-}$ molecules through α-, β-, and γ-phosphate oxygens, where the adenine base is protonated at N1 and the triphosphate group is fully deprotonated. No bonding interactions take place between the metal ion and the adenine base. The ATP molecule adopts the usual conformation for nucleotide in the Mg^{2+} and Ca^{2+} complexes, that is, *anti* base, C2′-*endo* sugar, and *gauche–gauche* about the C4′–C5′ bond (Table 5.4). The triphosphate chain is folded into left-handed helical form and the adenine base is away from the chelate ring — that is, it takes the Λ-*exo* configuration (Table 5.2).

These types of structures occur also for 3*d* divalent metal ions in isomorphous metal complexes of ATP with the general formula of $[M(HATP)_2]{\cdot}[M'(H_2O)_6]{\cdot}(Hdpa)_2{\cdot}nH_2O$, as described next.

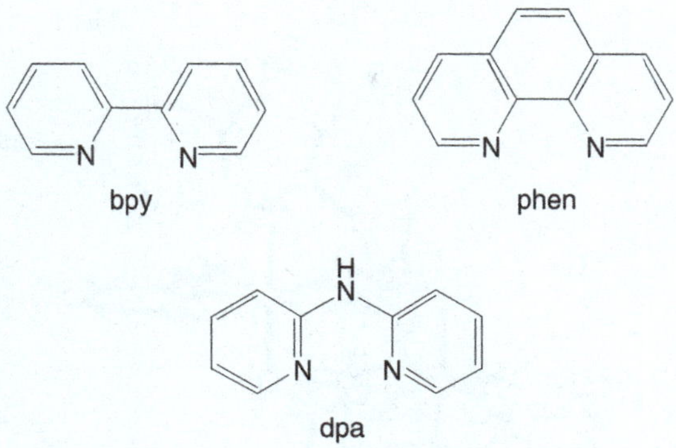

Figure 5.3 Chemical structures of aromatic amines, bpy, phen, and dpa.

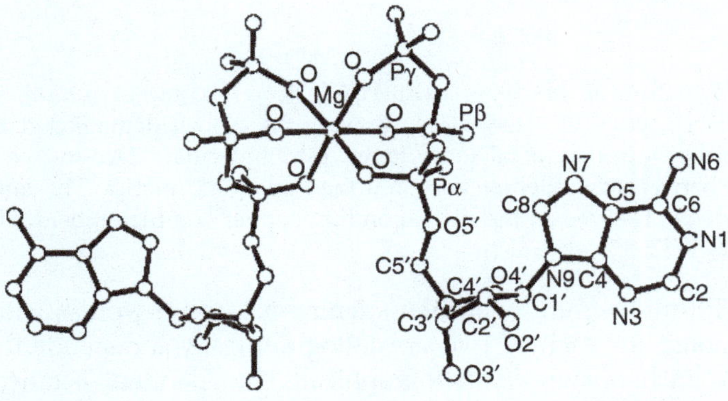

Figure 5.4 Structure of the $[Mg(HATP)_2]^{4-}$ anion,[21] showing Mg^{2+} coordination in which the Mg^{2+} ion resides on a twofold axis. Note the α,β,γ-tridentate chelation to the triphosphate group. The same coordination mode is observed in the corresponding calcium,[21] manganese,[20] cobalt,[20] zinc,[24] and cadmium[24] complexes.

5.3 *Structures of transition and heavy metal ion complexes*

A total number of six crystal structures of transition and heavy metal ion complexes with ATP have been reported. These are classified into two types: $[M(H_2ATP)(aromatic\ amine)]_2 \cdot nH_2O$ (M = Cu^{2+};[22] Zn^{2+}(ref.23)) and $[M(HATP)_2]\cdot[M'(H_2O)_6]\cdot(Hdpa)_2 \cdot nH_2O$ (M = Zn^{2+}, M′ = Mg^{2+} and Zn^{2+};[24] M = M′ = Mn^{2+};[20] M = M′ = Co^{2+};[20] M = Cd^{2+}, M′ = Ca^{2+}(ref.24)).

5.3.1 *Copper and zinc complexes of the type $[M(H_2ATP)(aromatic\ amine)]_2$*

A Cu^{2+} complex, $[Cu(H_2ATP)(phen)]_2 \cdot 7H_2O$ (phen = 1,10-phenanthroline; Figure 5.3),[22] and a Zn^{2+} complex, $[Zn(H_2ATP)(bpy)]_2 \cdot 4H_2O$ (bpy = 2,2′-bipyridyl; Figure 5.3),[23] belong to this type. Both complexes are isostructural and the Zn^{2+} complex is described here. Figure 5.5 shows the structure of the

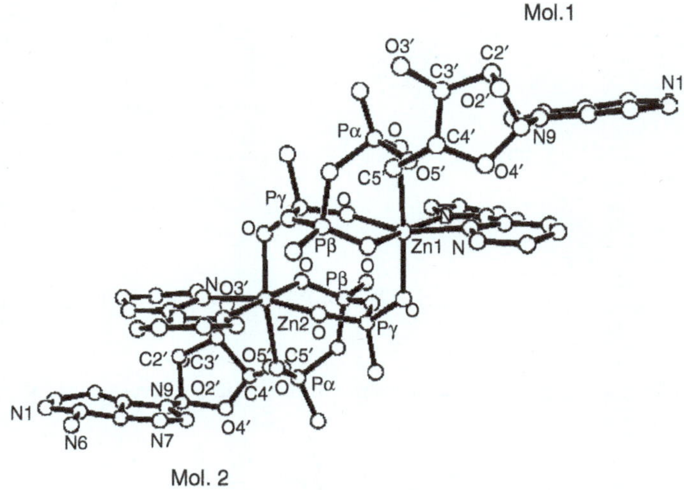

Figure 5.5 Structure of the dimeric $[Zn(H_2ATP)(bpy)]_2$ complex (Orioli, P. et al., *J. Am. Chem. Soc.*, 1981, 103, 4446–4452), showing the α,β,γ-tridentate chelation to the triphosphate group and metal ion-bridged intramolecular "face-to-face" stacking interaction between the adenine base and the bpy ligand planes. The same coordination mode is observed in the corresponding copper complex (Sheldrick, W.S., *Z. Naturforsch.*, 1982, 37b, 863–871).

$[Zn(H_2ATP)(bpy)]_2$ unit, where the adenine N1 and a γ-phosphate oxygen are protonated for each of two crystallographically independent H_2ATP^{2-} molecules. Each of two crystallographically independent octahedral Zn^{2+} ions binds to an ATP molecule through an α-, a β-, and a γ-phosphate oxygen, where the axial Zn-O(α) bonding is very weak (Table 5.3).

The other axial position is occupied by a γ-phosphate oxygen of the second ATP molecule, leading a dimeric structure with the formation of an eight-membered puckered ring in which a pseudocenter of symmetry exists at the midpoint of the Zn–Zn vector (except for optically active ribose moieties). The structure is stabilized by intramolecular stacking interaction between the adenine base moiety and the bpy bidentate ligand that occupies the remaining two basal positions of the Zn atom. The ATP molecules take preferred conformations in the Zn^{2+} and Cu^{2+} complexes: *anti* around the C1′–N9 glycosidic bond, C3′-*endo* at the ribose, and *gauche–gauche* around the C4′–C5′ bond. The phosphate chains are folded into right- and left-handed forms for Mol. 1 and Mol. 2, respectively, and the adenine bases are away from the chelate ring for Mol. 1 and Mol. 2; that is, Mol. 1 and Mol. 2 adopt Δ-*exo* and Λ-*exo* configurations, respectively (Table 5.2).

Sigel and coworkers originally studied interactions among metal ions, nucleotides, and aromatic amines in solution as a model for enzyme–metal ion–substrate interactions and demonstrated that metal ion-bridged ternary complexes are formed and their stability enhanced by intramolecular stacking interactions between the heterocyclic base and the purine moieties.[30]

These $[M(H_2ATP)(aromatic\ amine)]_2$ complexes[22,23] could support the formation of such ternary complexes in solution and serve as a model for enzyme–metal ion–nucleoside triphosphate complexes; $[Cu(HAMP)(bpy)]_2^{2+}$(ref.31) and $[Cu_4(HADP)_2(bpy)_4(H_2O)_2(NO_3)_2]^{2+}$(ref.22) ternary complexes also belong to this type. It should be noted that, although a significant number of crystal structures of ternary complexes of the type $[M(nucleotide)(aromatic\ amine)]_2$ are now available,[13] intramolecular "face-to-face" stacking interaction as observed here between the nucleotide base and the aromatic amine planes occurs solely for adenine nucleotides.

On the other hand, intramolecular "edge-to-face" interaction between the edge of the aromatic amine ring and the face of the nucleotide base occurs for other nucleotides, 5′-GMP, 5′-IMP, 5′-CMP, and 5′-UMP.[13] The presence of a face-to-face stacking interaction exclusively for adenine nucleotides in the solid state seems to be affected, at least in part, by the formation of adenine–adenine self-base-pairing in the crystal lattice; such a base–pair formation creates a larger plane and makes further stacking feasible.[13] In this regard, it is of interest to see whether intramolecular "face-to-face" stacking occurs also with NTPs other than ATP in future work.

5.3.2 *Zinc, manganese, cobalt, and cadmium complexes of the type $[M(HATP)_2]^{4-}$*

A total of four complexes of the type $[M(HATP)_2]\cdot[M'(H_2O)_6]\cdot(Hdpa)_2\cdot nH_2O$ have been reported: M = Zn^{2+}, M′ = Mg^{2+} (72%) and Zn^{2+} (28%), n = 12;[24] M = M′ = Mn^{2+}, n = 12;[20] M = M′ = Co^{2+}, n = 9;[20] M = Cd^{2+}, M′ = Ca^{2+}, n = 9.[24] These complexes are isomorphous to each other (Table 5.1) and also to $[Mg(HATP)_2]^{4-}$ and $[Ca(HATP)_2]^{4-}$,[20,21] whose structures are mentioned in Section 5.2.2 (see Figure 5.4). "Phosphate-only" metal bonding through α-, β-, and γ-phosphate oxygens of the two ATP molecules is observed (Table 5.2). ATP molecules assume the usual conformations: *anti* base, C2′-*endo*-C3′-*exo* ribose, and *gauche–gauche* about the C4′–C5′ bond (Table 5.4). The phosphate chains are in the folded conformation and can be described as Λ-*exo* configuration for all of this type of complex (Table 5.2).

5.4 *General structural features of adenosine 5′-triphosphate metal complexes*

5.4.1 *Molecular conformations of adenosine 5′-triphosphate*

As Table 5.4 shows, the concept of a "rigid" nucleotide[26] also holds for ATP. That is, it takes *anti* conformation around the C1′–N9 bond, C2′-*endo* or C3′-*endo* puckering modes in the ribose ring (C2′-*endo*-C3′-*exo* puckering can be regarded as belonging to a C2′-*endo* family), and *gauche–gauche* conformation around the exocyclic C4′–C5′ bond. This suggests that metal interactions with ATP molecules exert a minor effect on the conformations of ATP.

However, a deviation from this rule occurs in ribose puckering and around the exocyclic C4′–C5′ bond in $Na_2{\cdot}H_2ATP{\cdot}2H_2O$,[19] where one of two ATP molecules adopts unusual C4′-*endo* and *trans-gauche* conformations. The phosphate chains are fixed in the folded conformation by α, β, γ-tridentate chelation in all of the ATP–metal complexes, forming the P_α-P_β-P_γ angle of 85 to 100° (Table 5.4). The C4′–O5′–P(α)–O(α,β) chain is conformationally flexible and thus the triphosphate chain points toward the adenine base in $Na_2{\cdot}H_2ATP$[17,19] or away from the adenine base in $[M(HATP)_2]^{4-}$ (M = Mg^{2+},[21] Ca^{2+},[21] Mn^{2+},[20] Co^{2+},[20] Zn^{2+},[24] and Cd^{2+}(ref.24)), and $[M(H_2ATP)(\text{aromatic amine})]_2$ (M = Cu^{2+}(ref.22) and Zn^{2+}(ref.23)).

In all the ATP–metal complexes, the adenine base is protonated at N1 because their crystals have been prepared under acidic conditions (Table 5.1). In a future work, it would be desirable to see any N1-protonation effect on structural features of ATP and on interactions with metal ions, by comparing crystal structures with those of fully deprotonated ATP^{4-} complexes, which could be crystallized under physiological pH conditions.

5.4.2 *Structural types of adenosine 5′-triphosphate metal complexes and metal bonding modes*

A total of 10 ATP–metal complexes are categorized into three structural types: type (1) $Na_2{\cdot}H_2ATP$;[17,19] type (2) $[M(HATP)_2]^{4-}$ (M = Mg^{2+},[21] Ca^{2+},[21] Mn^{2+},[20] Co^{2+},[20] Zn^{2+},[24] and Cd^{2+}(ref.24)); and type (3) $[M(H_2ATP)(\text{aromatic amine})]_2$ (M = Cu^{2+} [22] and Zn^{2+}(ref.23)). In type (1), two kinds of metal bondings are observed: one is the simultaneous metal bonding to base and phosphate to form an intramolecular N7–M–O(γ-phosphate) chelation and the other is the metal bonding to phosphate only to form an α, β, γ-tridentate chelation (Figure 5.2), where the metal ion is bonded more weakly to the O(α) atom than to the O(β) and O(γ) atoms (Table 5.3). As noted in Section 5.2.1, metal bonding to the adenine base at N7 may be facilitated by the formation of an interligand hydrogen bond between the amino N6 and a phosphate oxygen (of a different ATP molecule) that ligates to the same metal ion.

In type (2), divalent alkaline earth and transition metal ions exhibit phosphate-only metal bonding to α-, β-, γ-phosphate oxygens of the two ATP molecules (Figure 5.4). The metal ion is coordinated more strongly to the O(γ) oxygen atom than to the O(β) and O(α) atoms (Table 5.3). In type (3), divalent transition metal ions in ternary metal complexes containing aromatic amine show phosphate-only metal bonding to form an α, β, γ-tridentate chelation (Figure 5.5). The M–O bond length increases in the order M–O(γ) < M–O(β) < M–O(α) (Table 5.3). These complexes are dimeric with the aromatic amine coordinating through the nitrogen donor and the ATP molecule bridging the two metal centers through the phosphate oxygens, forming a dimeric $(M\text{–}O\text{–}P\text{–}O)_2$ unit. The adenine base and the aromatic amine ligand are arranged so as to make intramolecular stacking interaction between them, forming a metal ion-bridged folded structure.

Transition metal ions usually bind to nucleotides through the base moiety — usually at N7 for purine and additionally at N1 for adenine — in addition to the phosphate group.[13] The phosphate-only metal bonding in the ternary nucleotide–metal ion–aromatic amine system is of special interest in that, due to its π-acceptor properties, the aromatic amine imparts discriminating properties to the metal ion.[32] In types (1) through (3), the formation of α, β, γ-tridentate chelation is common for alkali, alkaline earth, and transition metal ions. The α, β, γ-tridentate chelation imposes chirality on a helical chain, right (Δ)- or left (Λ)-handedness. Moreover, the nucleoside moiety can be arranged in two ways: pointing toward (*endo*) or away from (*exo*) the chelate ring.

Thus, four tridentate isomers are possible and, as Table 5.2 shows, all four configurations have been observed in ATP–metal complexes. It has been shown that at least some of the enzymes that use nucleotides or polynucleotides as substrates are selective for stereo and geometric isomers of metal–nucleotide complexes. Thus, the stereochemistry and geometry of the metal–nucleotide complex play important roles in metal–nucleotide phosphate interactions at enzyme-active sites.[25]

5.5 *Concluding remarks*

During the past more than three decades, X-ray crystallographic studies have played a major part in providing a wealth of well defined knowledge about three-dimensional structures of nucleic acid derivatives, including their metal complexes in the solid state. However, only a limited number of crystal structures are known for nucleoside triphosphates, all of which are those of adenosine 5′-triphosphate. Moreover, these complexes belong to any of only three structural types, as noted in Section 5.4.2. Clearly, in order to find and refine any more general structural aspects for ATP, especially for an ATP^{4-} species, further X-ray studies are needed, although crystallization of nucleoside triphosphate complexes has been particularly difficult due to the high negative charge of species such as ATP^{4-},[24] and to the hydrolysis of nucleoside triphosphate during crystallization. Of course, definitive X-ray studies of NTPs other than ATP are needed.

Acknowledgments

Gratitude is expressed to Professor Sigel at the University of Basel for offering the opportunity to write this chapter. The use of the Ortep-3 for Windows[33] program for drawing Figure 5.2, Figure 5.4, and Figure 5.5 is acknowledged.

Abbreviations

ATP general term for adenosine triphosphate

ATP^{4-} adenosine 5′-triphosphate
B base
bpy 2,2′-bipyridyl
CMP cytidine monophosphate
d *d* orbital
dpa 2,2′-dipyridylamine
GMP guanosine monophosphate
$HAMP^{-}$ monoprotonated adenosine monophosphate
$HADP^{2-}$ monoprotonated adenosine diphosphate
$HATP^{3-}$ monoprotonated adenosine triphosphate
H_2ATP^{2-} diprotonated adenosine triphosphate
Hdpa monoprotonated 2,2′-dipyridylamine
IMP inosine monophosphate
L ligand
M metal ion
M′ metal ion
NAD nicotinamide adenine dinucleotide
NTP general term for nucleoside triphosphate
phen 1,10-phenanthroline
P phosphate
R *R*-value
S sugar
UMP uridine monophosphate
V volume of the unit cell
W water ligand
Z number of molecules in the unit cell

References

1. Saenger, W. Metal ion binding to nucleic acids. In: Saenger, W., Ed. *Principles of Nucleic Acid Structure*. New York, Berlin: Springer–Verlag, 1984, 201–219.
2. Eichhorn, G.L. Complexes of polynucleotides and nucleic acids. In: Eichhorn, G.L., Ed. *Inorganic Chemistry*. Vol 2. Amsterdam: Elsevier, 1973, 1210–1243.
3. Spiro, T.G., Ed. *Nucleic Acid–Metal Ion Interactions*. Vol. 1 of *Metal Ions in Biology*. New York: John Wiley & Sons, 1980.
4. Sigel, H., Ed. *Metal Complexes as Anticancer Agents*. Vol. 11 of *Metal Ions in Biological Systems*. New York: Marcel Dekker, 1980.
5. Eichhorn, G.L. and Marzilli, L.G., Eds. *Advances in Inorganic Biochemistry*. Vol 3. New York: Elsevier, 1981.
6. Sigel, H. and Sigel, A., Eds. *Interactions among Metal Ions, Enzymes, and Gene Expression*. Vol 25 of *Metal Ions in Biological Systems*. New York: Marcel Dekker, 1989.
7. Hodgson, D.J. The stereochemistry of metal complexes of nucleic acid constituents. *Prog. Inorg. Chem.*, 1977, 23, 211–254.
8. Gellert, R.W. and Bau, R. X-ray structural studies of metal–nucleoside and metal–nucleotide complexes. *Metal Ions Biol. Syst.*, 1979, 8, 1–56.

9. Swaminathan, V. and Sundaralingam, M. The crystal structures of metal complexes of nucleic acids and their constituents. *CRC Crit. Rev. Biochem.*, 1979, 6, 245–336.
10. Cini, R. X-ray structural studies of adenosine 5′-triphosphate metal compounds. *Comments Inorg. Chem.*, 1992, 13, 1–34.
11. Aoki, K. Crystal structures of metal ion–nucleotide complexes. In: Saenger, W., Ed. *Landolt–Börnstein: Nukleinsäuren; Teilbandb; Kristallographische und Strukturelle Daten II.* New York: Springer–Verlag, 1989, 171–246.
12. Aoki, K. Metal binding effects on nucleic acid structures. In: Lehn, J.M., Ed. *Comprehensive Supramolecular Chemistry*, Vol. 5. Oxford: Pergamon Press, 1996, 249–294.
13. Aoki, K. General conclusions from solid state studies of nucleotide–metal ion complexes. *Metal Ions Biol. Syst.*, 1996, 32, 91–134.
14. Cooperman, B.S. The role of divalent metal ions in phosphoryl and nucleotidyl transfer. *Metal Ions Biol. Syst.*, 1976, 5, 79–126.
15. Sundaralingam, M. and Haromy, T.P. Structures of nucleic acids and their constituents. In: Saenger, W., Ed., *Landolt–Börnstein: Nukleinsäuren; Teilband a; Kristallographische und Strukturelle Daten I.* New York: Springer–Verlag, 1989, 22–254.
16. Kennard, O., Isaacs, N.W., Coppola, J.C., Kirby, A.J., Warren, S., Motherwell, W.D.S., Watson, D.G., Wampler, D.L., Chenery, D.H., Larson, A.C., Kerr, K.A., and Sanseverino, L.R.D. Three-dimensional structure of adenosine triphosphate. *Nature*, 1970, 225, 333–336.
17. Kennard, O., Isaacs, N.W., Motherwell, W.D.S., Coppola, J.C., Wampler, D.L., Larson, A.C., and Watson, D.G. The crystal and molecular structure of adenosine triphosphate. *Proc. R. Soc. Lond. A*, 1971, 325, 401–436.
18. Larson, A.C. Restrained refinement of disodium adenosine 5′-triphosphate trihydrate. *Acta Cryst.*, 1978, B34, 3601–3604.
19. Sugawara, Y., Kamiya, N., Iwasaki, H., Ito, T., and Satow, Y. Humidity-controlled reversible structure transition of disodium adenosine 5′-triphosphate between dihydrate and trihydrate in a single crystal state. *J. Am. Chem. Soc.*, 1991, 113, 5440–5445.
20. Cini, R., Sabat, M., Sundaralingam, M., Burla, M.C., Nunzi, A., Polidori, G., and Zanazzi, P.F. Interaction of adenosine 5′-triphosphate with metal ions. X-ray structure of ternary complexes containing Mg(II), Ca(II), Mn(II), Co(II), ATP and 2,2′-dipyridylamine. *J. Biomol. Struct. Dyn.*, 1983, 1, 633–637.
21. Cini, R., Burla, M.C., Nunzi, A., Polidori, G.P., and Zanazzi, P.F. Preparation and physico-chemical properties of the ternary complexes formed between adenosine 5′-triphosphoric acid, bis(2-pyridyl)amine, and divalent metal ions. Crystal and molecular structures of the compounds containing Mg^{II} and Ca^{II}. *J. Chem. Soc. Dalton Trans.*, 1984, 2467–2476.
22. Sheldrick, W.S. Preparation and structures of ternary copper (II) complexes of ADP and ATP [1]. Models for enzyme–metal ion–nucleoside polyphosphate complexes. *Z. Naturforsch.*, 1982, 37b, 863–871.
23. Orioli, P., Cini, R., Donati, D., and Mangani, S. Crystal and molecular structure of the ternary complex *bis*(adenosine 5′-triphosphato)(2,2′-bipyridine)zinc(II). *J. Am. Chem. Soc.*, 1981, 103, 4446–4452.
24. Cini, R. and Marzilli, L.G. Exploitation of crystalline architecture and solution data in the rational preparation of novel mixed-metal ATP complexes. *Inorg. Chem.*, 1988, 27, 1855–1856.

25. Klevickis, C. and Grisham, C.M. Phosphate–metal ion interactions of nucleotides and polynucleotides. *Metal Ions Biol. Syst.*, 1996, 32, 1–26.
26. Saenger, W. Defining terms for the nucleic acids. In: Saenger, W., Ed. *Principles of Nucleic Acid Structure*. New York, Berlin: Springer–Verlag, 1984, 9–28.
27. Szent–Györgi, A.G. *Bioenergetics*. New York: Academic Press, 1957, 64.
28. Sigel, H. Interactions of metal ions with nucleotides and nucleic acids and their constituents. *Chem. Soc. Rev.*, 1993, 22, 255–267.
29. Saenger, W., Reddy, B.S., Mülegger, K., and Weinann, G. X-ray study of the lithium complex of NAD^+. *Nature*, 1977, 267, 225–229.
30. Mitchell, P.R. and Sigel, H. Enhanced stability of ternary metal ion/adenosine 5′-triphosphate complexes. Cooperative effects caused by stacking interactions in complexes containing adenosine triphosphate, phenanthroline, and magnesium, calcium, or zinc. *J. Am. Chem. Soc.*, 1978, 100, 1564–1570.
31. Aoki, K. X-ray evidence for the metal ion bridged intra- and intermolecular stacking interactions between nucleotide bases and aromatic heterocyclic rings within the ternary complex $[Cu(5'\text{-AMP})(bpy)(H_2O)]_2 \cdot (NO_3)_2 \cdot 6H_2O$. *J. Am. Chem. Soc.*, 1978, 100, 7106–7108.
32. Sigel, H. Ternary Cu^{2+} complexes: stability, structure and reactivity. *Angew Chem. Int. Ed. Engl.*, 1975, 14, 394–402.
33. Farrugia, L.J. ORTEP-3 for Windows — a version of ORTEP-III with a graphical user interface (GUI). *J. Appl. Cryst.*, 1997, 30, 565.

chapter six

DNA polymerases and their interactions with DNA and nucleotides

Allison A. Johnson
Laboratory of Molecular Pharmacology, NCI, NIH, Bethesda, MD

Kevin A. Fiala
Department of Biochemistry, The Ohio State University, Columbus, OH

Zucai Suo*
Department of Biochemistry, The Ohio State University, Columbus, OH

Contents

* Corresponding author.

6.1 *Introduction*

The *Escherichia coli* (*E. coli*) DNA polymerase I was discovered in 1958 by Arthur Kornburg. Since then, the many DNA polymerases identified to catalyze replication, recombination, and DNA repair have been sorted into six superfamilies (select enzymes listed in Table 6.1):

- Family A polymerases are related to *E. coli* DNA polymerase I.
- Family B enzymes include the eukaryotic replicative polymerases.
- Family C polymerases are related to *E. coli* DNA polymerase III and are found exclusively in prokaryotes.
- Family D polymerases are involved in DNA replication in Euryarchaeota.[1]
- Family X includes enzymes related to eukaryotic polymerase β.
- Family Y polymerases are the recently discovered bypass polymerases found in all domains of life.

The focus of this chapter will be the DNA-dependent DNA polymerases: Klenow fragment (the large proteolytic fragment of *E. coli* DNA polymerase I); polymerase β; and T7 phage DNA polymerase, due to the wealth of information available defining polymerization and structures of these enzymes.

Table 6.1 Properties of Select Polymerases

Polymerase	Family	Function	Exonuclease	Fidelity
Pol[a] α (alpha)	B	Priming nuclear DNA synthesis	No	10^{-4}
Pol δ (delta)	B	Nuclear replication	Yes	10^{-4}–10^{-5}
Pol β (beta)	X	Base excision repair	No	10^{-4}
Pol γ (gamma)	A	Mitochondrial replication	Yes	10^{-5}–10^{-6}
Pol ε (epsilon)	B	Nuclear replication	Yes	10^{-5}
Pol ζ (zeta)	B	Translesion synthesis	No	10^{-4}–10^{-5}
Pol ι (iota)	Y	Translesion synthesis	No	10^{-1}–10^{-4}
Pol η (eta)	Y	Repair UV damage	No	10^{-2}–10^{-3}
Pol κ (kappa)	Y	Translesion synthesis	No	10^{-3}–10^{-4}
HIV-1 RT	A	Viral replication	Rnase H	10^{-4}
E. coli DNA Pol I	A	Bacterial replication	Yes	10^{-3}–10^{-4} (exo$^-$)
T7 DNA pol	A	Virus replication	Yes	10^{-5}–10^{-6}
E. coli DNA Pol II	B	Translesion synthesis, replisome reactivation	Yes	10^{-4}–10^{-5}
Pol θ (theta)	A	DNA repair	Yes	10^{-5}
Pol λ (lambda)	X	Base excision repair	No	10^{-3}–10^{-4}
Pol μ (mu)	X	Nonhomologous end joining	No	?
Pol σ (sigma)	X	Sister chromatid cohesion	Yes	?

[a]Polymerase is abbreviated pol.

The overall structures of T7 DNA polymerase, human DNA polymerase β, and *Sulfolobus solfataricus* P2 DNA polymerase IV (Dpo4) are illustrated in Figure 6.1 (derived from references 2 through 4). DNA polymerases resemble a right hand gripping the DNA and use three acidic amino acids and two metals to catalyze phosphoryl transfer. The structural differences between several lesion bypass polymerases and high-fidelity replicative polymerases will be addressed later in this chapter. The role of nucleotides during replication and the interactions occurring between nucleotides and polymerases will be discussed from a structural and mechanistic perspective for polymerases that have been extensively studied.

Nucleotides are linked together by polymerases in a template directed manner. Figure 6.2A and Figure 6.3 illustrate the polymerase metal chelation and primer extension, respectively. An existing primer containing a 3′-hydroxyl forms a phosphodiester bond with the 5′-phosphate of the incoming nucleotide, resulting in the release of pyrophosphate. This reaction is thermodynamically driven by the breaking of a weak phosphate–phosphate bond and the formation of a stable phosphodiester bond coupled with the entropy gain after the release of pyrophosphate.[5] The presence of a polymerase increases the rate of this reaction from extremely slow in solution to the incorporation of up to several hundred nucleotides per second within the polymerase active site.

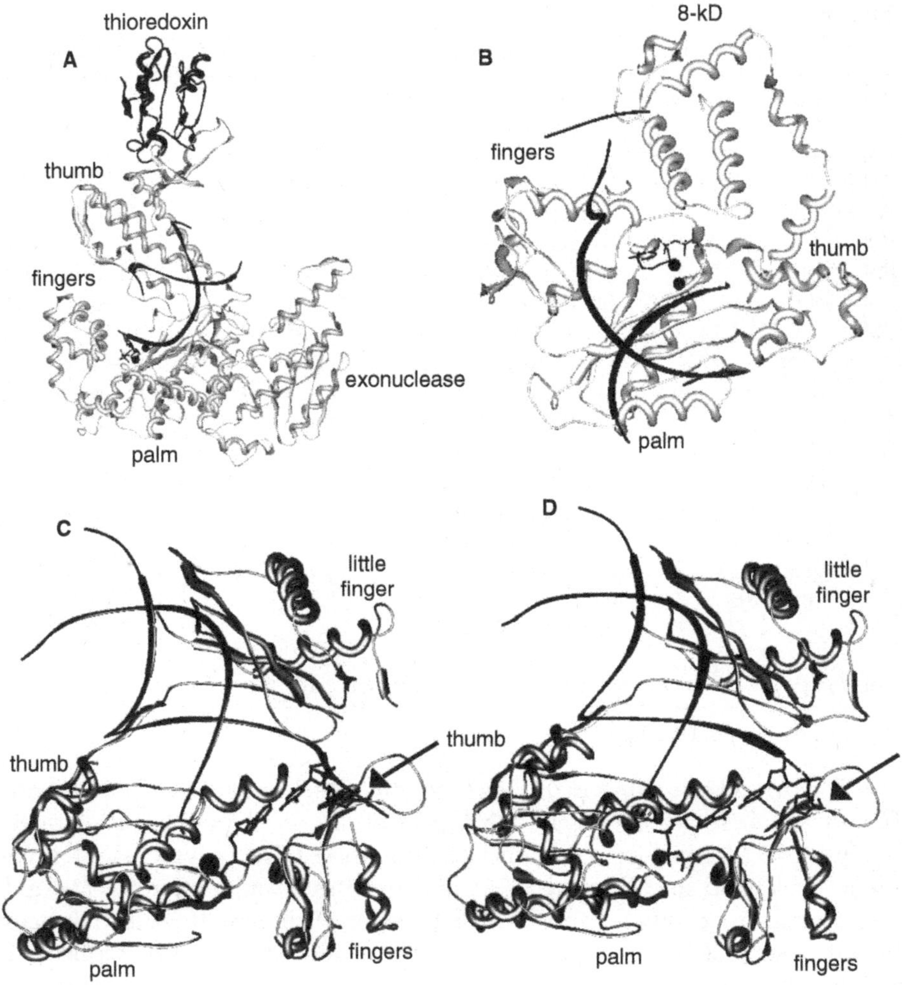

Figure 6.1 Polymerase structure comparison and metal chelation. A. T7 DNA polymerase crystallized with duplex DNA and the next dNTP. (From Doublie, S. et al., *Nature*, 1998, 391(6664), 251–258.) B. Human polymerase β crystallized with gapped DNA and the next dNTP. (From Sawaya, M.R. et al., *Biochemistry*, 1997, 36(37), 11205–11215.) C. Dpo4 crystallized with a DNA 13/18-mer duplex and the incoming matched ddADP. The arrow indicates the downstream template base tA4 is oriented away from the protein. D. Dpo4 in the ternary complex with a DNA 12/16-mer duplex and a mismatched ddGTP. The arrow indicates the downstream template base tT2 is oriented toward the same direction as the template base paired with the incoming ddGTP. E. The local active site structure of the Dpo4 structure in C. F.) The local active site structure of the Dpo4 structure in D. The four structures shown in C, D, E, and F are derived from Ling, H. et al., *Cell*, 2001, 107(1), 91–102. All structures contain divalent metal ions (filled black circles). (Continued)

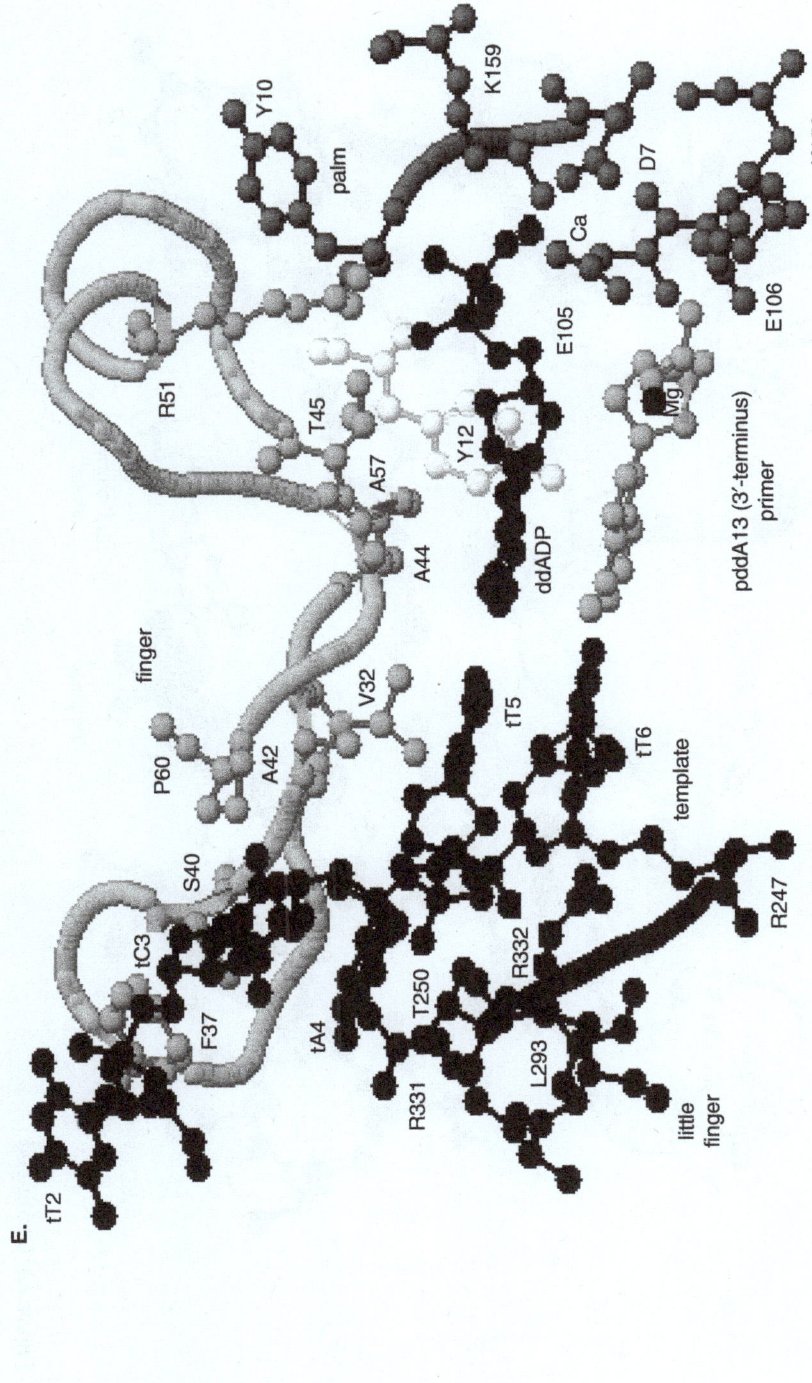

Figure 6.1 Continued.

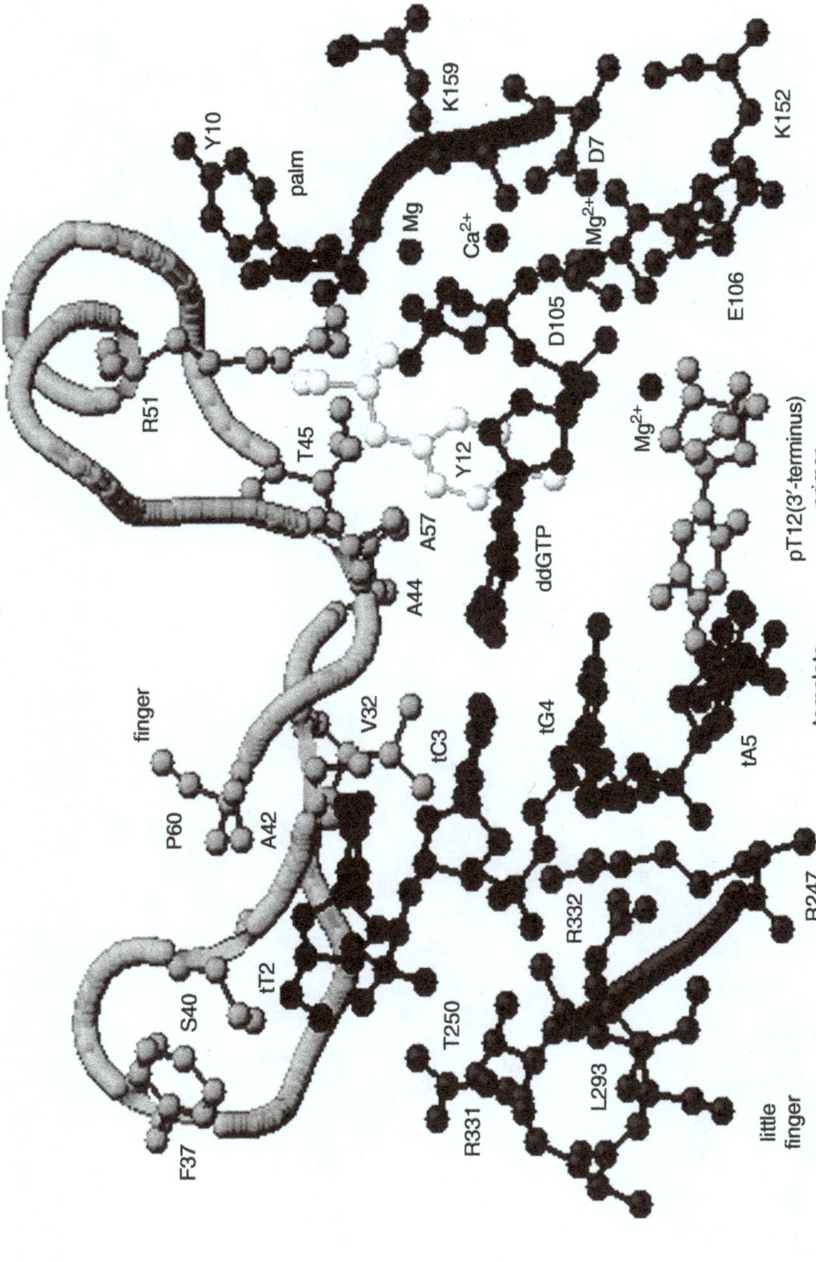

Figure 6.1 Continued.

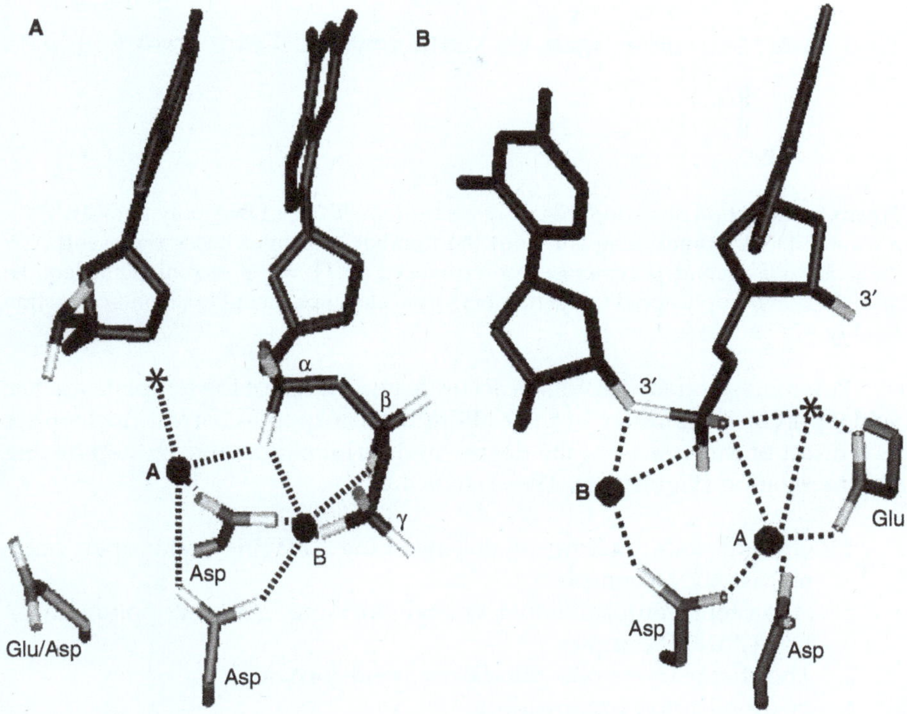

Figure 6.2 Transition states of DNA polymerization and excision. A. Transition state of DNA polymerization (From Doublie, S. et al., *Nature*, 1998, 391(6664), 251–258.) The 3′-terminal primer nucleotide, the next dNTP, the three acidic amino acids of the polymerase active site, and the two divalent metals are shown. The asterisk represents the 3′-hydroxyl of the primer because a dideoxy substrate was used for crystallization. The two divalent metal ions are labeled "A" and "B." B. Transition state cleavage by 3′-5′ exonuclease. (From Brautigam, C.A. et al., *Biochemistry*, 1999, 38(2), 696–704.) The two terminal bases of a single-stranded DNA, the three acidic amino acids of the exonuclease active site, and the two divalent metals are shown. The attacking hydroxyl ion is shown as an asterisk. The two divalent magnesium ions are labeled "A" and "B."

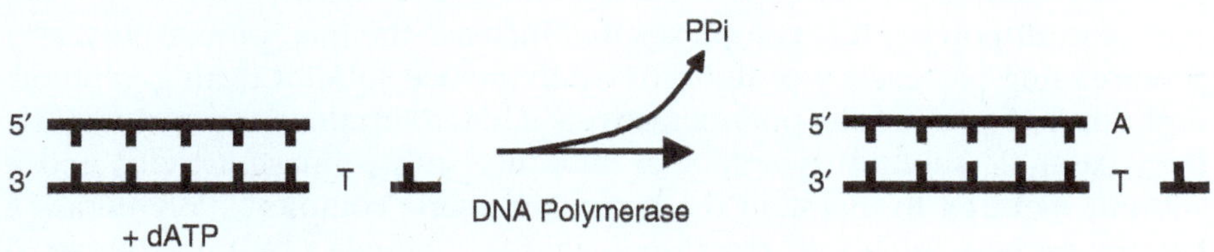

Figure 6.3 Polymerase extension reaction. A schematic diagram of single template-based nucleotide incorporation is illustrated.

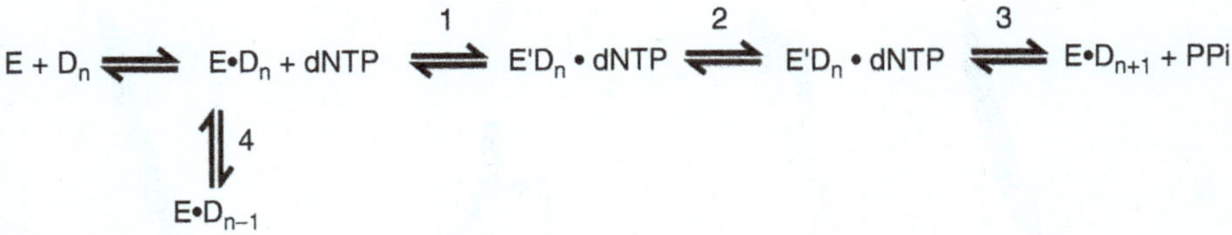

Figure 6.4 Reaction steps in DNA polymerization. "E" = a DNA polymerase; "D" = a duplex DNA primer/template with the number of primer bases represented by "n"; "E′" = a second polymerase conformation; dNTP = the incoming nucleotide; and "PPi" = pyrophosphate. The numbers indicate steps where nucleotide insertion fidelity can be affected.

Polymerase fidelity results in accurate replication of the template nucleic acids; polymerases differ in their ability to incorporate correct nucleotides as a result of differences in the degree of discrimination at each step during polymerization (Figure 6.4). These steps are:

1. Ground state binding of the incoming dNTP into the open polymerase/DNA complex
2. A protein conformational change forming a closed polymerase/DNA/dNTP complex
3. The chemistry of phosphodiester bond formation
4. Postreplication proofreading

In this chapter, we first describe several recently discovered polymerases belonging to different families. We then look at the roles of nucleotides in each of the preceding four steps and their interactions with polymerase active sites. Finally, the mechanistic and structural insights of DNA polymerization will be addressed in detail.

6.2 Human and recently discovered DNA polymerases

6.2.1 Eukaryotic replication

Eukaryotic DNA polymerases α, δ, and ε are members of the B family of polymerases. Polymerases β and γ belong to the X and A families of DNA polymerases and will be discussed separately. Polymerases α, δ, and ε are part of multipolypeptide complexes that increase the rate, processivity, and proofreading proficiency of the active polymerase subunit during genomic replication. For example, polymerase α is a heterotetrameric enzyme with a large catalytic subunit, two smaller subunits with primase activity, and a subunit required to maintain the heterotetrameric complex.[6] Polymerase ε has the highest fidelity of the three enzymes, followed by polymerases δ and α.[7] Polymerase α is primarily responsible for initiation of DNA synthesis and the priming of Okazaki fragments. Polymerases δ and ε catalyze replication of the leading and lagging strands, with polymerase δ most likely synthesizing the leading strand, although their specific roles are not explicitly defined. In general, the importance of each polymerase is indicated by the severity of mutant phenotypes. Mutagenesis studies indicate the function of polymerase δ is the most important, polymerase ε is intermediate, and polymerase α is the least important to eukaryotic replication.[8]

6.2.1.1 Polymerase α

Polymerase α synthesizes RNA–DNA and DNA–DNA primers on the leading and lagging strands during eukaryotic replication.[9] Polymerase α is able to initiate DNA synthesis using its primase activity, as well as catalyze processive replication with a polymerase activity. Polymerase α lacks associated 3′-5′ exonuclease activity, and association with the protein RPA may increase the processivity and fidelity of polymerase α replication.[10] Association of a noncatalytic subunit of polymerase α with SV40 large T antigen (helicase) may have implications for viral replication.

The conserved yeast polymerase α residue Gly952 is required for polymerase activity. Gly952 may reside on the polymerase α fingers domain that rotates from the open to closed conformation upon dNTP binding and may interact with the incoming dNTP.[11] Mutation of this residue indicates Gly952 is also required for high-fidelity incorporation of nucleotides.[11] In fact, the Gly952Ala mutant incorporates incorrect nucleotides with higher efficiency than correct nucleotides opposite template C, G, or T residues.[11] Although polymerase α is a high-fidelity polymerase, this observation resembles the characteristics of some bypass polymerases (discussed later) that have low intrinsic fidelity and insert the correct nucleotide with low efficiency.[11,12]

6.2.1.2 Polymerase δ

Human polymerase δ is the major replicative polymerase responsible for chromosomal replication in eukaryotic cells. Polymerase δ is a heterodimer composed of 125- and 50-kDa subunits.[13] The 125-kDa subunit contains the polymerase as well as 3′-5′ exonuclease activity, and the 50-kDa subunit is required for PCNA binding. Replication factor C and ATP are required for loading the sliding clamp PCNA onto DNA to facilitate polymerase δ replication.[14] In the presence of PCNA, polymerase δ processivity and overall activity are increased by up to 100-fold.[15,16] Mechanistic studies on polymerase δ purified from calf thymus show that polymerase δ replication[17] is similar to previously studied prokaryotic polymerases such as T7 DNA polymerase[18] and human polymerase γ, the only other eukaryotic replicative polymerase studied under pre-steady state conditions.[19,20]

Polymerase δ binds dNTPs tightly and incorporates them processively at a moderate rate.[17] PCNA increases the rate of incorporation and decreases the K_d value of the polymerase–DNA complex. Additionally, polymerase δ has a calculated processivity of an impressive 3500 dNTPs incorporated per single binding event. The overall fidelity of *Schizosaccharomyces pombe* polymerase δ was measured at 10^{-4} to 10^{-6}, but this may be an underestimation because this specific polymerase has a 3′-5′ exonuclease.[21] Deletion of the exonuclease activity results in an increase in spontaneous mutations.[22] Finally, the rate-limiting step during polymerization appears to be a step other then chemistry, probably a protein conformational change.

6.2.1.3 Polymerase ε

Polymerase ε consists of four subunits; the catalytic subunit includes polymerase and 3′-5′ exonuclease activities. Yeast polymerase ε has been overexpressed and purified as a four-subunit heterotetramer.[23] Deletion of the catalytic subunit from yeast is not lethal, indicating loss of polymerase ε function is compensated for by other polymerases.[24] Nonetheless, deletion of just the exonuclease activity of polymerase ε causes a mutator phenotype, thus indicating that polymerase ε is important for chromosomal replication.[22] PCNA and RFC are required for processive DNA synthesis *in vitro* but may not be required *in vivo*.[25]

6.2.1.4 Polymerase β

The family X member, eukaryotic polymerase β, fills gapped DNA during base excision repair. Damaged bases are hydrolyzed from a DNA strand by a DNA glycosylase, followed by backbone removal by phosphodiesterase or lyase.[3] The resulting one-nucleotide gap is repaired by polymerase β. This polymerase is composed of an N-terminal deoxyribosephosphate lyase activity domain and a C-terminal polymerase domain. Polymerase β lacks exonuclease activity and has fidelity only 200-fold higher than the discrimination provided by the difference in free energy of correct and incorrect base pairs in solution.[107] Some details of the polymerization mechanism and rate-limiting step of polymerase β are discussed later.

6.2.1.5 Polymerase γ

The A family member, eukaryotic polymerase γ, is composed of a large catalytic subunit and a dimer of smaller accessory subunits[26] and is responsible for replication of the mitochondrial genome. The simple enzyme structure of this polymerase has facilitated biochemical studies.[19,20,27,28] Polymerization catalyzed by polymerase γ is fairly accurate, with an average error rate of approximately 3×10^{-6}.[28] The 3′-5′ exonuclease contributes another 200-fold increase in fidelity when measured in the presence of dNTPs.[27] A crystal structure of the accessory subunit dimer has been solved,[26] but a crystal structure of the holoenzyme remains unsolved. The catalytic subunit contains the polymerase and 3′-5′ exonuclease activities of the enzyme. The accessory subunit of polymerase γ increases processivity by facilitating tighter binding of DNA and dNTP substrates and by increasing the rate of polymerization.[19]

Polymerase γ is susceptible to inhibition by nucleoside analogs used to treat AIDS; this inhibition may be among the causes of the many side effects of these drugs. For example, dideoxy-CTP, one of the highly toxic nucleoside analogs, is the most efficient of the FDA-approved nucleoside analogs to be incorporated into DNA by polymerase γ. Furthermore, once incorporated onto a primer 3′-terminus, ddCMP is removed very inefficiently by polymerase γ exonuclease activity.[29,30] Ultimately, a crystal structure will provide important insight into the interactions of polymerase γ with normal and analog substrates.

6.2.2 *Family D polymerases*

DNA replication in archaeal organisms is catalyzed by DNA polymerases from family D. These polymerases were identified initially from a hyperthermophilic archaeon, *Pyrococcus furiosus* (Pfu).[1] Two *Pfu* proteins encoded by tandem genes, DP1 and DP2, form a polymerase complex: the former is an accessory subunit, while the latter is the catalytic subunit. The two *Pfu* proteins are highly conserved in the Euryarchaeota. The homologs of DP2 share more than 50% amino acid conservation; the DP1 homologs possess more than 30% identity. However, these polymerases share little sequence homology to the polymerases from other families. The family D polymerases possess active 3′-5′ exonuclease activity. Their polymerase activity is sensitive to N-ethylmaleimide, but resistant to aphidicolin. The nucleotide incorporation catalyzed by family D polymerases has not been studied in detail to this point.

6.2.3 *Translesion polymerases*

Translesion DNA synthesis requires specialized DNA polymerases to accommodate the bulky lesions and aberrant base structures of damaged DNA. The members of the Y family of DNA polymerases, previously the UmuC/DinB/Rev1/Rad30 family, replicate undamaged DNA with low fidelity and poor processivity. These polymerases replicate across various DNA lesions, lack 3′-5′ exonuclease activity, and are known to increase mutagenesis. The error rates generated during replication of undamaged DNA is two to four orders of magnitude greater for bypass polymerases compared to replicative polymerases.[31]

Replication across DNA lesions often occurs with high fidelity. For example, polymerase ζ, a B family polymerase, and polymerase η incorporate two A residues opposite TT (6-4) photoproducts and TT *cis–syn* photodimers, respectively. Polymerase η is essential for correcting sunlight-induced DNA damage, and polymerases η and ζ and, possibly, polymerase ι are important for increasing immunoglobulin hypermutation. The bypass polymerases have little sequence similarity to higher fidelity replicative polymerases, yet their three-dimensional structures display the characteristic fingers, palm, and thumb domains (for example, see Dpo4 structures in Figure 6.1C and Figure 6.1D). The following is a short summary of several bypass polymerases.

6.2.3.1 Polymerase ζ (zeta)

Polymerase ζ is a B family polymerase related to polymerase δ, but is included here because of its bypass characteristics. This polymerase is relatively nonprocessive, with 50% dissociation after less than three dNTP extensions, although some events extend to 200 nucleotides.[32] Polymerase ζ is a translesion polymerase that works in concert with Rev1 protein. Rev1, a deoxycytidyl transferase that incorporates dCMP opposite abasic sites, was the first Y family enzyme identified. Polymerase δ then continues DNA synthesis from the mispaired C opposite the abasic site.[33,34] Rev1/polymerase ζ also can carry out translesion synthesis past TT (6-4) photoproducts *in vivo*.[35] Extension of terminal mismatches is catalyzed by polymerase 2- to >1000-fold greater compared to polymerase α.[36] The error rate for incorporation following a terminal mismatch is 10^{-1} to 10^{-2}, compared to a rate of 10^{-4} to 10^{-5} for extension from correctly paired termini.[37] This may be the major mutator function of polymerase ζ, because this enzyme causes 50 to 70% of spontaneous mutations in yeast.[36]

6.2.3.2 Polymerase η (eta)

Polymerase η (eta) was recently identified in yeast.[38] Mutations in the gene encoding polymerase η lead to a variant of xeroderma pigmentosum,[39,40] an inherited disorder characterized by sunlight hypersensitivity and high incidence of skin cancers and related to deficiencies in nucleotide excision repair. Polymerase η replicates through UV-induced *cis–syn* cyclobutane thymine–thymine dimers[38] and a variety of other DNA lesions (N-2-acetylaminofluroene (AAF)-modified guanine, 8-oxoguanine, abasic sites, and cisplatin-G), but is stalled by T–T (6-4) adducts.[41] Interestingly, replication through *cis–syn* TT dimers occurs with the same efficiency and fidelity as with undamaged DNA by human and yeast polymerase η.[42]

Polymerase η often inserts the correct nucleotide opposite a lesion. For example, dCTP is incorporated opposite modified dGMP residues[40,43] and dATP is incorporated opposite TT dimers.[44,45] A dCTP is also incorporated opposite 7,8-dihydro-8-oxoguanine, in contrast to eukaryotic replicative DNA polymerases.[46] Polymerization by polymerase η is relatively inaccurate during replication of undamaged DNA (~10^{-1} to 10^{-3}).[44,47,48] Misincorporation leads to polymerase η dissociation from the DNA, prohibiting extension of mismatched base pairs.[40] Base pairing components of the nucleotides involved at bypass replication sites are not affected by damaging chemical modifications, and perhaps polymerase η uses this base pairing information to maintain some accuracy during lesion bypass.[49] Therefore, polymerase η achieves some level of accurate replication by preferential incorporation of the correct nucleotide, as well as extension of correctly base-paired DNA.

6.2.3.3 Polymerase ι (iota)

Polymerase ι is encoded by the Rad30B gene, and its role within the cell is presently unknown. The fidelity of replication by polymerase ι differs with respect to the template base, with an error rate of 10^{-2} to 10^{-4} opposite a template A, G, or C.[47,50,51] Surprisingly, polymerase ι preferentially selects misincorporation of G opposite a template T,[47,50,51] even when the DNA lacks a lesion. Polymerase ι extends mismatches more efficiently than replicative polymerases and is unable to initiate incorporation from a DNA nick.[52] Polymerase ι has a limited processivity of one to three nucleotides per binding event, but efficiently extends a primer to fill DNA gaps of 7 to 10 nucleotides.[53,54]

Human polymerase ι (to facilitate misincorporation) and yeast polymerase ζ (to facilitate extension) are used to demonstrate coordinated activities for efficient bypass of a 6-4 pyrimidine–pyrimidone TT dimer (6-4 PP) and a synthetic abasic site.[37] This two-polymerase system may support the "two-step" model of error-prone lesion bypass in eukaryotes.[55,56] The *in vivo* role for polymerase ι may be to foster somatic hypermutation for the generation of antibody diversity, due to its template-T dependent insertion of G.[52] It is possible that polymerase ι and polymerase ζ are required in concert to traverse these lesions specifically.[52]

6.2.3.4 Polymerase κ (kappa)

Polymerase κ, the human ortholog of Dpo4 (see Section 6.4.3), is a product of the *DINB1* gene and a member of the Y family of DNA polymerases. Polymerase κ is a low-fidelity polymerase with observed misincorporation frequencies of 10^{-2} to 10^{-4}.[57] It is moderately processive, incorporating approximately 25 nucleotides or more per binding event.[58] Polymerase κ has demonstrated an ability to perform translesion synthesis on several aberrant primer-templates including substrates containing abasic sites, *N*-2-acetylaminofluorene (AAF)-adducts, 8-oxoguanine lesions, and (–)-*trans-anti*-benzo(a)pyrene-N^2-dG adducts. Interestingly, polymerase κ can extend the 3′-end of mismatched primer termini, but does so synergistically in the presence of polymerase δ, the human replicative polymerase.

Polymerase κ is expressed in a variety of tissues; however, in mouse embryos, detectable expression levels of polymerase κ have only been observed in the adrenal cortex.[59] It is suggested that polymerase κ may contribute to the bypass of oxidative DNA damage in the adrenal cortex, a site of steroid production and thus reactive oxygen species. Although the ultimate *in vivo* role of polymerase κ has not been definitively determined, its low base substitution fidelity and moderate processivity foster an increased proclivity toward mutagenesis.

6.3 Biochemical mechanism of DNA polymerization

6.3.1 Nucleotide binding: the role of hydrogen bonding

The incoming nucleotide forms hydrogen bonds with the template base and base stacks with the neighboring primer base. The correct base will properly hydrogen bond and in turn allow proper geometry of the triphosphate and ribose moieties. Base pairing is energetically favorable due to the entropic contributions of freed water molecules and the low cost of formation of the second and third hydrogen bonds.[5] Modified dNTPs lacking base pair or minor groove hydrogen bonding groups have been used to determine the contributions of hydrogen bonding to DNA polymerization.

Base pairing is not required for efficient dNTP incorporation by several DNA polymerases,[60] but hydrogen bonding between base pairs does increase the fidelity of dNTP incorporation.[5] Moreover, the A:T and G:C base pairs have similar free energies of base pair formation within the polymerase active site,[61] indicating that the number of hydrogen bonds between the base pairs is not critical to base pair formation. Overall, it appears that base pair hydrogen bonding is not required for DNA polymerization, but serves as an important initial check for correct base pair formation.

A correct base pair conforms to a specific polymerase active site geometry that is not optimal for an incorrect base pair. A "geometric selection" model of nucleotide insertion specificity was proposed to explain how active site geometry and electrostatic properties of the polymerase active site could increase replication fidelity.[62,63] The change in free energy ($\Delta\Delta G$) for correct compared to incorrect base pair formation increases from 1 to 3 kcal/mol observed in solution to 3 to 3.5 kcal/mol observed within the polymerase active site.[64]

The enthalpic and entropic contributions to free energy may change enough within the confines of the polymerase active site to account for polymerase fidelity. For example, exclusion of water and geometric constraints of the polymerase active site may combine to amplify the differences in base pairing free energy between aqueous solution and the active site environment.[65] Crystallization of polymerases in complex with DNA (or RNA) and nucleotide substrates and mutagenesis studies have indeed demonstrated that active site amino acids provide specific contacts ensuring the correct base pairing of incoming nucleotides.

Nucleotide analogs that retain base stacking properties but cannot form hydrogen bonds are incorporated by many polymerases; however, that incorporation has significantly lower fidelity compared to natural nucleotides,[5] indicating that hydrogen bonds are critical to polymerase fidelity. Interestingly, the base pairs formed by incorporated analogs lacking the ability to hydrogen bond have similar geometry to the canonical base pairs. It is thus clear that hydrogen bonds are essential to polymerase efficiency and fidelity.

6.3.2 Nucleotide binding: the role of base stacking

Base stacking is a stabilizing force in duplex DNA. Some studies have established that base stacking of DNA may be the dominant contributor to overall helix stability.[66,67] In fact, nucleotide analogs that lack the ability to form hydrogen bonds but fit well into the geometry of correctly base paired DNA and therefore contribute stacking to base pair selection appear sufficient for maintaining moderate replication fidelity.[5] A single unpaired base at the end of a DNA duplex stabilizes the duplex structure through stacking interactions[66]; moreover, half of the total free energy of a base pair may be derived from this base stacking.[5] Adenine is most often inserted opposite a template abasic site, and overall, the order of preference of insertion opposite an abasic site (A > G > T, C) corresponds to the bases' relative stacking ability.[66] Additionally, sequence context is important in selection of a base inserted opposite an abasic site[65]; a large base with stronger stacking than adenine is preferred by orders of magnitude for insertion at an abasic site.[66]

However, in the DNA polymerase active site, the contributions of base stacking at the position of the nascent base pair are influenced by active site amino acids. Many polymerases prevent stacking between the first template base and the neighboring 5′-base by bending the template DNA. Active site amino acids in Klenow fragment (Tyr766); Bacillus DNA polymerase (Tyr714); T7 DNA polymerase (Gly527); Taq DNA polymerase (Tyr671); and HIV-1 RT (Leu74), as well as other polymerases, perform the base stacking function required for stabilization of the template base opposite the incoming dNTP[68–70] or the base 5′ to the template base for His607 T7 of DNA polymerase and Phe61 of HIV-1 RT.[2,70]

Mutation of Klenow fragment Tyr766 to a nonstacking amino acid results in a significant decrease in polymerase fidelity.[71,72] Amino acid stacking also stabilizes the incoming base, as is the case for Tyr526 of T7 DNA polymerase, Phe667 of Taq DNA polymerase, and Tyr115 for HIV-1 RT.[2,69,70] Gapped and nicked DNA substrates in the polymerase β active site display a 90° bend in the template opposite the site of incorporation of the incoming dNTP[3]; the exposed bases on either side of the bend are stacked by His34 and Lys280. These observations indicate base stacking plays an important role in nucleotide binding and helix stabilization during DNA polymerization and that active site amino acids often provide substrate stabilization through stacking interactions.

6.3.3 *The rate-limiting step during DNA polymerization*

Selection of the correct base at the polymerase active site cannot be solely accounted for by the free energy of binding of correct compared to incorrect base pairs. The free energy difference of correct and incorrect nucleotides is approximately 1 to 3 kcal/mol.[64,73] This gain in free energy results from hydrogen bonding between base pairs and base stacking interactions between neighboring bases. The error frequency of one error out of 5 to 150 bases (10^{-1} to 10^{-2}) polymerized would result if the polymerase is solely attaching together base pairs formed with the said free energy difference.[61] However, the actual error frequencies of most polymerases are in the range of 10^{-3} to 10^{-6}.[62,73–75]

Kinetic studies were used to examine binding of correct and incorrect nucleotides by several polymerases. For T7 DNA polymerase, the K_d values for correct base pair formation are in the range of 20 μM, while the K_d values for incorrect base pair formation are in the range of 4 to 8 mM, contributing a factor of 200 to 400 to correct base pair formation.[76] Similar results were obtained for HIV-1 RT.[77] In contrast, the lack of selectivity in initial binding of nucleotides by Klenow fragment contributes to low fidelity of the enzyme.[78] The K_d values for correct compared to incorrect nucleotides are 5 and 12 μ*M*, respectively, indicating the Klenow fragment has low discrimination against incorrect nucleotide incorporation.[79] It was postulated that ground state nucleotide binding of properly base-paired nucleotides is required for incorporation of correct bases due to the higher fidelity of T7 DNA polymerase and HIV-1 RT compared to the Klenow fragment.[61] This suggests that the T7 DNA polymerase and HIV-1 RT active sites play a significant role in correct base pair formation compared to the Klenow fragment.

Two models have been suggested to explain the extraordinary fidelity of DNA polymerases. The *induced-fit model* suggests that polymerase fidelity is a result of conformational coupling in which the energy of nucleotide binding is used to drive a rate-limiting protein conformational change preceding a fast chemistry step.[61] The *rate-limiting transition state model* states that the free energy difference in chemical transition states between correct and incorrect base pairs dictates polymerase fidelity through a rate-limiting chemistry step.[80] The representative polymerases for these two models are T7 DNA polymerase (induced-fit model) and polymerase β (rate-limiting transition state model); in fact, differences in the mechanisms and structures of these two polymerases may prove each model to be true. The details of these two models are discussed next.

6.3.3.1 Induced-fit model of polymerase fidelity

According to the induced-fit model of polymerase fidelity, binding of dNTPs proceeds by a two-step mechanism: ground state binding followed by a rate-limiting formation of a ternary complex competent for catalysis.[61] The fidelity at this step is governed by ground state nucleotide binding in which the correct nucleotide is selected due to a favorable K_d for correct base pair formation.[76] Selection of the correct base pair occurs in part due to some of the intrinsic nucleotide binding energy being transferred to power a protein conformational change from an open to a closed conformation. The closed conformation has strict steric constraints governing the geometry of base pairing. This is reflected in a fast protein conformational change in the presence of correctly base paired nucleotides and a slower conformational change for an incorrect incoming nucleotide, due to polymerase sensing of proper base pair geometry.

Four kinetic experiments reported support the occurrence of a rate-limiting protein conformational change:[61]

- The rate of the chemical step should decrease with substitution of a nonbridging oxygen with sulfur at the α phosphorus of the incoming dNTP, which is defined as the elemental effect. An elemental effect of less than 4- to 11-fold is observed during correct dNTP incorporation by a polymerase, suggesting the observed rate did not define the chemistry step. Instead, an elemental effect of 4- to 11-fold is observed during incorrect dNTP incorporation, suggesting chemistry becomes rate limiting during misincorporation catalyzed by a polymerase. These conclusions are based on recent studies of the elemental effect of the cleavage of phosphate diesters.[81]

- The kinetic differences between a pulse-chase and an acid-quench experiment suggest the transient appearance of the closed-ternary complex containing tightly bound dNTP and DNA.[18]
- A change in protein fluorescence is observed to occur at the same rate as the maximum rate of dNTP incorporation.[61]
- The rate of dissociation of DNA from the closed state occurred at a 100-fold slower rate than from the open state.[77]

Interestingly, crystal structures of DNA polymerases and DNA in the absence and presence of dNTP differ in open and closed configurations of the active site.[2,3,68,70,82,83] In the closed complex, a polymerase active site makes sequence-independent contacts with the first few base pairs of DNA, including the nascent base pair, ensuring that correct base pairs are formed. The O helix of the polymerase fingers domain moves closer toward the incoming dNTP in the active site during formation of the closed complex. Mutagenesis has shown that interactions between the DNA/incoming nucleotide and several O helix amino acids are required for correct base selection and discrimination against dideoxynucleotides (discussed later). Together, the mechanistic and structural data support a rate-limiting conformational change that functions to increase polymerase fidelity.

6.3.3.2 *Rate-limiting transition state model of polymerase fidelity*

In contrast to the induced-fit model of nucleotide incorporation, it has been shown that the rate-limiting step of polymerization by polymerase β may be the chemistry step.[80] In this case, binding of the correct nucleotide by polymerase β is followed by a fast protein conformational change and a slower chemistry step, which governs polymerase fidelity. Protein conformational changes for polymerase β are observed before and after phosphodiester bond formation — the first demonstrated by the formation of the closed ternary complex and the second substantiated by the opening of the closed complex following catalysis.[84,85] Stopped-flow fluorescence experiments are used to show that the first conformational change is induced by Cr(III)•dTTP binding (inert substrate, divalent ion mimic),[84] and crystallography shows this conformational change is, in fact, the closed polymerase complex formation.[85]

A second crystal structure reveals the absence of electron density at the catalytic metal site when polymerase β is complexed with product in the closed conformation.[85] Mechanistically, this is interpreted to mean that the first Mg^{2+} ion facilitates proper dNTP binding and a fast protein conformational change.[84] The second Mg^{2+} ion binds the closed conformation, leading to the subsequent chemical reaction. Following release of the catalytic metal, a second protein conformational change occurs to allow reformation of the open polymerase complex. Interestingly, the rate of the first conformational change resulting in closed complex formation occurs 5- to 10-fold faster than the rate of the chemical reaction. The resulting conclusion is that, in contrast to T7 DNA polymerase, the chemical step is rate limiting for incorporation by polymerase β.[84,85]

The relationship between the rate-limiting chemistry step of polymerase β and polymerase fidelity can be explained by a correct fit model.[85] The selectivity of polymerase β relies on nascent base pair geometry to set up a transition state that is efficiently converted to product. The role of polymerase conformational changes is to convert the enzyme from a substrate-binding mode to a catalytic mode that does not play a role in fidelity. Instead, polymerase fidelity is a function of the difference in free energy between the chemical transition states of incorrect and correct base pairs.[80] The geometry and hydrogen bonding groups of Watson–Crick base pairs are similar for all correct base pairs, and these characteristics could be used for selection of the proper incoming dNTP at the transition state.[85]

It is likely that these two models can explain the fidelity observed for different types of DNA polymerases. Polymerase β, an X family member, plays an important role in gap-filling synthesis during base-excision repair in mammals. Polymerase β is not homologous with those replicative polymerases in family A, including T7 DNA polymerase. A crystal structure of HIV-1 RT complexed with DNA shows the YMDD motif, including two of the catalytic acidic amino acids, changes conformation upon DNA[86] and dNTP binding.[83] The YMDD motif resides on an unusual β-loop where the YMDD side chains project to the same side of the helix. A similar β-loop is observed in Bacillus[68] and T7 DNA polymerases,[2] as well as the polio virus RNA polymerase.[87]

In contrast, polymerase β does not have a similar loop or the corresponding active site conformational changes.[83] The movement of the YMDD motif is implicated in processivity and translocation— two events that differ significantly between polymerase β and replicative polymerases.[83] It is clear that mechanistic differences exist between polymerase β and family A DNA polymerases, and these differences could account for a change in the rate-limiting step and the active site contributions to fidelity during DNA polymerization.

6.3.4 Chemistry of polymerization

After nucleotide binding and closed complex formation, the polymerase is ready for the phosphodiester bond formation. The transition state of DNA polymerization is illustrated in Figure 6.2A. Burgers and Eckstein define the stereochemistry of *E. coli* DNA polymerase I interaction with dATP analogs containing phosphorothioate.[88] They suggest that:

- The β– and γ-phosphates of the nucleotide bind divalent metal in a bidentate manner.

- A positive charge on the enzyme neutralizes the negative charge of the α-phosphate.
- Nucleotidyl transfer occurs via an in-line attack of the primer 3′-hydroxyl upon the nucleotide α-phosphate, with subsequent release of pyrophosphate.

However, the positive charge on the enzyme in the second bullet is actually the second divalent metal ion bound at the active site. Subsequently, crystal structures demonstrate that polymerases catalyze replication using a two-metal-ion mechanism consistent with the model proposed for the 3′-5′ exonuclease of DNA polymerase I[89] (compare Figure 6.2A and Figure 6.2B, derived from Doublie et al.[2] and Brautigam et al.[90]). For DNA polymerases, both metals stabilize the structure and charge of the pentacovalent transition state.[91] The metal A activates the 3′-hydroxyl for attack of the α-phosphate by lowering the affinity of the hydroxyl for a proton. The metal B assists the pyrophosphate leaving group. Crystallization of human polymerase β with Cr(III) reveals that metal B binds with the incoming dNTP, facilitating protein conformation that is competent for metal A binding and subsequent catalysis.[85]

Structurally, the role of the polymerase active site is to bring together the reaction substituents in an alignment conducive for an efficient chemical reaction. Following dNTP binding and formation of the closed ternary complex, a nucleophilic 3′-hydroxyl of the primer displaces pyrophosphate from a dNTP, resulting in an increase in primer length by one nucleotide. The phosphoryl transfer reaction occurs through a semiassociative transition state in which the attacking 3′-hydroxyl of the primer and the pyrophosphate leaving group are positioned on opposite sides of an α-phosphate with trigonal bipyrimid configuration (Figure 6.5; reference 81).

Bonding of the reactive phosphate to the incoming nucleophile and outgoing leaving group is at a level between the larger amount of bonding observed for an associative transition state and the little bonding observed for a dissociative transition state.[81] The enzyme lowers the activation barrier by transition state stabilization facilitated by the presence of the two metal ions. Active site amino acids are required for positioning the metal ions to facilitate efficient polymerization. The roles of active site amino acids as depicted in Figure 6.2A may be incomplete. One of the acidic residues in rat polymerase β may facilitate deprotonation of the 3′-hydryoxyl group by acting as a general base.[92] Recent computer simulation of the polymerase transition state suggests that Asp654 of T7 DNA polymerase may also perform this function.[93]

6.3.5 *Translocation*

Translocation is important for processive replication. Many polymerases use accessory subunits, such as the association of PCNA with polymerase δ and thioredoxin with T7 DNA polymerase, to maintain the enzyme–DNA asso-

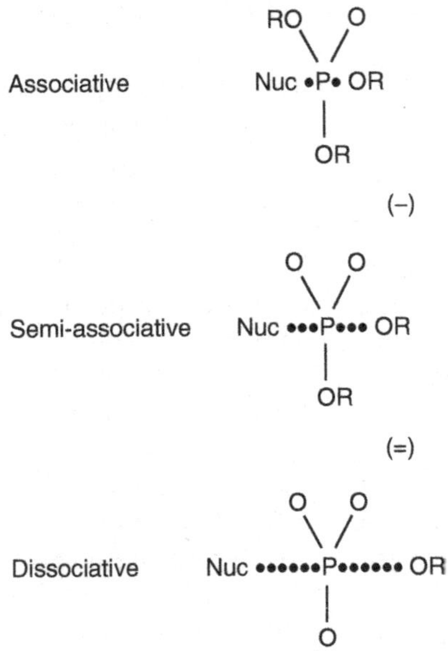

Figure 6.5 Different transition states. (From Herschlag, D. et al., *Biochemistry,* 1991, 30(20), 4844–4854.)

ciation and increase polymerase processivity. Translocation is not rate limiting to polymerization and kinetic parameters for this step have not been defined because translocation occurs too quickly to measure. Structural studies have shown the DNA minor groove becomes widened and shallow upon polymerase binding, facilitating protein interactions through the first four base pairs of duplex DNA. Minor groove contacts with active site amino acids of DNA polymerases are specifically important to fidelity and inhibition of translocation to allow error correction.[68] Mismatched base pairs are sensed throughout this region, as indicated by T7 DNA polymerase stalling,[61] thus allowing proofreading to correct the error. Efficient translocation is facilitated by correct base pair formation during polymerization.

Correct base pair formation may promote translocation because the energy released upon cleavage of the triphosphates of the incoming dNTP can be used to counteract the build-up of negative charges in the polymerase active site.[83] The energy-rich phosphodiester bond linking the α- and β-phosphates is hydrolyzed upon dNTP incorporation to stabilize the transition state and drive translocation.[94] The energy released by phosphodiester bond hydrolysis may be stored in the YMDD motif of HIV-1 RT, as indicated by movements of this domain during translocation.[83]

The YMDD motif may act as a "springboard" to push the DNA primer terminus forward for translocation. Translocation may also be facilitated by release of one divalent metal from the active site with pyrophosphate, following catalysis.[83] The local charge will become more negative due to loss of charge shielding by the metal, resulting in repulsion between negatively charged groups such as the acidic amino acids of the active site and the phosphate of the newly incorporated dNTP. Together, it appears that the energy of phosphodiester bond hydrolysis and the proximity of negative charges between the active site residues and the DNA promote translocation of the DNA.

6.3.6 3′-5′ Exonucleases

The 3′-5′ exonuclease activity of DNA polymerases enhances the fidelity of DNA synthesis by correcting misincorporated nucleotides. Of the eukaryotic polymerases, polymerase δ, polymerase γ, and polymerase ε have associated 3′-5′ exonucleases. Polymerases α and β, and the DNA lesion bypass polymerases do not have a 3′-5′ exonuclease activity, but autonomous 3′-5′ exonucleases (for example, TREX1 and WRN) might help maintain the accuracy of replicated DNA for these polymerases (for review see Shevelev and Hubscher[95]). Prokaryotes and viruses also have many 3′-5′ exonucleases; in particular, the 3′-5′ exonucleases of Klenow fragment and of T7 DNA polymerase will be described.

Kinetic studies have shown that partitioning of the enzyme–substrate complex among forward polymerization, proofreading, and dissociation controls the fidelity and efficiency of DNA polymerization.[61] When correct base pairs are formed, the rate of polymerization and the probability of extension are high. Some correctly base-paired DNA is transferred to the exonuclease active site for hydrolysis. This "cost" of removing mismatches means that 2 to 3% of correctly incorporated nucleotides are excised by polymerase I exonuclease and T7 DNA polymerase.[96–98] When an incorrect nucleotide is incorporated, the rate of polymerization slows, increasing the probability of dissociation or transfer of the DNA to the exonuclease active site for error correction.[98] Additionally, the stability of the DNA duplex decreases, facilitating transfer of the primer to the exonuclease active site.[99] The rate-limiting step of the exonuclease reaction is transfer of the single-stranded DNA from the polymerase to the exonuclease active site for T7 DNA polymerase[98] and is probably chemistry for the Klenow fragment. The observed excision rate increases in the presence of a mismatched base pair, favoring removal.

Even when they are part of the same polypeptide, the DNA polymerase and 3′-5′ exonuclease activities are spatially and functionally separated. For *E. coli* DNA polymerase I, the two active sites reside approximately 30 Å apart in crystal structures, on different domains of the Klenow fragment. Up to nine base pairs must melt for the DNA to move from the polymerase to the 3′-5′ exonuclease active site.[100] The polymerase and exonuclease active sites function independently and each can have DNA substrates bound to it at the same time.[101] DNA can transfer between the polymerase and exonuclease active sites with (Klenow fragment[79,97]) or without (T7 DNA polymerase[2]) dissociation from the enzyme.

Exonuclease active sites contact the three nucleotides at the 3′-terminus of the primer. Substitution of each of these three nucleotides by ribonucleotides showed that ribonucleotides are excised with the same efficiency as deoxynucleotides at the terminal base position.[102] Ribonucleotides positioned interior to the 3′-terminus are excised with reduced efficiency. Molecular modeling and mutagenesis suggested Klenow fragment residues Tyr423 and Met433/Asn420 sterically interfere with 2′-hydroxyl positioning at the second and third primer positions, respectively. Additionally, mutation of these residues to alanine eliminates or diminishes the excision rate reduction observed in the presence of ribonucleotides at the second or third primer positions. The rates of excision by T4 and T7 DNA polymerases are less adversely affected by ribonucleotides compared to Klenow fragment. Similar mutations in T4 DNA polymerase do not restore excision rates, probably because the excision step is not rate limiting.

6.3.7 *Chemistry of excision*

The crystal structure of Klenow fragment 3′-5′ exonuclease with dTMP product reveals the transition state organization of the active site.[89] The illustration of the Klenow exonuclease active site in Figure 6.2B is derived from Brautigam et al.[90] One metal is bound to the exonuclease active site (site A) in the absence of substrate. A second metal binds site B when single-stranded DNA (or dTMP) is present in the active site. The metal in site A is pentacoordinated with a distorted tetrahedral geometry by the carboxylate groups of three acidic residues, a substrate phosphate oxygen, and a water molecule positioned by active site amino acids. Metal B is coordinated octahedrally by the carboxylate of one acidic residue, two phosphate oxygens from the substrate, and three water molecules positioned by active site amino acids. Mutation of the one acidic residue directly coordinating metal B leads to inactivation of exonuclease function (e.g., Asp424Ala for Klenow fragment).[89] Accordingly, due to Asp424Ala mutagenesis, a second active site metal was not observed.

The transition state of excision[89,103] proceeds through an associative in-line mechanism in which the attacking hydroxyl (water molecule) approaches a position opposite the 3′-hydroxyl leaving group.[104] Metal A, Glu357, and Tyr497 exactly position the lone electron pair of the attacking hydroxyl ion towards the scissile phosphate.[89] Metal A activates the hydroxide ion for attack by acting as a Lewis acid and removing a proton from a water molecule. Metal B stabilizes the transition state or facilitates the release of the 3′-oxyanion leaving group. Many protein residues surrounding the exonuclease active site function to position the two metal ions, the single-stranded DNA, and the attacking hydroxide ion; as expected, these residues are conserved throughout 3′-5′ exonucleases.

6.3.8 *Exonuclease active site stabilization of single-stranded DNA*

Two crystal structures of Klenow fragment 3′-5′ exonuclease provide a view of the active site complexed with (1) one divalent metal and single-stranded

DNA in the exonuclease active site; and (2) two divalent metals and a product dTMP.[89] The single-stranded DNA complex shows that three B-form nucleotides fit into the hydrophobic pocket containing the exonuclease active site. This pocket most likely excludes duplex DNA due to steric interactions. The contacts between the DNA and protein are sequence independent and are important for stabilization of the single-stranded DNA for excision. Interactions between the two 3′-terminal bases and four amino acids hold the substrate terminus securely in the active site.

Phe473 stacks against the 3′-terminal base; Leu361 fits between the 3′-terminal and penultimate bases; Asn420 hydrogen bonds to the 4′-oxygen of the 3′-penultimate nucleotide; and Tyr431 stabilizes the sugar residues of the two 3′-terminal nucleotides. Interestingly, Leu361 disrupts base stacking between the two 3′-terminal nucleotides. In the product complex, the 3′-terminal base is positioned hydrophobically by Phe473 and Leu361, but there are no specific contacts with the thymine base. It is clear that Phe473 probably stabilizes and orients the single-stranded DNA within the active site by a stacking interaction absent in duplex DNA, perhaps aiding destabilization of duplex DNA. The mutant Leu361Ala is especially detrimental to exonuclease activity upon duplex DNA. Together, these residues stabilize single-stranded DNA before and after the excision reaction by compensating for stacking interactions present in duplex DNA.

6.4 Structural basis of DNA polymerization

6.4.1 Crystal structure overview

The crystal structures of several DNA polymerases complexed with duplex DNA and an incoming nucleotide substrate have been solved.[2,3,68,70,92] Four of these structures are illustrated in Figure 6.1A through Figure 6.1D (derived from references 2 through 4). Crystallization of these ternary complexes is facilitated by incorporation of a terminal ddNTP on the primer and inclusion of a second ddNTP to bind to the site of the nascent base pair. All DNA polymerases crystallized resemble a right hand with the fingers and thumb domains forming a cleft around the palm, which contains the conserved acidic residues of the active site.

Generally, the first several base pairs of duplex DNA within the polymerase active site are the A-form conformation. This conformation has a widened, shallow minor groove with decreased helical twist, facilitating protein contact with the DNA.[68] Sequence-independent hydrogen bonds form between the conserved thumb domain side chains and the purine N_3 and pyrimidine O_2 positions.[2,68] In fact, hydrogen bonding with these positions allows recognition of correctly base-paired DNA because of a twofold symmetry in G:C and A:T base pairs.[105] When a mismatch is present in the DNA, one or both of the mismatched bases rotate into the major groove,[106] disrupting minor groove interactions with the polymerase and signaling the polymerase to switch its activity from replication to proofreading.[2]

Bacillus DNA polymerase imposes a 90° bend in the template DNA next to the nascent base pair, making a conformational change before the chemistry step required for base pair hydrogen bond formation.[68] The sugar attached to the terminal base of the primer strand is in the C2′-endo conformation and is the only sugar within the first four base pairs that is in this typical B-form conformation. A similar bend in the DNA is observed for T7 DNA polymerase and HIV-1 RT, and for polymerase β complexed with gapped substrate, but not for Taq DNA polymerase.[2,3,69,70,82]

Nucleotide binding is guided by hydrogen bonding and base stacking with the terminal base pair of duplex DNA (discussed previously), interactions with active site amino acid residues, and stabilization by divalent metal ions. Specificity of base pairing is increased by steric interactions with the dNTP binding pocket. The geometry of a correct base pair is selected by active site steric complementarity that would not exist for incorrect base pairs.[2,68] Misincorporation is less likely because binding an incorrect base would cause improper alignment of the primer 3′-hydroxyl, thus prohibiting a subsequent chemical reaction.

Several amino acids have been identified as important for DNA polymerase fidelity. An important stacking interaction occurs between the terminal template base and Tyr714 of Bacillus DNA polymerase; mutation of the corresponding residue in the Klenow fragment (Tyr766) results in significantly lower replication fidelity.[71,72] Arg283 of human polymerase β contacts the minor groove of the template base of the nascent base pair, and mutation of this residue to alanine results in a 160-fold decrease in fidelity.[3,107] The Klenow fragment Arg668 makes an important minor groove contact with the first base of the primer during the rate-limiting step of polymerization.[108] Similar interactions are made by other DNA polymerases contributing to high-fidelity synthesis of DNA.

Interactions between the protein and the deoxyribose of the incoming dNTP also facilitate polymerization and selection of the correct substrate. A hydrophobic pocket formed by Tyr526, Glu480, and the strictly conserved Tyr530 surrounds the ribose of the incoming dNTP.[2] Ribonucleotides may be excluded from the T7 DNA polymerase and Klenow fragment active site due to a hydrophobic pocket, which cannot accommodate the 2′-hydroxyl of the ribonucleotides.[2,109] T7 DNA polymerase has low specificity against ddNTP incorporation, whereas Klenow fragment and Taq DNA polymerase discriminate against ddNTP incorporation.[110] The T7 DNA polymerase mutant Tyr526Phe exhibits a several thousand-fold increase in discrimination against ddNTPs, and relaxed discrimination for ddNTPs was observed in mutants containing the reciprocal mutations in the more selective Klenow fragment and Taq DNA polymerases.[2]

Two divalent metal ions bind at the polymerase active site and are important for triphosphate positioning and catalysis (Figure 6.2A). The primary function of metal B is to support binding and positioning of the substrate; metal A facilitates the catalytic reaction. Metal B positions the α,- β-, and γ-phosphates of the dNTP substrate in a tridentate configuration,[2,3,68] and metal A stabilizes the α-phosphate.[2,3] Both magnesium ions reduce the negative charge on the α-phosphate. Additionally, the triphosphate conformation is similar to that observed for free Na^{+}-ATP and Mg^{2+}-dNTP in solution, possibly offering significant entropic advantage to rapid polymerization.[111] A protein conformational change resulting in closed complex formation brings the dNTP α-phosphate closer to the primer 3′-hydroxyl.[3] Metal A stabilizes the pentacoordinated phosphate of the transition state and activates the attacking oxygen by acting as a Lewis acid, with a protein side chain (Asp256) acting as a proton acceptor.[92]

6.4.2 *Structural insights to reduced fidelity of HIV-1 RT*

HIV-1 RT functions as a DNA-dependent DNA polymerase and an RNA-dependent DNA polymerase, depending on the stage of the virus life cycle. Much effort has been devoted to understanding the mechanism and structure of RT to aid drug development. RT is an asymmetric heterodimer of p51/p66 subunits. The p66 subunit contains the polymerase and RNase H active sites. The p51 subunit is a truncated form of p66 lacking the RNase H domain. The tertiary structure of these two subunits is very different considering that they have identical sequences.[112] The catalytic subunit of RT is structurally similar to the palm–fingers–thumb organization of polymerases previously described, with the addition of a "connection" domain that links the RNase H domain to the rest of the polymerase.[112]

The dNTP binding pocket is of particular importance in HIV-1 RT crystal structures because many currently approved AIDS therapies are nucleoside analog RT inhibitors. The RT dNTP binding pocket will be discussed briefly here; for a more in-depth discussion on RT, see Chapter 8 in this book. As with other DNA and RNA polymerases, the fingers and thumb domains form a cleft containing the polymerase active site. The dNTP pocket accommodates the 3′-hydroxyl of an incoming dNTP, which is absent in nucleoside RT inhibitors; therefore, this pocket plays a role in RT NRTI resistance.

Upon dNTP binding, the fingers domain rotates toward the palm domain, bringing several amino acids into contact with the incoming dNTP,[70] including several sites that upon mutation confer resistance to nucleoside analogs. The incoming base stacks between the primer terminus and Arg72 and Gln151. The triphosphate is positioned by Lys65, Arg72, two main chain –NH groups, and the two metal ions. The metal ions are most likely octahedrally coordinated. Importantly, the 3′-hydroxyl fits in a pocket lined by Asp113, Tyr115, Phe116, and Gln151. The 3′-hydroxyl hydrogen bonds with Tyr115, and a Tyr115Phe mutant shows that the tyrosine prevents ribonucleotide incorporation.[113] Finally, HIV-1 RT lacks the O helix observed to increase fidelity for other polymerases,[112] possibly leading to the reduced fidelity of RT compared to T7 DNA polymerase.[76,77]

6.4.3 *Structural insights from bypass DNA polymerases*

Nearly all bypass polymerases belong to the Y family. These polymerases have low fidelity and low processivity on undamaged DNA, which suggests their potential displacement by high-fidelity, replicative polymerases shortly after the lesion is bypassed. Several crystal structures have illustrated characteristics of these polymerases that contribute to reduced replicative fidelity and propensity to traverse DNA lesions. In general, the active sites of these polymerases are more flexible, have greater solvent accessibility, and have more relaxed geometric constraints when compared to the high-fidelity, replicative polymerases. Important contacts between polymerase domains, the DNA minor groove, and the nascent base pair, are absent or reduced with respect to these bypass polymerases. It appears that these "loose" active sites allow these enzymes to replicate through aberrant substrates with low fidelity.

The crystal structure of the *S. acidocaldarius* DinB homolog (Dbh) shows the polymerase palm domain resembles that seen in other well characterized replicative polymerases; however, the fingers and thumb domains are structurally unique.[114,115] A full-length Dbh crystal structure showed that the C-terminal domain is structurally homologous to the polymerase-associated domain of polymerase η.[115] Additionally, the fingers domain appears in the closed conformation, which is a characteristic seen only following DNA and dNTP binding in other polymerases.[114] The active site of Dbh is quite accessible to solvent and the relatively few physical contacts between the polymerase and docked DNA allow for the possible accommodation of bulky DNA lesions in to the active site. The site of new base pair formation appears less restrictive compared to T7 DNA polymerase, possibly explaining the lower fidelity of Dbh.

Two full-length crystal structures of Dpo4, a DinB homolog of *S. solfataricus*, were recently obtained as complexes with duplex DNA and correct or incorrect incoming dNTP.[4] The overall structure of this polymerase is illustrated in Figure 6.1C and 6.1D. This enzyme is similar to polymerase η in that it can bypass *cis–syn* TT dimers. Dpo4 also exhibits the canonical right-handed structure — however, without the O helices of the fingers domain found in high-fidelity polymerases. The stubby fingers and thumb domains are smaller compared to those of high-fidelity polymerases, but Dpo4 has an additional "little fingers" domain that increases the interaction of this enzyme with DNA. The fingers domain of high-fidelity polymerases surrounds the nascent base pair and is important for mismatch discrimination. The nascent base pair has only limited contact with protein domains in the Dpo4 active site compared to replicative polymerases. The active site appears open and solvent accessible. Both of these factors may contribute significantly to the low-fidelity characteristic of bypass polymerases.

The crystal structure of Dpo4 shows several hydrogen bond and van der Waals contacts with the sugar and phosphodiester moieties of the B-form DNA in the active site. However, no interactions between Dpo4 and minor groove hydrogen bond acceptor or donor sites are observed, again in contrast to high-fidelity polymerases. In the presence of the next correct nucleotide, the downstream DNA template base tA4 is oriented away from the nascent base pair, which is similar to that observed in other DNA polymerases, and requires a rotation of over 90° in order to be positioned in the active site (Figure 6.1C and Figure 6.1E).

However, inclusion of a mismatched ddGTP in the crystallization mixture results in the observation that the 5′-unpaired template base tT2 is oriented toward the same direction as the base pairing template base and seems ready to slide into the active site (Figure 6.1D and Figure 6.1F). Moreover, Dpo4 can accommodate two adjacent template bases simultaneously in the active site so that the ddGTP forms a correct base pair with the next available template base tC3, rather than tG4 (Figure 6.1F). The lack of specific and numerous enzyme–DNA contacts, as well as the presence of two active site template bases, may have implications for a mechanism of *cis–syn* thymidine–thymidine dimer translesion synthesis, perhaps achieved through template slippage.

A crystal structure of the first 513 amino acids of yeast polymerase η[116] reveals striking differences between bypass and replicative DNA polymerases. Crystal structures of classic DNA polymerases, such as T7 DNA polymerase, strictly adhere to the fingers–thumb–palm subdomain structure. For polymerase η, the fingers and thumb domains are stubby. The O and O_1 helices, which are important for DNA binding and polymerase fidelity, are absent, although the total DNA binding surface is similar to other DNA polymerases due to the addition of a polymerase-associated domain. The novel polymerase-associated domain may compensate for the stubby fingers and thumb domains by mimicking an extra set of fingers.

Modeling of DNA and incoming dNTP reveals few contacts between the unpaired template 5′-end and the protein, and these bases remain in the active site. In contrast, the DNA is bent out of the active site at a 90° angle for T7 DNA polymerase.[2] The polymerase η active site is less restrictive to lesions such as TT dimers, which cannot bend due to the covalent linkage between the bases; in fact, inclusion of a second template base in the polymerase η active site might allow proper base pairing opposite 8-oxoG and TT dimers.[116]

6.5 *Inhibition of viral DNA polymerases by substrate analogs*

Because DNA polymerases play central roles in DNA replication, inhibition of viral DNA polymerases should slow down or stop viral infections. Most of the approved antiviral inhibitors are nucleoside analogs. They are chain terminators like AZT lacking the 3′-hydroxyl group, or are acyclic without a deoxyribose like acyclovir. These drugs are reviewed in detail in Chapter 11 in this book. Because pyrophosphate is a product formed during nucle-

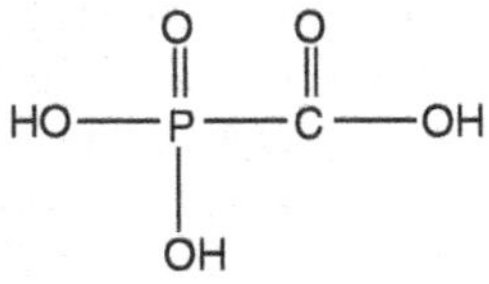

Figure 6.6 Chemical structure of phosphonoformic acid (Foscarnet).

otide incorporation, its analog could bind to and inhibit DNA polymerases. Phosphonoformic acid or Foscarnet (Figure 6.6), a pyrophosphate analog, is an antiherpes drug that binds to the site normally occupied by pyrophosphate and inhibits the catalytic cycle of a herpes DNA polymerase.[117]

6.6 Conclusion

DNA polymerases thus far identified are phylogenetically grouped into six families named A, B, C, D, X, and Y. Typical examples of these DNA polymerase families are summarized briefly in this chapter. Polymerases from different families or even from the same family have varying fidelity. High-fidelity polymerases catalyze genome replication and low-fidelity polymerases repair or bypass DNA lesions. The fidelity of a replicative polymerase follows an induced-fit model: a nucleotide binds to a binary complex of the polymerase and DNA to form a loose or "open" ternary complex, followed by a rate-limiting protein conformational change to form a tight or "closed" ternary complex. The ground state binding step is primarily governed by the intrinsic properties of the nascent base pair: the hydrogen bonds and base stacking. The protein conformational change is limited by the strict steric constraints of the polymerase active site and the geometry of base pairing.

In contrast, the fidelity of DNA polymerase β follows the rate-limiting transition state model; the protein conformational change following the "open" ternary complex formation is faster than the subsequent chemistry step, and the free energy difference in chemical transition states between correct and incorrect base pairs dictates this polymerase fidelity. The crystal structures of replicative polymerases in the presence of DNA and nucleotides reveal tight active sites that are dissimilar to the active sites of DNA lesion bypass polymerases observed thus far. The loose and solvent-accessible active site and lack of fidelity checking structural elements, such as the O helix, form the structural basis for the bypass polymerases to traverse various DNA lesions. The central role of viral DNA polymerases in genome replication has led to intensive drug discovery, and antiviral nucleoside and pyrophosphate analogs have been developed.

References

1. Cann, I.K. and Y. Ishino. Archaeal DNA replication: identifying the pieces to solve a puzzle. *Genetics*, 1999, 152(4), 1249–1267.
2. Doublie, S., S. Tabor, A.M. Long, C.C. Richardson, and T. Ellenberger. Crystal structure of a bacteriophage T7 DNA replication complex at 2.2-Å resolution. *Nature*, 1998, 391(6664), 251–258.
3. Sawaya, M.R., R. Prasad, S.H. Wilson, J. Kraut, and H. Pelletier. Crystal structures of human DNA polymerase beta complexed with gapped and nicked DNA: evidence for an induced fit mechanism. *Biochemistry*, 1997, 36(37), 11205–11215.
4. Ling, H., F. Boudsocq, R. Woodgate, and W. Yang. Crystal structure of a Y-family DNA polymerase in action: a mechanism for error-prone and lesion-bypass replication. *Cell*, 2001, 107(1), 91–102.
5. Kool, E.T. Hydrogen bonding, base stacking, and steric effects in dna replication. *Annu. Rev. Biophys. Biomol. Struct.*, 2001, 30, 1–22.
6. Hubscher, U, G Maga, and S Spadari. Eukaryotic DNA polymerases. *Annu. Rev. Biochem.*, 2002, 71, 133–163.
7. Thomas, D.C., J.D. Roberts, R.D. Sabatino, T.W. Myers, C.K. Tan, K.M. Downey, A.G. So, R.A. Bambara, and T.A. Kunkel. Fidelity of mammalian DNA replication and replicative DNA polymerases. *Biochemistry*, 1991, 30(51), 11751–11759.
8. Pavlov, Y.I., P.V. Shcherbakova, and T.A. Kunkel. *In vivo* consequences of putative active site mutations in yeast DNA polymerases alpha, epsilon, delta, and zeta. *Genetics*, 2001, 159(1), 47–64.
9. Ng, L., C.K. Tan, K.M. Downey, and P.A. Fisher. Enzymologic mechanism of calf thymus DNA polymerase delta. *J. Biol. Chem.*, 1991, 266(18), 11699–11704.
10. Maga, G., I. Frouin, S. Spadari, and U. Hubscher. Replication protein A as a "fidelity clamp" for DNA polymerase alpha. *J. Biol. Chem.*, 2001, 276(21), 18235–18242.
11. Limsirichaikul, S., M. Ogawa, A. Niimi, S. Iwai, T. Murate, S. Yoshida, and M. Suzuki. The Gly-952 residue of *Saccharomyces cerevisiae* DNA polymerase alpha is important in discriminating correct deoxyribonucleotides from incorrect ones. *J. Biol. Chem.*, 2003, 278(21), 19079–19086.
12. Beard, W.A., D.D. Shock, B.J. Vande Berg, and S.H. Wilson. Efficiency of correct nucleotide insertion governs DNA polymerase fidelity. *J. Biol. Chem.*, 2002, 277(49), 47393–47398.
13. Lee, M.Y., C.K. Tan, K.M. Downey, and A.G. So. Further studies on calf thymus DNA polymerase delta purified to homogeneity by a new procedure. *Biochemistry*, 1984, 23(9), 1906–1913.
14. Hindges, R. and U. Hubscher. DNA polymerase delta, an essential enzyme for DNA transactions. *Biol. Chem.*, 1997, 378(5), 345–362.
15. Tan, C.K., C. Castillo, A.G. So, and K.M. Downey. An auxiliary protein for DNA polymerase delta from fetal calf thymus. *J. Biol. Chem.*, 1986, 261(26), 12310–12316.
16. Prelich, G., M. Kostura, D.R. Marshak, M.B. Mathews, and B. Stillman. The cell-cycle regulated proliferating cell nuclear antigen is required for SV40 DNA replication *in vitro*. *Nature*, 1987, 326(6112), 471–475.

17. Einolf, H.J. and F.P. Guengerich. Kinetic analysis of nucleotide incorporation by mammalian DNA polymerase delta. *J. Biol. Chem.*, 2000, 275(21), 16316–16322.
18. Patel, S.S., I. Wong, and K.A. Johnson. Pre-steady-state kinetic analysis of processive DNA replication including complete characterization of an exonuclease-deficient mutant. *Biochemistry*, 1991, 30(2), 511–525.
19. Johnson, A.A., Y. Tsai, S.W. Graves, and K.A. Johnson. Human mitochondrial DNA polymerase holoenzyme: reconstitution and characterization. *Biochemistry*, 2000, 39(7), 1702–1708.
20. Graves, S.W., A.A. Johnson, and K.A. Johnson. Expression, purification, and initial kinetic characterization of the large subunit of the human mitochondrial DNA polymerase. *Biochemistry*, 1998, 37(17), 6050–6058.
21. Chen, X., S. Zuo, Z. Kelman, M. O'Donnell, J. Hurwitz, and M.F. Goodman. Fidelity of eucaryotic DNA polymerase delta holoenzyme from *Schizosaccharomyces pombe*. *J. Biol. Chem.*, 2000, 275(23), 17677–17682.
22. Morrison, A. and A. Sugino. The 3′ → 5′ exonucleases of both DNA polymerases delta and epsilon participate in correcting errors of DNA replication in *Saccharomyces cerevisiae*. *Mol. Gen. Genet.*, 1994, 242(3), 289–296.
23. Chilkova, O., B.H. Jonsson, and E. Johansson. The quarternary structure of DNA polymerase epsilon from *S. cerevisiae*. *J. Biol. Chem.*, 2003.
24. Kesti, T., K. Flick, S. Keranen, J.E. Syvaoja, and C. Wittenberg. DNA polymerase epsilon catalytic domains are dispensable for DNA replication, DNA repair, and cell viability. *Mol. Cell*, 1999, 3(5), 679–685.
25. Burgers, P.M. *Saccharomyces cerevisiae* replication factor C. II. Formation and activity of complexes with the proliferating cell nuclear antigen and with DNA polymerases delta and epsilon. *J. Biol. Chem.*, 1991, 266(33), 22698–22706.
26. Carrodeguas, J.A., K. Theis, D.F. Bogenhagen, and C. Kisker. Crystal structure and deletion analysis show that the accessory subunit of mammalian DNA polymerase gamma, Pol gamma B, functions as a homodimer. *Mol. Cell*, 2001, 7(1), 43–54.
27. Johnson, A.A. and K.A. Johnson. Exonuclease proofreading by human mitochondrial DNA polymerase. *J. Biol. Chem.*, 2001, 276(41), 38097–38107.
28. Johnson, A.A. and K.A. Johnson. Fidelity of nucleotide incorporation by human mitochondrial DNA polymerase. *J. Biol. Chem.*, 2001, 276(41), 38090–38096.
29. Johnson, A.A., A.S. Ray, J. Hanes, Z. Suo, J.M. Colacino, K.S. Anderson, and K.A. Johnson. Toxicity of antiviral nucleoside analogs and the human mitochondrial DNA polymerase. *J. Biol. Chem.*, 2001, 276(44), 40847–40857.
30. Feng, J.Y., A.A. Johnson, K.A. Johnson, and K.S. Anderson. Insights into the molecular mechanism of mitochondrial toxicity by AIDS drugs. *J. Biol. Chem.*, 2001, 276(26), 23832–23837.
31. Friedberg, E.C., R. Wagner, and M. Radman. Specialized DNA polymerases, cellular survival, and the genesis of mutations. *Science*, 2002, 296(5573), 1627–1630.
32. Nelson, J.R., C.W. Lawrence, and D.C. Hinkle. Thymine–thymine dimer bypass by yeast DNA polymerase zeta. *Science*, 1996, 272(5268), 1646–1649.
33. Nelson, J.R., C.W. Lawrence, and D.C. Hinkle. Deoxycytidyl transferase activity of yeast REV1 protein. *Nature*, 1996, 382(6593), 729–731.

34. Lin, W., H. Xin, Y. Zhang, X. Wu, F. Yuan, and Z. Wang. The human REV1 gene codes for a DNA template-dependent dCMP transferase. *Nucleic Acids Res.*, 1999, 27(22), 4468–4475.
35. Nelson, J.R., P.E. Gibbs, A.M. Nowicka, D.C. Hinkle, and C.W. Lawrence. Evidence for a second function for *Saccharomyces cerevisiae* Rev1p. *Mol. Microbiol.*, 2000, 37(3), 549–554.
36. Lawrence, C.W. and V.M. Maher. Eukaryotic mutagenesis and translesion replication dependent on DNA polymerase zeta and Rev1 protein. *Biochem. Soc. Trans.*, 2001, 29(Pt 2), 187–191.
37. Johnson, R.E., M.T. Washington, L. Haracska, S. Prakash, and L. Prakash. Eukaryotic polymerases iota and zeta act sequentially to bypass DNA lesions. *Nature*, 2000, 406(6799), 1015–1019.
38. Johnson, R.E., S. Prakash, and L. Prakash. Efficient bypass of a thymine–thymine dimer by yeast DNA polymerase, pol iota. *Science*, 1999, 283(5404), 1001–1004.
39. Johnson, R.E., C.M. Kondratick, S. Prakash, and L. Prakash. hRAD30 mutations in the variant form of xeroderma pigmentosum. *Science*, 1999, 285(5425), 263–265.
40. Masutani, C., R. Kusumoto, S. Iwai, and F. Hanaoka. Mechanisms of accurate translesion synthesis by human DNA polymerase eta. *Embo J.*, 2000, 19(12), 3100–3109.
41. Wang, Z. Translesion synthesis by the UmuC family of DNA polymerases. *Mutat. Res.*, 2001, 486(2), 59–70.
42. Prakash, S. and L. Prakash. Translesion DNA synthesis in eukaryotes: a one- or two-polymerase affair. *Genes Dev.*, 2002, 16(15), 1872–1883.
43. Vaisman, A., C. Masutani, F. Hanaoka, and S.G. Chaney. Efficient translesion replication past oxaliplatin and cisplatin GpG adducts by human DNA polymerase eta. *Biochemistry*, 2000, 39(16), 4575–4580.
44. Washington, M.T., R.E. Johnson, S. Prakash, and L. Prakash. Fidelity and processivity of *Saccharomyces cerevisiae* DNA polymerase eta. *J. Biol. Chem.*, 1999, 274(52), 36835–36838.
45. Johnson, R.E., M.T. Washington, S. Prakash, and L. Prakash. Bridging the gap: a family of novel DNA polymerases that replicate faulty DNA. *Proc. Natl. Acad. Sci. USA*, 1999, 96(22), 12224–12226.
46. Haracska, L., S.L. Yu, R.E. Johnson, L. Prakash, and S. Prakash. Efficient and accurate replication in the presence of 7,8-dihydro-8-oxoguanine by DNA polymerase eta. *Nat. Genet.*, 2000, 25(4), 458–461.
47. Johnson, R.E., M.T. Washington, S. Prakash, and L. Prakash. Fidelity of human DNA polymerase eta. *J. Biol. Chem.*, 2000, 275(11), 7447–7450.
48. Matsuda, T., K. Bebenek, C. Masutani, F. Hanaoka, and T.A. Kunkel. Low fidelity DNA synthesis by human DNA polymerase eta. *Nature*, 2000, 404(6781), 1011–1013.
49. Livneh, Z. DNA damage control by novel DNA polymerases: translesion replication and mutagenesis. *J. Biol. Chem.*, 2001, 276(28), 25639–25642.
50. Tissier, A., J.P. McDonald, E.G. Frank, and R. Woodgate. pol iota, a remarkably error-prone human DNA polymerase. *Genes Dev.*, 2000, 14(13), 1642–1650.
51. Zhang, Y., F. Yuan, X. Wu, and Z. Wang. Preferential incorporation of G opposite template T by the low-fidelity human DNA polymerase iota. *Mol. Cell Biol.*, 2000, 20(19), 7099–7108.

52. Tissier, A., E.G. Frank, J.P. McDonald, A. Vaisman, A.R. Fernandez de Henestrosa, F. Boudsocq, M.P. McLenigan, and R. Woodgate. Biochemical characterization of human DNA polymerase iota provides clues to its biological function. *Biochem. Soc. Trans.*, 2001, 29(part 2), 183–187.
53. Poltoratsky, V., C.J. Woo, B. Tippin, A. Martin, M.F. Goodman, and M.D. Scharff. Expression of error-prone polymerases in BL2 cells activated for Ig somatic hypermutation. *Proc. Natl. Acad. Sci. USA*, 2001, 98(14), 7976–7981.
54. Frank, E.G., A. Tissier, J.P. McDonald, V. Rapic–Otrin, X. Zeng, P.J. Gearhart, and R. Woodgate. Altered nucleotide misinsertion fidelity associated with pol iota-dependent replication at the end of a DNA template. *Embo J.*, 2001, 20(11), 2914–2922.
55. Tissier, A., E.G. Frank, J.P. McDonald, S. Iwai, F. Hanaoka, and R. Woodgate. Misinsertion and bypass of thymine–thymine dimers by human DNA polymerase iota. *Embo J.*, 2000, 19(19), 5259–5266.
56. Woodgate, R. Evolution of the two-step model for UV-mutagenesis. *Mutat. Res.*, 2001, 485(1), 83–92.
57. Zhang, Y., F. Yuan, H. Xin, X. Wu, D.K. Rajpal, D. Yang, and Z. Wang. Human DNA polymerase kappa synthesizes DNA with extraordinarily low fidelity. *Nucleic Acids Res.*, 2000, 28(21), 4147–4156.
58. Ohashi, E., T. Ogi, R. Kusumoto, S. Iwai, C. Masutani, F. Hanaoka, and H. Ohmori. Error-prone bypass of certain DNA lesions by the human DNA polymerase kappa. *Genes Dev.*, 2000, 14(13), 1589–1594.
59. Velasco–Miguel, S., J.A. Richardson, V.L. Gerlach, W.C. Lai, T. Gao, L.D. Russell, C.L. Hladik, C.L. White, and E.C. Friedberg. Constitutive and regulated expression of the mouse Dinb (Pol kappa) gene encoding DNA polymerase kappa. *DNA Repair* (Amst), 2003, 2(1), 91–106.
60. Morales, J.C. and E.T. Kool. Varied molecular interactions at the active sites of several DNA polymerases: nonpolar nucleoside isosteres as probes. *J. Am. Chem. Soc.*, 2000, 122(6), 1001–1007.
61. Johnson, K.A. Conformational coupling in DNA polymerase fidelity. *Annu. Rev. Biochem.*, 1993, 62, 685–713.
62. Echols, H. and M.F. Goodman. Fidelity mechanisms in DNA replication. *Annu. Rev. Biochem.*, 1991, 60, 477–511.
63. Sloane, D.L., M.F. Goodman, and H. Echols. The fidelity of base selection by the polymerase subunit of DNA polymerase III holoenzyme. *Nucleic Acids Res.*, 1988, 16(14A), 6465–6475.
64. Aboul–Ela, F., D. Koh, I. Tinoco, Jr., and F.H. Martin. Base–base mismatches. Thermodynamics of double helix formation for dCA3XA3G + dCT3YT3G (X, Y = A,C,G,T). *Nucleic Acids Res.*, 1985, 13(13), 4811–4824.
65. Goodman, M.F. Hydrogen bonding revisited: geometric selection as a principal determinant of DNA replication fidelity. *Proc. Natl. Acad. Sci. USA*, 1997, 94(20), 10493–10495.
66. Bommarito, S., N. Peyret, and J. SantaLucia, Jr. Thermodynamic parameters for DNA sequences with dangling ends. *Nucleic Acids Res.*, 2000, 28(9), 1929–1934.
67. Petersheim, M. and D.H. Turner. Base-stacking and base-pairing contributions to helix stability: thermodynamics of double-helix formation with CCGG, CCGGp, CCGGAp, ACCGGp, CCGGUp, and ACCGGUp. *Biochemistry*, 1983, 22(2), 256–263.

68. Kiefer, J.R., C. Mao, J.C. Braman, and L.S. Beese. Visualizing DNA replication in a catalytically active *Bacillus* DNA polymerase crystal. *Nature,* 1998, 391(6664), 304–307.
69. Eom, S.H., J. Wang, and T.A. Steitz. Structure of Taq ploymerase with DNA at the polymerase active site. *Nature,* 1996, 382(6588), 278–281.
70. Huang, H., R. Chopra, G.L. Verdine, and S.C. Harrison. Structure of a covalently trapped catalytic complex of HIV-1 reverse transcriptase: implications for drug resistance. *Science,* 1998, 282(5394), 1669–1675.
71. Carroll, S.S., M. Cowart, and S.J. Benkovic. A mutant of DNA polymerase I (Klenow fragment) with reduced fidelity. *Biochemistry,* 1991, 30(3), 804–813.
72. Bell, J.B., K.A. Eckert, C.M. Joyce, and T.A. Kunkel. Base miscoding and strand misalignment errors by mutator Klenow polymerases with amino acid substitutions at tyrosine 766 in the O helix of the fingers subdomain. *J. Biol. Chem.,* 1997, 272(11), 7345–7351.
73. Loeb, L.A. and T.A. Kunkel. Fidelity of DNA synthesis. *Annu. Rev. Biochem.,* 1982, 51, 429–457.
74. Kunkel, T.A. and K. Bebenek. Recent studies of the fidelity of DNA synthesis. *Biochim. Biophys. Acta,* 1988, 951(1), 1–15.
75. Kunkel, T.A. and K. Bebenek. DNA replication fidelity. *Annu. Rev. Biochem.,* 2000, 69, 497–529.
76. Wong, I., S.S. Patel, and K.A. Johnson. An induced-fit kinetic mechanism for DNA replication fidelity: direct measurement by single-turnover kinetics. *Biochemistry,* 1991, 30(2), 526–537.
77. Kati, W.M., K.A. Johnson, L.F. Jerva, and K.S. Anderson. Mechanism and fidelity of HIV reverse transcriptase. *J. Biol. Chem.,* 1992, 267(36), 25988–25997.
78. Eger, B.T., R.D. Kuchta, S.S. Carroll, P.A. Benkovic, M.E. Dahlberg, C.M. Joyce, and S.J. Benkovic. Mechanism of DNA replication fidelity for three mutants of DNA polymerase I: Klenow fragment KF(exo+), KF(polA5), and KF(exo-). *Biochemistry,* 1991, 30(5), 1441–1448.
79. Kuchta, R.D., V. Mizrahi, P.A. Benkovic, K.A. Johnson, and S.J. Benkovic. Kinetic mechanism of DNA polymerase I (Klenow). *Biochemistry,* 1987, 26(25), 8410–8417.
80. Showalter, A.K. and M.D. Tsai. A reexamination of the nucleotide incorporation fidelity of DNA polymerases. *Biochemistry,* 2002, 41(34), 10571–10576.
81. Herschlag, D., J.A. Piccirilli, and T.R. Cech. Ribozyme-catalyzed and nonenzymatic reactions of phosphate diesters: rate effects upon substitution of sulfur for a nonbridging phosphoryl oxygen atom. *Biochemistry,* 1991, 30(20), 4844–4854.
82. Li, Y., S. Korolev, and G. Waksman. Crystal structures of open and closed forms of binary and ternary complexes of the large fragment of *Thermus aquaticus* DNA polymerase I: structural basis for nucleotide incorporation. *Embo J.,* 1998, 17(24), 7514–7525.
83. Sarafianos, S.G., A.D. Clark, Jr., K. Das, S. Tuske, J.J. Birktoft, P. Ilankumaran, A.R. Ramesha, J.M. Sayer, D.M. Jerina, P.L. Boyer, S.H. Hughes, and E. Arnold. Structures of HIV–1 reverse transcriptase with pre- and post-translocation AZTMP-terminated DNA. *Embo J.,* 2002, 21(23), 6614–6624.
84. Zhong, X., S.S. Patel, and M.D. Tsai. DNA polymerase beta. 5. Dissecting the functional roles of the Twl metal ions with Cr(III)dTTP. *J. Am. Chem. Soc.,* 1998, 120, 235–236.

85. Arndt, J.W., W. Gong, X. Zhong, A.K. Showalter, J. Liu, C.A. Dunlap, Z. Lin, C. Paxson, M.D. Tsai, and M.K. Chan. Insight into the catalytic mechanism of DNA polymerase beta: structures of intermediate complexes. *Biochemistry*, 2001, 40(18), 5368–5375.
86. Ding, J., K. Das, Y. Hsiou, S.G. Sarafianos, A.D. Clark, Jr., A. Jacobo–Molina, C. Tantillo, S.H. Hughes, and E. Arnold. Structure and functional implications of the polymerase active site region in a complex of HIV-1 RT with a double-stranded DNA template-primer and an antibody Fab fragment at 2.8-A resolution. *J. Mol. Biol.*, 1998, 284(4), 1095–1111.
87. Hansen, J.L., A.M. Long, and S.C. Schultz. Structure of the RNA-dependent RNA polymerase of poliovirus. *Structure*, 1997, 5(8), 1109–1122.
88. Burgers, P.M. and F. Eckstein. A study of the mechanism of DNA polymerase I from *Escherichia coli* with diastereomeric phosphorothioate analogs of deoxyadenosine triphosphate. *J. Biol. Chem.*, 1979, 254(15), 6889–6893.
89. Beese, L.S. and T.A. Steitz. Structural basis for the 3′-5′ exonuclease activity of *Escherichia coli* DNA polymerase I: a two metal ion mechanism. *Embo J.*, 1991, 10(1), 25–33.
90. Brautigam, C.A., S. Sun., J.A. Piccirilli, and T.A. Steitz. Structures of normal single-stranded DNA and deoxyribo-3′-S-phosphorothiolates bound to the 3′-5′ exonucleolytic active site of DNA polymerase I from *Escherichia coli*. *Biochemistry*, 1999, 38(2), 696–704.
91. Steitz, T.A. A mechanism for all polymerases. *Nature*, 1998, 391(6664), 231–232.
92. Pelletier, H., M.R. Sawaya, A. Kumar, S.H. Wilson, and J. Kraut. Structures of ternary complexes of rat DNA polymerase beta, a DNA template-primer, and ddCTP. *Science*, 1994, 264(5167), 1891–1903.
93. Florian, J., M.F. Goodman, and A. Warshel. Computer simulation studies of the fidelity of DNA polymerases. *Biopolymers*, 2003, 68(3), 286–299.
94. Patel, P.H., A. Jacobo–Molina, J. Ding, C. Tantillo, A.D. Clark, Jr., R. Raag, R.G. Nanni, S.H. Hughes, and E. Arnold. Insights into DNA polymerization mechanisms from structure and function analysis of HIV-1 reverse transcriptase. *Biochemistry*, 1995, 34(16), 5351–5363.
95. Shevelev, I.V. and U. Hubscher. The 3′-5′ exonucleases. *Nat. Rev. Mol. Cell Biol.*, 2002, 3(5), 364–376.
96. Fersht, A.R., J.W. Knill–Jones, and W.C. Tsui. Kinetic basis of spontaneous mutation. Misinsertion frequencies, proofreading specificities and cost of proofreading by DNA polymerases of *Escherichia coli*. *J. Mol. Biol.*, 1982, 156(1), 37–51.
97. Joyce, C.M. How DNA travels between the separate polymerase and 3′-5′-exonuclease sites of DNA polymerase I (Klenow fragment). *J. Biol. Chem.*, 1989, 264(18), 10858–10866.
98. Donlin, M.J., S.S. Patel, and K.A. Johnson. Kinetic partitioning between the exonuclease and polymerase sites in DNA error correction. *Biochemistry*, 1991, 30(2), 538–546.
99. Steitz, T.A., L. Beese, P.S. Freemont, J.M. Friedman, and M.R. Sanderson. Structural studies of Klenow fragment: an enzyme with two active sites. *Cold Spring Harbor Symp. Quant. Biol.*, 1987, 52, 465–471.
100. Ollis, D.L., P. Brick, R. Hamlin, N.G. Xuong, and T.A. Steitz. Structure of large fragment of *Escherichia coli* DNA polymerase I complexed with dTMP. *Nature*, 1985, 313(6005), 762–766.

101. Catalano, C.E. and S.J. Benkovic. Inactivation of DNA polymerase I (Klenow fragment) by adenosine 2′,3′-epoxide 5′-triphosphate: evidence for the formation of a tight-binding inhibitor. *Biochemistry,* 1989, 28(10), 4374–4382.
102. Lin, T.C., C.X. Wang, C.M. Joyce, and W.H. Konigsberg. 3′-5′ Exonucleolytic activity of DNA polymerases: structural features that allow kinetic discrimination between ribo- and deoxyribonucleotide residues. *Biochemistry,* 2001, 40(30), 8749–8755.
103. Freemont, P.S., J.M. Friedman, L.S. Beese, M.R. Sanderson, and T.A. Steitz. Cocrystal structure of an editing complex of Klenow fragment with DNA. *Proc. Natl. Acad. Sci. USA,* 1988, 85(23), 8924–8928.
104. Gupta, A.P. and S.J. Benkovic. Stereochemical course of the 3′-5′-exonuclease activity of DNA polymerase I. *Biochemistry,* 1984, 23(24), 5874–5881.
105. Seeman, N.C., J.M. Rosenberg, and A. Rich. Sequence-specific recognition of double helical nucleic acids by proteins. *Proc. Natl. Acad. Sci. USA,* 1976, 73(3), 804–808.
106. Hunter, W.N., T. Brown, N.N. Anand, and O. Kennard. Structure of an adenine–cytosine base pair in DNA and its implications for mismatch repair. *Nature,* 1986, 320(6062), 552–555.
107. Beard, W.A., W.P. Osheroff, R. Prasad, M.R. Sawaya, M. Jaju, T.G. Wood, J. Kraut, T.A. Kunkel, and S.H. Wilson. Enzyme–DNA interactions required for efficient nucleotide incorporation and discrimination in human DNA polymerase beta. *J. Biol. Chem.,* 1996, 271(21), 12141–12144.
108. Spratt, T.E. Identification of hydrogen bonds between *Escherichia coli* DNA polymerase I (Klenow fragment) and the minor groove of DNA by amino acid substitution of the polymerase and atomic substitution of the DNA. *Biochemistry,* 2001, 40(9), 2647–2652.
109. Joyce, C.M. Choosing the right sugar: how polymerases select a nucleotide substrate. *Proc. Natl. Acad. Sci. USA,* 1997, 94(5), 1619–1622.
110. Tabor, S. and C.C. Richardson. A single residue in DNA polymerases of the *Escherichia coli* DNA polymerase I family is critical for distinguishing between deoxy- and dideoxyribonucleotides. *Proc. Natl. Acad. Sci. USA,* 1995, 92(14), 6339–6343.
111. Sarafianos, S.G., K. Das, J. Ding, P.L. Boyer, S.H. Hughes, and E. Arnold. Touching the heart of HIV-1 drug resistance: the fingers close down on the dNTP at the polymerase active site. *Chem. Biol.,* 1999, 6(5), R137–146.
112. Kohlstaedt, L.A., J. Wang, J.M. Friedman, P.A. Rice, and T.A. Steitz. Crystal structure at 3.5-Å resolution of HIV-1 reverse transcriptase complexed with an inhibitor. *Science,* 1992, 256(5065), 1783–1790.
113. Gao, H.Q., P.L. Boyer, S.G. Sarafianos, E. Arnold, and S.H. Hughes. The role of steric hindrance in 3TC resistance of human immunodeficiency virus type-1 reverse transcriptase. *J. Mol. Biol.,* 2000, 300(2), 403–418.
114. Zhou, B.L., J.D. Pata, and T.A. Steitz. Crystal structure of a DinB lesion bypass DNA polymerase catalytic fragment reveals a classic polymerase catalytic domain. *Mol. Cell,* 2001, 8(2), 427–437.
115. Silvian, L.F., E.A. Toth, P. Pham, M.F. Goodman, and T. Ellenberger. Crystal structure of a DinB family error-prone DNA polymerase from *Sulfolobus solfataricus. Nat. Struct. Biol.,* 2001, 8(11), 984–989.

116. Trincao, J., R.E. Johnson, C.R. Escalante, S. Prakash, L. Prakash, and A.K. Aggarwal. Structure of the catalytic core of *S. cerevisiae* DNA polymerase eta: implications for translesion DNA synthesis. *Mol. Cell,* 2001, 8(2), 417–426.
117. Coen, D.M. Viral DNA polymerases in *DNA Replication in Eukaryotic Cells,* M.L. Depamphilis, Ed. 1996, Cold Spring Harbor Laboratory Press: Cold Spring Harbor, NY, 495– 523.

chapter seven

RNA polymerases

Wissam Abdel–Gawad
The Biophysics Program, The Ohio State University, Columbus, OH

Seng-Lai Tan
Lilly Research Laboratories, Eli Lilly & Company, Indianapolis, IN

Zucai Suo*
Department of Biochemistry, The Ohio State University, Columbus, OH

Contents

* Corresponding author.

7.1 *Introduction*

Gene transcription is a highly coordinated process catalyzed by multisubunit RNA polymerases. Although the fundamental principles of transcription are conserved among prokaryotes and eukaryotes, the latter utilizes much more complex mechanisms. Unlike those of prokaryotes, eukaryotic monocistronic mRNA transcripts contain relatively complex regulatory sequences and are localized in different subcellular compartments and subject to processing to yield mature transcripts.

Three DNA-dependent RNA polymerases are responsible for the synthesis of eukaryotic mRNAs. The most important is RNA polymerase II (RNA Pol II), which is involved in the transcription of all protein genes and most small nuclear (sn) RNA genes. RNA Pol II from a typical eukaryotic cell contains up to 15 polypeptide subunits and has a molecular mass up to 600 kDa. Despite the complexity and huge size of eukaryotic RNA polymerases, important insights into various aspects of the transcription mechanisms have been gained from comparison of the three-dimensional structure of a bacterial and a yeast RNA polymerase bound to nucleic acids.

Interestingly, most viral RNA polymerases are characterized as single subunit DNA-dependent or RNA-dependent RNA polymerases. Viral DNA-dependent RNA polymerases typically comprise one subunit and do not share an apparent sequence or structural homology with multisubunit DNA-dependent RNA polymerases.

This is not the case with viral RNA-dependent RNA polymerases (RdRp), which share functional similarities with eukaryotic DNA-dependent RNA polymerases and carry out replication of viral RNA genomes and transcription into mRNA. Limited by viral genetic coding capacity, viral RdRps inevitably act in concert with other viral and host factors. Thus, a better understanding of the structure, function, and regulation of the viral RdRp relative to eukaryotic DNA-dependent RNA polymerases, as well as of their interactions with other viral and host factors, is essential to the development of viral RdRp-specific inhibitors as potential antiviral therapeutic agents. In this chapter, we will first introduce typical examples of DNA-dependent RNA polymerases and RNA-dependent RNA polymerases and then summarize the mechanistic, structural, and biological features of the RNA polymerases.

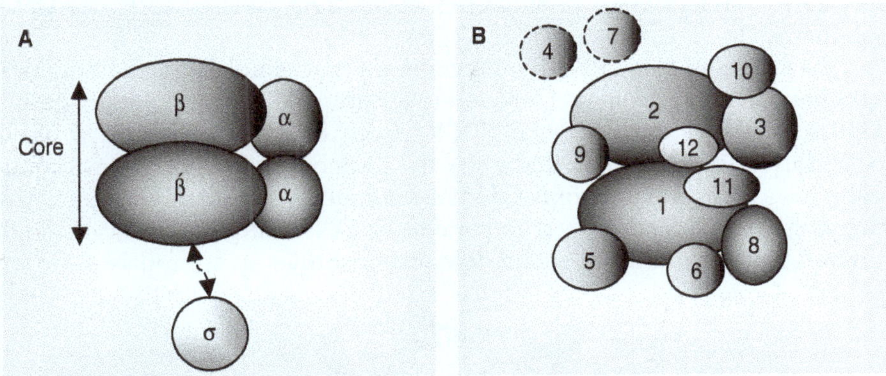

Figure 7.1 Schematic architecture of RNA polymerase subunits: (A) in *E. coli* and (B) RNA Pol II in yeast. The crystal structure of RNA Pol II from yeast lacks subunits 4 and 7 indispensable for transcription, (see Figure 7.3 for crystal structure). The arrow in (A) indicates that the σ subunit is dissociable. (Drawn by the authors using Microsoft Word drawing tools.)

7.2 DNA-dependent RNA polymerases

7.2.1 E. coli DNA-dependent RNA polymerase

An *E. coli* DNA-dependent RNA polymerase comprises five subunits: two α subunits and one for each β, β′, and σ subunit. Beta (151 kDa) and β′ (156 kDa) are significantly larger than the α (37 kDa) (Figure 7.1A). Seven different forms of σ subunits, also known as the σ factors, have been identified: σ^{70}, σ^{N}, σ^{S}, σ^{H}, σ^{F}, σ^{E}, and σ^{FecI}, with molecular weights ranging from 28 to 70 kDa. The holoenzyme-containing subunit, σ^{70} ($E\sigma^{70}$), is responsible for transcription of the most housekeeping genes during the exponential growth phase. Some specific genes expressed during the stationary phase are transcribed by $E\sigma^{S}$.[1]

The σ subunit acts as a specificity factor and plays an important role in recognizing the transcriptional initiation site; it also possesses helicase activity that unwinds the DNA. Nucleotide synthesis is catalyzed by the four subunits comprising two α subunits, β, and β′ subunit, collectively known as the core polymerase. The term "holoenzyme" refers to a complete and fully functional enzyme, which includes the core polymerase and the σ factor.

7.2.2 Viral DNA-dependent RNA polymerases

Bacteriophage T7 RNA polymerase (T7 RNAP) is an ideal case study of the DNA-dependent RNA polymerase family due to the solution of several T7 RNAP structures that led to significant understanding of the functional aspects of its transcription.[2–7] T7 RNAP consists of one subunit and does not share an apparent sequence or structural homology with multisubunit DNA-dependent RNA polymerases (in bacteria and eukaryotes). However, T7 RNAP shares structural homology to mitochondrial and chloroplast RNA polymerase and to DNA polymerases of family A (related to *E. coli* DNA polymerase I).

The structure of T7 RNAP morphologically resembles the anatomy of a right hand with subdomains referred to as "thumb," "fingers," and "palm" (Figure 7.2 shows the T7 RNAP/DNA "open" complex[4]). The active site is located in a deep pocket bounded by the thumb, fingers, and palm subdomains, as well as the N-terminal domain common to RNA polymerases. The fingers domain offers contacts to handle the DNA template strand through its DNA binding motifs; the palm domain constitutes an appropriate surface

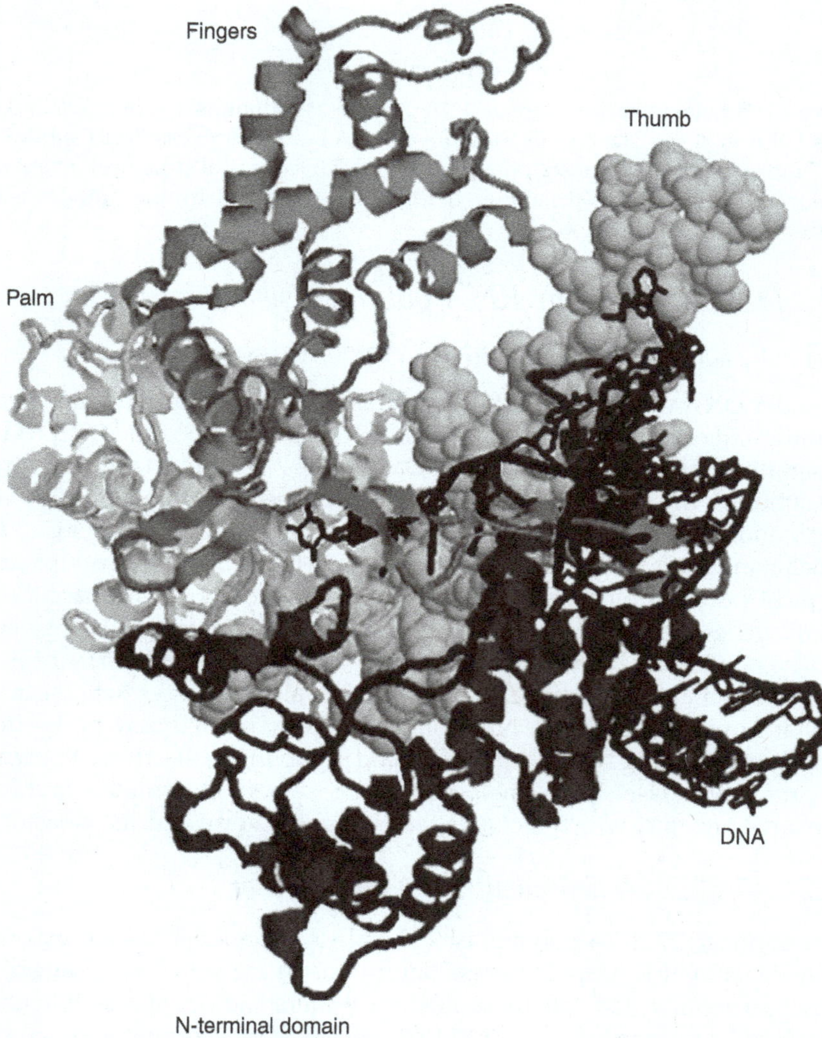

Figure 7.2 Crystal structure of T7 RNAP/DNA open complex (PDB 1CEZ [Cheetham, G.M. et al., *Nature*, 399, 80–83, 1999] Protein Explorer).

to accommodate DNA, incoming ribonucleotide triphosphate (rNTP), and two catalytic Mg^{2+} ions. The thumb domain stabilizes the elongation complex and the N-terminal domain plays an important role in DNA recognition, formation of the transcription bubble, and transcription cycling via conformational change.

7.2.3 Eukaryotic DNA-dependent RNA polymerases

Unlike prokaryotes, eukaryotes have several RNA polymerases: a mitochondrial polymerase (and a chloroplast polymerase in plants) and three nuclear polymerases. We will focus on the nuclear RNA polymerases, which consist of three distinct classes: I, II, and III. RNA Pol I is localized in the nucleolus and is responsible for the synthesis of a transcript called preribosomal RNA that contains the precursor for rRNA genes (except 5S rRNA). RNA Pol II, which is present in the nucleoplasm, is involved in the transcription of all protein genes and most snRNA genes. RNA Pol III is also localized in the nucleoplasm — specifically, outside the nucleolus — and functions to transcribe 5S rRNA, tRNA, U6 snRNA, and some small RNA precursors.

In yeast, each DNA-dependent RNA polymerase consists of two large subunits and 12 smaller subunits encoded by the *RPB1* to *RPB12* genes (Figure 7.1B and Figure 7.3).[8,9] Extensive structural conservation takes place among the subunits of eukaryotic RNA Pol II as evidenced by six subunits of human RNA Pol II that can functionally replace their yeast counterparts.[10] The Rpb1, Rpb2, Rpb3, and Rpb11 subunits of RNA Pol II are homologous to subunits of RNA Pol I and III. Rpb5, Rpb6, Rpb8, Rpb10, and Rpb12 are common to all three RNA polymerases. Only Rpb4, Rpb7, and Rpb9 are unique to RNA Pol II. Thus, RNA polymerases are assembled from common as well as class-specific subunits. Ten of the yeast genes encoding RNA Pol II subunits are essential for cell viability, whereas deletion of the *RPB4* or the *RPB9* gene confers conditional growth phenotypes. Mutations in *RPB9* apparently affect start site selection.

The two largest RNA Pol II subunits, Rpb1 (~200 kDa) and Rpb2 (~150 kDa), are the most highly conserved subunits and are homologous to the *E. coli* subunit β′ and β subunits, respectively. Rpb1 has an unusual structural feature not found in prokaryotes, specifically a long C-terminal "tail" known as the C-terminal domain (CTD; discussed later). Not surprisingly, the structural similarity between Rpb1/β′ and Rpb2/β extends to functional similarity. Rpb1 and β′ are involved in DNA binding, whereas Rpb2 and β bind nucleotide substrates. Rpb1 and Rpb2 participate in the catalytic site of the RNA polymerase.

Rpb3 (45 kDa) is related to the α subunit of bacterial RNA polymerase based on partial amino acid sequence similarity, size similarity, identical subunit stoichiometry (two per molecule), and assembly defects associated with mutations in either subunit.[11] Limited mutational analysis revealed that Rpb3 is involved in RNA Pol II assembly.[12] Rpd4 (32 kDa) also shares sig-

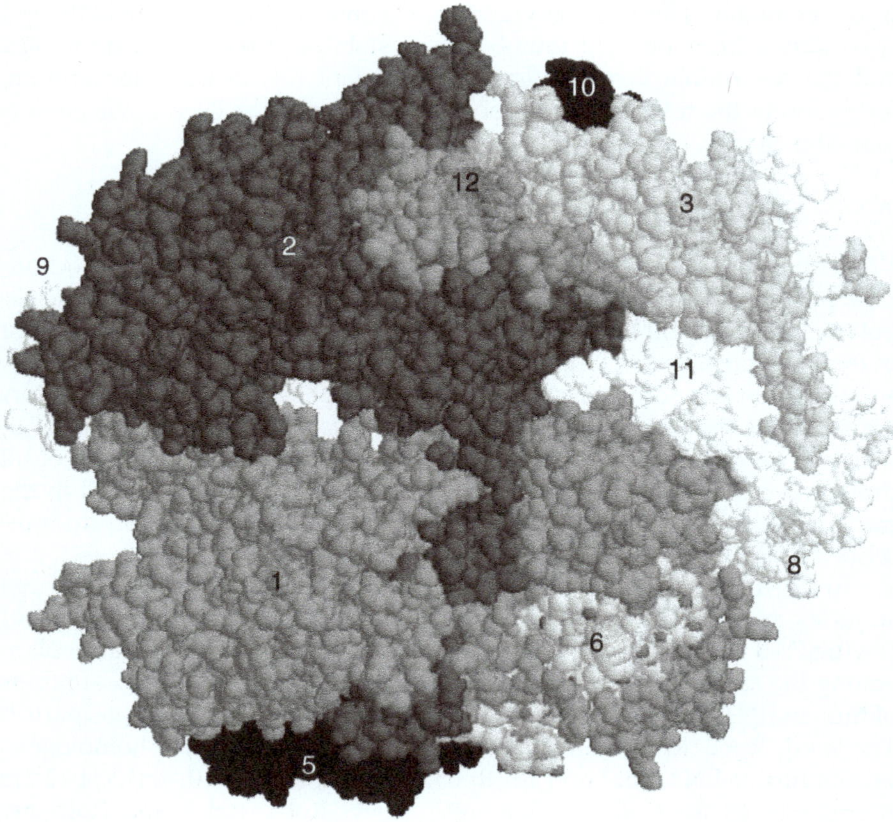

Figure 7.3 Crystal structure of the yeast RNA Pol II (PDB 1I5O [Cramer, P. et al., *Science*, 292, 1863–1876, 2001] Protein Explorer). Subunits are numbered from 1 to 12. The structure lacks subunits 4 and 7.

nificant sequence similarity with the σ factor of *E. coli* and thus is thought to be involved with promoter recognition.

Rpb7 is functionally related to Rpb4 and both subunits can be dissociated from RNA Pol II. RNA Pol II purified from an *rpb4* null mutant lacks Rpb7. This form of RNA Pol II is indistinguishable from wild-type RNA Pol II in an *in-vitro* elongation assay but is inactive in promoter-directed transcription initiation. Furthermore, an inactive RNA Pol II with a defective form of Rpb1 could complement the Rpb7-less form of RNA Pol II *in vitro*. These results demonstrate that Rpb4 and Rpb7 function in transcription initiation and suggest that they can shuttle between RNA Pol II molecules.

The 2.8-Å crystal structure of yeast RNA Pol II alone[13] and of the 3.3-Å crystal structure of the elongation complex have been solved by Kornberg and coworkers.[14] The crystal structure of yeast RNA Pol II lacks the critical subunits Rbp4 and Rbp7; both subunits have not been amenable to crystallization. The CTD structure is disordered and not observed in the molecular coordinates.

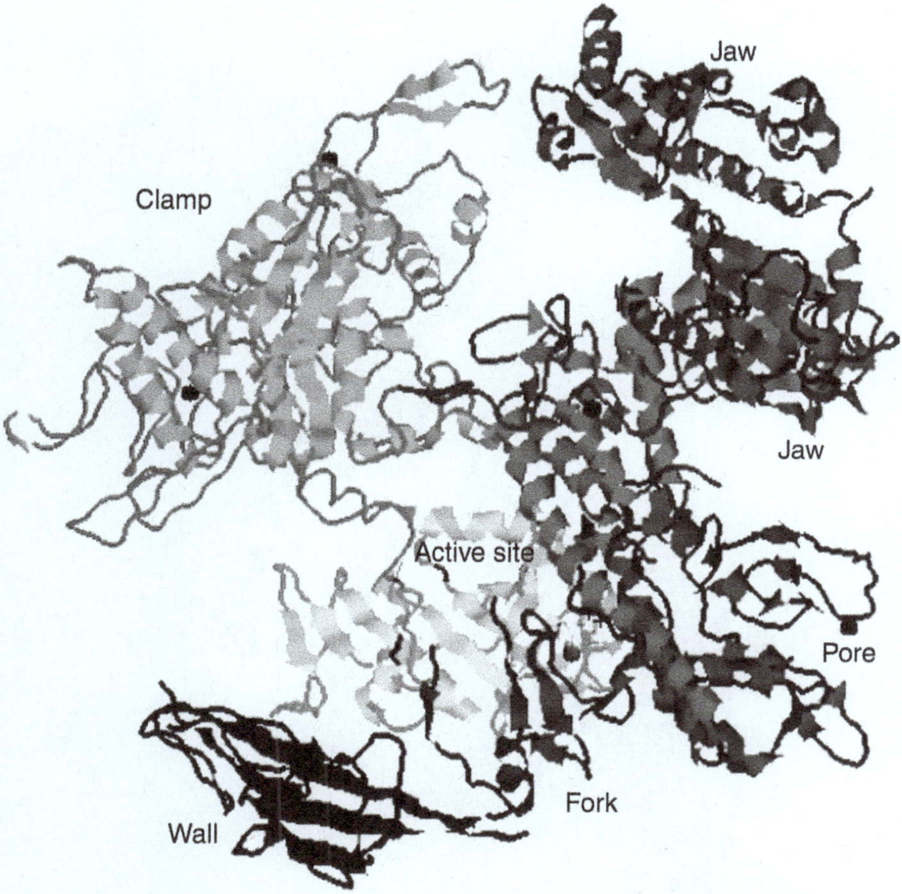

Figure 7.4 Crystal structure of the yeast RNA Pol II (PDB 1I50 [Cramer, P. et al., *Science*, 292, 1863–1876, 2001] Protein Explorer).

The general structural features of yeast RNA Pol II are displayed in Figure 7.4. Three border regions of a large cleft exist and are termed "clamp," "jaws," and "wall." A bridging helix is located in Rbp1 (residues 831 to 843) and separates the active site and the next downstream base in the template DNA strand. Near the bridging helix a pore traverses the molecule with the active site magnesium situated in the pore. The pore is the only entry for rNTP substrates because the RNA/DNA hybrid blocks entry of nucleotides from above the elongation complex.

The clamp domain plays crucial roles in the process of transcription. Figure 7.5 and Figure 7.6 show the clamp domain in two conformations: "open" in the case of enzyme structure[13] and "closed" in the elongation complex structure,[14] respectively. The closing of the clamp in the elongation complex involves a shift of some amino acid residues by as much as 30 Å. As a result of clamp closure, downstream DNA and RNA/DNA hybrid in

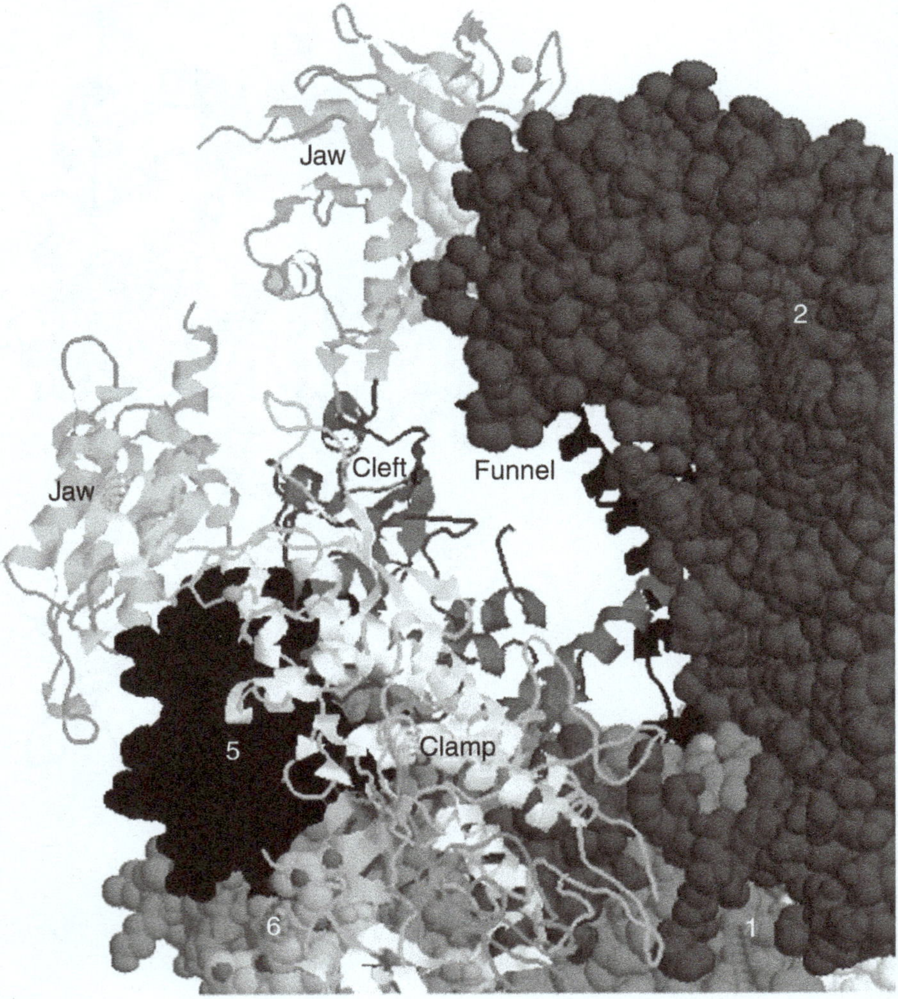

Figure 7.5 Crystal structure of the yeast RNA Pol II (PDB 1I50 [Cramer, P. et al., *Science*, 292, 1863–1876, 2001] Protein Explorer) showing the clamp domain in the open conformation.

the elongation complex are unable to dissociate from the enzyme molecule, leading to high processivity. The clamp therefore may function as a built-in processivity element.

Evidence from molecular modeling also indicates that the clamp domain plays important roles in DNA strand separation and as the DNA exit domain formed by two structural features: the rudder and the lid loops.[15] The rudder, lid, and zipper are three loops that protrude from the clamp domain. The rudder loop may separate RNA from DNA at the upstream end of the hybrid. The lid and zipper loops might maintain the upstream end of the transcription bubble (Figure 7.7).

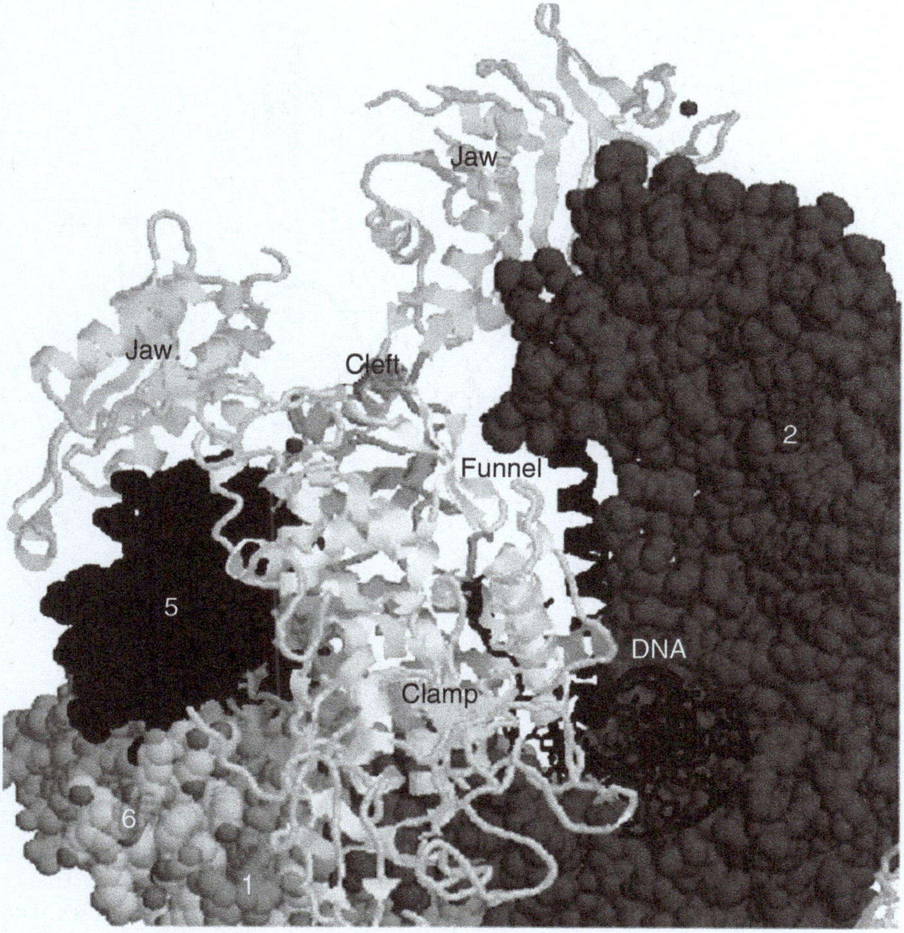

Figure 7.6 Crystal structure of the yeast RNA Pol II elongation complex (PDB 1I6H [Gnatt, A.L. et al., *Science*, 292, 1876–1882, 2001] Protein Explorer) showing the clamp domain in the closed conformation.

The jaw domain consists of subunits Rpb1 and Rpb9. This domain is suggested to play an important role in binding and stabilizing the position of downstream DNA. Mobility of the jaws would allow them to open or close on the downstream DNA, providing DNA with higher degrees of freedom for movement and increasing the B-factors of DNA/RNA hybrid. The wall border is involved in direct binding of nucleic acid. Additionally, a structural feature, referred to as the "funnel," is located below the active site and serves with the pore as an entry for the rNTP (Figure 7.7).

7.2.4 Mechanism of DNA-dependent RNA polymerases

Synthesis of RNA exhibits several features synonymous with DNA replication. For example, RNA synthesis requires accurate and efficient initiation

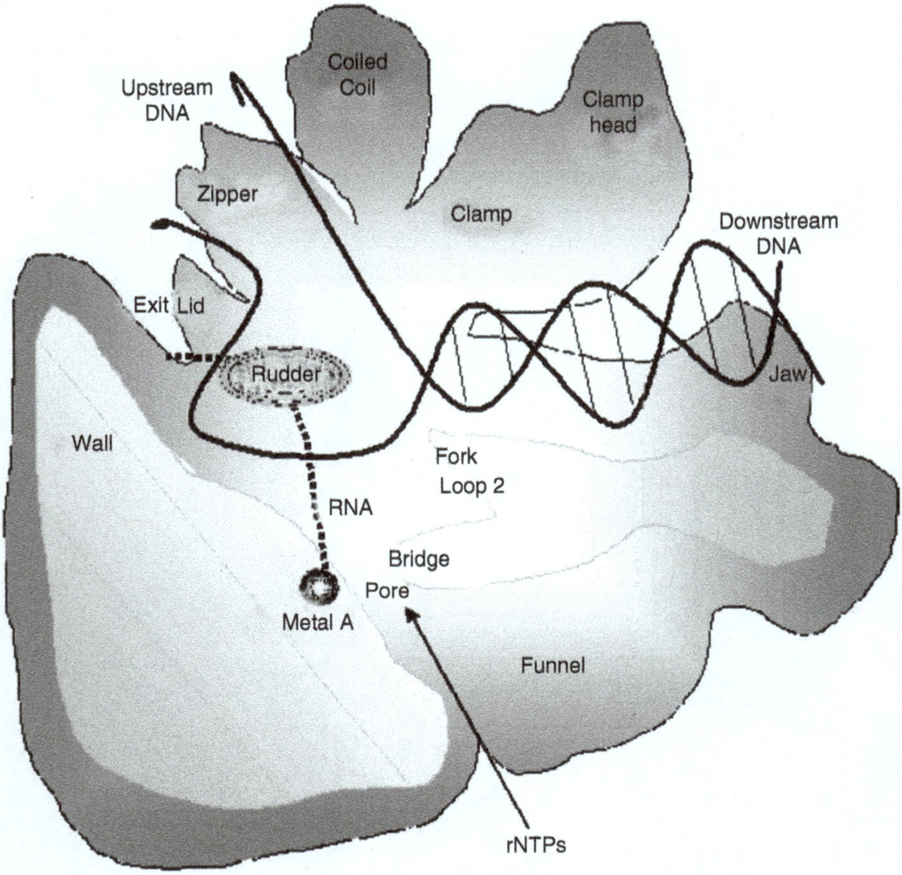

Figure 7.7 Schematic architecture of the yeast RNA Pol II elongation complex. The RNA transcript is shown in a dashed curve. The solvent-exposed stretches of DNA template and RNA transcript were disordered in the crystal structure. (Redrawn from Cramer, P., *Curr. Opin. Struct. Biol.*, 12, 89–97, 2002, by the authors using Microsoft Word drawing tools).

and elongation proceeds in the 5′ → 3′ direction (i.e., the polymerase moves along the template strand of DNA in the 3′ → 5′ direction). The RNA 3′-hydroxyl group acts as a nucleophile, attacking the α-phosphate of the incoming ribonucleoside triphosphate and releasing pyrophosphate. In incorporating the correct nucleotides, the polymerases follow Watson–Crick base pairing requirements. However, transcription exhibits several features distinct from DNA replication. Unlike DNA replication, transcription does not require a primer to initiate RNA synthesis. Additionally, transcription initiates in prokaryotes and eukaryotes from many more sites than DNA replication and occurs when RNA polymerases recognize specific DNA sequences called "promoters." Furthermore, RNA synthesis requires distinct and accurate termination.

RNA polymerases proceed at a rate much slower than the replicative DNA polymerases (approximately 50 to100 bases/sec for RNA synthesis vs. ~1000 bases/sec for DNA synthesis). RNA polymerases have high processivity (i.e., the average number of nucleotides incorporated before a polymerase dissociates) because RNA polymerases cannot resume the synthesis of the prematurely released RNA without starting over. On the other hand, DNA polymerases vary widely in processivity.

Although lacking the 3′ → 5′ proofreading exonuclease activity that can be found in replicative DNA polymerases, RNA polymerases can detect the mismatched base pairs within the DNA/RNA hybrid. The distortion of the hybrid by misincoporation results in the reverse movement of the RNA polymerase, known as "backtracking."[16] However, the removal of a misincorporated base requires extrinsic cleavage factors (Gre proteins in bacteria and SII/TFIIS in the Pol II system). RNA polymerases make about one error for every 10^4 to 10^5 ribonucleotides incorporated in the RNA strand; the error rate for replicative DNA polymerases is about one mistake per 10^6 to 10^8 nucleotides. The fidelity of RNA polymerases is thus much lower than that of replicative DNA polymerases. This is tolerable because the aberrant RNA molecules can simply be turned over while new correct molecules are made. Moreover, a mistake in an RNA transcript is of less consequence to cell survival than a mistake in the permanent codes stored in DNA.

7.2.4.1 E. coli RNA polymerase

The holoenzyme of *E. coli* DNA-dependent RNA polymerase is composed of the core enzyme and the transiently associated σ subunit (see Section 7.2.1 and Figure 7.1A). The σ subunit is required for accurate initiation of transcription by providing the polymerase with the proper cues when a start site has been encountered. The efficiency of DNA and the RNA polymerase binding required for the basal level of transcription in *E. coli* is affected by the variation in specific consensus sequences on the promoter and the spacing between these sequences. In prokaryotic and eukaryotic transcription, the first incorporated ribonucleotide is a purine triphosphate. Nucleotide addition continues until specific termination signals are encountered.

Following termination, the core polymerase dissociates from the template. The core and the σ subunit can then reassociate to form the holoenzyme again, ready to initiate another round of transcription. Transcriptional termination occurs in *E. coli* by factor-dependent and factor-independent means. Two structural features of all *E. coli* factor-independent terminating genes have been identified. One is the presence of two symmetrical GC-rich segments capable of forming a stem-loop structure in the RNA and the second is a downstream A-rich sequence in the template. The formation of stem loop in the RNA destabilizes the association between the RNA polymerase and the DNA template. This is further destabilized by the weaker nature of the AU base pairs formed between the template and the RNA following the stem loop. This leads to dissociation of the polymerase and termination of transcription.[17] The transcription of most *E. coli* genes is terminated by this method.

Factor-dependent termination requires the recognition of termination sequences by the termination protein, *rho* (ρ factor). The ρ factor is a homohexameric protein that has an RNA-binding domain and an ATP-binding domain, capable of catalyzing ATP-coupled translocation of a polymerase along an RNA transcript.[18] The factor recognizes and binds to sequences in the 3′ portion of the RNA. This binding destabilizes the interactions between the RNA polymerase and the DNA template, leading to dissociation of the polymerase and termination of transcription.

7.2.4.2 T7 phage RNA polymerase

In bacteriophage, sequence-specific recognition of the duplex promoter is accomplished by specific T7 RNAP/DNA interactions, which involve the major and minor grooves of DNA (see Section 7.2.6 for more detail). In their crystal structure model of T7 RNAP/DNA initiation complex, Cheetham et al. suggested the reasons why T7 RNAP can initiate transcription from a single priming nucleotide depending on structural perspectives.[4] First, RNAP binds specifically to the promoter and positions the template at the enzyme active site. Second, the aromatic side chain of Trp422 stacks on the template base T-1 (Figure 7.8), bending the template strand to orient the template base position at +1 for the priming ribonucleotide. Finally, His811 is shown to form hydrogen bonding with the 2′-OH of the priming nucleotide that preferentially selects priming ribonucleotides over deoxyribonucleotides.

During the early stages of transcription, T7 RNAP forms an unstable initiation complex that synthesizes and releases RNA transcripts that are two to eight nucleotides in length. This phenomenon is known as the "abortive phase of transcription" and is a general feature of all RNA polymerases. One of the major factors that help to trigger the transition from abortive initiation to elongation is the stability of interaction between the heteroduplex and its binding site.

Two main models have been proposed to explain how transcription initiation and elongation dynamics proceed by T7 RNAP. The first model, called "protein inchworming," requires that the DNA recognition activity and polymerase activity of the enzyme reside on different domains and move independently of each other as RNA elongates. The second model, known as the "DNA scrunching" model, keeps the enzyme structure unchanged but requires that the template and the non-template DNA strands be progressively accommodated to accumulate the transcription bubble in the active site.

However, three recent structures have been reported to end this debate. The first is a 2.4-Å resolution structure of the T7 RNAP initiation complex captured transcribing an RNA trinucleotide from a 17 base pair promoter DNA containing a 5-nucleotide single strand extension, reported by Cheetham and Steitz.[5] In support of the DNA scrunching model, the structure exhibited that the template strand has accumulated into the active site

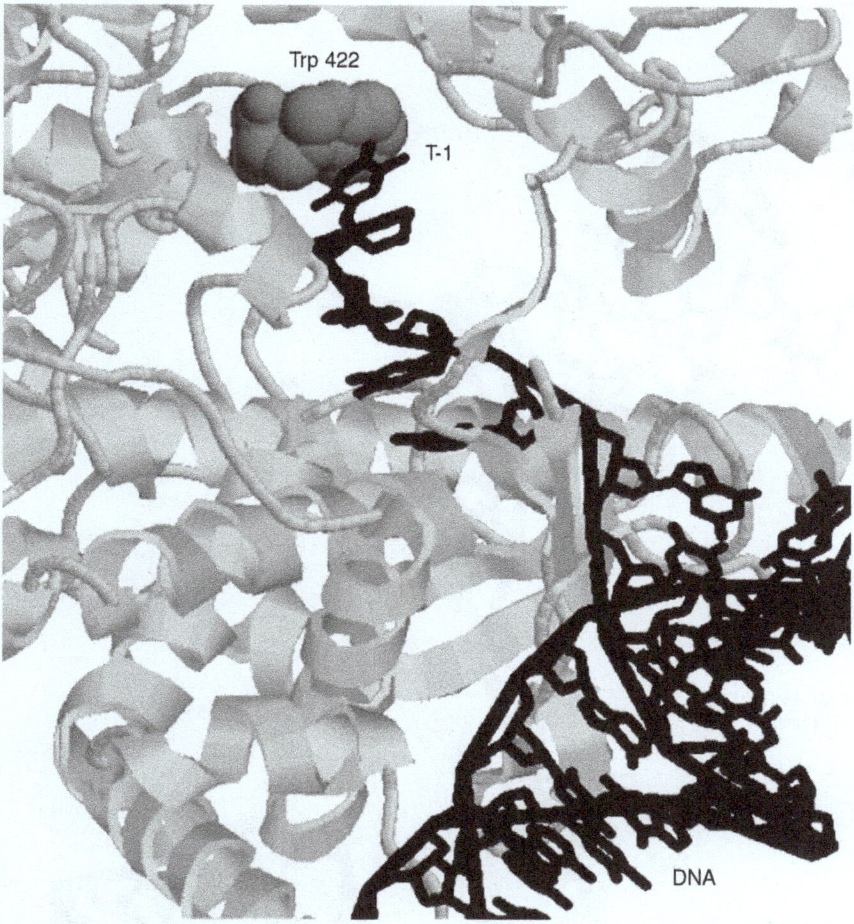

Figure 7.8 Crystal structure of T7 RNAP/DNA open complex (PDB 1CEZ [Cheetham, G.M. et al., *Nature*, 399, 80–83, 1999] Protein Explorer). The aromatic side chain of Trp422 stacks on the template base thymine –1.

pocket to place positions +3 and +4 of the template strand at the catalytic site. Furthermore, the structure showed that His784 and Gly542 form a steric gate to the incoming nucleotide, which allows T7 RNAP to incorporate ribonucleotides in preference of deoxyribonucleotides (Figure 7.9). Biochemical evidence indicates that Tyr639 plays an important role in this type of nucleotide specificity during elongation. Site-directed mutagenesis of Tyr639Phe reduces the T7 RNAP nucleotide preference[19] and double mutant Tyr639Phe, Ser641Ala allowed its use for preparative synthesis of hybrid DNA/RNA–polynucleotide products for use as templates for reverse transcriptase (the so-called mixed polynucleotide).[20]

Although it provided valuable structural information about the sequence-specific recognition and RNA transcription in the early initiation phase, Cheetham and Steitz's structure of the T7 RNAP initiation complex

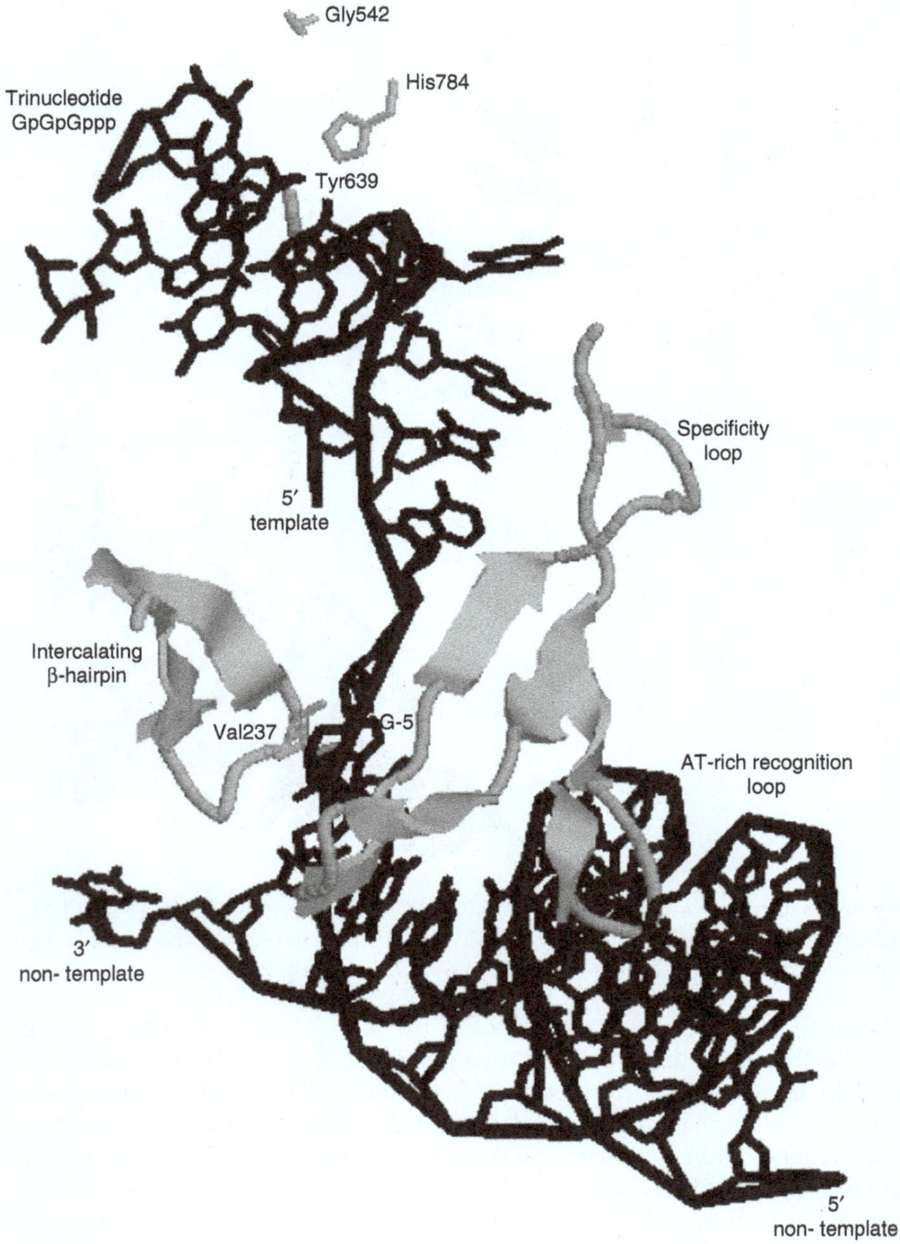

Figure 7.9 Crystal structure of the T7 RNAP/DNA initiation complex: binding motifs and structural features (PDB 1QLN [Cheetham, G.M. and Steitz, T.A., *Science*, 286, 2305–2309, 1999] Protein Explorer).

could not present conclusive evidence for the mechanism by which T7 RNAP extends RNA transcript in the late initiation and the elongation phases. The final mechanism has yet to be proven experimentally.

Recently, a second crystal structure has been solved by the Yin and Steitz.[6] The 2.1-Å resolution structure of the T7 RNAP elongation complex with 30 base pairs of duplex DNA contains both a transcription bubble and a 17-nucleotide RNA transcript.[6] These researchers reported that the N-terminal domain changed its shape and tertiary structure substantially, compared to the structure of the initiation complex. As a consequence, the promoter-binding site is destroyed. Two new conformational features are created: a channel that accommodates the heteroduplex at the active site and an exit tunnel through which an RNA transcript can pass (Figure 7.10). These structural features show that T7 RNAP changes progressively to accommodate the RNA/DNA hybrid as it grows which, in turn, is compelling evidence for elongation by the polymerase inchworming mechanism.

This evidence has been simultaneously supported by the 2.9-Å crystal structure of the elongation complex, consisting of T7 RNAP, 8 bp RNA/DNA hybrid and 10 bp of downstream DNA, reported by Tahirov et al.[7] Thus, the crystal structures discussed here present a dynamic picture of how T7 RNAP transcribes RNA; RNA extends in the abortive phase by the DNA scrunching mechanism while transcription elongation proceeds by the RNAP inchworming mechanism.

7.2.4.3 Eukaryotic RNA polymerases

In eukaryotes, an assembly of polymerase and general transcription factors (at least six TFs: TFIIA, B, D, E, F, and H) is required for RNA Pol II to bind a promoter. Figure 7.11 summarizes the sequence of events of transcription initiation and elongation as a simplified model of transcription by eukaryotic RNA Pol II *in vivo*.[21] First, TATA-binding protein (TBP) binds to the TATA box. The TATA box is the TATAAA sequence present 25 to 30 base pairs (bp) upstream of the transcription start site (note that an initiator element (Inr) is defined as a discrete core promoter element and is functionally similar to the TATA box). The two elements act synergistically when separated by 25 to 30 bp, but act independently when separated by more than 30 bp.

Next, TFIIB binds TBP. This binding is stabilized by TFIIA or TFIID. The DNA/TBP/TFIIB complex is next bound by another complex consisting of TFIIF and RNA Pol II, forming the open complex. TFIIF helps target RNA Pol II to its promoter by reducing nonspecific polymerase/DNA binding. Finally, TFIIE and TFIIH bind and create the closed complex. TFIIH possesses DNA–helicase activity that promotes the unwinding of DNA near the RNA start site. This process requires ATP hydrolysis and can be performed by the kinase activity of TFIIH, which results in phosphorylation of RNA Pol II at several places in the CTD. The function of TFIIH extends from mediating initiation to playing an essential role in nucleotide-excision repair of the DNA lesions. The genetic loss of TFIIH subunits causes genetic diseases

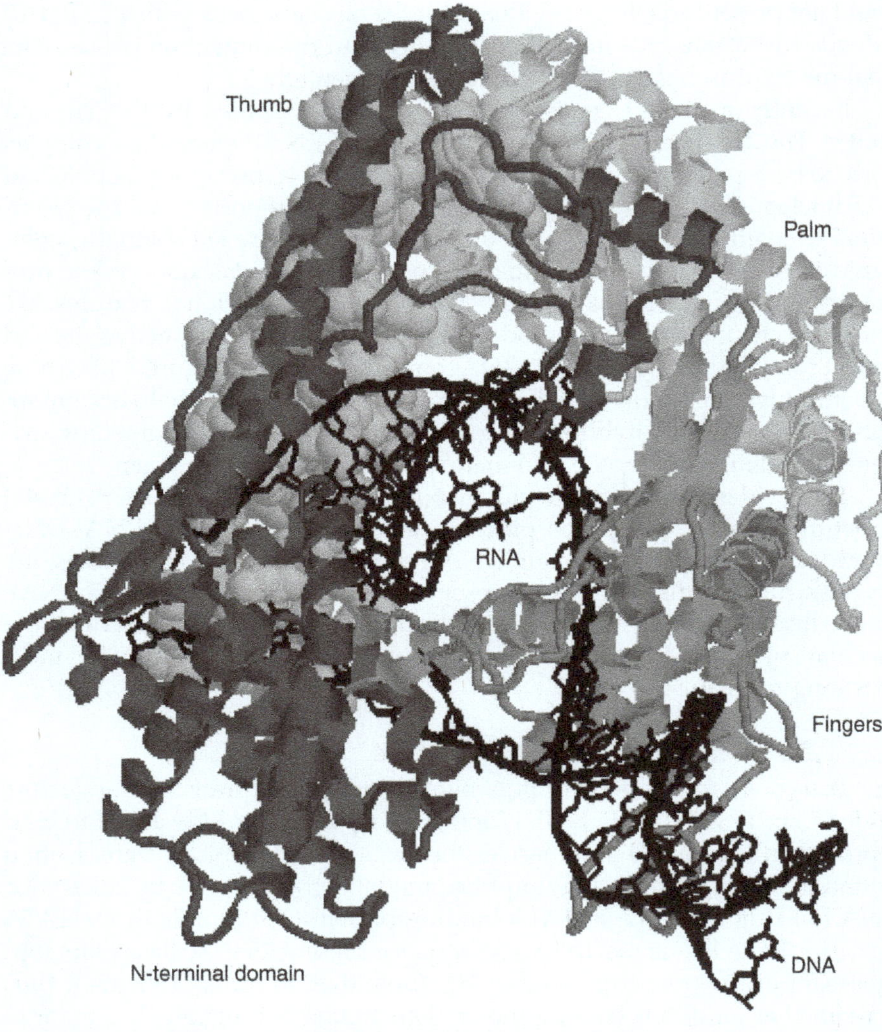

Figure 7.10 Crystal structure of the T7 RNAP elongation complex. Two conformational features are exhibited: a channel that accommodates the heteroduplex in the active site and an exit tunnel through which the RNA transcript can pass (PDB 1MSW [Yin, W.S. and Steitz, T.A., *Science*, 298, 1387–1395, 2002] Protein Explorer).

(xeroderma pigmentosa is an example; for more details about the RNA Pol II core promoter and protein regulators, see Smale[22] and Shilatifard et al.[23]).

The RNA Pol II phosphorylation is followed by a conformational change in the overall complex, thus triggering initiation of transcription. During the synthesis of the initial 60 to 70 nucleotides of RNA, TFIIE and, subsequently, TFIIH are released and RNA Pol II enters the elongation phase. During elongation, TFIIF remains bound to RNA Pol II and affects the activity of the enzyme. After elongation is completed, RNA Pol II is dephosphorylated by

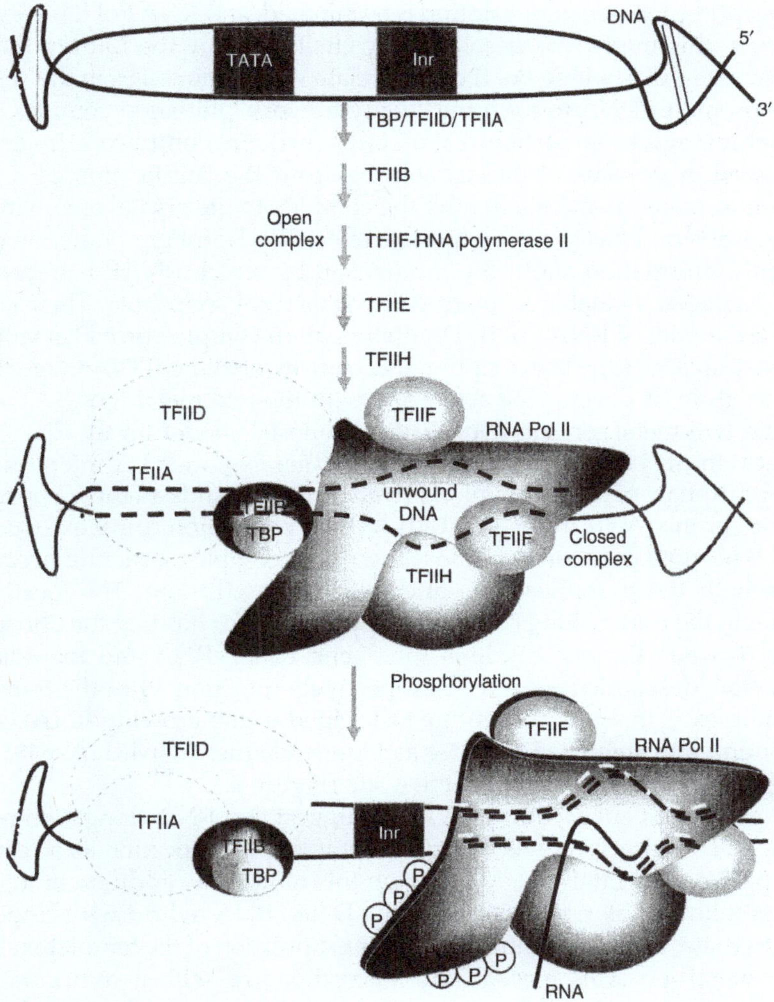

Figure 7.11 Transcription of RNA Pol II promoter. First, the TATA-binding protein (TBP) binds to the TATA box. Then, TFIIB binds TBP; this binding is stabilized by TFIIA or TFIID. DNA/TBP/TFIIB complex is next bound by another complex consisting of TFIIF and RNA Pol II, forming the open complex. Finally, TFIIE and TFIIH bind to create the close complex. TFIIH has kinase activity that phosphorylates RNA Pol II at several places in CTD. The RNA Pol II phosphorylation is followed by a conformational change in the overall complex and the transcription initiation is triggered. During the synthesis of the initial 60 to 70 nucleotides of RNA, first TFIIE and subsequently TFIIH are released and RNA Pol II enters the elongation phase. During elongation, TFIIF remains bound to RNA Pol II, affecting the activity of the enzyme. Finally, RNA Pol II is dephosphorylated by phosphatases and recycled (not shown). (This figure was redrawn by the authors from Nelson, D.L. and Cox, M.M., in *Lehninger Principles of Biochemistry,* chap. 26, 987–990, 2000, New York: Worth Publishers, Incorporated, using Microsoft Word drawing tools.)

various phosphatases, transcription is terminated, and RNA Pol II is recycled. However, this model has been recently challenged by the finding that the holoenzyme exists where Pol II can associate with various TFs in the absence of the promoter DNA to form the "ready to work" initiation complex.[24,25]

Achieving X-ray structures of large protein complexes invariably encounters numerous obstacles, starting from the purification of a large amount of material and stabilizing the complex in the crystal to solving the phase problem. Due to the significant efforts by Kornberg and coworkers, in-depth information about the mechanism by which RNAP II transcribes DNA has been available through X-ray structural snapshots. Their crystal structure model of RNA Pol II/DNA elongation complex provides valuable information about the transcription mechanism, enzyme/DNA interactions, and structure of the enzyme active site with the two metal ions.[14]

The two-metal ion mechanism is a proposed model for all RNA polymerases. In the case of single subunit polymerases, metal A interacts with the α-phosphate of the incoming rNTP and metal B binds all three phosphate moieties of the rNTP. In the yeast RNA Pol II elongation complex structure, metal B has low occupancy in the transcribing complex structure because it leaves with the pyrophosphate after nucleotide addition. The location of metal A in the transcribing complex is appropriate for binding the phosphate group between the nucleotide at the 3′-end of the RNA and the adjacent nucleotide, designated +1 and –1, respectively (position +1 in the transcribing complex is that of a nucleotide just added to the growing RNA). Metal A is bound and stabilized by three aspartate residues, Asp481, Asp483, and Asp485, located at the enzyme active site (Figure 7.12).

Results from eukaryotic RNA Pol II suggest that RNA elongation occurs by the DNA scrunching mechanism (a template translocation as a result of RNA polymerase catalysis).[26] Following this model, the addition of a nucleotide is followed by translocation of the DNA/RNA hybrid with respect to the active site, which accommodates the next position of the template relative to the metal ion. This mechanism is accompanied with a conformational change in the bridging helix: from a "straight" state to a "bend" state and recycled back to a straight state (see Figure 7.12 for the straight conformation). The bridging helix is also involved in the RNA Pol II proofreading and repair mechanisms. After incorporation of an incorrect nucleotide in the RNA transcript, the bridging helix separates RNA from DNA at the downstream edge of the hybrid during backtracking, mediating the removal of an RNA stretch by transcript cleavage factors.

7.2.5 Regulation of RNA polymerases

7.2.5.1 E. coli and T7 RNA polymerases

The activities of the bacterial and eukaryotic transcription initiation complexes are regulated in response to environmental, developmental, and cell-type signals. In most cases, regulation is mediated by factors that bind to specific

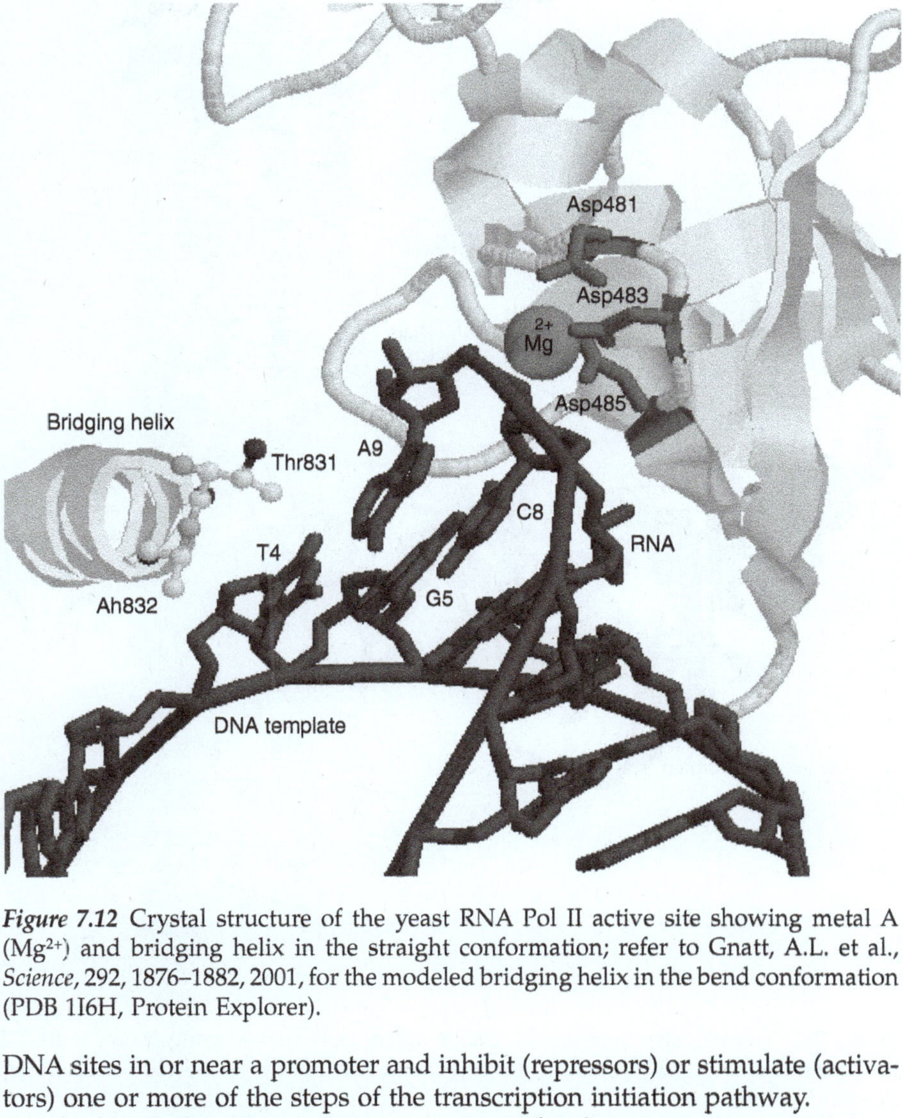

Figure 7.12 Crystal structure of the yeast RNA Pol II active site showing metal A (Mg^{2+}) and bridging helix in the straight conformation; refer to Gnatt, A.L. et al., *Science*, 292, 1876–1882, 2001, for the modeled bridging helix in the bend conformation (PDB 1I6H, Protein Explorer).

DNA sites in or near a promoter and inhibit (repressors) or stimulate (activators) one or more of the steps of the transcription initiation pathway.

Cyclic AMP receptor protein is an example of transcription activators in *E. coli*. When *E. coli* is grown in the absence of sugars other than glucose, the binding of cyclic AMP receptor protein activates transcription of more than 20 RNA transcripts that code for the enzymes that metabolize these sugars. The 3-Å resolution crystal structure of the cyclic AMP receptor protein and DNA complex suggests that the cyclic AMP receptor protein induces a 90° hairpin bend in the DNA that might facilitate the contacts between the RNA polymerase and DNA.[27]

In bacteriophage T7, T7 lysozyme inhibits T7 RNAP by producing an allosteric conformational change in the enzyme (Figure 7.13). This conformational change modifies the necessary active site structure required for

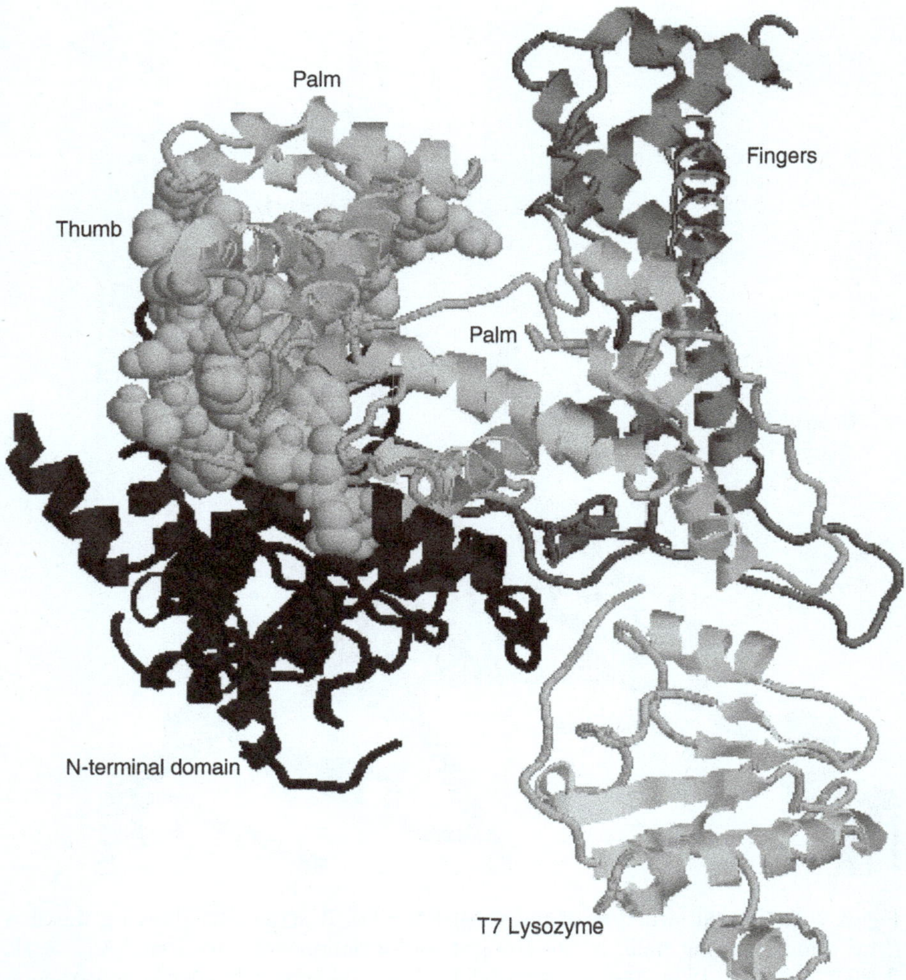

Figure 7.13 Crystal structure of T7 phage RNA polymerase complexed with T7 lysozyme (PDB 1ARO [Jeruzalmi, D. and Steitz, T.A., *Embo J.*, 17, 4101–4113, 1998] Protein Explorer).

substrate binding. The 2.8-Å resolution crystal structure of the T7 RNAP/T7lysozyme complex reveals that the C-terminus dipeptide Phe882–Ala883 plays a crucial role in this mechanism.[3] In the absence of bound promoter, Phe882–Ala883 can be cleaved in solution by carboxypeptidase A. During transcription initiation, this terminus is bound to a hydrophobic pocket beneath the palm domain, rearranging the catalytic pocket at the active site to stably bind rNTP and RNA. However, Phe882–Ala883 adopts an extended conformation in the TRNAP/T7 lysozyme complex that, in turn, destabilizes the conformation of the active site.

7.2.5.2 RNA Pol II

The ultimate goal of any transcriptional control process is to govern the recruitment and activity of RNA polymerase II. Unless Pol II can initiate transcription at the 5′ end of a gene and successfully transcribe the complete message, a protein product cannot be produced. Not surprisingly, therefore, Pol II is a target for regulation.

7.2.5.2.1 Carboxy-terminal repeat domain. The carboxy-terminal domain performs important regulatory roles in all major steps of mRNA formation, including transcription initiation and elongation, capping, splicing, and 3′-end processing (reviewed in Proudfoot et al.[28] and Hampsey[29]). The CTD of the largest RNA Pol II subunit comprises tandem repeats of a heptapeptide consensus sequence Tyr-Ser-Pro-Thr-Ser-Pro-Ser, which is highly conserved with varied repeat lengths among eukaryotic organisms. The repeat length appears to increase with increasing genome complexity, with 27 repeats in yeast Rpb1, 34 repeats in the *C. elegans* CTD, 45 repeats in the Drosophila CTD, and 52 repeats in the mouse and human CTD.

The CTD is differentially phosphorylated by multiple kinases during the transcription cycle.[30,31] RNA Pol II *in vivo* has two forms: the first, IIO, is extensively phosphorylated at the CTD and found in the elongating complex, whereas form IIA is hypophosphorylated and preferentially enters the preinitiation complex.[32] Form IIA is converted to IIO concomitant with or shortly after initiation, suggesting that CTD phosphorylation regulates transcription initiation.[33] Although several kinases have been reported to mediate phosphorylation of the RNA Pol II CTD (reviewed in Prelich[34]) — including Cdk7/Kin28, Cdk8/Srb10, Cdk9/P-TEFb, CTDK-1, and Sgv1/Bur1 — it remains to be determined whether these CTD kinases affect different steps in the transition from initiation to elongation.

Much less is known about the dephosphorylation process of the CTD. The CTD phosphatase called Fcp1 is regulated by TFIIB and TFIIF. The RAP74 subunit of TFIIF stimulates Fcp1 phosphatase activity, whereas TFIIB inhibits the stimulatory activity of TFIIF.[34–36] The role of Fcp1 in counteracting CTD kinase activity remains to be elucidated within the context of the transcription cycle.

7.2.5.2.2 Ubiquitylation. Ubiquitylation of RNA Pol II also plays a role in transcriptional control. When eukaryotic cells are exposed to DNA-damaging agents, one of their first priorities is to repair transcriptionally active genes.[37,38] This is achieved, in part, when active RNA Pol II stalled at a DNA lesion is ubiquitylated and presumably destroyed by the proteasome. The removal of RNA Pol II is then followed by the coordinated recruitment of the DNA-repair machinery and repair of the damaged DNA. Thus, in this protective strategy known as the transcription-coupled repair (TCR) mechanism, cells are able to use Pol II to probe for DNA damage and shut down expression of the genetic information until damaged DNA segments are corrected.

How does the proteasome machinery recognize transcriptionally active, but stalled, RNA polymerase molecules? One way is through phosphorylation events in the CTD of the largest subunit of Pol II. Because the phosphorylation status of the RNA Pol II CTD changes during transcription and CTD phosphorylation promotes RNA Pol II ubiquitylation,[38] it has been suggested that a specific pattern of CTD phosphorylation shown by a stalled polymerase could function as a direct signal for ubiquitylation.

7.2.6 Protein/DNA interactions and binding motifs

Direct binding of various amino acid residues of a polymerase to an RNA/DNA hybrid is essential for positioning of nucleic acids at the active site and stability of the complex. The most common protein/DNA binding motifs adopt an α-helical secondary structure, which is inserted in the major groove of the DNA.

Discrimination between the T7 promoter and other DNA sequences by T7 RNAP is directly accomplished by a less common antiparallel β-strand motif inserted into the DNA major groove through a hydrogen-bonding network (Figure 7.9). This β-motif is known as a specificity loop (residues 739 to 770) and is located at the insertion between the polymerase fingers and palm domains. Furthermore, the N-terminal domain of T7 RNAP indirectly recognizes a specific AT-rich sequence by inserting a flexible surface loop (residues 93 to 101) into the minor groove.

In addition to its role in sequence-specific recognition, the N-terminal domain facilitates transcription bubble formation through the intercalating β-hairpin loop (Figure 7.9). This loop plays an important role in melting the promoter by stacking Val237 on base G-5. All these structural features have been reported in the crystal structure of the DNA/T7 RNAP initiation complex.[5]

Elongation complexes, on the other hand, display different DNA binding motifs due to the protein conformational change. For example, a loop (residues 593 to 610) and a helix–turn–helix fragment (residues 647 to 675) form a channel that accommodates the nontemplate strand and extends to the template strand exit.[7] Furthermore, the specificity loop moves sideways in the elongation complex to become part of the RNA transcript exit tunnel.[6]

Eukaryotic RNA polymerases have little affinity for their promoters because DNA recognition is almost entirely performed by the action of activator proteins. Most polymerase/DNA interactions are therefore nonspecific in nature. For yeast RNA Pol II, the stability of the polymerase/DNA elongation complex is determined by a few structural and biochemical factors. The clamp closure determines the basal level of elongation complex stability. As the DNA/RNA hybrid reaches 9 to 10 bases, the RNA and DNA contact the rudder and lid loops, which separate the DNA and RNA (Figure 7.7). The growing RNA chain can then reach and bind the RNA-binding channel at a length of 13 to 15 bases. Finally, the transcript exits the RNA-binding channel at a length of 18 to 20 bases.

The increased contact of the single-stranded RNA transcript to the RNA exit channel increases the stability to the complex. DNA duplex unwinding is performed by specific protein interactions. Two elements, known as "forks," are suggested to maintain the transcription bubble in the yeast RNA Pol II elongation complex. Fork loop 1 is in the vicinity of Rpb 2 residue 470 and fork loop 2 is in the Rpb2 cleft domain (Figure 7.7). The structure of DNA helix unwinding is prohibited by the disordering of the nontemplate strand as well as the disorder in several residues of fork loop 2. The crystal structure of the yeast RNA Pol II elongation complex reported the burying of 3400 Å^2 of surface area in protein/nucleic acid interface. Although the crystal structure exhibits such strong DNA-binding characteristics, these interactions do not interfere with translocation of nucleic acid during transcription.

The mobility of DNA is allowable for three reasons:[14]

- Most of the interactions are with the sugar-phosphate backbone with a significant lack of base specificity.
- Several amino acid side chains interact with two phosphate groups along the DNA backbone simultaneously, which may reduce the free energy of translocation.
- Formation of a positively charged electrostatic feature called the "second shell" around the hybrid at a distance 4 to 8 Å may attract the hybrid without restraining its movement.

In the yeast RNA Pol II active site, polymerase/DNA nonspecific interactions play an important role in transcription. Nonspecific interactions at the active site take the form of van der Waals contacts between the enzyme and various nucleotide bases at the end of the hybrid, position +1, through Thr831 and Ala832 of the Rbp1 bridging helix (Figure 7.12). The alignment of the nucleotide base at position +1 is believed to be stabilized by these interactions after the bridging helix undergoes a conformational change (back from the bend state to the straight state) and DNA simultaneously undergoes the coupled translocation at the active site.

7.3 Viral RNA-dependent RNA polymerases

7.3.1 Poliovirus polymerase

Poliovirus, the prototype picornavirus , is among the most diverse and oldest known viruses and is closely related to several medically important viruses, including rhinoviruses, hepatitis A virus, coxsackieviruses, and echoviruses. RNA replication by poliovirus RNA Pol is distinct from eukaryotic RNA Pol II in that it does not involve a nuclear phase, occurring within large protein complexes associated with the membranes of virus-induced cytoplasmic vesicles.[39]

Polovirus replication is a two-step process, beginning with the synthesis of a complementary minus strand. That strand, in turn, serves as a template for the production of plus-strand RNA. The replication initiation in poliovirus RNA synthesis is the covalent linkage of the 5′-phosphate of UMP to the hydroxyl group of a tyrosine in a terminal protein VPg (a small protein covalently linked to the 5′-end of the plus-stranded poliovirus genomic RNA that initiates transcription of the poliovirus genome).[40,41] Although viral replication occurs in large protein complexes *in vivo*, purified poliovirus RNA polymerase alone ($3D^{pol}$) is capable of elongating primed template RNAs *in vitro*.[42]

Highly purified poliovirus polymerase, a 52-kDa protein, shows a high degree of cooperativity in terms of RNA binding and template utilization, suggesting a possible role of polymerase/polymerase interactions for efficient RNA replication.[43,44] Data in support of polymerase/polymerase interactions have also come from interactions observed in the yeast two-hybrid system,[45] from glutaraldehyde cross-linking,[44] and from the three-dimensional structure of the polymerase.[46] At present, the exact composition of the polio replication complex is not clear. In addition to $3D^{pol}$, this complex may contain viral vectors 3AB, 3CD, and 3B. 3AB and 3CD are required primarily to establish an initiation complex, possibly by recruiting $3D^{pol}$.[47] 3AB, the precursor of the VPg, is an RNA-binding protein, capable of interacting with $3D^{pol}$,[48,49] that possibly stimulates elongation of nascent RNA[50] or enhances the efficiency of primer utilization.[51]

The X-ray crystal structure of the Mahoney type 1 poliovirus polymerase[52] shows that this polymerase has a typical topology observed for reverse transcriptases and other DNA polymerases and can be compared with a cupped right hand with fingers, palm, and thumb subdomains (Figure 7.14). The polymerase molecules interact within the crystal lattice via two extensive polymerase/polymerase interfaces, referred to as Interface I and Interface II (Figure 7.14 shows interfaces I and II in light gray and black colors, respectively).

These interactions are much more extensive than those typically observed for crystal-packing interactions, with Interface I alone burying a total of 2180 Å^2 of solvent-accessible surface area. Amino acid residue mutations predicted to disrupt interface I are lethal to the virus and reduce RNA-binding affinity.[53,54] The X-ray structural observation is further confirmed by the results of electron microscopy, which show the purified $3D^{pol}$ forms planar and tubular oligomeric arrays.[55] Membranous vesicles isolated from poliovirus-infected cells contain structures consistent with the presence of two-dimensional polymerase arrays on their surfaces during infection.[55] Thus, host cytoplasmic membranes may function as physical foundations for two-dimensional polymerase arrays, conferring the advantages of surface catalysis to viral RNA replication.[55]

7.3.2 *HCV RNA-dependent RNA polymerase NS5B*

Hepatitis C (HCV) virus possesses an RNA-dependent RNA polymerase NS5B, which is essential for viral replication and infectivity.[56] NS5B is a

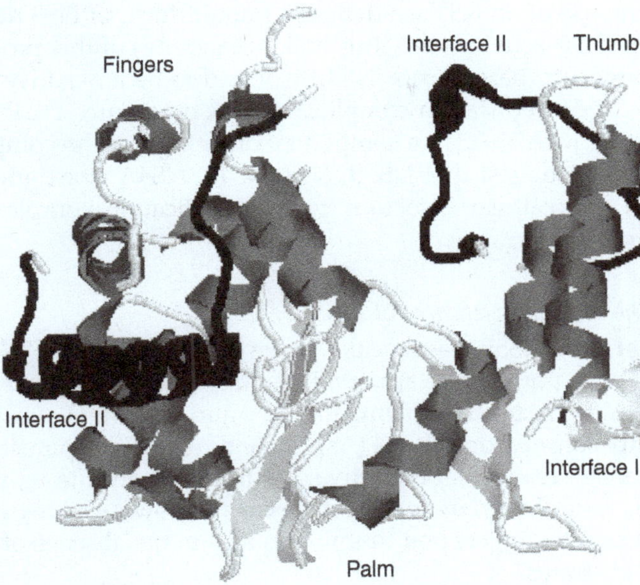

Figure 7.14 Crystal structure of the RNA-dependent RNA polymerase of poliovirus (PDB 1RDR [Hansen, J.L. et al., Long, A.M., and Schultz, S.C., *Structure*, 5, 1109–1122, 1997] Protein Explorer). Interface I is shown as a light gray ribbon in the thumb domain. The black color ribbons in the thumb and the fingers domains form interface II.

65-kDa membrane-associated phosphoprotein.[57] Sequence analysis reveals four motifs including a GDD motif that is conserved among polymerases from positive-stranded RNA viruses.[58,59] NS5B copies long heteropolymeric templates with an estimated elongation rate of 150 to 200 nucleotides per minute at 22°C.[60]

Due to detection of dimer-sized, hairpin-like RNA molecules using large heteropolymeric templates derived from the HCV RNA genome, HCV replication was first proposed to go through a "fold-back" priming mechanism.[59,61] However, NS5B has recently been shown to synthesize a full-length HCV genome *in vitro* in a template-dependent and primer-independent manner.[62–64] Thus, NS5B does not require a separate primer or a folded-back 3′-template end as a primer. Instead, it recognizes the specific sequences and structures at the 3′-termini of the plus and minus strands of the HCV RNA genome for the initiation of RNA synthesis.[62,65] NS5B uses mono-, di-, or trinucleotides efficiently to initiate *de novo* RNA synthesis.[66] Interestingly, it has been shown that a high concentration of rGTP stimulates RNA synthesis by NS5B by up to two orders of magnitude.[67]

NS5B has been found to interact physically and functionally with several HCV proteins, including NS3 (a protease/helicase),[68] NS4B (a hydrophobic protein of unknown function),[68] NS5A (containing an interferon sensitivity determining region),[69] and core protein.[70] NS3 dramatically modulates template recognition by NS5B and changes the synthetic products generated by this enzyme.[68] The use of an NTPase-deficient mutant form of NS3 demonstrates that the NTPase activity (and thus helicase activity) of this protein is specifically required for these effects. NS4B is found to be a negative regulator of the NS3/NS5B replication complex.[68] NS5A modulates HCV replication by interaction with NS5B as a component of the replication complex.[69] Overall, these results suggest that NS5B, NS3, NS4B, NS4A, core, and possibly, other viral and cellular cofactors, form a replication complex that catalyzes HCV replication.

7.3.2.1 The crystal structure of NS5B

A detailed view of HCV NS5B was revealed by its crystal structures.[71,72] HCV NS5B adopts a typical polymerase structure that resembles the human right hand with subdomains termed palm, fingers, and thumb. Palm motif includes the highly conserved residues (Asp220 and Asp318) responsible for the nucleotide transfer reaction at the active site. Interestingly, the active site of HCV NS5B is completely in the crystal structure owing to extensive interactions between the fingers and thumb domains in the absence of RNA and nucleotide substrates.

Figure 7.15 shows the thumb/fingers interdomain linkage through two insertion loops: short loop (loop 1, light gray color, residues 139 to 160) and longer loop (loop 2, black color, residues 11 to 45). The fingers and thumb domains are therefore not free to undergo a conformational change independent of each other. Residue Asp225 plays a crucial role in the selection of rNTP over dNTP by forming hydrogen bonds with ribose 2′-hydroxyl group of the incoming rNTP. The C-terminal residues of 545 to 562 — particularly the hydrophobic residues Leu547, Trp550, and Phe551 — occupy a putative RNA-binding cleft in this polymerase and seem to regulate the polymerase activity.[73]

On the basis of the closed tertiary structures of HCV NS5B in the absence of nucleic acids, this RdRp is predicted to be able to accommodate the template–primer duplex without undergoing global conformation changes, indicating that the general structure is probably preserved during the reaction pathway. This is in contrast to other previously determined polymerase structures in which the intersubdomain contacts are relatively flexible and can undergo large-scale subdomain movement (e.g., in T7 DNA-dependent RNAP). However, the observed closed conformation of NS5B could be brought about due to crystal packing. More studies are thereby needed to confirm the prediction.

Crystal structures of HCV NS5B of complexed with rNTPs and divalent metal ions in the absence of an RNA substrate has been reported by Bressanelli et al. (at resolutions: 1.7 Å for NS5B/rGTP complex and 1.85 Å for NS5B/rUTP complex).[74] Four main structure–function features have been revealed by these crystal structures (see Figure 7.16):

- A specific site for the rGTP found precisely at the finger/thumb connection (referred to as the rGTP specific site)

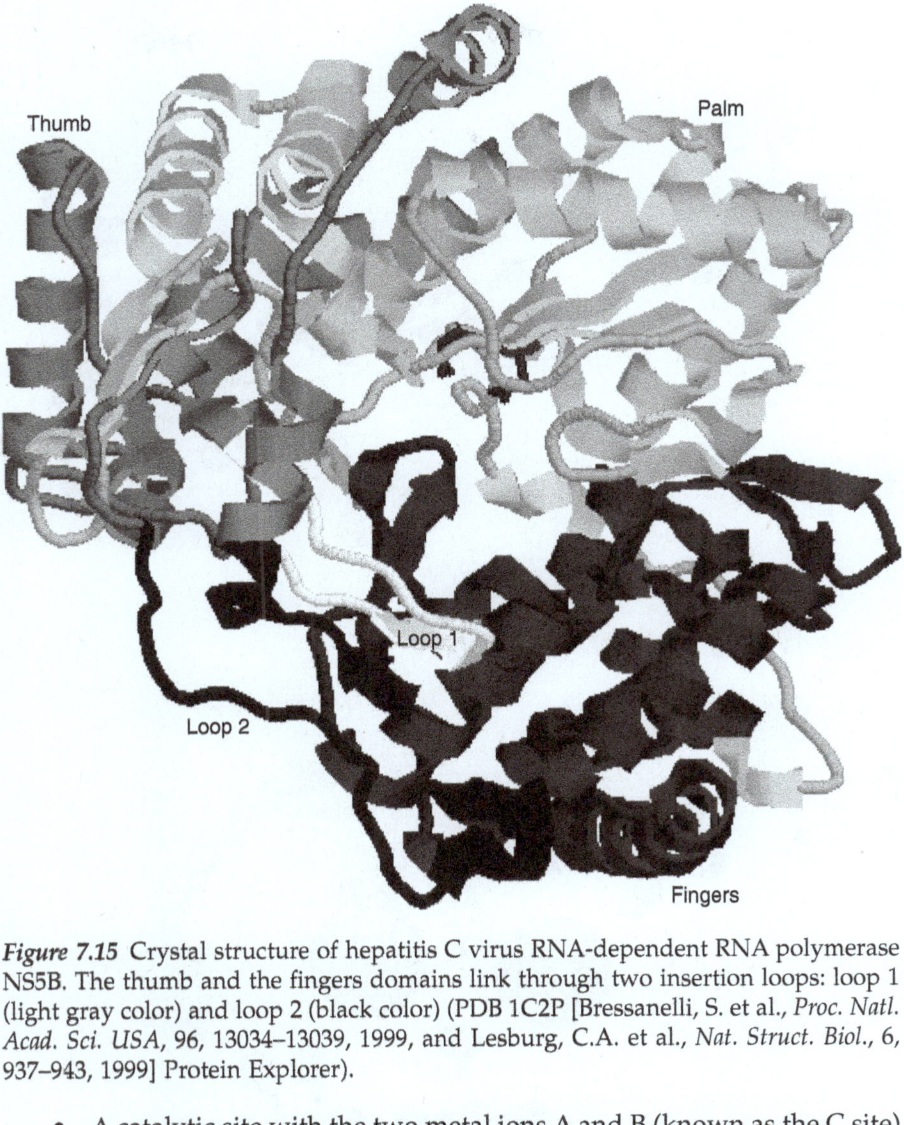

Figure 7.15 Crystal structure of hepatitis C virus RNA-dependent RNA polymerase NS5B. The thumb and the fingers domains link through two insertion loops: loop 1 (light gray color) and loop 2 (black color) (PDB 1C2P [Bressanelli, S. et al., *Proc. Natl. Acad. Sci. USA*, 96, 13034–13039, 1999, and Lesburg, C.A. et al., *Nat. Struct. Biol.*, 6, 937–943, 1999] Protein Explorer).

- A catalytic site with the two metal ions A and B (known as the C site)
- Two additional rNTP binding sites (referred to as the P site and the I site)

Interestingly, the structure shows the presence of three distinct nucleotide-binding sites at the enzyme active center in the absence of an RNA substrate.

7.3.2.1.1 The NS5B-rGTP specific site. The crystal structure reported a specific site for rGTP binding and showed no occupancy of this site by other ribonucleotides (Figure 7.16A). A kinetic analysis of NS5B showed that

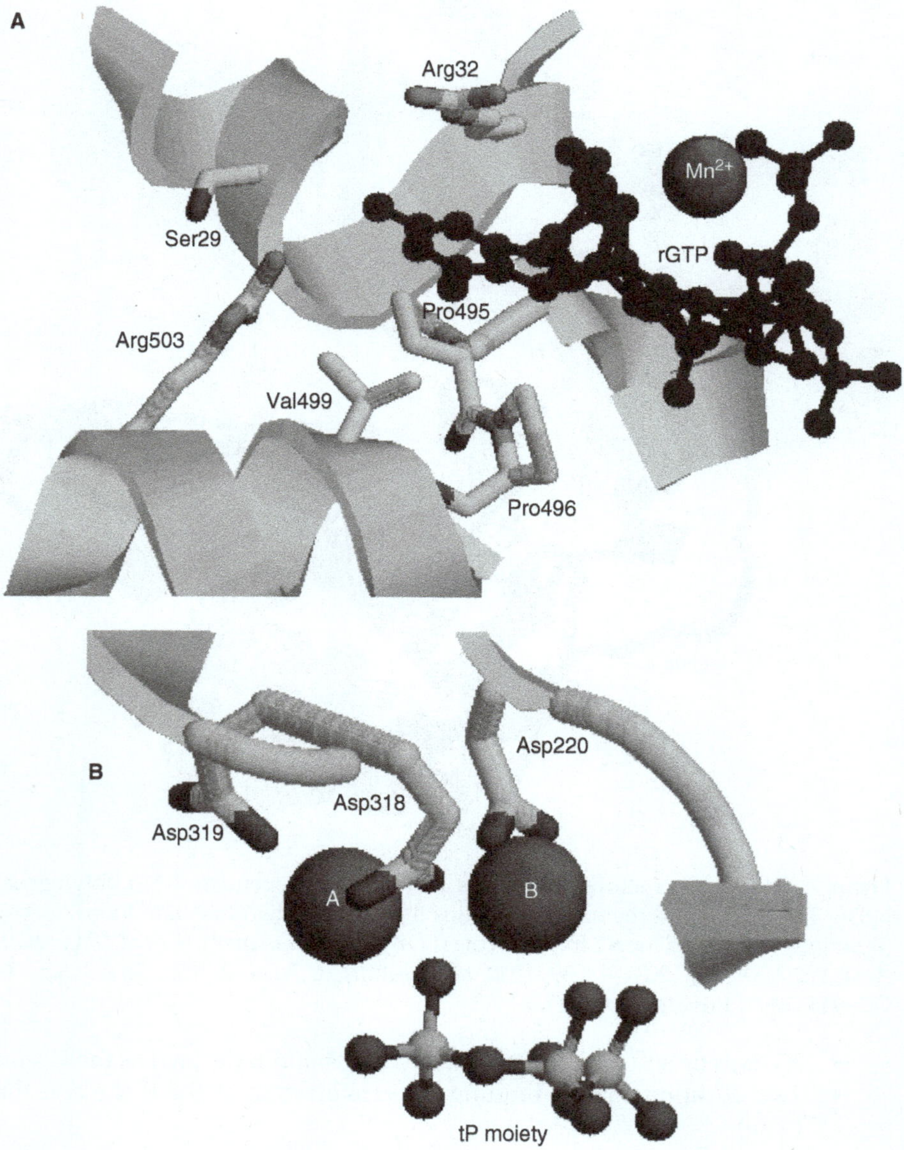

Figure 7.16 Crystal structure of HCV RNA-dependent RNA polymerase NS5B complexed with rNTP. (A) The rGTP specific site. (B) The C site with rGTP triphosphate moiety (tP) and two Mn^{2+} ions, labeled A and B. (C) The C, P, and I sites with metal ions and rUTP bound to the C site and the triphosphate moiety bound to P and I sites. (PDBs 1GX5 and 1GX6 [Bressanelli, S. et al., *J. Virol.*, 76, 3482–3492, 2002] Protein Explorer)

Continued.

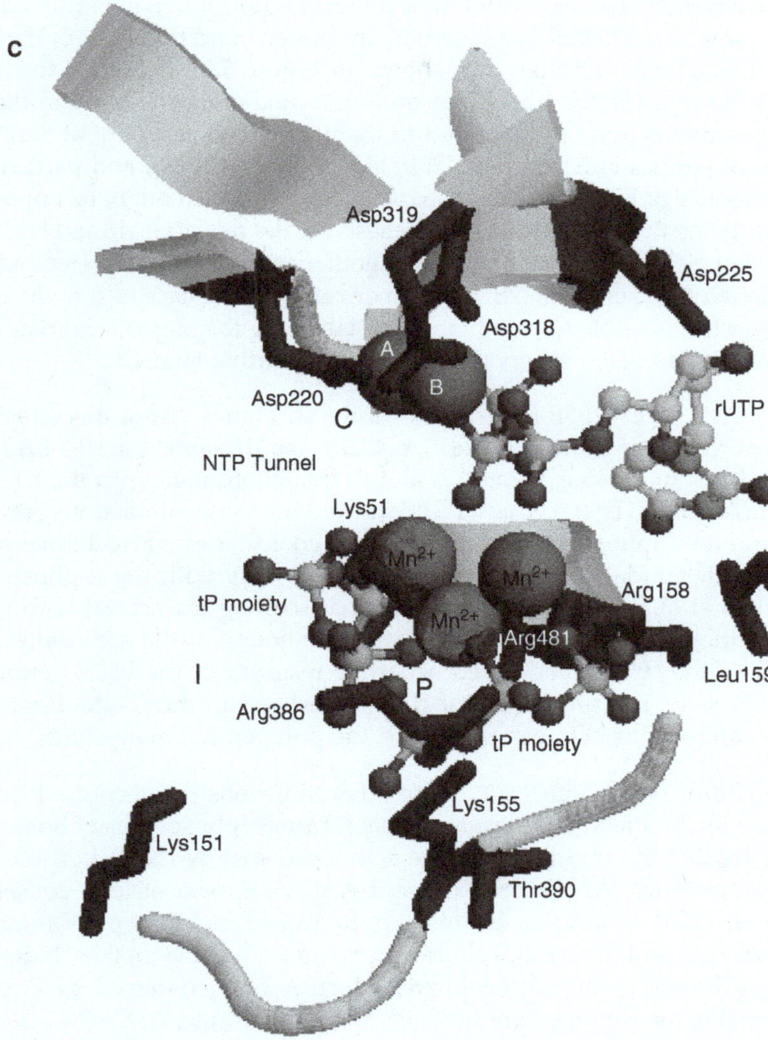

Figure 7.16 Continued.

this site is 70 to 80% occupied at 1 m*M* rGTP.[74] In this site, the contact between rGTP and the enzyme is mediated by six amino acids: four from the thumb domain (Pro495, Pro496, Val499, and Arg503) and two from the fingertips (Arg32 and Ser 29).

The ring of Pro495 and the side chain of Val499 provide a platform (via hydrophobic and van der Waals interactions, respectively) against which the base is aligned. Pro496 makes lateral van der Waals contacts with rGTP. The surface residue, Arg32, seems to be a crucial specificity determinant. In the absence of rGTP, Arg32 is highly mobile. The freedom of movement of Arg32 is restricted upon rGTP binding due to bidentate hydrogen bonding between the 2′-OH group of the ribose and the N2 position of the guanine base (the latter case is mediated by a water molecule and is not shown in Figure 7.16A). Furthermore, Arg503 makes a bidentate hydrogen bond to the N2 of the base via another water molecule (not shown in Figure 7.16A) and to the main chain carbonyl of Ser29. The water molecule mediating the bond to the N2 of the guanine is hydrogen bonded to the main chain carbonyl of Ser29.

Recent studies indicate that NS5B, like poliovirus RdRp and particularly in the presence of RNA, can oligomerize *in vitro*,[75–77] which might be important for modulating the polymerase. Mutagenesis studies reveal Glu18 and His502,[77] as well as Leu30,[76] are critical for the oligomerization that is a prerequisite to RdRp activity. The fact that NS5B oligomerization takes place next to the rGTP site suggests a possible role of this site in stabilizing the oligomeric complex.[74] However, these *in-vitro* observations need to be further studied.

7.3.2.1.2 The NS5B-C site. As shown in Figure 7.16B, this site comprised of three aspartate residues (Asp220, Asp318, and Asp319) bridging the two Mn^{2+} metal ions, named A and B, that coordinate with the triphosphate moiety of rGTP (rATP and rCTP showed the same interactions between NS5B and the triphosphate moiety). Metal B coordinates with all three phosphate of nucleotide, whereas metal A interacts only with the α-phosphate. Bressanelli et al. reported, interestingly, that soaking the crystal with rUTP revealed the density of a complete nucleotide bound at the active site (specifically, to the strictly conserved aspartate residues in the HCV genotype; Figure 7.16C).[74] The uracil base of rUTP is ordered in the C site due to the specific interactions between uracil and the polypeptide main chain.

7.3.2.1.3 The NS5B-P site. The three Mn^{2+} ions between the P and C sites may likely screen the charge repulsion from triphosphates of both sites, thus increasing the occupancy of the active site area by rNTP. In the P site, the three residues, Arg158, Ser367, and Arg386, appear strictly conserved among all HCV genotypes. In this site, hydrogen bonding predominately mediates the protein–triphosphate interactions. Arg158 makes bidentate hydrogen bonds to the α-phosphate, which is also contacted by Arg394. Additionally, the β-phosphate interacts with Ser367 and Arg386 via hydrogen bonding and Thr390 hydrogen bonds to the γ-phosphate (Figure 7.16C).

7.3.2.1.4 The NS5B-I site. The I site is located further towards the NTP tunnel. In contrast to hydrogen bonding in the P site, electrostatic interactions are the predominating forces that hold the I site in place. Specifically, the triphosphate moiety of rUTP shows no hydrogen bonding to the side chains of I site residues. This observation led Bressanelli et al.[74] to suggest that this site might hold the nucleotides transiently. In all HCV genotypes, the five side chains of the residues interacting with rUTP (Arg48, Lys51, Lys151, Lys155, and Arg158) appear strictly conserved (Figure 7.16C).

7.3.2.2 *Anti-NS5B drug development*

The current therapy for HCV infection is a combination of ribavirin and interferon-α. Ribavirin is a broad-spectrum antiviral nucleoside analog (Fig-

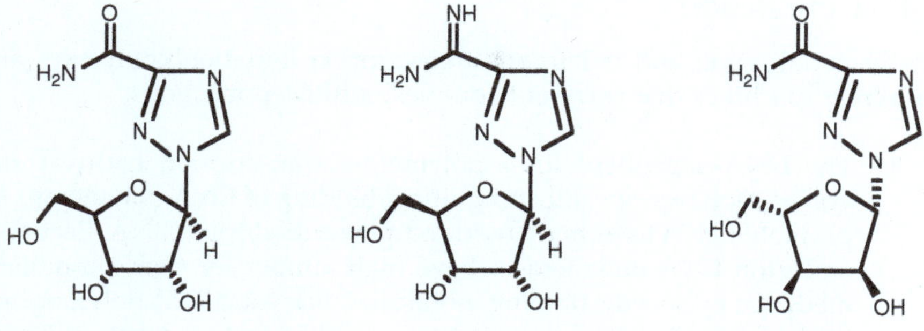

Figure 7.17 Chemical structures of the three antiviral nucleoside analogs: ribavirin, viramidine, and levovirin.

ure 7.17). The 5′-monophosphate of ribavirin (RMP) has been shown to inhibit inosine monophosphate dehydrogenase (IMPDH), causing the depletion of intracellular GTP pools. RMP has also been proposed to inhibit viral transcription[78] and guanylyltransferase activity.[79] Recently, substantial evidence has suggested ribavirin is an RNA mutagen because its 5′-triphosphate (RTP) has been shown to be incorporated into viral RNA by NS5B, which base pairs equally well with a uracil or a cytosine via two hydrogen bonds through rotation of its exocyclic carboxyamide moiety.[80,81]

Driven by the clinical success of nucleoside-analog inhibitors of HIV reverse transcriptase and the efficacy of ribavirin, significant efforts have been devoted to finding effective anti-HCV nucleoside analogs that act as powerful mutagens or as chain terminators. Some progress has been achieved recently. Two derivatives of ribavirin, Levovirin (the *L*-enantiomer the ribavirin) and Viramidine (the 3-carboxamidine derivative), have similar antiviral functions of ribavirin but appear to be less toxic[82] (Figure 7.17). Alternatively, the 5′-triphosphates of 2′-C-methyladenosine and 2′-O-methylcytidine have been found to inhibit NS5B-catalyzed RNA synthesis *in vitro* and in subgenomic HCV-replicon assays in a manner competitive with substrate nucleoside triphosphates. NS5B is able to incorporate these two nucleotide analogs into RNA, but is subsequently impaired in its ability to extend the incorporated analog by addition of the next nucleotide.[83]

In addition, a series of diketobutanoic acids have been shown to inhibit NS5B activity *in vitro* and in the HCV-replicon assay. These compounds apparently interfere with the binding of the phosphoryl groups of rNTPs at the active site of NS5B, thereby preventing the formation of phosphodiester bonds catalyzed by the polymerase. In the search for alternative anti-HCV inhibitors, *N,N*-disubstituted phenylalanines,[84] heterocyclic derivatives,[85] monoclonal antibodies,[86] and high-affinity and specific RNA aptamers of NS5B[87] have been found to inhibit the RdRp activity of NS5B by different mechanisms. Hopefully, more potent and selective anti-HCV agents will be developed in the near future.

7.4 Conclusion

The bacterial, viral, and eukaryotic transcription initiation complexes are molecular machines that carry out complex, multistep reactions:

1. The DNA-dependent RNA polymerase transcription pathway involves sequence-specific recognition: binding of RNA polymerase to promoter DNA to form a closed complex with duplex DNA. Bacterial and viral RNA polymerases have high affinity for their promoters mediated by specific binding motifs. In eukaryotic RNA polymerases this step involves the binding of transcription factors, which enhances the protein affinity to DNA
2. Melting the DNA duplex through specific DNA/protein interaction (hydrophobic and/or van der Waals interactions in nature) and isomerization through several intermediates to form an open complex with an ~14-nucleotide region of melted single-stranded DNA surrounding the transcription start
3. Abortive cycles of synthesis and release of 2- to 8-nucleotide RNA oligomers as an initial transcribing complex
4. Upon synthesis of a 9-nucleotide RNA oligomer, isomerization to break protein/DNA interactions between RNA polymerase and the promoter (in the case of viral and bacterial DNA-dependent RNA polymerases) and to break, or weaken, protein/protein interactions between RNA polymerase and initiation factor(s) (in the case of eukaryotic DNA-dependent RNA polymerases), resulting in an elongation complex that processively translocates along DNA and extends the RNA product

Different polymerases extend RNA transcript using different mechanisms. T7 RNAP elongates RNA by protein inchworming, whereas yeast Pol II follows the DNA scrunching mechanism.

Viral RdRps show different mechanisms of replication and transcription. On one hand, the RdRp activity of HCV resides in the membrane-associated nonstructural protein NS5B. HCV NS5B can initiate RNA replication from a single rNTP; the structure of HCV NS5B alone shows the "closed" conformation. The structure of NS5B in complex with rNTP exhibits three distinct nucleotide binding sites at the enzyme active site in the absence of an RNA substrate; one of the three sites is specific for rGTP.

On the other hand, the RdRp of poliovirus is in an "open" conformation in the crystal structure. The replication initiation in poliovirus RNA synthesis is the covalent linkage of the 5′-phosphate of UMP to the hydroxyl group of a tyrosine in the terminal protein VPg (3B). Apparently, the kinetic and structural characteristics of RdRp exhibit high tendency of oligomerization and cooperativity. Nevertheless, a detailed understanding of the mechanism of transcription by viral RdRp is prohibited by the lack of sufficient structural information about the conformational transitions from initiation to termination.

Each step in the transcription pathway appears to involve conformational changes in the RNA polymerase and/or promoter DNA. Issues for ongoing work involve defining the structure of the complex at each step of transcription (initiation, elongation, and termination), defining the conformational transitions, and defining the kinetics of the transitions. These studies will set the stage for comprehensive understanding of the mechanism of gene transcription.

Abbreviations

CTD carboxy-terminal domain
HCV hepatitis C virus
RdRp RNA-dependent RNA polymerase
RNA Pol II RNA polymerase II
rNTP ribonucleotide triphosphate
TBP TATA binding protein
TF transcription factor
T7 RNAP T7 RNA polymerase

References

1. Ishihama, A. (2000) Functional modulation of *Escherichia coli* RNA polymerase, *Annu. Rev. Microbiol.*, 54, 499–518.
2. Sousa, R., Chung, Y.J., Rose, J.P., and Wang, B.C. (1993) Crystal structure of bacteriophage T7 RNA polymerase at 3.3-A resolution, *Nature*, 364, 593–599.
3. Jeruzalmi, D. and Steitz, T.A. (1998) Structure of T7 RNA polymerase complexed to the transcriptional inhibitor T7 lysozyme, *Embo J.*, 17, 4101–4113.
4. Cheetham, G.M., Jeruzalmi, D., and Steitz, T.A. (1999) Structural basis for initiation of transcription from an RNA polymerase-promoter complex, *Nature*, 399, 80–83.
5. Cheetham, G.M. and Steitz, T.A. (1999) Structure of a transcribing T7 RNA polymerase initiation complex, *Science*, 286, 2305–2309.
6. Yin, W.S. and Steitz, T.A. (2002) Structural basis of the transition from initiation to elongation transcription in T7 RNA polymerase, *Science*, 298, 1387–1395.
7. Tahirov, T., Temiakov, D., Anlkin, M., Patlan, V., McAllister, W.T., Vassylyv, D.G., and Yokoyama, S. (2002) Structure of a T7 RNA polymerase elongation complex at 2.9-Å resolution, *Nature*, 420, 43–50.
8. Sakurai, H. and Ishihama, A. (2001) Transcription organization and mRNA levels of the genes for all 12 subunits of the fission yeast RNA polymerase II, *Genes Cells*, 6, 25–36.
9. Cramer, P. (2002) Multisubunit RNA polymerases, *Curr. Opin. Struct. Biol.*, 12, 89–97.
10. McKune, K., Moore, P.A., Hull, M.W., and Woychik, N.A. (1995) Six human RNA polymerase subunits functionally substitute for their yeast counterparts, *Mol. Cell. Biol.*, 15, 6895–6900.

11. Pati, U.K. (1994) Human RNA polymerase II subunit hRPB14 is homologous to yeast RNA polymerase I, II, and III subunits (AC19 and RPB11) and is similar to a portion of the bacterial RNA polymerase alpha subunit, *Gene*, 145, 289–292.
12. Mitobe, J., Mitsuzawa, H., and Ishihama, A. (2001) Functional analysis of RNA polymerase II Rpb3 mutants of the fission yeast *Schizosaccharomyces pombe*, *Curr. Genet.*, 39, 210–221.
13. Cramer, P., Bushnell, D.A., and Kornberg, R.D. (2001) Structural basis of transcription: RNA polymerase II at 2.8 angstrom resolution, *Science*, 292, 1863–1876.
14. Gnatt, A.L., Cramer, P., Fu, J., Bushnell, D.A., and Kornberg, R.D. (2001) Structural basis of transcription: an RNA polymerase II elongation complex at 3.3-A resolution, *Science*, 292, 1876–1882.
15. Gnatt, A. (2002) Elongation by RNA polymerase II: structure–function relationship, *Biochim. Biophys. Acta*, 1577, 175–190.
16. Nudler, E., Mustaev, A., Lukhtanov, E., and Goldfarb, A. (1997) The RNA–DNA hybrid maintains the register of transcription by preventing backtracking of RNA polymerase, *Cell*, 89, 38–41.
17. Adhya, S., Sarkar, P., Valenzuela, D., and Maitra, U. (1979) Termination of transcription by Escherichia coli RNA polymerase: influence of secondary structure of RNA transcripts on rho-independent and rho-dependent termination, *Proc. Natl. Acad. Sci. USA*, 76, 1613–1617.
18. Richardson, J.P. (2002) Rho-dependent termination and ATPases in transcript termination, *Biochim. Biophys. Acta*, 1577, 251–260.
19. Sousa, R. and Padilla, R. (1995) A mutant T7 RNA polymerase as a DNA polymerase, *Embo J.*, 14, 4609–4621.
20. Gudima, S.O., Kazantseva, E.G., Kostyuk, D.A., Shchaveleva, I.L., Grishchenko, O.I., Memelova, L.V., and Kochetkov, S.N. (1997) Deoxyribonucleotide-containing RNAs: a novel class of templates for HIV-1 reverse transcriptase, *Nucleic Acids Res.*, 25, 4614–4618.
21. Nelson, D.L. and Cox, M.M. (2000) RNA metabolism, in *Lehninger Principles of Biochemistry* (chap. 26), 987–990. New York: Worth Publishers, Incorporated.
22. Smale, S.A.J.K. (2003) The RNA polymerase II core promoter, *Annu. Rev. Biochem.*, 72, 449–479.
23. Shilatifard, A., Conaway, R.C. and JW Conaway, J.W. (2003) The RNA polymerase II elongation complex, *Annu. Rev. Biochem.*, 72, 693–715.
24. Koleske, A.J. and Young, R.A. (1994) An RNA polymerase II holoenzyme responsive to activators, *Nature*, 368, 466–469.
25. Ossipow, V., Tassan, J.P., Nigg, E.A., and Schibler, U. (1995) A mammalian RNA polymerase II holoenzyme containing all components required for promoter-specific transcription initiation, *Cell*, 83, 137–146.
26. Cook, P.R. (1999) The organization of replication and transcription, *Science*, 284, 1790–1795.
27. Schultz, S.C., Shields, G.C., and Steitz, T.A. (1991) Crystal structure of a CAP–DNA complex: the DNA is bent by 90°, *Science*, 253, 1001–1007.
28. Proudfoot, N.J., Furger, A., and Dye, M.J. (2002) Integrating mRNA processing with transcription, *Cell*, 108, 501–512.
29. Hampsey, M. (1998) Molecular genetics of the RNA polymerase II general transcriptional machinery, *Microbiol. Mol. Biol. Rev.*, 62, 465–503.

30. Kim, J.B. and Sharp, P.A. (2001) Positive transcription elongation factor B phosphorylates hSPT5 and RNA polymerase II carboxyl-terminal domain independently of cyclin-dependent kinase-activating kinase, *J. Biol. Chem.*, 276, 12317–12323.
31. Cadena, D.L. and Dahmus, M.E. (1987) Messenger RNA synthesis in mammalian cells is catalyzed by the phosphorylated form of RNA polymerase II, *J. Biol. Chem.*, 262, 12468–12474.
32. Dahmus, M.E. (1996) Reversible phosphorylation of the C-terminal domain of RNA polymerase II, *J. Biol. Chem.*, 271, 19009–19012.
33. O'Brien, T., Hardin, S., Greenleaf, A., and Lis, J.T. (1994) Phosphorylation of RNA polymerase II C-terminal domain and transcriptional elongation, *Nature*, 370, 75–77.
34. Prelich, G. (2002) RNA polymerase II carboxy-terminal domain kinases: emerging clues to their function, *Eukaryotic Cell*, 1, 153–162.
35. Chambers, R.S., Wang, B.Q., Burton, Z.F., and Dahmus, M.E. (1995) The activity of COOH-terminal domain phosphatase is regulated by a docking site on RNA polymerase II and by the general transcription factors IIF and IIB, *J. Biol. Chem.*, 270, 14962–14969.
36. Chambers, R.S. and Dahmus, M.E. (1994) Purification and characterization of a phosphatase from HeLa cells which dephosphorylates the C-terminal domain of RNA polymerase II, *J. Biol. Chem.*, 269, 26243–26248.
37. van den Boom, V., Jaspers, N.G., and Vermeulen, W.(2002) When machines get stuck — obstructed RNA polymerase II: displacement, degradation or suicide, *Bioessays*, 24, 780–784.
38. Ratner, J.N., Balasubramanian, B., Corden, J., Warren, S.L., and Bregman, D.B. (1998) Ultraviolet radiation-induced ubiquitination and proteasomal degradation of the large subunit of RNA polymerase II. Implications for transcription-coupled DNA repair, *J. Biol. Chem.*, 273, 5184–5189.
39. Bienz, K., Egger, D., Pfister, T., and Troxler, M. (1992) Structural and functional characterization of the poliovirus replication complex, *J. Virol.*, 66, 2740–2747.
40. Ambros, V. and Baltimore, D. (1978) Protein is linked to the 5′ end of poliovirus RNA by a phosphodiester linkage to tyrosine, *J. Biol. Chem.*, 253, 5263–5266.
41. Rothberg, P.G., Harris, T.J., Nomoto, A., and Wimmer, E. (1978) O4-(5′-uridylyl)tyrosine is the bond between the genome-linked protein and the RNA of poliovirus, *Proc. Natl. Acad. Sci. USA*, 75, 4868–4672.
42. Van Dyke, T.A. and Flanegan, J.B. (1980) Identification of poliovirus polypeptide P63 as a soluble RNA-dependent RNA polymerase, *J. Virol.*, 35, 732–740.
43. Beckman, M.T. and Kirkegaard, K. (1998) Site size of cooperative single-stranded RNA binding by poliovirus RNA-dependent RNA polymerase, *J. Biol. Chem.*, 273, 6724–6670.
44. Pata, J.D., Schultz, S.C., and Kirkegaard, K. (1995) Functional oligomerization of poliovirus RNA-dependent RNA polymerase, *RNA*, 1, 466–477.
45. Hope, D.A., Diamond, S.E., and Kirkegaard, K. (1997) Genetic dissection of interaction between poliovirus 3D polymerase and viral protein 3AB, *J. Virol.*, 71, 9490–8.
46. Pathak, H.B., Ghosh, S.K., Roberts, A.W., Sharma, S.D., Yoder, J.D., Arnold, J.J., Gohara, D.W., Barton, D.J., Paul, A.V., and Cameron, C.E. (2002) Structure–function relationships of the RNA-dependent RNA polymerase from poliovirus (3Dpol). A surface of the primary oligomerization domain functions in capsid precursor processing and VPg uridylylation, *J. Biol. Chem.*, 277, 31551–31562.

47. Andino, R., Rieckhof, G.E., Achacoso, P.L., and Baltimore, D. (1993) Poliovirus RNA synthesis utilizes an RNP complex formed around the 5′-end of viral RNA, *Embo J.*, 12, 3587–3598.
48. Xiang, W., Cuconati, A., Paul, A.V., Cao, X., and Wimmer, E. (1995) Molecular dissection of the multifunctional poliovirus RNA-binding protein 3AB, *RNA*, 1, 892–904.
49. Hope, D.A., Diamond, S.E., and Kirkegaard, K. (1997) Genetic dissection of interaction between poliovirus 3D polymerase and viral protein 3AB, *J. Virol.*, 71, 9490–9498.
50. Paul, A.V., Cao, X., Harris, K.S., Lama, J., and Wimmer, E. (1994) Studies with poliovirus polymerase 3Dpol. Stimulation of poly(U) synthesis *in vitro* by purified poliovirus protein 3AB, *J. Biol. Chem.*, 269, 29173–29181.
51. Richards, O.C. and Ehrenfeld, E. (1998) Effects of poliovirus 3AB protein on 3D polymerase-catalyzed reaction, *J. Biol. Chem.*, 273, 12832–12840.
52. Hansen, J.L., Long, A.M., and Schultz, S.C. (1997) Structure of the RNA-dependent RNA polymerase of poliovirus, *Structure*, 5, 1109–1122.
53. Diamond, S.E. and Kirkegaard, K. (1994) Clustered charged-to-alanine mutagenesis of poliovirus RNA-dependent RNA polymerase yields multiple temperature-sensitive mutants defective in RNA synthesis, *J. Virol.*, 68, 863–876.
54. Hobson, S.D., Rosenblum, E.S., Richards, O.C., Richmond, K., Kirkegaard, K., and Schultz, S.C. (2001) Oligomeric structures of poliovirus polymerase are important for function, *Embo J.*, 20, 1153–1163.
55. Lyle, J.M., Bullitt, E., Bienz, K., and Kirkegaard, K. (2002) Visualization and functional analysis of RNA-dependent RNA polymerase lattices, *Science*, 296, 2218–2222.
56. Kolykhalov, A.A., Mihalik, K., Feinstone, S.M., and Rice, C.M. (2000) Hepatitis C virus-encoded enzymatic activities and conserved RNA elements in the 3′ nontranslated region are essential for virus replication *in vivo*, *J. Virol.*, 74, 2046–51.
57. Hwang, S.B., Park, K.J., Kim, Y.S., Sung, Y.C., and Lai, M.M. (1997) Hepatitis C virus NS5B protein is a membrane-associated phosphoprotein with a predominantly perinuclear localization, *Virology*, 227, 439–446.
58. Poch, O., Sauvaget, I., Delarue, M., and Tordo, N. (1989) Identification of four conserved motifs among the RNA-dependent polymerase encoding elements, *Embo J.*, 8, 3867–3874.
59. Lohmann, V., Korner, F., Herian, U., and Bartenschlager, R. (1997) Biochemical properties of hepatitis C virus NS5B RNA-dependent RNA polymerase and identification of amino acid sequence motifs essential for enzymatic activity, *J. Virol.*, 71, 8416–8428.
60. Lohmann, V., Roos, A., Korner, F., Koch, J.O., and Bartenschlager, R. (1998) Biochemical and kinetic analyses of NS5B RNA-dependent RNA polymerase of the hepatitis C virus, *Virology*, 249, 108–118.
61. Behrens, S.E., Tomei, L., and De Francesco, R. (1996) Identification and properties of the RNA-dependent RNA polymerase of hepatitis C virus, *Embo J.*, 15, 12–22.
62. Oh, J.W., Ito, T., and Lai, M.M. (1999) A recombinant hepatitis C virus RNA-dependent RNA polymerase capable of copying the full-length viral RNA, *J. Virol.*, 73, 7694–7702.

63. Sun, X.L., Johnson, R.B., Hockman, M.A., and Wang, Q.M. (2000) *De novo* RNA synthesis catalyzed by HCV RNA-dependent RNA polymerase, *Biochem. Biophys. Res. Commun.*, 268, 798–803.
64. Oh, J.W., Sheu, G.T., and Lai, M.M. (2000) Template requirement and initiation site selection by hepatitis C virus polymerase on a minimal viral RNA template, *J. Biol. Chem.*, 275, 17710–17717.
65. Cheng, J.C., Chang, M.F., and Chang, S.C. (1999) Specific interaction between the hepatitis C virus NS5B RNA polymerase and the 3′ end of the viral RNA, *J. Virol.*, 73, 7044–7049.
66. Shim, J.H., Larson, G., Wu, J.Z., and Hong, Z. (2002) Selection of 3′-template bases and initiating nucleotides by hepatitis C virus NS5B RNA-dependent RNA polymerase, *J. Virol.*, 76, 7030–9.
67. Lohmann, V., Overton, H., and Bartenschlager, R. (1999) Selective stimulation of hepatitis C virus and pestivirus NS5B RNA polymerase activity by GTP, *J. Biol. Chem.*, 274, 10807–10815.
68. Piccininni, S., Varaklioti, A., Nardelli, M., Dave, B., Raney, K.D., and McCarthy, J.E. (2002) Modulation of the hepatitis C virus RNA-dependent RNA polymerase activity by the nonstructural (NS) 3 helicase and the NS4B membrane protein, *J. Biol. Chem.*, 277, 45670–45679.
69. Shirota, Y., Luo, H., Qin, W., Kaneko, S., Yamashita, T., Kobayashi, K., and Murakami, S. (2002) Hepatitis C virus (HCV) NS5A binds RNA-dependent RNA polymerase (RdRP) NS5B and modulates RNA-dependent RNA polymerase activity, *J. Biol. Chem.*, 277, 11149–11155.
70. Uchida, M., Hino, N., Yamanaka, T., Fukushima, H., Imanishi, T., Uchiyama, Y., Kodama, T., and Doi, T. (2002) Hepatitis C virus core protein binds to a C-terminal region of NS5B RNA polymerase, *Hepatol. Res.*, 22, 297–306.
71. Bressanelli, S., Tomei, L., Roussel, A., Incitti, I., Vitale, R.L., Mathieu, M., De Francesco, R., and Rey, F.A. (1999) Crystal structure of the RNA-dependent RNA polymerase of hepatitis C virus, *Proc. Natl. Acad. Sci. USA*, 96, 13034–13039.
72. Lesburg, C.A., Cable, M.B., Ferrari, E., Hong, Z., Mannarino, A.F., and Weber, P.C. (1999) Crystal structure of the RNA-dependent RNA polymerase from hepatitis C virus reveals a fully encircled active site, *Nat. Struct. Biol.*, 6, 937–943.
73. Adachi, T., Ago, H., Habuka, N., Okuda, K., Komatsu, M., Ikeda, S., and Yatsunami, K. (2002) The essential role of C-terminal residues in regulating the activity of hepatitis C virus RNA-dependent RNA polymerase, *Biochim. Biophys. Acta*, 1601, 38–48.
74. Bressanelli, S., Tomei, L., Rey, F.A., and De Francesco, R. (2002) Structural analysis of the hepatitis C virus RNA polymerase in complex with ribonucleotides, *J. Virol.*, 76, 3482–3492.
75. Wang, Q.M., Hockman, M.A., Staschke, K., Johnson, R.B., Case, K.A., Lu, J., Parsons, S., Zhang, F., Rathnachalam, R., Kirkegaard, K., and Colacino, J.M. (2002) Oligomerization and cooperative RNA synthesis activity of hepatitis C virus RNA-dependent RNA polymerase, *J. Virol.*, 76, 3865–3872.
76. Labonte, P., Axelrod, V., Agarwal, A., Aulabaugh, A., Amin, A., and Mak, P. (2002) Modulation of hepatitis C virus RNA-dependent RNA polymerase activity by structure-based site-directed mutagenesis, *J. Biol. Chem.*, 277, 38838–38846.

77. Qin, W., Luo, H., Nomura, T., Hayashi, N., Yamashita, T., and Murakami, S. (2002) Oligomeric interaction of hepatitis C virus NS5B is critical for catalytic activity of RNA-dependent RNA polymerase, *J. Biol. Chem.*, 277, 2132–2137.
78. Eriksson, B., Helgstrand, E., Johansson, N.G., Larsson, A., Misiorny, A., Noren, J.O., Philipson, L., Stenberg, K., Stening, G., Stridh, S., and Oberg, B. (1977) Inhibition of influenza virus ribonucleic acid polymerase by ribavirin triphosphate, *Antimicrob. Agents Chemother.*, 11, 946–951.
79. Goswami, B.B., Borek, E., Sharma, O.K., Fujitaki, J., and Smith, R.A. (1979) The broad spectrum antiviral agent ribavirin inhibits capping of mRNA, *Biochem. Biophys. Res. Commun.*, 89, 830–836.
80. Crotty, S., Maag, D., Arnold, J.J., Zhong, W., Lau, J.Y., Hong, Z., Andino, R., and Cameron, C.E. (2000) The broad-spectrum antiviral ribonucleoside ribavirin is an RNA virus mutagen, *Nat. Med.*, 6, 1375–1379.
81. Crotty, S., Cameron, C.E., and Andino, R. (2001) RNA virus error catastrophe: direct molecular test by using ribavirin, *Proc. Natl. Acad. Sci. USA*, 98, 6895–6900.
82. Watson, J. (2002) Prospects for hepatitis C virus therapeutics: levovirin and viramidine as improved derivatives of ribavirin, *Curr. Opin. Invest. Drugs*, 3, 680–683.
83. Carroll, S.S., Tomassini, J.E., Bosserman, M., Getty, K., Stahlhut, M.W., Eldrup, A.B., Bhat, B., Hall, D., Simcoe, A.L., LaFemina, R., Rutkowski, C.A., Wolanski, B., Yang, Z., Migliaccio, G., De Francesco, R., Kuo, L.C., MacCoss, M., and Olsen, D.B. (2003) Inhibition of hepatitis C virus RNA replication by 2′-modified nucleoside analogs, *J. Biol. Chem.*, 278, 11979–11984.
84. Chan, L., Reddy, T.J., Proulx, M., Das, S.K., Pereira, O., Wang, W., Siddiqui, A., Yannopoulos, C.G., Poisson, C., Turcotte, N., Drouin, A., Alaoui–Ismaili, M.H., Bethell, R., Hamel, M., L'Heureux, L., Bilimoria, D., and Nguyen–Ba, N. (2003) Identification of N,N-disubstituted phenylalanines as a novel class of inhibitors of hepatitis C NS5B polymerase, *J. Med. Chem.*, 46, 1283–1285.
85. Gu, B., Johnston, V.K., Gutshall, L.L., Nguyen, T.T., Gontarek, R.R., Darcy, M.G., Tedesco, R., Dhanak, D., Duffy, K.J., Kao, C.C., and Sarisky, R.T. (2003) Arresting initiation of hepatitis C virus RNA synthesis using heterocyclic derivatives, *J. Biol. Chem.*, 278, 16602–16607.
86. Moradpour, D., Bieck, E., Hugle, T., Wels, W., Wu, J.Z., Hong, Z., Blum, H.E., and Bartenschlager, R. (2002) Functional properties of a monoclonal antibody inhibiting the hepatitis C virus RNA-dependent RNA polymerase, *J. Biol. Chem.*, 277, 593–601.
87. Kumar, P.K., Machida, K., Urvil, P.T., Kakiuchi, N., Vishnuvardhan, D., Shimotohno, K., Taira, K., and Nishikawa, S. (1997) Isolation of RNA aptamers specific to the NS3 protein of hepatitis C virus from a pool of completely random RNA, *Virology*, 237, 270–282.

chapter eight

Reverse transcriptases

Louise Wang
Institute of Cellular and Molecular Biology, University of Texas at Austin, TX

Zucai Suo*
Department of Biochemistry, Ohio State University, Columbus, OH

Contents

* Corresponding author.

8.1 Introduction

Reverse transcriptase (RT) has been studied for the past three decades after its discovery in 1970 by Baltimore[1] and Temin.[2] RTs are biochemically distinct from other DNA polymerases in two aspects. First, RTs copy a single-strand viral RNA genome into double-strand proviral DNA using RNA and DNA as a template, unlike DNA polymerases, which can only copy DNA templates to synthesize new DNA strands. The flexible RT active site can accommodate DNA and RNA substrates to fulfill the requisite primer/template alignment during different stages of reverse transcription. Second, RTs lack the 3′→5′ exonuclease activity of replicative DNA polymerases[3] that is responsible for proofreading and correcting mismatched nucleotides. Therefore, the fidelity of DNA synthesis is substantially lower for RTs in comparison to the fidelity of most replicative DNA polymerases. Error-prone polymerization essentially accounts for the characteristic hypermutability of retroviruses and hepadnaviruses, which ultimately confers drug resistance.

In this chapter, the reverse transcriptase of human immunodeficiency type 1 virus (HIV-1 RT) will be discussed most extensively due to the abundance of information available for this enzyme. RTs of other origins will only be briefly described to contrast with HIV-1 RT. RTs will be reviewed in the aspects of structure, function, mechanism, and evolution with respect to their roles in antiviral drug research.

8.2 Evolution of RT-containing viruses

RT can rapidly evolve and increase the diversity of genetic populations via error-prone replication. Homologous elements of RTs are encoded in the genomes of many retroviruses[1] such as human immunodeficiency virus (HIV); hepadnaviruses such as hepatitis B virus (HBV)[4,5]; plant and animal DNA viruses such as cauliflower mosaic viruses[4]; retrotransposons[6–8]; and mitochondrial group II introns.[9] Retroviral and hepadnaviral RTs have been the main focus of RT research because HIV and HBV cause worldwide pandemic diseases spread through transfusion of blood and/or body fluid.

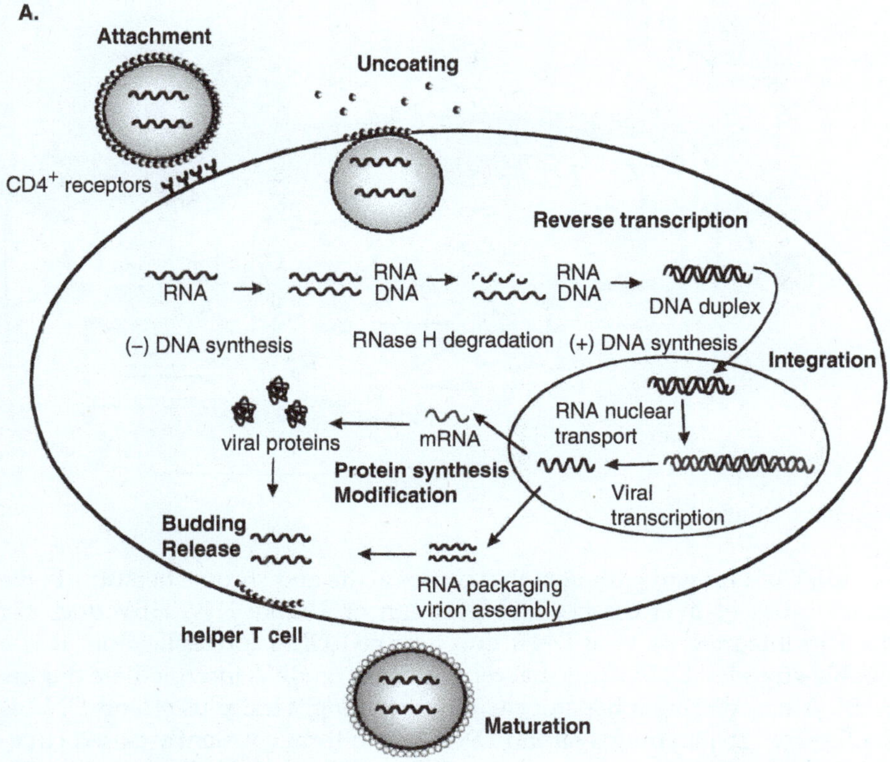

Figure 8.1 Schematic illustrations of viral life cycles. A) HIV fuses with a CD4+ cell via surface receptors. Viral RNA is reverse transcribed into DNA, and integrated into the host genes. The viral DNA remains latent until it is transcribed into mRNA and then translated into viral proteins. The viral RNA and viral proteins are assembled into a mature virus at the membrane, which is then released to infect other host cells. B) Infectious HBV attaches to host cellular receptors and releases nucleocapsids that move into the host cell nucleus. The partially double-stranded HBV DNA is converted to covalently closed-circular form (cccDNA), which is the template for synthesis of HBV proteins. The pregenomic RNA (pgRNA) binds core or capsid protein and polymerase (P) proteins to form RNA packaged nucleocapsids. The pgRNA is reverse transcribed into dsDNA by the polymerase. The viral DNA is redirected to the nucleus or coated by surface glycoproteins in the Golgi and endoplasmic reticulum (ER) before being exported as enveloped virions. Continued.

HIV is the etiological agent of the acquired immunodeficiency syndrome (AIDS).[10] HIV infects CD4+ helper T-cells of the immune system (Figure 8.1A). After being engulfed by the host cells, the viral single-stranded RNA is copied into double-stranded proviral DNA by HIV RT. The proviral DNA is integrated into the host genome by the HIV integrase. Host cellular polymerases transcribe the proviral DNA into viral RNA, which is subsequently translated into viral proteins by cellular translational machinery. Virions are assembled and bud from the cell membrane to infect other host cells.

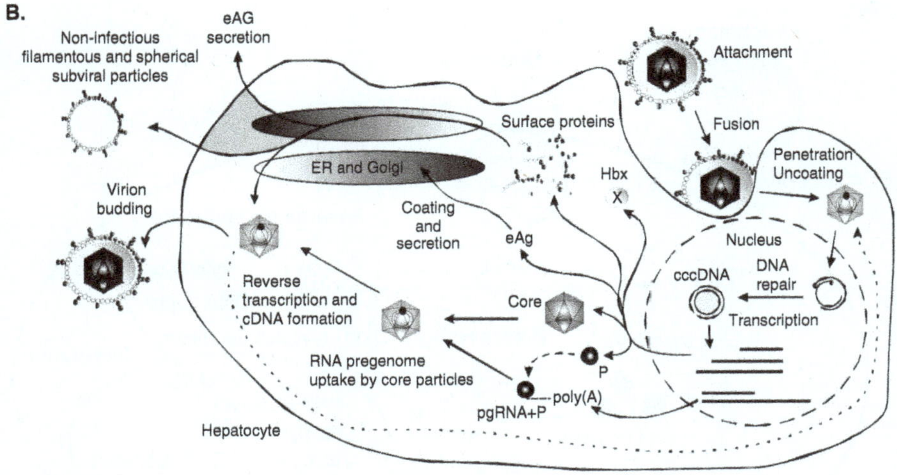

Figure 8.1 Continued.

HBV is a human pathogen that causes acute and chronic hepatitis B, the major cause of liver disease and liver cancer. Unlike HIV, HBV does not need to integrate its viral DNA into host cell DNA for replication. It is a double-stranded DNA virus that replicates via an RNA intermediate (Figure 8.1B). After entering a hepatocyte and uncoating, viral plus-strand DNA is synthesized to join minus-strand DNA and to form covalently closed circular DNA (cccDNA) in the nucleus. The cccDNA serves as a template for transcription by host RNA polymerase II to pregenomic RNA (pgRNA), which is translated to yield HBV proteins. The pgRNA is subsequently reverse transcribed to minus-strand and plus-strand DNA by HBV polymerase, as the pgRNA is concomitantly degraded. Once the genome is translated and packaged into a mature core, the core buds into the lumen of endoplasmic reticulum, secretes through cellular membrane, and infects new hepatocytes.

De novo mutations are generated by inherently promiscuous reverse transcription during the viral life cycle (Figure 8.1), thus introducing base substitutions, frame shifts, genetic rearrangements, and hypermutations.[11,12] HIV has been shown to contain a high level of genetic variation.[13,14] HIV RT has a higher error rate than those of the RTs of avian myeloblastosis virus (AMV) and murine leukemia virus (MLV).[15] Thus, the fidelity of HIV-1 RT is a major contributing factor to the strain diversity.[15–17] Each single nucleotide variant on a branch in the HIV phylogenetic tree is deemed a unique and viable virus.[18] There are 296 drug-resistant mutations that have been recorded in the HIV database from Los Alamos National Laboratory (http://hiv-web.lanl.gov). With a rate of less than one mutation per genome per cycle,[18] the hypermutation of HIV is considered the Achilles' heel of anti-AIDS therapy. Drug-resistant strains appear in a matter of weeks after the treatment begins.[19]

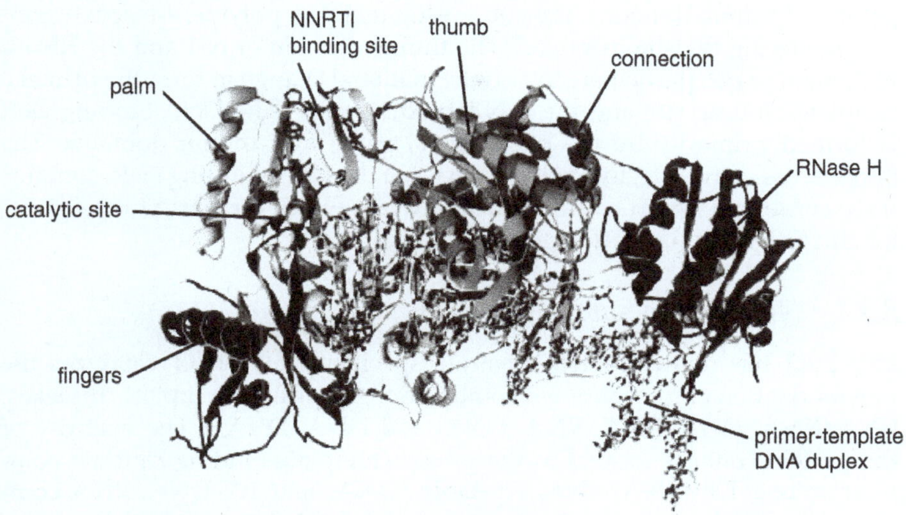

Figure 8.2 HIV-1 RT heterodimer in catalytic complex with double-stranded DNA. The five domains of p66 are labeled and p51 is in the lightest gray color. Binding sites for NNRTIs and NRTIs on RT are distinguishable from each other. The NNRTI binding pocket is outlined by the side chains of the hydrophobic residues composing that region. The NRTI binding site is highlighted by the active site conserved residues, D110, D185, and D186, in the space-filled model. The primer/template duplex is illustrated in the ball-and-stick model. This image was generated with 1RTD.pdb (28) using Swiss PDB viewer and POV-Ray. (From Huang, H.F. et al., *Science*, 282(5394), 1669–1675, 1998.)

8.3 Structures of reverse transcriptase

HIV-1 RT has attracted the attention of scientists from academic institutions and industry because of the urgency to treat AIDS patients. As a result, HIV-1 RT is the best-characterized retroviral RT in terms of structure and function. HIV-1 RT is a heterodimer of two tightly associated polypeptides, a 66 kDa subunit (p66) and a 51 kDa subunit (p51). The p66 subunit consists of five distinct domains (Figure 8.2). Four domains, comprising the N-terminal 440 amino acid residues, are referred to as the fingers, palm, thumb, and connection[20] and form the polymerase core in the general shape of a right hand. The fifth p66 subunit domain is the ribonuclease H (RNase H) domain made up of the C-terminal 120 amino acid residues. This domain is responsible for the hydrolytic cleavage of the RNA template in an RNA-DNA duplex. The p51 subunit is formed by proteolytic cleavage of the C-terminal 120 amino acids of the p66 subunit by HIV protease[21] and hence contains the four domains of the polymerase core while lacking the RNase H domain.

The two subunits have identical N-terminal sequences. However, the orientations of the domains are different in the p66 and p51 subunits, such that only the p66 subunit possesses a functional polymerase active site in its palm domain. In the p51 subunit, the connection domain lies between its palm and thumb domains, thus obscuring its DNA polymerase active site[20] and rendering the site inactive.[22] The thumb domain of p51 and the RNase H domain of p66 participate in a conformational change to form the primer/template-binding site and the tRNA-binding site.[23] The DNA binding cleft is formed primarily by the p66 fingers, palm, and thumb domains. The polymerase active site, located at the base of the DNA-binding cleft, contains three conserved aspartic acid residues, two of which are part of the characteristic YXDD motif of RT structures.

8.3.1 Nucleic acid substrate binding

HIV-1 RT has nonrestrictive substrate specificity. HIV-1 RT catalyzes the polymerization of four different combinations of primer/template duplexes: RNA/RNA, DNA/RNA, RNA/DNA, and DNA/DNA.[24] The majority of the protein contacts located in the primer/template-binding cleft are common among RT–tRNA/RNA, RT–DNA/DNA, and RT–DNA/RNA complexes, except for a few contacts between the p66 fingers domain and the viral tRNA/RNA that are not detected with RT–DNA/DNA.[25] The "RNase H primer grip," consisting of amino acids that interact with the DNA primer strand, may control the interactions between the RNA template and the RNase H active site, depending on the sequence and structure of the nucleic acid.[26] Another important role for the RNase H domain is to bind divalent cations within the RNase H active site, which neutralizes the negative charges of residues in the RNase H region and thereby stabilizes the association of RT with primer/template.[27]

Crystallographic studies of RT-DNA complexes reveal conformational changes upon DNA binding. In the RT–DNA binary complex, DNA binding along a groove stretching from the polymerase active site to the RNase H active site[28] increases the flexibility of RT[29] and causes a rotation of the thumb domain of p66[30] towards the primer/template (Figure 8.3), in concert with conformational changes in the YMDD motif.[31] The DNA primer terminus is in the vicinity of the catalytic aspartic triad in the palm domain of p66. Hydrophobic interactions and possible hydrogen bonds help stabilize the polymerase–DNA complex.[32] DNA base pairs close to the active site form a type A-like structure with a widened minor groove, similar to what has been observed with other DNA polymerases.[33] Six to seven base pairs away from the polymerase active site, the bound DNA exhibits a 40° bend[34] and transitions to B-form DNA structure.

8.3.2 Ternary complex formation

The formation of the ternary RT–DNA–dNTP complex resembles ternary complexes of other DNA polymerases.[35–37] A binding pocket is formed by closure of the fingers domain towards the palm and thumb domains to accommodate the 3′-hydroxyl group of an incoming nucleotide. The distal portion of the fingers domain bends inward to the palm domain and the

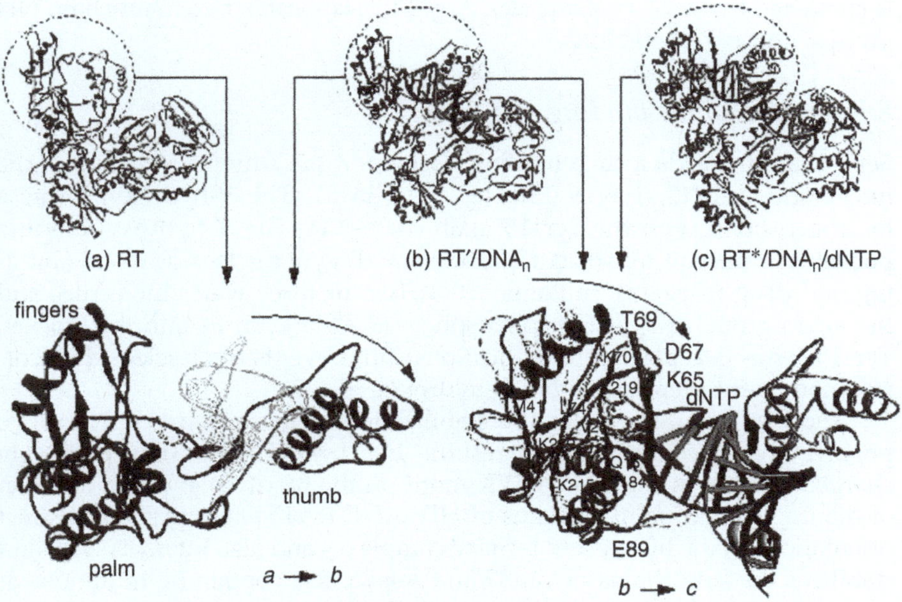

Figure 8.3 Conformational changes inside HIV-1 RT p66 subunit during DNA polymerization. (a) unliganded form of RT; (b) binary complex RT′/DNA$_n$; (c) ternary complex RT*/DNA$_n$/dNTP; (a → b) rotation of the p66 thumb domain away from the fingers domain induced by DNA binding (see the arrow); (b → c) hinge movement of the fingers domain to close down to the polymerase active site at the floor of the palm domain induced by nucleotide binding (see the arrow). The thumb domain in a → b is in light gray in the absence of DNA, and dark gray in the presence of DNA. The fingers domain in b → c is in light gray in the absence of dNTP, and dark gray in the presence of dNTP. RT′ denotes the enzyme form after the conformational change induced by DNA binding. RT* represents the enzyme form after the conformational change induced by nucleotide binding. The protein and DNA backbones are shown in ribbon diagrams. The DNA template and primer are in dark gray and light gray, respectively. (Courtesy of Drs. K. Das, S. Sarafianos, and E. Arnold, Center for Advanced Biotechnology and Medicine, Rutgers University, Piscataway, New Jersey.)

polymerase active site (see Figure 8.3), inducing a repositioning of the primer 3′-terminus with respect to the rest of the polymerase active site. The palm domain shifts slightly to rearrange the side chains of the active site residues.

The incoming nucleotide also brings in two divalent metal ions, one catalytic and the other for nucleotide binding, to coordinate the oxygen ions of the triphosphate of the nucleotide; the side chains of strictly conserved Asp185 and Asp110; and the backbone carbonyl oxygen of Val111. The metal ions are positioned closely to the 3′-hydroxyl group of the primer terminus. The conformational change induced upon nucleotide binding specifically positions Asp110 for metal chelation. The nucleotide base stacks against the primer terminus and residues Arg72 and Gln151. The triphosphate portion is contacted by Lys65 (γ phosphate), Arg72 (α phosphate), two main-chain NH groups, and two metal ions.

8.3.3 *Nucleotide binding site*

Several amino acids form a binding pocket for the 3′-hydroxyl group of the nucleotide: Asp113, Tyr115, Phe116, and Gln151. The 3′-hydroxyl forms a hydrogen bond with the Tyr115 main-chain-NH. The 3′-hydroxyl-binding pocket is important for structural understanding of nucleoside analog binding and drug resistance mutants of RT. Two or three water molecules and the azido group of AZT 5′-tripohosphate (AZTTP) can fit into this pocket. Tyr115 contacts 2′ position modifications and prevents ribonucleoside incorporation by interference with a 2′-hydroxyl.[28,33]

The last requirement for the stabilization of the ternary complex is a positive charge conferred from residue 154. Lys154 is the only positively charged residue in the VLPQGWK motif on the β8-αE loop at the junction of the fingers and palm domains of HIV-1 RT. Lys154 assumes two distinct orientations in the binary and ternary complexes and also interacts with and stabilizes the side chains of Glu89 and Asp86. RTs containing mutations at position 154 are found to be resistant to nucleoside analog drugs,[38] which validates the significance of a positive charge on residue 154.

8.3.4 *Translocation*

The aromatic residues in the vicinity of the polymerase active site are speculated to be involved in transitions between different stages of the catalytic process. During DNA polymerization and translocation, conformational changes occur at the "primer grip" (the β12–β13 hairpin of p66 palm domain). Translocation refers to the process in which elongated primer/template DNA is relocated from the nucleotide-binding site (pretranslocation complex N) to the priming site (post-translocation complex P).[39]

Structural elements that help move nucleic acid and protein along the translocation track during polymerization include the primer grip, template grip, and helices αH and αI of the p66 thumb.[31] Two antiparallel α-helices (H and I) of the thumb domain interact in the region of DNA bend. Whereas α-helix I interacts with the sugar-phosphate backbone of the template strand, α-helix H is partially embedded in the widened DNA minor groove and interacts primarily with the primer strand. Alanine substitutions in helix H of the thumb domain have been found to reduce frameshift fidelity and lower the processivity of HIV-1 RT.[40]

Functional analysis coupled with molecular dynamics modeling identified five residues of the minor groove binding track that are important for DNA binding and frameshift fidelity. These residues protrude into the minor groove over a stretch of the second through the sixth base pair from the 3′-hydroxyl primer terminus and provide important RT interactions with primer/template during processive synthesis.[41] Changes in the nucleic acid structure and sequences induce mutational hot spots[42,43] and strong sites for termination of processive synthesis.[44]

8.3.5 RT fidelity determinants

Exploitation of *in vivo* fidelity and the structure–function relationship between various RT domains has identified several amino acid motifs present among retroviral RTs. Such fidelity determinants include:

- Residues of the Tyr-X-Asp-Asp (YXDD) motif
- The nucleotide binding site, including the residues lining the pocket and the residues directly contacting the nucleotide substrate[45]
- The α–helix H of the thumb domain[40]
- The conserved Leu-Pro-Gln-Gly (LPQG) motif
- The primer grip comprising residues 227 to 235, which positions the primer terminus for nucleophilic attack on an incoming nucleotide[34,46]
- The conserved RNase H primer grip[26]
- The finger domain

8.3.6 YMDD motif

Despite the high frequency of mutations, the sequences, Tyr-(Met, Val or Leu)-Asp-Asp (YXDD boxes), are found in the RTs of many retroviruses,[47] cauliflower mosaic viruses, and hepatitis B viruses. The YMDD loop of HIV-1 RT participates in translocation as a springboard (Figure 8.4), projecting the primer terminus from the N to the P site after deoxynucleoside monophosphate incorporation.[39] Mutations in the YXDD boxes have been associated with a decrease in enzymatic activity and viral infectivity; changes in the orientation of the primer in the primer/template complex; and resistance to nucleoside analogs.[48–52] The crystal structure of HIV-1 RT complexed with a primer/template substrate and an incoming dTTP has identified six amino acid residues that contact dTTP.[28] Four of these six residues are from the YMDD motif, and three of these residues have been shown to alter the *in vivo* fidelity of RT.[53,54]

8.3.7 Structural differences in other reverse transcriptases

Although related to HIV-1 RT, the RTs of avian sarcoma-leukosis virus (ASV) and moloney murine leukemia virus (MoMLV) can be distinguished by the sequences of their polymerase and RNase H domains,[55] as well as their overall subunit organization. MoMLV RT is a 74-kDa monomer in solution, but may dimerize upon nucleic acid binding.[56] MoMLV RT contains a polymerase and an RNase H domain. In contrast to MoMLV RT, ASV RT is a heterodimer. The smaller subunit (α) contains a polymerase and an RNase H domain. The larger subunit (β) contains three domains: a polymerase, an RNase H domain, and an integrase.[57–59]

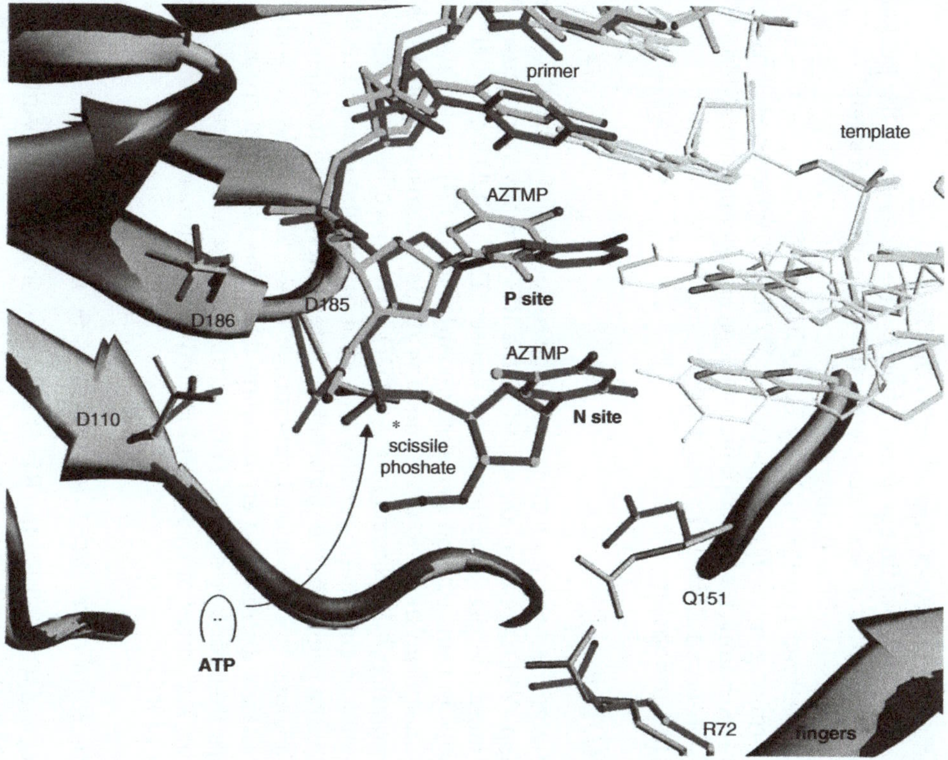

Figure 8.4 Ribbon representation of the superimposed polymerase active sites of complexes P and N (alignment based on p66 residues 107 ± 112 and 155 ± 215). Color scheme: primer strands and side chains of complex P (light gray), and those of complex N (dark gray). Differences between the P and N sites are visualized by the displacement of the YMDD loop and the position of AZTMP. (Modified from Sarafianos, S.G. et al., *EMBO J.*, 21(23), 6614–6624, 2002.)

8.4 Functions of reverse transcriptase

Reverse transcriptase basically has three consecutive catalytic activities (Figure 8.1A). First, it synthesizes a minus-strand DNA by RNA-dependent DNA polymerization, which is initiated by an RNA primer. Second, the RNase H domain of reverse transcriptase degrades the RNA template from the RNA–DNA heteroduplex[60] as minus-strand DNA synthesis proceeds. Finally, it synthesizes the plus-strand DNA by DNA-dependent DNA polymerization using a DNA primer.[61–63]

8.4.1 The polymerase and RNase H activities

The complete process of reverse transcription involves five steps (Figure 8.5). In the first step, the synthesis of (–) strong-stop DNA initiates from the 3′ end of a primer tRNA partially annealed to the primer binding site (PBS) region located near the 5′ end of the (+) genomic RNA template. DNA synthesis stops when RT encounters the 7-methyl-guanine cap. The second step involves the first strand transfer. In this step, the 5′ end of the genomic RNA is degraded by the RNase H activity of RT, and (–) strong-stop DNA is translocated from the 5′ end to the 3′ end of the genomic RNA complementary to the 3′ end of (–) strong-stop DNA.

In the third step, the (–) strong-stop DNA synthesis resumes into the pbs region, whereas the (+) synthesis is initiated from the polypurine tract (PPT) fragment of genomic RNA created by RNase H degradation of the genomic RNA. Primer tRNA is also removed by RNase H in this step. The fourth step involves the second strand transfer. In this step, the PBS region at the 3′ ends of (+) strong-stop DNA hybridizes to its complementary copy, pbs, of the (–) strand DNA, generating a circular molecule. The PPT RNA primer is removed by the RNase H activity of RT. In the last step, (–) and (+) DNA synthesis continues, thus completing the synthesis of the intact double-stranded preintegrative DNA.[56]

8.4.2 RNase H cleavage

The polymerase and RNase H active sites of HIV-1 RT are separated by 19 to 20 base pairs of DNA/RNA heteroduplex when the template does not contain a secondary structure.[20,34,64–68] The primer/template can be productively bound to RNase H while nonproductively bound to the polymerase. RNase H activity is not required for RT to read through the secondary structure of the RNA template,[64] although the polymerase and RNase H activities in HIV-1 RT are coupled by the movement of the enzyme along the template during the polymerization process. This correlation is elucidated by the overlap of RNA cleavage sites and pause sites for primer elongation through an RNA hairpin on the template.[64]

RNase H cleavage can be nonspecific or specific. During retrovirus replication, the RNase H domain of RT carries out nonspecific hydrolysis that

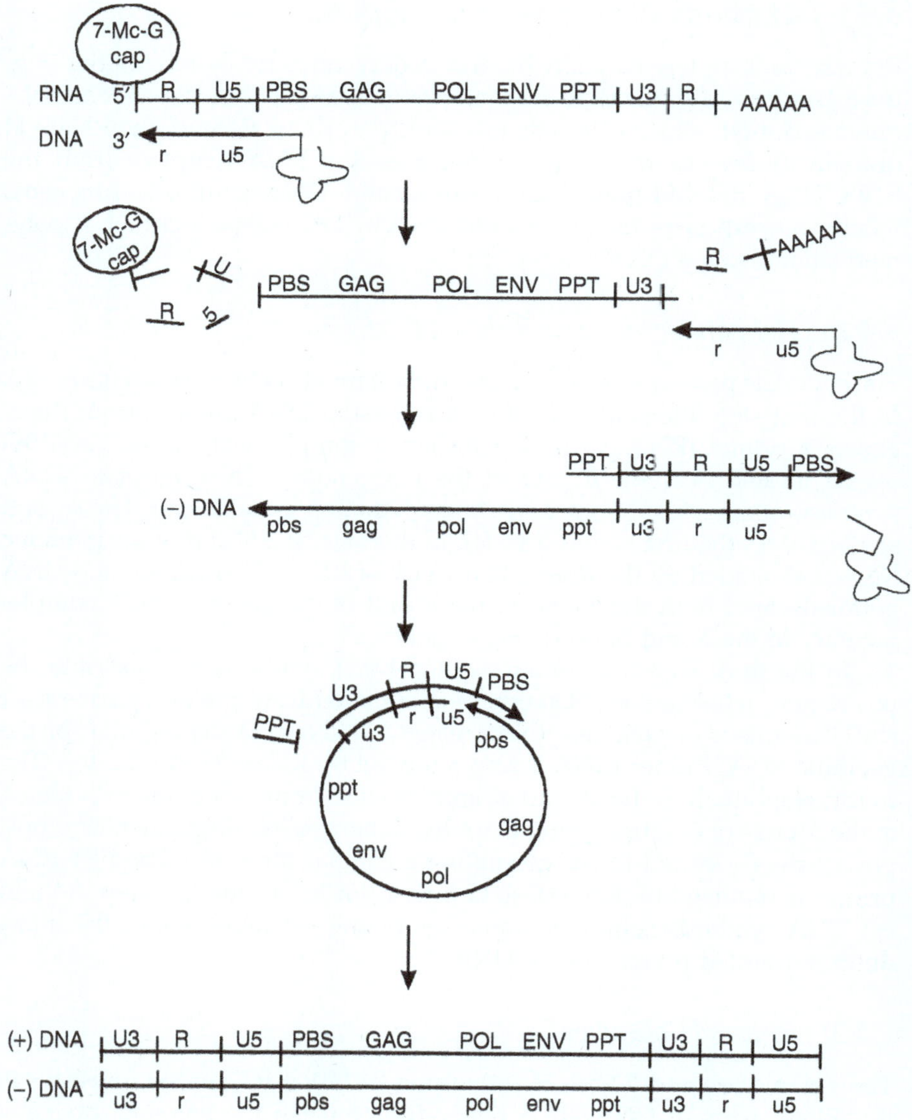

Figure 8.5 The process of reverse transcription. Dark and thicker lines are used to represent DNA strands while light and thinner lines are used for RNA. Arrowheads represent the 3′ ends of DNA that will be elongated in subsequent steps. Uppercase letters indicate (+) strand regions while lowercase letters indicate (–) strand regions. The drawings are not to scale. There are five steps: (1) the synthesis of (–) strand strong-stop DNA; (2) the first strand transfer reaction; (3) the synthesis of (+) strand strong-stop DNA; (4) the second strand transfer reaction; (5) completion of the synthesis of (+) and (–) strand DNA.

removes RNA from the DNA/RNA replication intermediate, making nascent DNA available for a second round of synthesis.[69] Specific RNase cleavage is required in the following steps:

1. Release of the cognate tRNA primer from (–) DNA prior to second strand transfer[70,71]
2. Excision of the 3′ end PPT of the RNA genome from the replication intermediate and subsequent use of PPT to prime (+) strand DNA synthesis
3. Removal of the PPT primer from (+) strand DNA by a precise cleavage at the PPT/U3 junction, generating a correct viral DNA end for subsequent integration[72]

The PPT is resistant to RNase H cleavage. The PPT-selective processing is induced by inherent structural distortions resulting from the PPT sequence.[73]

The specificity of RNase H cleavage is governed by the structure and the proper positioning of the DNA/RNA primer/template duplex in the p66 subunit, relative to the RNase H active site.[74] A series of HIV-1 RT p66 residues within the connection domain and RNase H domain interact with the DNA primer strand and may control the trajectory of the RNA template relative to the RNase H catalytic center. This region is referred to as the "RNase H primer grip" and is conserved between retroviral and bacterial RNases H.[26] Substitution of the conserved residues in the "RNase H primer grip" region reduces the polymerization-independent cleavage activity significantly.[75] Residues in the RNase H primer grip, the thumb domain, and, particularly, the minor groove binding track[76] are crucial for the unique interactions between RT and PPT that create the correct positioning for precise RNase H cleavage.

8.5 Kinetic mechanism of reverse transcriptase

The kinetic parameters of HIV-1 RT have been measured by transient state single nucleotide incorporation assays.[77,78] For the examination of individual steps of a reaction pathway, transient state (also known as pre-steady-state) kinetic methods are preferred over the steady-state kinetic analyses. Steady-state kinetic analyses do not address the individual steps of polymerization. Moreover, the observed steady-state turnover rate, k_{cat}, is usually dominated by the slow dissociation rate of DNA from a polymerase rather than the actual polymerization rate.

Individual reaction steps, including the binding of DNA and an incoming nucleotide; the putative protein conformational change; the chemistry step; and the product dissociation step, can only be resolved by implementing pre-steady-state kinetic methods. These methods allow direct measurement of each elementary step in a reaction mechanism while the enzyme is examined as a stoichiometric reactant. Pre-steady-state kinetic studies using

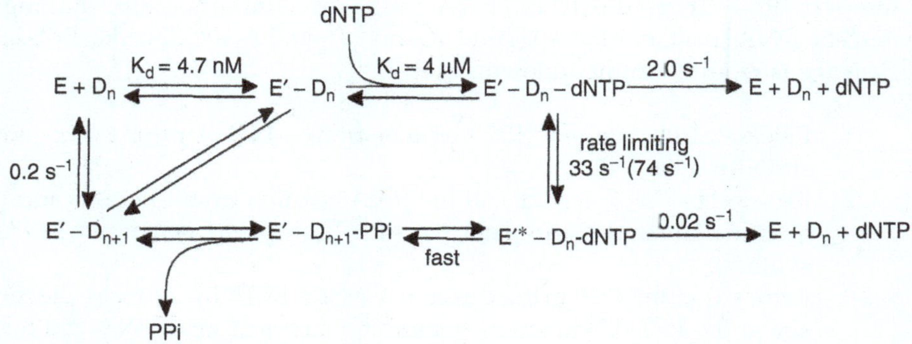

Figure 8.6 A schematic diagram for the polymerization pathway of HIV-1 reverse transcriptase. Kinetic measurements using an RNA template are indicated in parentheses; all others are determined using a duplex DNA/DNA primer/template. E = enzyme. D_n = substrate DNA. D_{n+1} = product DNA with an elongated primer. dNTP = the incoming nucleotide. E′ = the form of RT after a conformational change induced by DNA-binding. E′* = the transitional state of RT after the second conformational change upon nucleotide binding. PPi = pyrophosphate. (Data reported by Kati, W.M. et al., *J. Biol.Chem.*, 267(36), 25988–25997, 1992.)

synthetic primer/template molecules and nucleotides have determined the polymerization mechanism of RT and the rate constants for the elementary steps in the reaction pathway.[77]

The elucidated mechanistic pathway for RT polymerization (Figure 8.6) is similar to other polymerases and involves the following steps:

1. Tight binding of a primer/template duplex substrate to the polymerase heterodimer (E), which induces a conformational change of the enzyme (E′-D_n)[79,80]
2. Binding of next correct nucleotide (dNTP) and divalent metal ions required for catalysis to the enzyme/DNA binary complex (E′-D_n), which forms a ground-state "open" ternary complex of enzyme/DNA/nucleotide (E′-D_n-dNTP), a diffusion-limited initial association
3. A rate-limiting conformational change, which is a relatively slow isomerization to a "closed" ternary complex (E′*-D_n-dNTP) that properly positions the nucleotide for catalysis[78]
4. A fast chemical incorporation step involving the formation of a phosphodiester bond between the 3′-hydroxyl primer terminus and the α-phosphate of the nucleotide (E′-D_{n+1}-PPi), followed by the release of pyrophosphate (PPi)
5. Processive synthesis whereby the elongated primer/template DNA is translocated from complex N to complex P,[39,77,81–83] after which the enzyme continues to polymerize or alternatively exhibits distributive synthesis where the enzyme dissociates from the primer/template DNA

Following dissociation, the polymerization process restarts via the formation of the initial binary complex (E′-D_n).

The binding of RT to a primer/template DNA substrate is rather complicated, involving a protein conformational change. Rothwell et al. reported different distinct binding modes of E-DNA binary complexes in solution.[84] The first mode is observed when the DNA substrate is bound at a site distal to the DNA-binding tract, rendering a nonproductive E-DNA complex. The second mode is observed when the nonproductive complex isomerizes to form a productive E′-DNA complex, which can successfully bind and incorporate an incoming nucleotide.[84,85]

8.6 RNA-dependent vs. DNA-dependent DNA polymerization

The difference between RNA- and DNA-dependent DNA polymerization has been elucidated by pre-steady-state kinetic methods.[77] Single nucleotide incorporation assays have been previously conducted using HIV-1 RT and synthetic primer/template duplexes. Transient state kinetic parameters for each type of primer/template demonstrate a faster polymerization rate (66 s^{-1} compared to 20 s^{-1}) and a slower steady state rate (0.06 s^{-1} compared to 0.18 s^{-1}) for RNA-dependent DNA polymerization vs. DNA-dependent DNA polymerization.[86] In short, the elongated DNA/DNA product is synthesized at a slower rate but released faster from the ternary complex than DNA/RNA product. The fidelity of DNA synthesis by HIV-1 RT is 10- to 60-fold higher with an RNA template than with a DNA template (discussed later). The protein conformational change following nucleotide binding is the rate-limiting step for RNA- and DNA-dependent DNA polymerization.[77,78,83,87]

8.7 RNA secondary structure switching

The secondary structure of an RNA or DNA template stalls RT during polymerization.[64,88–91] The DNA/RNA and DNA/DNA substrates predominantly bind at the polymerase active site of RT in a nonproductive state at strong pause sites and in a productive state at nonpause sites. Pause sites are segments of templates in which secondary structures such as hairpin stems prevent the DNA substrate from binding productively to RT, consequently stalling the polymerization process. Suo and Johnson proposed a model for the motion of RT walking along the RNA secondary structure at 37°C.[64] Although remaining associated with RT, it is postulated that the nonproductively bound DNA/RNA is slowly converted to a productive state upon melting of the next stem base pair (Figure 8.7). Interestingly, the RNA secondary structure changes from one hairpin to a new one when HIV-1 RT reads through the stem region of the first hairpin structure. Moreover, HIV-1 RT facilitates such RNA secondary structure switching.[64,88–90] This model

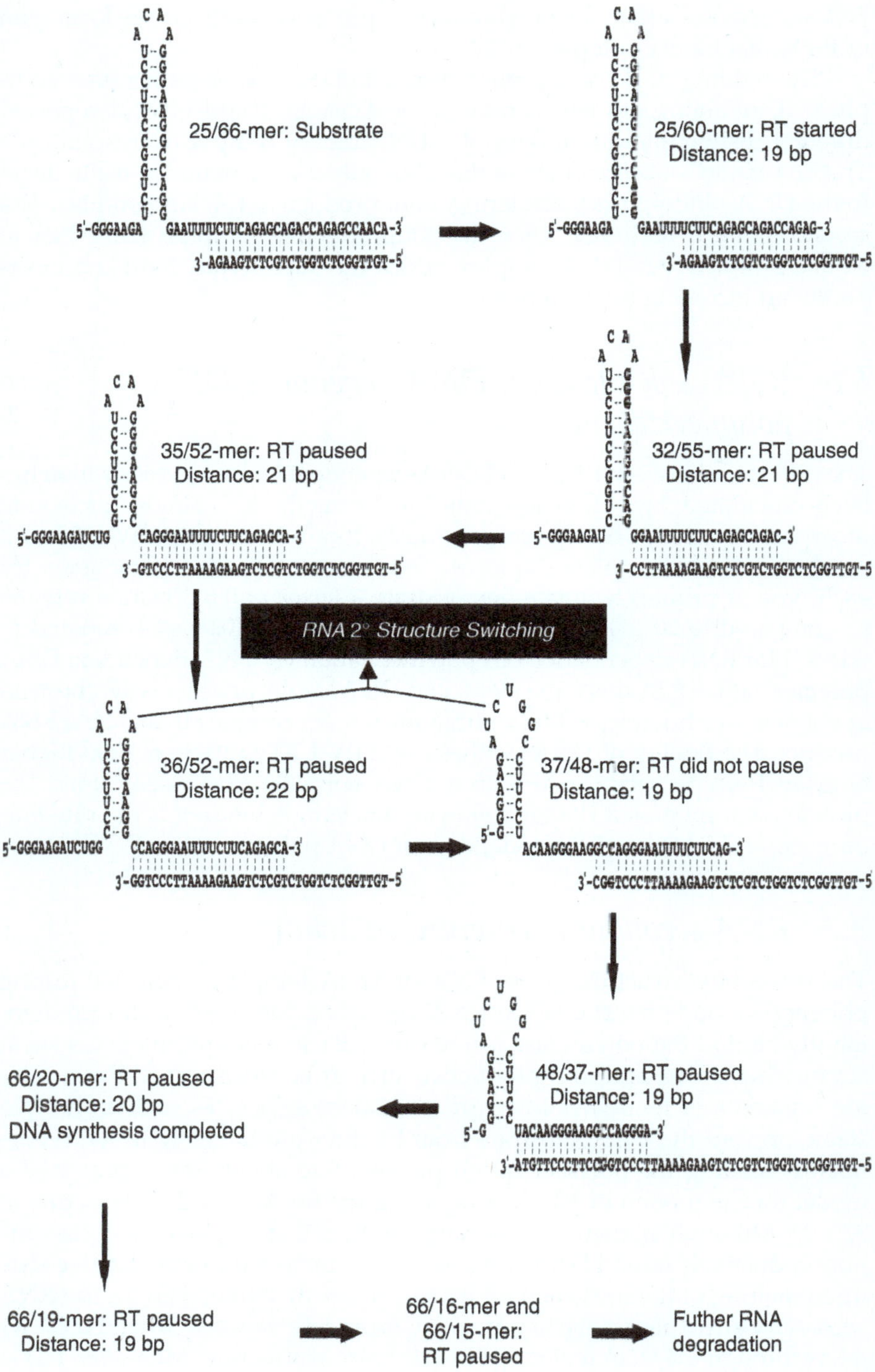
25/66-mer: Substrate
25/60-mer: RT started
Distance: 19 bp
32/55-mer: RT paused
Distance: 21 bp
35/52-mer: RT paused
Distance: 21 bp
RNA 2° Structure Switching
36/52-mer: RT paused
Distance: 22 bp
37/48-mer: RT did not pause
Distance: 19 bp
48/37-mer: RT paused
Distance: 19 bp
66/20-mer: RT paused
Distance: 20 bp
DNA synthesis completed
66/19-mer: RT paused
Distance: 19 bp
66/16-mer and
66/15-mer:
RT paused
Futher RNA
degradation

Figure 8.7

Table 8.1 Fidelity of HIV-1 RT with 25/45 Primer/Template

	DNA template[a]				RNA template[b]			
dNTP	K_d (μM)	k_{pol} (s^{-1})	k_{pol}/K_d ($\mu M^{-1}s^{-1}$)	Fidelity[c]	K_d (μM)	k_{pol} (s^{-1})	k_{pol}/K_d ($\mu M^{-1}s^{-1}$)	Fidelity[c]
dATP	4	33	8.25	—	14	74	5.28	—
dGTP	1010	4.8	4.75×10^{-3}	1,740	1000	0.2	2.0×10^{-4}	26,000
dCTP	1240	0.52	4.19×10^{-4}	19,700	1100	0.03	3.0×10^{-5}	176,000
dTTP	840	0.41	4.88×10^{-4}	16,900	700	0.003	4.9×10^{-6}	1,100,000

[a] Kati, W.M. et al., *J. Biol.Chem.*, 267(36), 25988–25997, 1992.

[b] Kerr, S.G. and Anderson, K.S., *Biochemistry*, 36(46), 14056–14063, 1997.

[c] Calculated as $[(k_{pol}/K_d)_{correct} + (k_{pol}/K_d)_{incorrect}]/(k_{pol}/K_d)_{incorrect}$.

demonstrates a dynamic RNA or DNA template secondary structure during replication by RT.

8.8 *Fidelity of reverse transcriptase*

The fidelity of a polymerase is a measure of the frequency of incorporation of incorrect nucleotides by the enzyme. RT performs the remarkable task of reversing the flow of genetic information, although it is rather erroneous due to the lack of 3′ to 5′ exonuclease proofreading activity and the flexible accommodation of DNA and RNA templates. The fidelity of a retroviral polymerase depends on its protein sequence, base pair geometry, flexibility of the nucleotide-binding pocket[28,54,92,93] and the contact between amino acid residues and nucleotides.[45] Mutation rates vary among different RTs. For example, the mutation rate is estimated to be 10^{-2} to 10^{-5} errors per cycle in HIV through a host cell.[11,94,95] HIV-1 and HBV RT have almost identical mutation rates[96] and AMV and MLV RT have 10-fold and 18-fold higher fidelity than HIV-1 RT, respectively.[97]

Higher fidelity of DNA synthesis by HIV-1 RT is observed with an RNA template[86] in comparison to a DNA template.[77] There can be several explanations for this difference in fidelity. Misaligned intermediates are formed less frequently with RNA templates than with DNA templates.[98] Topology and thermodynamic stability of DNA/DNA and DNA/RNA duplexes vary in a sequence-dependent manner.[99,100] In addition, the secondary structure

***Figure 8.7* (See figure, facing page.)** A proposed model of RT traversing the RNA secondary structure at 37°C. During the concurrent processes of DNA polymerization and RNA cleavage, RT pauses at several sites. The RNA secondary structures change their locations by melting and reforming base pairs. The completion of DNA synthesis further degrades RNA template. As a result of the mobile RNA secondary structure, the distance between the 3′ terminus of DNA primer and the RNA cleavage site ranges from 19 to 22 base pairs. (Suo, Z.C. and Johnson, K.A., *Biochemistry*, 36(41), 12468–12476, 1997.)

of a template might affect the efficiency of nucleotide incorporation.[64,88–91] Kerr and Anderson have reported extremely slow rates for incorporation of incorrect nucleotides during RNA-directed DNA polymerization.[86]

Pre-steady state kinetic methods have been used to estimate the fidelity of HIV-1 RT with DNA or RNA as a template (see Table 8.1). With a DNA template, mismatched nucleotide incorporations opposite a template base thymine display about a 250-fold increase of the equilibrium dissociation constant (K_d) and a 7- to 80-fold decrease of the maximal incorporation rate (k_{pol}), relative to the incorporation of matched deoxyadenosine 5′-triphosphate (dATP). The estimated fidelity factors are 1740 for discrimination against dG:dT, 19,700 for dC:dT, and 16,900 for dT:dT mismatch.[77]

In contrast, RNA-dependent DNA replication is much more selective because the maximum rates of misincorporation and the binding affinities of incorrect nucleotides are substantially lower.[86] In the presence of an RNA template, the maximum incorporation rates for dGTP, dCTP, and dTTP opposite a template uridine (rUTP) are 370- to 24,700-fold slower than the k_{pol} of the correct nucleotide (dATP).[86] At the same time, the binding affinity of mismatched dGTP, dCTP, and dTTP is 50- to 80-fold lower than that of the correct nucleotide dATP.[86] The much slower incorporation rate of a mismatched nucleotide with an RNA template than with a DNA template leads to higher enzyme fidelity. With a template uridine, the fidelity is estimated to be 26,000 for discrimination against dGTP, 176,000 for dCTP, and 1×10^6 for dTTP.[86] The fidelity rate with a DNA/RNA primer/template substrate is 10- to 100-fold higher than the fidelity rate with a DNA/DNA primer/template substrate.

Three checkpoints ensure discrimination against non-Watson–Crick base pairing: initial nucleotide binding in the "open" form; induced-fit conformational change to the "closed" form; and catalytic incorporation.[33,92,101] For example, residue Gln151 interacts with the 3′-hydroxyl group of the incoming nucleotide to secure the binding of correct nucleotides to HIV-1 RT[102] and thereby distinguishes the sugar moieties of the incoming nucleotides from those of the incorrect nucleotides.[103] Even if an incorrect nucleotide escapes the first checkpoint, non-Watson–Crick base pairing between the template and the incoming nucleotide will decrease the rate of the conformational change.[92] Lastly, rigid and bulky nucleotides that constrain the polymerase active site drastically decrease the incorporation rates and the binding affinities.[101] The overall size and shape of a nucleotide substrate and the nucleotide binding pocket around the nascent base pair at the polymerase active site are significant factors in nucleotide selectivity.

8.9 Applications of nucleotides in the research of reverse transcriptase

Structural derivatives of nucleosides and nucleotides have been designed as retroviral inhibitors (see also Chapter 11 in this book) in antiretroviral ther-

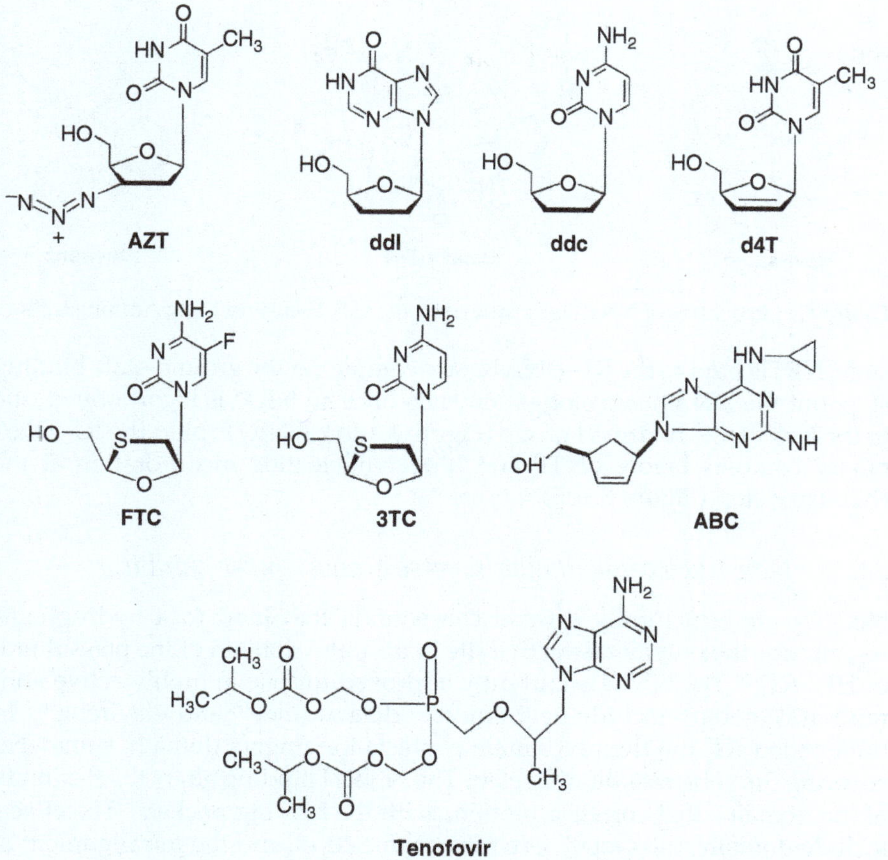

Figure 8.8 Structures of NRTIs approved by the U.S. Food and Drug Administration.

apy. Two classes of RT inhibitors are used in the clinical treatment of AIDS: nucleoside/nucleotide analog reverse transcriptase inhibitors (NRTIs), and nonnucleoside analog reverse transcriptase inhibitors (NNRTIs). Some NRTIs are also being used for hepatitis B virus infection. These two types of RT inhibitors and HIV protease inhibitors are used in combination therapies to achieve maximum potency at a tolerable level of toxicity.

NRTIs and NNRTIs differ in their chemical structures (Figure 8.8 and Figure 8.9); sites of RT binding (Figure 8.2); and inhibition mechanisms (Figure 8.10). NRTIs are structural derivatives of nucleosides or nucleotides with modified functional groups. NNRTIs are often aromatic compounds that are highly hydrophobic. Because of their close resemblance to nucleotides, the 5′-triphosphates of NRTIs directly compete against incoming nucleotides and bind at the nucleotide-binding sites on HIV-1 RT and HBV RT.

Unlike NRTIs, NNRTIs enter a pocket adjacent to the active site that is lined with hydrophobic residues. Binding of NNRTIs to this hydrophobic pocket induces conformational changes that lock the enzyme into an inactive

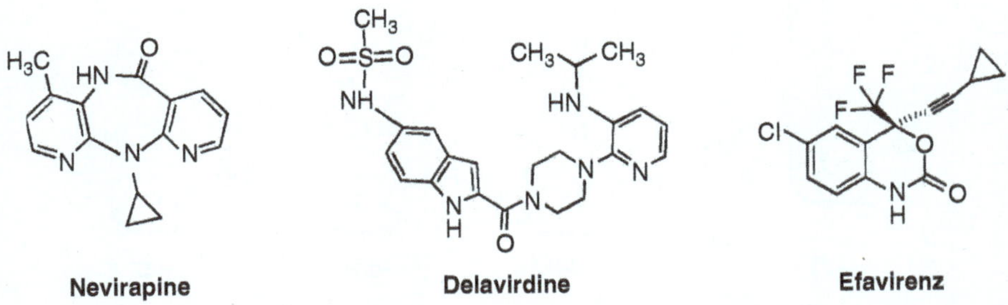

Figure 8.9 Structures of NNRTIs approved by the U.S. Food and Drug Administration.

state. NRTIs bind to the RT–DNA binary complex at the ground-state binding step, but the DNA chain elongation ends once an NRTI is incorporated due to the lack of a 3′-hydroxyl group (Figure 8.10A). NNRTIs bind the RT–DNA binary complex before dNTP and hinder nucleotide incorporation at the chemistry step (Figure 8.10B).

8.9.1 Non-nucleoside analog reverse transcriptase inhibitors

NNRTIs are structurally diverse compounds that bind to a hydrophobic region near the polymerase active site in the palm domain of the p66 subunit of HIV RT.[104] The NNRTIs currently approved for use in highly active antiretroviral therapy include nevirapine,[105] delavirdine,[106] and efavirenz.[107] In unliganded RT, the thumb domain contacts the fingers domain, somewhat covering the substrate-binding cleft. The NNRTI binding alters the flexibility of the domains that correlate motions with the binding pocket.[29] The effects include domain shifts relative to the enzyme core[30] and the misalignment of the conserved residues of the polymerase active site (Asp110, Asp185, and Asp186) relative to the catalytically required Mg^{2+} ions.[78]

Esnouf et al. suggest that NNRTIs inhibit RT by locking the polymerase active site into an inactive conformation, reminiscent of the conformation observed in the inactive p51 subunit.[104] Temiz and Bahar speculate that nevirapine redirects the domain movement upon binding while efavirenz destroys the p66 thumb flexibility.[108] Rotation and stiffening of the p66 thumb domain skew the orientation of the primer/template with respect to the polymerase and RNase H active sites of RT. As a result, NNRTIs mechanistically block the chemical reaction of the DNA polymerization, without affecting the nucleotide binding or the nucleotide-induced conformational change.[78,87]

Pre-steady state kinetic studies[78] have revealed that NNRTIs inhibit HIV-1 RT by stalling the chemical reaction (step 3′, Figure 8.10B), without affecting the conformational change (step 2′, Figure 8.10B).[78] Consequently, the chemical reaction of the enzyme–DNA–inhibitor–nucleotide complex becomes the rate-limiting step in the polymerization pathway. These inhibitors do not compete with primer/template or nucleotide substrates for binding to the enzyme. Once bound, the dissociation rate of the inhibitor

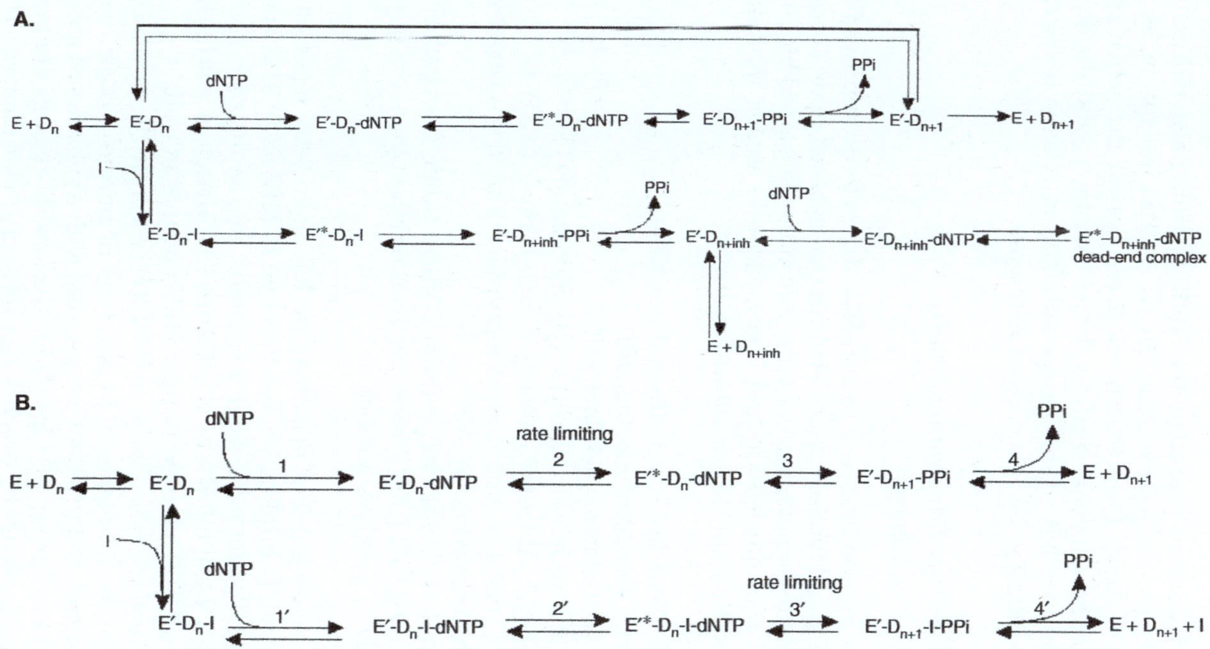

Figure 8.10 Schemes for the inhibition mechanisms of HIV-1 RT by (A) NRTIs, and (B) NNRTIs. Notations are the same as in Figure 8.6. NRTIs bind at the active site as natural nucleotides and terminate the primer elongation by forming the dead-end E′*-D$_{n+inh}$-dNTP complex, which cannot continue the chemical reaction due to lack of 3′-hydroxyl group in the primer of D$_{n+inh}$. NNRTIs inhibit the chemical reaction (step 3′), but do not affect nucleotide binding (step 1′) or the nucleotide-induced conformational change (step 2′). (Spence, R.A. et al., *Science*, 267(5200), 988–993, 1995.)

from the enzyme–DNA–inhibitor complex is rather slow relative to the rate of the polymerization reaction. Spence et al. suggest that the binding of the NNRTIs switches the positions of the carboxyl groups of the conserved aspartic acid residues and resultantly reduces the rate of the Mg^{2+}-dependent chemical reaction.[78]

The structural distortions caused by the NNRTI binding also reduce the affinity of HIV-1 RT for the primer/template as well as the specificity of RNase H cleavage.[109] Aside from the chemical reaction at the active site, other reverse transcription events, such as strand transfer, strand displacement, or recognition of tRNA primer in the initiation of reverse transcription,[51] could also be affected by this class of inhibitors.

8.9.2 Nucleoside/nucleotide analog reverse transcriptase inhibitors

Nucleoside analogs are the first type of anti-HIV drugs to be approved by the U.S. Food and Drug Administration. Their design is based upon success with acyclovir in the treatment of herpes virus infections developed by Elion et al.[110] Currently, seven nucleoside and one nucleotide analog inhibitors have been approved by the FDA for clinical treatment:

- Zidovudine (3′-azido-3′-deoxythymidine, AZT)
- Didanosine (2′,3′-dideoxyinosine, ddI)
- Zalcitabine (2′,3′-dideoxycytidine, ddC)
- Stavudine (2′,3′-didehydro-2′-dideoxythymidine, d4T)
- Lamivudine (2′,3′-dideoxy-3′-thiacytidine (–)isomer of 3TC)
- Abacavir ((1S-cis)-4-[2-amino-6-(cyclopropylamino)-9H-purine-9-yl]-2-cyclopentene-1-methanol, ABC)
- Emtricitabine ((–)-2′,3′-dideoxy-5-fluoro-3′-thiacytidine, Coviracil, FTC)
- Tenofovir (9-[(R)-2-[[Bis[[(isopropoxycarbonyl)oxy]methoxy]phosphinyl]methoxy]propyl]adenine)

More recently, retroviral and hepadnaviral RTs have been reported to have a similar nucleotide-binding pocket structure to HIV RT.[111] This is not surprising because HIV and HBV share a common step in their life cycles: conversion of their RNA into DNA by RT (Figure 8.1). Therefore, nucleoside analogs like 3TC and FTC are used to treat HBV patients as well.

All nucleoside analogs are believed to inhibit HIV via an identical mechanism. NRTIs are synthetic analogs of pyrimidines or purines, consisting of nitrogenous bases associated to a deoxyribose ring with a substituent replacing the natural hydroxyl group at 3′-position. Nucleoside analogs are phosphorylated to 5′-triphosphates before they can act as chain terminators. They prevent the formation of 3′,5′-phosphodiester linkages between the last nucleotide of the primer and the incoming nucleotide, thereby terminating the elongation of a nascent viral DNA chain. Factors controlling the efficacy of these inhibitors include uptake, transport, metabolism, incorporation, and drug degradation[112] — all of which contribute to variation among cell-, tissue-, and organ-specific toxicity levels.[113]

Unlike the nucleoside analogs, tenofovir disooproxil fumarate is an acyclic nucleoside phosphonate derivative and a water-soluble prodrug (Figure 8.8). It contains two ester moieties that mask the phosphonate group and increase the oral and cellular bioavailability of the drug. It is hydrolyzed to PMPA (R-9-(2-phosphonylmethoxypropyl)adenine) and phosphorylated to form PMPA diphosphate (PMPA-DP), which then acts as an HIV reverse transcriptase inhibitor.[114] The high bioavailability of PMPA is partially attributed to the fact that only two phosphorylation steps are required to activate PMPA *in vivo* while three phosphorylation steps are required to intracellularly activate other NRTIs.

As a potent inhibitor of HIV-1 RT, PMPA-DP has substrate specificity constants (k_{pol}/K_d) similar to the nucleotide analog, ddATP, and only fivefold lower than the natural nucleotide substrate, dATP.[115] Meanwhile, PMPA-DP is highly selective against T7 phage DNA polymerase,[115] human mitochondrial DNA polymerase,[116] rat DNA polymerase β, and, most likely, human DNA polymerase β.[115] Once incorporated into DNA, PMPA is less likely to be removed from it by pyrophosphorolysis and/or by nucleotide-dependent chain-terminator removal than AZT and d4T.[117]

8.9.3 Cellular toxicity of NRTIs

Toxicity is one major drawback to AIDS treatment with NRTIs[118,119] because of nonspecific inhibition of host cellular DNA polymerases α, δ, ε, DNA primase,[120] and human mitochondrial DNA polymerase.[3] The human mitochondrial DNA polymerase is responsible for mitochondrial DNA replication and repair. The toxicity of a nucleoside analog depends on the frequency of incorporation of the chain terminator and the kinetics of excision by the proofreading exonuclease in a host cellular DNA polymerase, with respect to the minimal rate of DNA replication required to sustain growth.[121] The toxicity indices of NRTIs are correlated with the specificity constants (k_{pol}/K_d) of incorporation by mitochondrial DNA polymerase, varying more than 500,000-fold in the order of magnitude: ddC > ddI (metabolized to ddA) > d4T >> (–)3TC > PMPA > AZT > ABC (metabolized to CBV).[116] The higher toxicities of d4T, ddC, and ddI (ddA) are attributed to their 13- to 36-fold tighter binding with respect to the normal nucleotide.

The combination of efficient incorporation and ineffective exonuclease removal is considered the cause for the high toxicities of dideoxy nucleoside analogs, ddC and ddI.[116] Furthermore, the stereochemical structure of a nucleoside analog plays an important role in its potency and toxicity. For instance, (+) 3TC-TP is a more potent inhibitor for human mitochondrial DNA polymerase than the FDA-approved (–) isomer, with tighter binding and faster incorporation into a DNA primer/template duplex.[122]

8.9.4 Drug resistance mechanisms

Sequence comparisons between RT variants demonstrated a correlation between amino acid residue substitutions or insertions and the development of drug-resistant strains of HIV-1. To date, all known HIV and HBV resistance mutations to antiviral drugs have been located at conserved residues near the active sites.[52,123–125] Most of the hypermutable conserved residues make direct contacts with antiviral drugs. Multiple mutations occurring in one NRTI-resistant strain of HIV can be at up to five positions: A62V, V75I, F77L, F116Y, and Q151M.[126] The resistance accrues as the number of mutation sites rises.

Molecular modeling studies and structural analysis of nucleoside analogs complexed with wild-type and mutated RT unravel the mechanisms of drug resistance conferred by mutations. Two distinct mechanisms have been elucidated for resistance of HIV-1 RT to NRTIs: (1) mutations hinder the incorporation of a nucleoside analog; and (2) mutations enhance the excision of a nucleoside analog after it has been incorporated onto viral DNA.

8.9.4.1 NRTI resistance through inhibition of incorporation

Clustered resistance mutations correlate with the chemical structures of the NRTI or the NNRTI.[28,51,127] NRTI-resistant mutations interfere with the binding or reactivity of the nucleoside analogs relative to natural nucleotides. Most NRTI mutations are likely to reposition the primer/template, whereas other mutations hinder the inhibitor binding by directly changing interactions between RT and an inhibitor or by indirectly altering residues proximal to the nucleotide-binding pocket.

Point mutations in the vicinity of the incoming nucleotide can affect the side chain contact and metal ion coordination, and thus the orientation of the incoming nucleotide. For example, Arg72 in the nucleotide-binding pocket forms hydrogen bonds with the nucleotide and is responsible for pyrophosphate removal while its guanidium group stacks with the base of the nucleotide. A mutation of R72A renders HIV-1 RT deficient in pyrophosphorolysis. Gln151 recognizes the 3′-hydroxyl and β-phosphate of the incoming nucleotide and interacts with residues that directly contact the nucleotide or affect the position of important structural elements. The Q151M mutation disrupts the hydrogen-bond network and therefore generates drug resistance to nucleoside analog inhibitors.

Because the YMDD loop pushes the primer terminus from the N to the P site (see Section 8.4) after dNTP incorporation during translocation,[39] Met184 (the second amino acid of the YMDD motif) is important for the positioning of the 3′-primer terminus and the binding of nucleotide. The M184I, M184V, M184T, and M184A mutations develop resistance by three mechanisms:

- Steric hindrance between side chains of β–branched residue at position 184 and the ring of the β–L-enantiomer of NRTI, specifically 3TC,

so that the strained binding between the inhibitor and HIV-1 RT is salt sensitive and not catalytically competent[52,128]
- Reorientation of the primer/template[33]
- Enhancement of ATP binding, pyrophosphorolysis, and excision of the analog from the 3′-end of a primer[129]

The 3TC-resistant HIV-1 RT^{M184V} drastically reduces incorporation efficiency by weakening the binding for the triphosphate of 3TC, yet its fidelity is increased 2.4-fold relative to wild-type RT. The incorporation of a normal nucleotide, dCTP, is similar between the wild-type and the mutant HIV-1 RT during DNA-dependent DNA polymerization, but is 3.5-fold less efficient by the mutant during RNA-dependent DNA polymerization. The incorporation of the triphosphate of 3TC is 146- and 117-fold less efficient in the mutant HIV-1 RT for DNA- and RNA-dependent DNA polymerization, respectively, compared to wild type RT.[130] 3TC reverses AZT-resistant strains to AZT-sensitive strains,[131] which supports the combination regimen of 3TC and AZT in clinical treatment.

8.9.4.2 NRTI resistance through analog excision

The excision reaction is mechanistically the reverse of the chemical step of a polymerization reaction, requiring a pyrophosphate donor, the ATP β,γ-phosphate moiety, joined to AZTMP at the 3′ primer terminus by RT to form a dinucleoside tetraphosphate.[129,132–135] The AZTMP-terminated primer is more often in the N site, which is the substrate for excision, because the azido group imposes steric strain (Figure 8.4) and destabilizes the ternary complex in the P site. The excision of an AZTMP-terminated primer is specific to complex N.[39] The ATP β,γ-phosphate moiety is proximal to the scissile phosphate of the AZTMP-terminated primer.

The nucleotide base makes aromatic π–π interactions with the Tyr215 ring of the AZT-resistant enzyme, which is stabilized by the Trp210 indole ring in an L210W mutant RT. [135,136] The excision of nucleoside analogs, including ddTMP and d4TMP from DNA, is aggravated by amino acid substitutions, such as T215Y, and insertions in the fingers domain. Gel shift analysis supports a mechanism in which insertions in the fingers domain destabilize the ternary complex, allowing the end of the primer to gain access to the N site to prepare for excision.[137]

8.9.4.3 NNRTI resistance mechanism

Like NRTIs, NNRTIs induce mutations in the vicinity of their binding pocket in HIV-1 RT. Single amino acid substitutions include L74I/V; V75I/L; A98G; L100I; K101E/I/Q; K103N/Q/R/T; V106A/I; V108I; E138K; T139I; G141E; V179D/E; Y181C/I; Y188C/H/L; V189I; G190E/Q; P225H; P236L; and V233E.[138–149] The side chains of mutated amino acid residues prevent RT from binding to the inhibitors. The hydrophobic and electrostatic interactions between the enzyme and NNRTI are altered. Switched amino acids can cause rearrangement of residues inside the NNRTI binding pocket and thus repositioning of NNRTIs.[150,151] These mutational factors contribute to weaker binding of NNRTIs to HIV RT, which leads to drug resistance.

8.9.5 *The future of antiviral drugs*

In order to improve adherence, tolerability, long-term toxicity, and drug- and cross-resistance in the current regimens, structural derivatives of nucleoside and nucleotide analogs are being tested to search for more specific but less toxic antiviral drugs. Essentially, the side chains of approved drugs are being modified to increase selectivity towards HIV-1 RT and reduce toxicity towards cellular DNA polymerases.[152] Among all the investigational NRTIs, two are active against both HIV and HBV: amdoxovir (DAPD, diaminopurine dioxolane) and ACH-126443 (Beta-L-Fd4C).[153–156]

Amdoxovir is an analog of guanosine that is deaminated *in vivo* to DXG (dioxolane guanine). It is active against some nucleoside-resistant viruses but less potent to the K65R or L74V mutant.[157] Recently, Feng et al. and Feng and Anderson report that L-nucleosides are generally less toxic while their activity is comparable to their corresponding D-nucleosides.[122,158] This leads to a testing of the L-derivatives of current nucleoside/nucleotide analogs. ACH-126,443 emerged from the massive screening of 2′,3′-dideoxy(dd)- and 2′,3′-didehydro-2′,3′-dideoxy(D4)-L-nucleoside analogs. It has been reported to be the most potent anti-HBV agent to date.[159]

Aside from reverse transcriptase inhibitors, scientists are exploring new ways to stop HIV viral replication using synthetic nucleotides. New classes of HIV drug candidates include nucleic acid aptamers as template-analog inhibitors; short interference RNA oligos; ribozymes; antisense RNA oligos; and decoy RNA oligos. Template analogs and decoy RNA oligos directly compete against DNA for binding to HIV RT.

Nucleic acid aptamers bring great hopes for overcoming the hypermutation of HIV-1 RT. Burke et al. and Fisher et al. report that HIV-1 RT may not be able to mutate to aptamer-resistant forms without deteriorating its own viral replication activity.[160,161] Bacterially expressed RNA aptamers are designed for inhibiting HIV-1 RT inside bacterial cells and additionally display potent antiviral activity in human cell culture.[162] Pseudoknot RNA aptamers show substantially tighter binding than the natural substrate, exceptional specificity, and a broad range of working conditions.[163] Virion-encapsulated template analog RT inhibitors (small RNA molecules) can predispose virions for inhibition immediately upon entry.[164] DNA aptamers against RNase H diminish the *in vitro* RNase H activity as well as viral replication of HIV-1 in infected human cells.[165]

Short interference RNAs (siRNAs) are double-strand RNA segments of 21 to 23 nucleotides that guide mRNA degradation in a sequence-specific fashion to shut down gene expression. The HIV-1 life cycle can be interrupted by siRNA technology at multiple stages. Currently, mRNA targets for siRNAs include the HIV-1 cellular receptor CD4; the viral structural Gag protein; the green fluorescence protein substituted for the Nef regulatory protein[166];

and the *rev* regulatory gene.[167] Silencing expression of these proteins can reduce the viral entry into human cells, prevent viral genomic RNA accumulation and virion uncoating and packaging, or extinguish the replication of progeny virus inside the host genome. In short, siRNAs can inhibit pre- and/or postintegration infection events in the HIV-1 life cycle.

Other RNA-based gene therapies include ribozyme,[168,169] antisense, and decoy inhibition.[169] Ribozymes bind to and cleave RNA. Antisense RNA oligos interact and base pair with structured RNA, resulting in the disruption of an RNA structural motif crucial for protein recognition or the degradation of RNA/RNA duplexes. Decoys are short structured synthetic RNA oligos that mimic the viral RNA and compete with natural RNA ligands.

8.10 Conclusions

This chapter has described the origin, structure, function, and mechanism of reverse transcriptases in various viruses. The discovery of reverse transcriptase has shed light on the mechanism of viral replication, especially the unique genetic conversion from RNA to DNA. The establishment of a kinetic mechanism of polymerization by reverse transcriptase has provided a kinetic basis to evaluate anti-HIV/anti-HBV drugs, as well as to characterize mutant strains of HIV. The effects of each RT drug-resistant amino acid change have been studied to help improve the design of antiviral drugs.

The reverse transcriptases of HIV and HBV are the highlight of this chapter because they are major drug targets for therapeutic treatments. A significant amount of effort has been devoted toward the study of these two viruses in order to counter the worldwide AIDS and hepatitis epidemics. The similarity between the structures of the nucleotide-binding pockets of HIV and HBV reverse transcriptases has allowed for the treatment of hepatitis B patients with nucleoside analogs initially developed for the treatment of AIDS patients. However, the therapeutic use of nucleoside analogs is limited by cellular toxicity, lack of activity in some cell types, and susceptibility to viral resistance. The development of viral drug resistance also limits the potency of non-nucleoside analogs toward HIV-1 RT.

In order to address current limitations, scientists are seeking alternative therapeutics such as template analogs, short interference RNAs, and nucleic acid aptamers. The spectrum of antiretroviral and antihepadnaviral nucleoside/nucleotide analogs is being expanded as new approaches are continually emerging. Mechanistic evaluation of antiviral drugs remains an important tool for improvements in the antiviral regimens. Further *in vivo* work using more diverse experimental systems is required for a more complete assessment of these new drugs.

Abbreviations

AIDS acquired immunodeficiency syndrome
AMV avian myeloblastosis virus

ASLV avian sarcoma-leukosis virus;
dATP 2′-deoxyadenine-5′-triphosphate
dCTP 2′-deoxycytidine-5′-triphosphate
dGTP 2′-deoxyguanosine-5′-triphosphate
dTTP 2′-deoxythymidine-5′-triphosphate
HBV hepatitis B virus
HIV human immunodeficiency virus
MLV murine leukemia virus
MoMLV moloney murine leukemia virus
PBS primer binding site
pgRNA pregenomic RNA
PPi pyrophosphate
PPT polypurine tract
RNase H ribonuclease H
RT reverse transcriptase
siRNA short interference RNA

References

1. Baltimore, D. Milestones in biological research — discovery of the reverse transcriptase. *FASEB J.*, 9(15), 1660–1663, 1995.
2. Temin, H.M. and Mizutani, S. RNA-dependent DNA polymerase in virions of Rous sarcoma virus (Reprinted from *Nature* vol. 226, 1211–1213, 1970). Reviews in *Med. Virol.*, 8(1), 6–8, 1998.
3. Gray, H. and Wong, T.W. Purification and Identification of Subunit Structure of the Human Mitochondrial–DNA polymerase. *J. Biological Chem.*, 267(9), 5835–5841, 3-25-1992.
4. Toh, H., Hayashida, H., and Miyata, T. Sequence homology between retroviral reverse transcriptase and putative polymerases of hepatitis-B virus and cauliflower mosaic-virus. *Nature*, 305(5937), 827–829, 1983.
5. Miller, R.H. and Robinson, W.S. Common evolutionary origin of hepatitis-B virus and retroviruses. *Proc. Natl. Acad. Sci. USA*, 83(8), 2531–2535, 1986.
6. Temin, H.M. Origin of retroviruses from cellular moveable genetic elements. *Cell*, 21(3), 599–600, 1980.
7. Inouye, S., Saigo, K., Yamada, K., and Kuchino, Y. Identification and nucleotide-sequence determination of a potential primer transfer-RNA for reverse transcription of a Drosophila retrotransposon, 297. *Nucl. Acids Res.*, 14(7), 3031–3043. 4-11-1986.
8. Yuki, S., Ishimaru, S., Inouye, S., and Saigo, K. Identification of genes for reverse transcriptase-like enzymes in 2 Drosophila retrotransposons, 412 and gypsy — a rapid detection method of reverse transcriptase genes using Yxdd box probes. *Nucl. Acids Res.*, 14(7), 3017–3030, 4-11-1986.
9. Michel, F. and Lang, B.F. Mitochondrial class-II introns encode proteins related to the reverse transcriptases of retroviruses. *Nature*, 316(6029), 641–643, 1985.
10. Blattner, W., Gallo, R.C., and Temin, H.M. HIV causes aids. *Science*, 241(4865), 515, 7-29-1988.

11. Pathak, V.K. and Temin, H.M. Broad-spectrum of *in vivo* forward mutations, hypermutations, and mutational hotspots in a retroviral shuttle vector after a single replication cycle — substitutions, frameshifts, and hypermutations. *Proc. Natl. Acad. Sci. USA*, 87(16), 6019–6023, 1990.
12. Preston, B.D. Reverse transcriptase fidelity and HIV-1 variation. *Science*, 275(5297), 228–229, 1-10-1997.
13. Wain–Hobson, S. HIV genome variability *in vivo*. *AIDS*, 3 Suppl. 1, S13–S18. 1989.
14. Wain–Hobson, S. Running the gamut of retroviral variation. *Trends Microbiol.*, 4, 135–141, 1996.
15. Preston, B.D. and Loeb, L.A. Error-prone DNA polymerization by human immunodeficiency virus (HIV) reverse transcriptase. *Proc. Am. Assoc. Cancer Res.*, 29, 443. 1988.
16. Preston, B.D., Poiesz, B.J., and Loeb, L.A. Fidelity of HIV-1 reverse transcriptase. *Science*, 242(4882), 1168–1171, 11-25-1988.
17. Preston, B.D. Error-prone retrotransposition: rime of the ancient mutators. *Proc. Natl. Acad. Sci. USA*, 93(15), 7427–7431, 7-23-1996.
18. Mansky, L.M. and Temin, H.M. Lower *in-vivo* mutation rate of human immunodeficiency virus type-1 than that predicted from the fidelity of purified reverse transcriptase. *J. Virol.*, 69(8), 5087–5094, 1995.
19. Fischer, M., Hafner, R., Schneider, C., Trkola, A., Joos, B., Joller, H., Hirschel, B., Weber, R., and Gunthard, H.F. HIV RNA in plasma rebounds within days during structured treatment interruptions. *AIDS*, 17(2), 195–199, 1-24-2003.
20. Kohlstaedt, L.A., Wang, J., Friedman, J.M., Rice, P.A., and Steitz, T.A. Crystal structure at 3.5 angstrom resolution of HIV-1 reverse transcriptase complexed with an inhibitor. *Science*, 256(5065), 1783–1790, 6-26-1992.
21. Lightfoote, M.M., Coligan, J.E., Folks, T.M., Fauci, A.S., Martin, M.A., and Venkatesan, S. Structural characterization of reverse transcriptase and endonuclease polypeptides of the acquired immunodeficiency syndrome retrovirus. *J. Virol.*, 60(2), 771–775, 1986.
22. Legrice, S.F.J., Naas, T., Wohlgensinger, B., and Schatz, O. Subunit-selective mutagenesis indicates minimal polymerase-activity in heterodimer-associated p51 HIV-1 reverse transcriptase. *EMBO J.*, 10(12), 3905–3911, 1991.
23. Morris, M.C., Berducou, C., Mery, J., Heitz, F., and Divita, G. The thumb domain of the P51-subunit is essential for activation of HIV reverse transcriptase. *Biochemistry*, 38(46), 15097–15103, 11-16-1999.
24. Arts, E.J. and Le Grice, S.F. Interaction of retroviral reverse transcriptase with template–primer duplexes during replication. *Prog. Nucl. Acid Res. Mol. Biol.*, 58, 339–393, 1998.
25. Kvaratskhelia, M., Miller, J.T., Budihas, S.R., Pannell, L.K., and Le Grice, S.F. Identification of specific HIV-1 reverse transcriptase contacts to the viral RNA:tRNA complex by mass spectrometry and a primary amine selective reagent. *Proc. Natl. Acad. Sci. USA*, 99(25), 15988–15993, 12-10-2002.
26. Sarafianos, S.G., Das, K., Tantillo, C., Clark, A.D., Ding, J., Whitcomb, J.M., Boyer, P.L., Hughes, S.H., and Arnold, E. Crystal structure of HIV-1 reverse transcriptase in complex with a polypurine tract RNA:DNA. *EMBO J.*, 20(6), 1449–1461, 3-15-2001.
27. Cristofaro, J.V., Rausch, J.W., Le Grice, S.F.J., and DeStefano, J.J. Mutations in the ribonuclease H active site of HIV-RT reveal a role for this site in stabilizing enzyme–primer–template binding. *Biochemistry*, 41(36), 10968–10975, 9-10-2002.

28. Huang, H.F., Chopra, R., Verdine, G.L., and Harrison, S.C. Structure of a covalently trapped catalytic complex of HIV-I reverse transcriptase: Implications for drug resistance. *Science*, 282(5394), 1669–1675, 11-27-1998.
29. Madrid, M., Lukin, J.A., Madura, J.D., Ding, J., and Arnold, E. Molecular dynamics of HIV-1 reverse transcriptase indicates increased flexibility upon DNA binding. *Proteins*, 45(3), 176–182, 11-15-2001.
30. Rodgers, D.W., Gamblin, S.J., Harris, B.A., Ray, S., Culp, J.S., Hellmig, B., Woolf, D.J., Debouck, C., and Harrison, S.C. The structure of unliganded reverse transcriptase from the human immunodeficiency virus type-1. *Proc. Natl. Acad. Sci. USA*, 92(4), 1222–1226, 2-14-1995.
31. Ding, J., Das, K., Hsiou, Y., Sarafianos, S.G., Clark, A.D., Jr., Jacobo–Molina, A., Tantillo, C., Hughes, S.H., and Arnold, E. Structure and functional implications of the polymerase active site region in a complex of HIV-1 RT with a double-stranded DNA template-primer and an antibody Fab fragment at 2.8-A resolution. *J. Mol. Biol.*, 284(4), 1095–1111, 12-11-1998.
32. Beard, W.A., Bebenek, K., Darden, T.A., Li, L., Prasad, R., Kunkel, T.A., and Wilson, S.H. Vertical-scanning mutagenesis of a critical tryptophan in the minor groove binding track of HIV-1 reverse transcriptase — molecular nature of polymerase-nucleic acid interactions. *J. Biological Chem.*, 273(46), 30435–30442, 11-13-1998.
33. Sarafianos, S.G., Das, K., Ding, J.P., Boyer, P.L., Hughes, S.H., and Arnold, E. Touching the heart of HIV-1 drug resistance: the fingers close down on the dNTP at the polymerase active site. *Chem. Biol.*, 6(5), R137–R146, 1999.
34. Jacobomolina, A., Ding, J.P., Nanni, R.G., Clark, A.D., Lu, X.D., Tantillo, C., Williams, R.L., Kamer, G., Ferris, A.L., Clark, P., Hizi, A., Hughes, S.H., and Arnold, E. Crystal-structure of human-immunodeficiency-virus type-1 reverse transcriptase complexed with double-stranded DNA at 3.0 angstrom-resolution shows bent DNA. *Proc. Natl. Acad. Sci. USA*, 90(13), 6320–6324, 7-1-1993.
35. Sawaya, M.R., Prasad, R., Wilson, S.H., Kraut, J., and Pelletier, H. Crystal structures of human DNA polymerase beta complexed with gapped and nicked DNA: evidence for an induced fit mechanism. *Biochemistry*, 36(37), 11205–11215, 9-16-1997.
36. Brautigam, C.A. and Steitz, T.A. Structural and functional insights provided by crystal structures of DNA polymerases and their substrate complexes. *Curr. Opin. Struct. Biol.*, 8(1), 54–63, 1998.
37. Doublie, S., Tabor, S., Long, A.M., Richardson, C.C., and Ellenberger, T. Crystal structure of a bacteriophage T7 DNA replication complex at 2.2-A resolution. *Nature*, 391(6664), 251–258, 1-15-1998.
38. Sharma, B., Kaushik, N., Upadhyay, A., Tripathi, S., Singh, K., and Pandey, V.N. A positively charged side chain at position 154 on the beta8-alphaE loop of HIV-1 RT is required for stable ternary complex formation. *Nucl. Acids Res.*, 31(17), 5167–5174, 9-1-2003.
39. Sarafianos, S.G., Clark, A.D., Das, K., Tuske, S., Birktoft, J.J., Ilankumaran, P., Ramesha, A.R., Sayer, J.M., Jerina, D.M., Boyer, P.L., Hughes, S.H., and Arnold, E. Structures of HIV-1 reverse transcriptase with pre- and post-translocation AZTMP-terminated DNA. *EMBO J.*, 21(23), 6614–6624, 12-2-2002.

40. Bebenek, K., Beard, W.A., Casasfinet, J.R., Kim, H.R., Darden, T.A., Wilson, S.H., and Kunkel, T.A. Reduced frameshift fidelity and processivity of HIV-1 reverse transcriptase mutants containing alanine substitutions in helix-H of the thumb subdomain. *J. Biological Chem.*, 270(33), 19516–19523, 8-18-1995.
41. Bebenek, K., Beard, W.A., Darden, T.A., Li, L.P., Prasad, R., Luxon, B.A., Gorenstein, D.G., Wilson, S.H., and Kunkel, T.A. A minor groove-binding track in reverse transcriptase. *Nat. Struct. Biol.*, 4(3), 194–197, 1997.
42. Bebenek, K., Abbotts, J., Wilson, S.H., and Kunkel, T.A. Error-prone polymerization by HIV-1 reverse transcriptase — contribution of template-primer misalignment, miscoding, and termination probability to mutational hot-spots. *J. Biological Chem.*, 268(14), 10324–10334, 5-15-1993.
43. Bebenek, K., Abbotts, J., Roberts, J.D., Wilson, S.H., and Kunkel, T.A. Specificity and mechanism of error-prone replication by human immunodeficiency virus-1 reverse transcriptase. *J. Biological Chem.*, 264(28), 16948–16956, 10-5-1989.
44. Abbotts, J., Bebenek, K., Kunkel, T.A., and Wilson, S.H. Mechanism of HIV-1 Reverse transcriptase — termination of processive synthesis on a natural DNA — template is influenced by the sequence of the template-primer stem. *J. Biological Chem.*, 268(14), 10312–10323, 5-15-1993.
45. Halvas, E.K., Svarovskaia, E.S., and Pathak, V.K. Role of murine leukemia virus reverse transcriptase deoxyribonucleoside triphosphate-binding site in retroviral replication and *in vivo* fidelity. *J. Virol.*, 74(22), 10349–10358, 2000.
46. Gutierrez–Rivas, M. and Menendez–Arias, L. A mutation in the primer grip region of HIV-1 reverse transcriptase that confers reduced fidelity of DNA synthesis. *Nucl. Acids Res.*, 29(24), 4963–4972, 12-15-2001.
47. Johnson, M.S., McClure, M.A., Feng, D.F., Gray, J., and Doolittle, R.F. Computer analysis of retroviral pol genes: assignment of enzymatic functions to specific sequences and homologies with nonviral enzymes. *Proc. Natl. Acad. Sci. USA*, 83(20), 7648–7652, 1986.
48. Boyer, P.L. and Hughes, S.H. Analysis of mutations at position 184 in reverse transcriptase of human immunodeficiency virus type 1. *Antimicrob. Agents Chemother.*, 39(7), 1624–1628, 1995.
49. Tantillo, C., Ding, J., Jacobo–Molina, A., Nanni, R.G., Boyer, P.L., Hughes, S.H., Pauwels, R., Andries, K., Janssen, P.A., and Arnold, E. Locations of anti-AIDS drug binding sites and resistance mutations in the three-dimensional structure of HIV-1 reverse transcriptase. Implications for mechanisms of drug inhibition and resistance. *J. Mol. Biol.*, 243(3), 369–387, 10-28-1994.
50. Wakefield, J.K., Jablonski, S.A., and Morrow, C.D. *In vitro* enzymatic activity of human immunodeficiency virus type 1 reverse transcriptase mutants in the highly conserved YMDD amino acid motif correlates with the infectious potential of the proviral genome. *J. Virol.*, 66(11), 6806–6812, 1992.
51. Hsiou, Y., Ding, J., Das, K., Clark, A.D., Jr., Hughes, S.H., and Arnold, E. Structure of unliganded HIV-1 reverse transcriptase at 2.7 A resolution: implications of conformational changes for polymerization and inhibition mechanisms. *Structure*, 4(7), 853–860, 7-15-1996.
52. Boyer, P.L., Gao, H.Q., Clark, P.K., Sarafianos, S.G., Arnold, E., and Hughes, S.H. YADD mutants of human immunodeficiency virus type 1 and Moloney murine leukemia virus reverse transcriptase are resistant to lamivudine triphosphate (3TCTP) *in vitro*. *J. Virol.*, 75(14), 6321–6328, 2001.

53. Martin–Hernandez, A.M., Gutierrez–Rivas, M., Domingo, E., and Menendez–Arias, L. Mispair extension fidelity of human immunodeficiency virus type 1 reverse transcriptases with amino acid substitutions affecting Tyr115. *Nucl. Acids Res.*, 25(7), 1383–1389, 4-1-1997.
54. Harris, D., Kaushik, N., Pandey, P.K., Yadav, P.N.S., and Pandey, V.N. Functional analysis of amino acid residues constituting the dNTP binding pocket of HIV-1 reverse transcriptase. *J. Biological Chem.*, 273(50), 33624–33634, 12-11-1998.
55. Boyer, P.L., Gao, H.Q., Frank, P., Clark, P.K., and Hughes, S.H. The basic loop of the RNase H domain of MLV RT is important both for RNase H and for polymerase activity. *Virology*, 282(1), 206–213, 3-30-2001.
56. Telesnitsky, A. and Goody, R.S. Strong-stop strand transfer during reverse transcription. In: Skalka, A.M. and Goff, S.P. *Reverse Transcriptase*, 49–83, 1993. New York, Cold Spring Harbor Laboratory Publications.
57. Hizi, A., Leis, J.P., and Joklik, W.K. The RNA-dependent DNA polymerase of avian sarcoma virus B77. Binding of viral and nonviral ribonucleic acids to the alpha, beta2, and alphabeta forms of the enzyme. *J. Biol. Chem.*, 252(19), 6878–6884, 10-10-1977.
58. Coffin, J.M., Hughes, S.H., and Varmus, H.E. *Retroviruses*. 1997. New York, Cold Spring Harbor Laboratory Press.
59. Hughes, S.H., Arnold, E., and Hostomsky, Z. RNase H of retroviral reverse transcritpases. In: Toulme, J.J. and Crouch, R.J. *Ribonucleases H*. 195–224. 2003. Paris, INSERM editions.
60. Watson, K.F., Schendel, P.L., Rosok, M.J., and Ramsey, L.R. Model RNA-directed DNA synthesis by avian myeloblastosis virus DNA polymerase and its associated Rnase-H. *Biochemistry*, 18(15), 3210–3219, 1979.
61. Goff, S.P. Retroviral reverse transcriptase — synthesis, structure, and function. *J. Acquired Immune Deficiency Syndr. Hum. Retrovirol.*, 3(8), 817–831, 1990.
62. Huber, H.E., Mccoy, J.M., Seehra, J.S., and Richardson, C.C. Human immunodeficiency virus-1 reverse transcriptase — template binding, processivity, strand displacement synthesis, and template switching. *J. Biological Chem.*, 264(8), 4669–4678, 3-15-1989.
63. Varmus, H., Swanstrom, R., Weiss, R, Teich, N., Varmus, H., and Coffin, J. *RNA Tumor Viruses*, 369–512, 1984. New York, Cold Spring Harbor Laboratory Publications.
64. Suo, Z.C. and Johnson, K.A. Effect of RNA secondary structure on RNA cleavage catalyzed by HIV-1 reverse transcriptase. *Biochemistry*, 36(41), 12468–12476, 10-14-1997.
65. Ding, J., Jacobomolina, A., Nanni, R.G., Lu, X., Williams, R.L., Clark, A.D., Hughes, S.H., and Arnold, E. Structure of HIV-1 reverse transcriptase/DsDNA/Fab complex — protein–DNA interactions and structure of the polymerase active-site. *Biophys. J.*, 64(2), A350, 1993.
66. Hermann, T., Meier, T., Gotte, M., and Heumann, H. The "helix clamp" in HIV-1 reverse transcriptase: a new nucleic acid-binding motif common in nucleic acid polymerases. *Nucl. Acids Res.*, 22(22), 4625–4633, 11-11-1994.
67. Divita, G., Baillon, J.G., Rittinger, K., Chermann, J.C., and Goody, R.S. Interface peptides as structure-based human immunodeficiency virus reverse transcriptase inhibitors. *J. Biological Chem.*, 270(48), 28642–28646, 12-1-1995.
68. Wohrl, B.M., Tantillo, C., Arnold, E., and Le Grice, S.F. An expanded model of replicating human immunodeficiency virus reverse transcriptase. *Biochemistry*, 34(16), 5343–5356, 4-25-1995.

69. Molling, K., Bolognesi, D.P., Bauer, H., Busen, W., Plassmann, H.W., and Hausen, P. Association of viral reverse transcriptase with an enzyme degrading the RNA moiety of RNA–DNA hybrids. *Nat. New Biol.*, 234(51), 240–243, 12-22-1971.
70. Smith, C.M., Smith, J.S., and Roth, M.J. RNase H requirements for the second strand transfer reaction of human immunodeficiency virus type 1 reverse transcription. *J. Virol.*, 73(8), 6573–6581, 1999.
71. Smith, C.M., Potts, W.B., Smith, J.S., and Roth, M.J. RNase H cleavage of tRNA(Pro) mediated by M-MuLV and HIV-1 reverse. *Virology*, 229(2), 437–446, 3-17-1997.
72. Champoux, J.J. Roles of ribonuclease H in reverse transcription. In: Skalka, A.M. and Goff, S.P., Eds. *Reverse Transcriptase*. 103–177. 1993. New York, Cold Spring Harbor Labratory Publications.
73. Kvaratskhelia, M., Budihas, S.R., and Le Grice, S.F.J. Pre-existing distortions in nucleic acid structure aid polypurine tract selection by HIV-1 reverse transcriptase. *J. Biological Chem.*, 277(19), 16689–16696, 5-10-2002.
74. Gao, H.Q., Boyer, P.L., Arnold, E., and Hughes, S.H. Effects of mutations in the polymerase domain on the polymerase, RNase H and strand transfer activities of human immunodeficiency virus type 1 reverse transcriptase. *J. Mol. Biol.*, 277(3), 559–572. 4-3-1998.
75. Rausch, J.W., Lener, D., Miller, J.T., Julias, J.G., Hughes, S.H., and Le Grice, S.F.J. Altering the RNase H primer grip of human immunodeficiency virus reverse transcriptase modifies cleavage specificity. *Biochemistry*, 41(15), 4856–4865, 4-16-2002.
76. Powell, M.D., Beard, W.A., Bebenek, K., Howard, K.J., Le Grice, S.F.J., Darden, T.A., Kunkel, T.A., Wilson, S.H., and Levin, J.G. Residues in the alpha H and alpha I helices of the HIV-1 reverse transcriptase thumb subdomain required for the specificity of RNase H-catalyzed removal of the polypurine tract primer. *J. Biological Chem.*, 274(28), 19885–19893, 7-9-1999.
77. Kati, W.M., Johnson, K.A., Jerva, L.F., and Anderson, K.S. Mechanism and fidelity of HIV reverse transcriptase. *J. Biol.Chem.*, 267(36), 25988–25997, 12-25-1992.
78. Spence, R.A., Kati, W.M., Anderson, K.S., and Johnson, K.A. Mechanism of inhibition of HIV-1 reverse transcriptase by nonnucleoside inhibitors. *Science*, 267(5200), 988–993, 2-17-1995.
79. Divita, G., Muller, B., Immendorfer, U., Gautel, M., Rittinger, K., Restle, T., and Goody, R.S. Kinetics of interaction of Hiv Reverse transcriptase with primer template. *Biochemistry*, 32(31), 7966–7971, 8-10-1993.
80. Kruhoffer, M., Urbanke, C., and Grosse, F. 2-Step binding of HIV-1 Reverse transcriptase to nucleic-acid substrates. *Nucl. Acids Res.*, 21(17), 3943–3949, 8-25-1993.
81. Reardon, J.E. Human immunodeficiency virus reverse transcriptase – steady state and pre-steady state kinetics of nucleotide incorporation. *Biochemistry*, 31(18), 4473–4479, 5-12-1992.
82. Reardon, J.E. Human immunodeficiency virus reverse transcriptase — a kinetic analysis of RNA-dependent and DNA-dependent DNA polymerization. *J. Biological Chem.*, 268(12), 8743–8751, 4-25-1993.
83. Hsieh, J.C., Zinnen, S., and Modrich, P. Kinetic mechanism of the DNA-dependent DNA polymerase activity of human immunodeficiency virus reverse transcriptase. *J. Biological Chem.*, 268(33), 24607–24613, 11-25-1993.

84. Rothwell, P.J., Berger, S., Kensch, O., Felekyan, S., Antonik, M., Wohrl, B.M., Restle, T., Goody, R.S., and Seidel, C.A.M. Multiparameter single-molecule fluorescence spectroscopy reveals heterogeneity of HIV-1 reverse transcriptase: primer/template complexes. *Proc. Natl. Acad. Sci. USA,* 100(4), 1655–1660, 2-18-2003.
85. Wohrl, B.M., Krebs, R., Goody, R.S., and Restle, T. Refined model for primer/template binding by HIV-1 reverse transcriptase: pre-steady state kinetic analyses of primer/template binding and nucleotide incorporation events distinguish between different binding modes depending on the nature of the nucleic acid substrate. *J. Mol. Biol.,* 292(2), 333–344, 9-17-1999.
86. Kerr, S.G. and Anderson, K.S. RNA-dependent DNA replication fidelity of HIV-1 reverse transcriptase: evidence of discrimination between DNA and RNA substrates. *Biochemistry,* 36(46), 14056–14063, 11-18-1997.
87. Rittinger, K., Divita, G., and Goody, R.S. Human immunodeficiency virus reverse transcriptase substrate-induced conformational changes and the mechanism of inhibition by non-nucleoside inhibitors. *Proc. Natl. Acad. Sci. USA,* 92(17), 8046–8049, 8-15-1995.
88. Suo, Z.C. and Johnson, K.A. Effect of RNA secondary structure on the kinetics of DNA synthesis catalyzed by HIV-1 reverse transcriptase. *Biochemistry,* 36(41), 12459–12467, 10-14-1997.
89. Suo, Z.C. and Johnson, K.A. DNA secondary structure effects on DNA synthesis catalyzed by HIV-1 reverse transcriptase. *J. Biological Chem.,* 273(42), 27259–27267, 10-16-1998.
90. Suo, Z.C. and Johnson, K.A. RNA secondary structure switching during DNA synthesis catalyzed by HIV-1 reverse transcriptase. *Biochemistry,* 36(48), 14778–14785, 12-2-1997.
91. Suo, Z.C. and Johnson, K.A. The effect of RNA secondary structure on the kinetics of polymerization catalyzed by HIV-1 reverse transcriptase. *Biochemistry,* 35(28), 9, 7-16-1996.
92. Johnson, K.A. Conformational coupling in DNA polymerase fidelity. *Annu. Rev. Biochem.,* 62, 685–713, 1993.
93. Kaushik, N., Chowdhury, K., Pandey, V.N., and Modak, M.J. Valine of the YMDD motif of Moloney murine leukemia virus reverse transcriptase: role in the fidelity of DNA synthesis. *Biochemistry,* 39(17), 5155–5165, 5-2-2000.
94. Boutabout, M., Wilhelm, M., and Wilhelm, F.X. DNA synthesis fidelity by the reverse transcriptase of the yeast retrotransposon Ty1. *Nucl. Acids Res.,* 29(11), 2217–2222, 6-1-2001.
95. Leider, J.M., Palese, P., and Smith, F.I. Determination of the mutation rate of a retrovirus. *J. Virol.,* 62(9), 3084–3091, 1988.
96. Park, S.G., Kim, Y., Park, E., Ryu, H.M., and Jung, G. Fidelity of hepatitis B virus polymerase. *Eur. J. Biochemistry,* 270(14), 2929–2936, 2003.
97. Roberts, J.D., Bebenek, K., and Kunkel, T.A. The accuracy of reverse transcriptase from HIV-1. *Science,* 242(4882), 1171–1173, 11-25-1988.
98. Boyer, J.C., Bebenek, K., and Kunkel, T.A. Unequal human immunodeficiency virus type-1 reverse transcriptase error rates with RNA and DNA templates. *Proc. Natl. Acad. Sci. USA,* 89(15), 6919–6923, 8-1-1992.
99. Fedoroff, O.Y., Salazar, M., and Reid, B.R. Structure of a DNA–RNA hybrid duplex — why Rnase-H does not cleave pure RNA. *J. Mol. Biol.,* 233(3), 509–523, 10-5-1993.

100. Sugimoto, N., Nakano, S., Katoh, M., Matsumura, A., Nakamuta, H., Ohmichi, T., Yoneyama, M., and Sasaki, M. Thermodynamic parameters to predict stability of RNA/DNA hybrid duplexes. *Biochemistry*, 34(35), 11211–11216, 9-5-1995.
101. Cramer, J., Strerath, M., Marx, A., and Restle, T. Exploring the effects of active site constraints on HIV-1 reverse transcriptase DNA polymerase fidelity. *J. Biological Chem.*, 277(46), 43593–43598, 11-15-2002.
102. Weiss, K.K., Bambara, R.A., and Kim, B. Mechanistic role of residue Gln(151) in error-prone DNA synthesis by human immunodeficiency virus type 1 (HIV-1) reverse transcriptase (RT) — pre-steady state kinetic study of the Q151N HIV-1 RT mutant with increased fidelity. *J. Biological Chem.*, 277(25), 22662–22669, 6-21-2002.
103. Kaushik, N., Talele, T.T., Pandey, P.K., Harris, D., Yadav, P.N.S., and Pandey, V.N. Role of glutamine 151 of human immunodeficiency virus type-1 reverse transcriptase in substrate selection as assessed by site-directed mutagenesis. *Biochemistry*, 39(11), 2912–2920, 3-21-2000.
104. Esnouf, R., Ren, J., Ross, C., Jones, Y., Stammers, D., and Stuart, D. Mechanism of inhibition of HIV-1 reverse transcriptase by non-nucleoside inhibitors. *Nat. Struct. Biol.*, 2(4), 303–308, 1995.
105. Merluzzi, V.J., Hargrave, K.D., Labadia, M., Grozinger, K., Skoog, M., Wu, J.C., Shih, C.K., Eckner, K., Hattox, S., Adams, J. Inhibition of HIV-1 replication by a non-nucleoside reverse transcriptase inhibitor. *Science*, 250(4986), 1411–1413, 12-7-1990.
106. Romero, D.L., Busso, M., Tan, C.K., Reusser, F., Palmer, J.R., Poppe, S.M., Aristoff, P.A., Downey, K.M., So, A.G., Resnick, L. Non-nucleoside reverse transcriptase inhibitors that potently and specifically block human immunodeficiency virus type 1 replication. *Proc. Natl. Acad. Sci. USA*, 88(19), 8806–8810, 10-1-1991.
107. Young, S.D., Britcher, S.F., Tran, L.O., Payne, L.S., Lumma, W.C., Lyle, T.A., Huff, J.R., Anderson, P.S., Olsen, D.B., Carroll, S.S. L-743, 726 (DMP-266): a novel, highly potent non-nucleoside inhibitor of the human immunodeficiency virus type 1 reverse transcriptase. *Antimicrob. Agents Chemother.*, 39(12), 2602–2605, 1995.
108. Temiz, N.A. and Bahar, I. Inhibitor binding alters the directions of domain motions in HIV-1 reverse transcriptase. *Proteins*, 49(1), 61–70, 10-1-2002.
109. Palaniappan, C., Fay, P.J., and Bambara, R.A. Nevirapine alters the cleavage specificity of ribonuclease H of human immunodeficiency virus 1 reverse transcriptase. *J. Biol. Chem.*, 270(9), 4861–4869, 3-3-1995.
110. Elion, G.B., Furman, P.A., Fyfe, J.A., de Miranda, P., Beauchamp, L., and Schaeffer, H.J. Selectivity of action of an antiherpetic agent, 9-(2-hydroxyethoxymethyl) guanine. *Proc. Natl. Acad. Sci. USA*, 74(12), 5716–5720, 1977.
111. Beck, J., Vogel, M., and Nassal, M. dNTP versus NTP discrimination by phenylalanine 451 in duck hepatitis B virus P protein indicates a common structure of the dNTP-binding pocket with other reverse transcriptases. *Nucl. Acids Res.*, 30(7), 1679–1687, 4-1-2002.
112. Li, X. and Chan, W.K. Transport, metabolism and elimination mechanisms of anti-HIV agents. *Adv. Drug Deliv. Rev.*, 39(1–3), 81–103, 10-18-1999.
113. Lewis, W., Day, B.J., and Copeland, W.C. Mitochondrial toxicity of NRTI antiviral drugs: an integrated cellular perspective. *Nat. Rev. Drug Discov.*, 2(10), 812–822, 2003.

114. Robbins, B.L., Srinivas, R.V., Kim, C., Bischofberger, N., and Fridland, A. Anti-human immunodeficiency virus activity and cellular metabolism of a potential prodrug of the acyclic nucleoside phosphonate 9-R-(2-phosphonomethoxypropyl)adenine (PMPA), bis(isopropyloxymethylcarbonyl)PMPA. *Antimicrob. Agents Chemother.*, 42(3), 612–617, 1998.
115. Suo, Z.C. and Johnson, K.A. Selective inhibition of HIV-1 reverse transcriptase by an antiviral inhibitor, (R)-9-(2-phosphonylmethoxypropyl)adenine. *J. Biological Chem.*, 273(42), 27250–27258, 10-16-1998.
116. Johnson, A.A., Ray, A.S., Hanes, J., Suo, Z.C., Colacino, J.M., Anderson, K.S., and Johnson, K.A. Toxicity of antiviral nucleoside analogs and the human mitochondrial DNA polymerase. *J. Biological Chem.*, 276(44), 40847–40857, 11-2-2001.
117. Naeger, L.K., Margot, N.A., and Miller, M.D. Tenofovir (PMPA) is less susceptible to pyrophosphorolysis and nucleotide-dependent chain-terminator removal than zidovudine or stavudine. *Nucleosides Nucleotides Nucl. Acids*, 20(4–7), 635–639, 2001.
118. Richman, D.D., Fischl, M.A., Grieco, M.H., Gottlieb, M.S., Volberding, P.A., Laskin, O.L., Leedom, J.M., Groopman, J.E., Mildvan, D., Hirsch, M.S. The toxicity of azidothymidine (AZT) in the treatment of patients with AIDS and AIDS-related complex. A double-blind, placebo-controlled trial. *N. Engl. J. Med.*, 317(4), 192–197, 7-23-1987.
119. Yarchoan, R., Pluda, J.M., Thomas, R.V., Mitsuya, H., Brouwers, P., Wyvill, K.M., Hartman, N., Johns, D.G., and Broder, S. Long-term toxicity/activity profile of 2′,3′-dideoxyinosine in AIDS or AIDS-related complex. *Lancet*, 336(8714), 526–529, 9-1-1990.
120. Nickel, W., Austermann, S., Bialek, G., and Grosse, F. Interactions of azidothymidine triphosphate with the cellular DNA polymerases alpha, delta, and epsilon and with DNA primase. *J. Biol. Chem.*, 267(2), 848–854, 1-15-1992.
121. Anderson, K.S. Perspectives on the molecular mechanism of inhibition and toxicity of nucleoside analogs that target HIV-1 reverse transcriptase. *Biochem. Biophys. Acta — Mol. Basis Dis.*, 1587(2–3), 296–299, 7-18-2002.
122. Feng, J.Y., Johnson, A.A., Johnson, K.A., and Anderson, K.S. Insights into the molecular mechanism of mitochondrial toxicity by AIDS drugs. *J. Biol. Chem.*, 276(26), 23832–23837, 6-29-2001.
123. Delaney, W.E., IV, Locarnini, S., and Shaw, T. Resistance of hepatitis B virus to antiviral drugs: current aspects and directions for future investigation. *Antivir Chem. Chemother.*, 12(1), 1–35, 2001.
124. Shafer, R.W., Stevenson, D., and Chan, B. Human immunodeficiency virus reverse transcriptase and protease sequence database. *Nucl. Acids Res.*, 27(1), 348–352, 1-1-1999.
125. Soriano, V. and de Mendoza, C. Genetic mechanisms of resistance to protease inhibitors and entry inhibitors. *HIV Clin. Trials*, 3(3), 249–257, 2002.
126. Ueno, T., Shirasaka, T., and Mitsuya, H. Enzymatic characterization of human immunodeficiency virus type 1 reverse transcriptase resistant to multiple 2′,3′-dideoxynucleoside 5′-triphosphates. *J. Biol. Chem.*, 270(40), 23605–23611, 10-6-1995.
127. Chong, Y., Borroto–Esoda, K., Furman, P.A., Schinazi, R.F., and Chu, C.K. Molecular mechanism of DApd/DXG against zidovudine- and lamivudine-drug resistant mutants: a molecular modelling approach. *Antivir Chem. Chemother.*, 13(2), 115–128, 2002.

128. Gao, H.Q., Boyer, P.L., Sarafianos, S.G., Arnold, E., and Hughes, S.H. The role of steric hindrance in 3TC resistance of human immunodeficiency virus type-1 reverse transcriptase. *J. Mol. Biol.*, 300(2), 403–418, 7-7-2000.
129. Boyer, P.L., Sarafianos, S.G., Arnold, E., and Hughes, S.H. Mechanisms of nucleoside analogue resistance. *Antiviral Ther.*, 7, S32, 2002.
130. Feng, J.Y. and Anderson, K.S. Mechanistic studies examining the efficiency and fidelity of DNA synthesis by the 3TC-resistant mutant (184V) of HIV-1 reverse transcriptase. *Biochemistry*, 38(29), 9440–9448, 7-20-1999.
131. Larder, B.A., Kemp, S.D., and Harrigan, P.R. Potential mechanism for sustained antiretroviral efficacy of AZT-3Tc combination therapy. *Science*, 269(5224), 696–699, 8-4-1995.
132. Arion, D., Kaushik, N., McCormick, S., Borkow, G., and Parniak, M.A. Phenotypic mechanism of HIV-1 resistance to 3′-azido-3′-deoxythymidine (AZT): increased polymerization processivity and enhanced sensitivity to pyrophosphate of the mutant viral reverse transcriptase. *Biochemistry*, 37(45), 15908–15917, 11-10-1998.
133. Meyer, P.R., Matsuura, S.E., So, R.G., and Scott, W.A. Unblocking of chain-terminated primer by HIV-1 reverse transcriptase through a nucleotide-dependent mechanism. *Proc. Natl. Acad. Sci. USA*, 95(23), 13471–13476, 11-10-1998.
134. Meyer, P.R., Matsuura, S.E., Mian, A.M., So, A.G., and Scott, W.A. A mechanism of AZT resistance: An increase in nucleotide-dependent primer unblocking by mutant HIV-1 reverse transcriptase. *Mol. Cell*, 4(1), 35–43, 1999.
135. Boyer, P.L., Sarafianos, S.G., Arnold, E., and Hughes, S.H. Selective excision of AZTMP by drug-resistant human immunodeficiency virus reverse transcriptase. *J. Virol.*, 75(10), 4832–4842, 2001.
136. Meyer, P.R., Matsuura, S.E., Tolun, A.A., Pfeifer, I., So, A.G., Mellors, J.W., and Scott, W.A. Effects of specific zidovudine resistance mutations and substrate structure on nucleotide-dependent primer unblocking by human immunodeficiency virus type I reverse transcriptase. *Antimicrob. Agents Chemother.*, 46(5), 1540–1545, 2002.
137. Boyer, P.L., Sarafianos, S.G., Arnold, E., and Hughes, S.H. Nucleoside analog resistance caused by insertions in the fingers of human immunodeficiency virus type 1 reverse transcriptase involves ATP-mediated excision. *J. Virol.*, 76(18), 9143–9151, 2002.
138. Richman, D.D. HIV drug resistance. *Annu. Rev. Pharmacol. Toxicol.*, 33, 149–164, 1993.
139. Balzarini, J., Karlsson, A., Perez–Perez, M.J., Vrang, L., Walbers, J., Zhang, H., Oberg, B., Vandamme, A.M., Camarasa, M.J., and De Clercq, E. HIV-1-specific reverse transcriptase inhibitors show differential activity against HIV-1 mutant strains containing different amino acid substitutions in the reverse transcriptase. *Virology*, 192(1), 246–253, 1993.
140. Balzarini, J., Karlsson, A., Vandamme, A.M., Perez–Perez, M.J., Zhang, H., Vrang, L., Oberg, B., Backbro, K., Unge, T., and San Felix, A. Human immunodeficiency virus type 1 (HIV-1) strains selected for resistance against the HIV-1-specific (2′,5′-bis-O-(tert-butyldimethylsilyl)-3′-spiro-5″-[4″-amino-1″, 2″-oxathiole-2″,2″-dioxide)]-beta-D-pentofurano syl (TSAO) nucleoside analogues retain sensitivity to HIV-1-specific nonnucleoside inhibitors. *Proc. Natl. Acad. Sci. USA*, 90(15), 6952–6956, 8-1-1993.

141. Saag, M.S., Emini, E.A., Laskin, O.L., Douglas, J., Lapidus, W.I., Schleif, W.A., Whitley, R.J., Hildebrand, C., Byrnes, V.W., and Kappes, J.C. A short-term clinical evaluation of L-697,661, a non-nucleoside inhibitor of HIV-1 reverse transcriptase. L-697, 661 Working Group. *N. Engl. J. Med.*, 329(15), 1065–1072, 10-7-1993.
142. Byrnes, V.W., Sardana, V.V., Schleif, W.A., Condra, J.H., Waterbury, J.A., Wolfgang, J.A., Long, W.J., Schneider, C.L., Schlabach, A.J., and Wolanski, B.S. Comprehensive mutant enzyme and viral variant assessment of human immunodeficiency virus type 1 reverse transcriptase resistance to non-nucleoside inhibitors. *Antimicrob. Agents Chemother.*, 37(8), 1576–1579, 1993.
143. Balzarini, J., Weeger, M., Camarasa, M.J., De Clercq, E., and Uberla, K. Sensitivity/resistance profile of a simian immunodeficiency virus containing the reverse transcriptase gene of human immunodeficiency virus type 1 (HIV-1) toward the HIV-1-specific non-nucleoside reverse transcriptase inhibitors. *Biochem. Biophys. Res. Commun.*, 211(3), 850–856, 6-26-1995.
144. Balzarini, J., Perez–Perez, M.J., Velazquez, S., San Felix, A., Camarasa, M.J., De Clercq, E., and Karlsson, A. Suppression of the breakthrough of human immunodeficiency virus type 1 (HIV-1) in cell culture by thiocarboxanilide derivatives when used individually or in combination with other HIV-1-specific inhibitors (i.e., TSAO derivatives). *Proc. Natl. Acad. Sci. USA*, 92(12), 5470–5474, 6-6-1995.
145. Balzarini, J. and De Clercq, E. Analysis of inhibition of retroviral reverse transcriptase. *Methods Enzymol.*, 275, 472–502, 1996.
146. Buckheit, R.W., Jr., Kinjerski, T.L., Fliakas–Boltz, V., Russell, J.D., Stup, T.L., Pallansch, L.A., Brouwer, W.G., Dao, D.C., Harrison, W.A., and Schultz, R.J. Structure activity and cross-resistance evaluations of a series of human immunodeficiency virus type-1-specific compounds related to oxathiin carboxanilide. *Antimicrob. Agents Chemother.*, 39(12), 2718–2727, 1995.
147. Buckheit, R.W., Jr., Fliakas–Boltz, V., Yeagy–Bargo, S., Weislow, O., Mayers, D.L., Boyer, P.L., Hughes, S.H., Pan, B.C., Chu, S.H., and Bader, J.P. Resistance to 1-([2-hydroxyethoxy)methyl]-6-(phenylthio)thymine derivatives is generated by mutations at multiple sites in the HIV-1 reverse transcriptase. *Virology*, 210(1), 186–193, 6-20-1995.
148. Buckheit, R.W., Jr., Fliakas–Boltz, V., Decker, W.D., Roberson, J.L., Stup, T.L., Pyle, C.A., White, E.L., McMahon, J.B., Currens, M.J., and Boyd, M.R. Comparative anti-HIV evaluation of diverse HIV-1-specific reverse transcriptase inhibitor-resistant virus isolates demonstrates the existence of distinct phenotypic subgroups. *Antiviral Res.*, 26(2), 117–132, 1995.
149. Pelemans, H., Esnouf, R., Dunkler, A., Parniak, M.A., Vandamme, A.M., Karlsson, A., De Clercq, E., Kleim, J.P., and Balzarini, J. Characteristics of the Pro225His mutation in human immunodeficiency virus type 1 (HIV-1) reverse transcriptase that appears under selective pressure of dose-escalating quinoxaline treatment of HIV-1. *J. Virol.*, 71(11), 8195–8203, 1997.
150. Ren, J., Milton, J., Weaver, K.L., Short, S.A., Stuart, D.I., and Stammers, D.K. Structural basis for the resilience of efavirenz (DMP-266) to drug resistance mutations in HIV-1 reverse transcriptase. *Structure Fold. Des.*, 8(10), 1089–1094, 10-15-2000.

151. Lindberg, J., Sigurdsson, S., Lowgren, S., Andersson, H.O., Sahlberg, C., Noreen, R., Fridborg, K., Zhang, H., and Unge, T. Structural basis for the inhibitory efficacy of efavirenz (DMP-266), MSC194 and PNU142721 towards the HIV-1 RT K103N mutant. *Eur. J. Biochem.*, 269(6), 1670–1677, 2002.
152. Marx, A., Amacker, M., Stucki, M., Hubscher, U., Bickle, T.A., and Giese, B. 4′-Acylated thymidine 5′-triphosphates: a tool to increase selectivity towards HIV-1 reverse transcriptase. *Nucl. Acids Res.*, 26(17), 4063–4067, 9-1-1998.
153. Gish, R.G., Leung, N.W., Wright, T.L., Trinh, H., Lang, W., Kessler, H.A., Fang, L., Wang, L.H., Delehanty, J., Rigney, A., Mondou, E., Snow, A., and Rousseau, F. Dose range study of pharmacokinetics, safety, and preliminary antiviral activity of emtricitabine in adults with hepatitis B virus infection. *Antimicrob. Agents Chemother.*, 46(6), 1734–1740, 2002.
154. Schinazi, R.F., McMillan, A., Cannon, D., Mathis, R., Lloyd, R.M., Peck, A., Sommadossi, J.P., St Clair, M., Wilson, J., and Furman, P.A. Selective inhibition of human immunodeficiency viruses by racemates and enantiomers of cis-5-fluoro-1-(2-(hydroxymethyl)-1,3-oxathiolan-5-yl]cytosine. *Antimicrob. Agents Chemother.*, 36(11), 2423–2431, 1992.
155. Furman, P.A., Jeffrey, J., Keifer, L.L., Feng, J.Y., Anderson, K.S., Borroto–Esoda, K., Hill, E., Copeland, W.C., Chu, C.K., Sommadossi, J.P., Liberman, I., Schinazi, R.F., and Painter, G.R. Mechanism of action of 1-beta-D-2,6-diaminopurine dioxolane, a prodrug of the human immunodeficiency virus type 1 inhibitor 1-beta-D-dioxolane guanosine. *Antimicrob. Agents Chemother.*, 45(1), 158–165, 2001.
156. Patel, J. and Mitra, A.K. ACH-126443 Achillion/Yale University. *Curr. Opin. Invest. Drugs*, 3(11), 1580–1584, 2002.
157. Mewshaw, J.P., Myrick, F.T., Wakefield, D.A., Hooper, B.J., Harris, J.L., McCreedy, B., and Borroto–Esoda, K. Dioxolane guanosine, the active form of the prodrug diaminopurine dioxolane, is a potent inhibitor of drug-resistant HIV-1 isolates from patients for whom standard nucleoside therapy fails. *J. Acquired Immune Deficiency Syndr.*, 29(1), 11–20, 1-1-2002.
158. Feng, J.Y. and Anderson, K.S. Mechanistic studies comparing the incorporation of (+) and (–) isomers of 3TCTP by HIV-1 reverse transcriptase. *Biochemistry*, 38(1), 55–63, 1-5-1999.
159. Chen, S.H. Comparative evaluation of L-Fd4C and related nucleoside analogs as promising antiviral agents. *Curr. Med. Chem.*, 9(9), 899–912, 2002.
160. Burke, D.H., Scates, L., Andrews, K., and Gold, L. Bent pseudoknots and novel RNA inhibitors of type 1 human immunodeficiency virus (HIV-1) reverse transcriptase. *J. Mol. Biol.*, 264(4), 650–666, 12-13-1996.
161. Fisher, T.S., Joshi, P., and Prasad, V.R. Mutations that confer resistance to template-analog inhibitors of human immunodeficiency virus (HIV) type 1 reverse transcriptase lead to severe defects in HIV replication. *J. Virol.*, 76(8), 4068–4072, 2002.
162. Nickens, D.G., Patterson, J.T., and Burke, D.H. Inhibition of HIV-1 reverse transcriptase by RNA aptamers in *Escherichia coli*. *RNA*, 9(9), 1029–1033, 2003.
163. Kensch, O., Connolly, B.A., Steinhoff, H.J., McGregor, A., Goody, R.S., and Restle, T. HIV-1 reverse transcriptase-pseudoknot RNA aptamer interaction has a binding affinity in the low picomolar range coupled with high specificity. *J. Biol. Chem.*, 275(24), 18271–18278, 6-16-2000.

164. Joshi, P. and Prasad, V.R. Potent inhibition of human immunodeficiency virus type 1 replication by template analog reverse transcriptase inhibitors derived by SELEX (systematic evolution of ligands by exponential enrichment). *J. Virol.*, 76(13), 6545–6557, 2002.
165. Andreola, M.L., Pileur, F., Calmels, C., Ventura, M., Tarrago–Litvak, L., Toulme, J.J., and Litvak, S. DNA aptamers selected against the HIV-1 RNase H display *in vitro* antiviral activity. *Biochemistry*, 40(34), 10087–10094, 8-28-2001.
166. Novina, C.D., Murray, M.F., Dykxhoorn, D.M., Beresford, P.J., Riess, J., Lee, S.K., Collman, R.G., Lieberman, J., Shankar, P., and Sharp, P.A. siRNA-directed inhibition of HIV-1 infection. *Nat. Med.*, 8(7), 681–686, 2002.
167. Lee, N.S., Dohjima, T., Bauer, G., Li, H., Li, M.J., Ehsani, A., Salvaterra, P., and Rossi, J. Expression of small interfering RNAs targeted against HIV-1 rev transcripts in human cells. *Nat. Biotechnol.*, 20(5), 500–505, 2002.
168. James, H.A. and Gibson, I. The therapeutic potential of ribozymes. *Blood*, 91(2), 371–382, 1-15-1998.
169. Dorman, N. and Lever, A.M. RNA-based gene therapy for HIV infection. *HIV Med.*, 2(2), 114–122, 2001.

chapter nine

Base-modified nucleoside triphosphates: chemistry and biological applications

Stephen G. Will
Department of Core Chemistry, Roche Molecular Diagnostics, Alameda, CA

Veeraiah Bodepudi*
Department of Core Chemistry, Roche Molecular Diagnostics, Alameda, CA

Contents

* Corresponding author.

9.1 Introduction

This review is intended to cover the synthesis and use of base-modified nucleoside triphosphates (BM-NTPs). The BM-NTPs covered in this review are predominantly synthetic in origin, although a selection of rare analogs that occur in nature will also be discussed.

The application of nucleic acids as therapeutic agents in antisense and triplex technologies — and as molecular biology tools in the field of diagnostics — inspired researchers to study various nucleobase, sugar, and phosphate backbone modifications of the naturally occurring nucleic acids. As a result, several new inventions largely responsible for the successful sequencing of the human genome, such as polymerase chain reaction (PCR) and DNA sequencing technologies, were developed. They are also widely used techniques in molecular diagnostics. BM-NTPs were critical components of the success of this endeavor. In addition, a tremendous amount of work published in the last three decades has improved understanding of nucleic acids with various chemical modifications.

However, this chapter is dedicated to base-modified nucleoside triphosphates and their applications in biotechnology only. The use of some standard natural triphosphates in biotechnological applications will also be discussed briefly.

Base-Modified Nucleoside Triphosphates consist of analogs of the naturally occurring DNA and RNA monomers. These analogs confer a blend of physical and chemical properties that allow a diverse set of applications of these molecules at the interface between chemistry and biology, from the areas of biochemical tools through medicinal applications. Among many other uses, they offer the researcher ways to probe some basic enzymatic processes and they provide unique and highly selective methods of labeling nucleic acids.

In order to be effective substrates in technologically useful biological processes, the BM-NTPs must in some way mimic the properties of naturally occurring NTPs. The basic structural elements of the NTPs should be substantially maintained to ensure biological recognition, although any base modification is inevitably going to modify the structure of the nucleoside on an atomic or electronic level. These modifications may, in some cases, allow for binding of the analog to an enzyme's active site, but may block the biological activity of the enzyme involved.

This is the key mode of action of the cytotoxic nucleoside analogs based on the 5-fluoro-pyrimidine nucleosides; the analogs bind very strongly to the active site of kinases, but cannot serve as substrates for the second step of triphosphate synthesis due to stereoelectronic effects of the small but highly electronegative fluorine atom. In this case, the analog is recognized by the enzyme, which implies that the fluorine atom at the 5-position and the stereoelectronic perturbation induced are not intimately involved in the recognition process. The binding of the analog MP to the enzyme inhibits the *de novo* synthesis of 2′-deoxythymidine-triphosphate, resulting in the cytotoxic effect. The 5-fluorocytidine analog exhibits the same effect, although it's believed that conversion to the 5-fluorouridine analog is required for acitivity.

9.2 *Nucleic acid chemistry, H-bonding structure*

The hydrogen-bonding pattern of natural nucleic acid bases is the key to the complementary recognition of alternating nucleic acid strands and of the nucleoside triphosphate components used to make them. This pattern of hydrogen bond donor and acceptor functionality results in strong, relatively stable, and unambiguous pairing of the naturally occurring bases. By examining structural variants in which the pattern of hydrogen bonding is modulated through the change of the positions of the nitrogen atoms in the bases or through modification of the type of exocyclic substituent presented for base pairing in specific locations within the bases, the combination of specificity and strength in natural bases can be tuned to meet certain specific tasks in biological and technological applications.

For example, the normal 3-hydrogen bonds, formed between guanine and cytidine in DNA, can be reduced to 2-hydrogen bonds by the substitution of deoxyinosine for dG in oligonucleotides. Deoxyinosine's lack of the exocyclic N^2 amino group of the guanine base results in loss of strength of hydrogen bonding, which results in the generation of weaker specific DNA duplexes; in this case, it also results in the possibility of formation of significantly strong structures with classically noncomplementary bases. This propensity is used in nature, whereby the "wobble" position of the third position of the anticodon loop in tRNA may be substituted by inosine, which reduces the specificity by which the anticodon loop hybridizes with its target sequence. The weaker hydrogen bonding associated with dITP has been exploited in PCR applications, in which the denaturation temperature of PCRs can be significantly reduced over conventional PCRs due to the attendant lowering of the T_{den} of the dI containing PCR product strands.[1]

9.3 *Chemistry*

Modified nucleosides have been found in ribonucleic acids (RNAs)[2,3] and deoxyribonucleic acids (DNAs)[2,3] due to the methylation of nucleic acids at various bases. In 1948, Hotchkiss discovered the first modified nucleobase, 5-methylcytosine, in a sample of calf-thymus DNA, presumably due to DNA methyl transferase activity.[4] Since then much work has been done in the area of DNA methylation, revealing it to be an important process contributing to transcriptional regulation in living cells.

Cytosine methylation within CpG dinucleotides has been implicated with the regulation of gene expression in higher organisms.[5] When incorporated into DNA, 5-methyl-2′-deoxycytidine (5-Me-dC) induced apoptosis,[6] gene silencing,[7–11] and other transcriptional regulation changes and controls.[12–16] Another methylated nucleoside analog is 7-methylguanosine, iso-

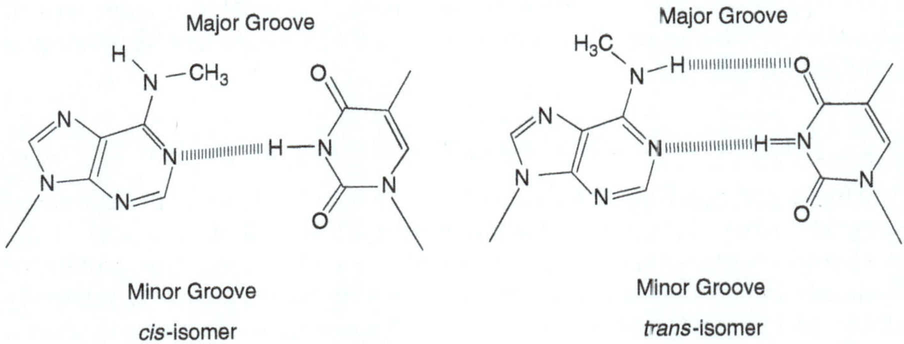

Figure 9.1 Rotational isomers of N^6-Me-dA and their H-bonding in Watson-Crick base pairing.

lated by Dunn in 1963 as a free base from acid hydrolysates of tRNA from pig liver.[17] The positive charge on 7-methylguanosine (N^7-Me-G) might stabilize certain regions of the tRNA structure[18] and also may play an important role in translation.[19–23] Similarly, due to its preferential *cis* conformation, N^6-methyl-2′-deoxyadenosine (N^6-Me-dA) could affect local nucleic acid conformation because N^6-Me-dA analogs exist in solution primarily in the *cis* conformation (about 96% of the time) rather than in the *trans* conformation.[24]

In the *cis* conformation, the methyl group of N^6-Me-dA is located on the N^1 side of the base and thus prevents the formation of the second hydrogen bond in a normal Watson–Crick base pairing (Figure 9.1). By contrast, in the *trans* conformation the methyl group is on the N^7 side of the base and thereby favorable to a normal A–T base pair. Because of these *cis* and *trans* rotational isomers, substitution with N^6-Me-dA can lower the stability of the resulting base pair in duplexes.[25,26] Conversely, the substitution of 5-Me-dCTP in place of dCTP results in an increase in stability of ds DNA.[c] Both N^6-Me-dATP and 5-Me-dCTP have found use in PCR applications.[27]

Formycin and pseudouridine analogs are C-nucleosides in which the normal N base is replaced by the more stable C–C bond between sugar and base. Formycin and pseudouridine participate in hydrogen bonding with G and A, respectively, in a normal Watson–Crick base pairing. In triple-helix oligonucleotides, or triplexes, (Figure 9.2), the association of the third strand is through the formation of specific hydrogen bonds with one of the two antiparallel strands in a Watson–Crick duplex. These new H-bonding patterns do not form in normal DNA duplex structure. The stability of the third strand is weaker than the double strand (Watson–Crick) interactions and thus dissociates first. Ts′o and his group reported that use of pseudouridine in place of uridine in selected places can improve the stability of triple helices.[28]

Reactive oxygen species such as hydroxyl radicals react with nucleic acids in cellular systems and can result in oxidatively modified nucleotides such as 8-oxo-2′-deoxyguanosine (8-oxo-dG); 8-oxo-2′-deoxyadenosine (8-oxo-dA); hydroxymethyl-2′-deoxyuridine (HMDU); or

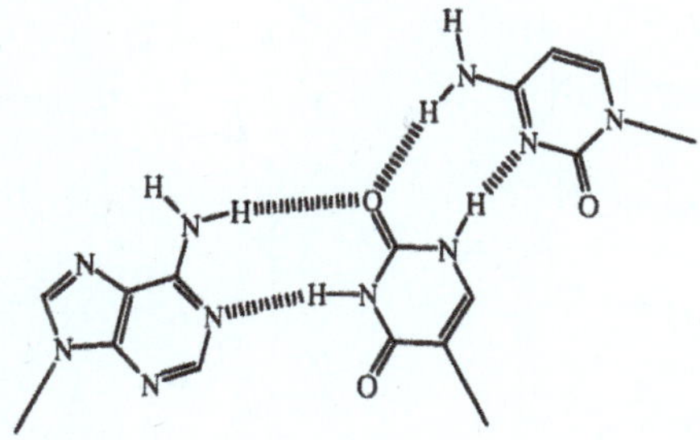

Figure 9.2 H-Bonding of A·pseudoU·C triad.

5-hydroxy-2′-deoxyuridine. These oxidatively modified nucleotides are mutagenic in nature and are believed to play a significant role in carcinogenesis. To understand the biophysical and biochemical properties of these nucleotides better, the analogs were synthesized and studied in DNA and RNA polymerase activity, fidelity assays,[J,K] as well as activity towards kinases,[L,M] and phosphohydrolases.[N]

Base modifications can be envisioned based on the understanding of the thermodynamic properties of nucleic acids. One mode of such base modification is based on the enhanced stability of DNA/RNA, through extra hydrogen bonding. The purine analog 2, 6-diaminopurine-2′-deoxyribonucleoside-5′-triphosphate (DAP), has an extra 2′-amine group (Figure 9.3), which renders an additional hydrogen bond in the minor groove of DNA (in standard Watson and Crick base pairing) with thymidine, thereby leading to the stabilization of duplex.[29,30] The increase in the melting temperature of a duplex with DAP is dependent on the sequence and the length of the oligonucleotide. A comprehensive review on structural properties of diaminopurine was published in 1998.[31]

Another approach to improving stability of DNA/RNA complexes included strengthening the hydrogen bonds between bases and/or increasing the base pair stacking interactions. Such compounds included pyrazolo[3,4-d]pyrimidine[32–35]; 7-halo-8-aza-7-deaza-2′-deoxyguanosine[36–38]; and 8-aza-7-deaza-2′-deoxyguanosine.[32,36,38] These modifications improved the

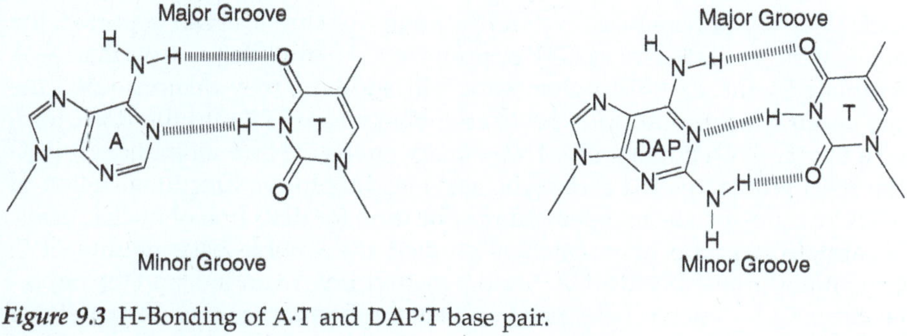

Figure 9.3 H-Bonding of A·T and DAP·T base pair.

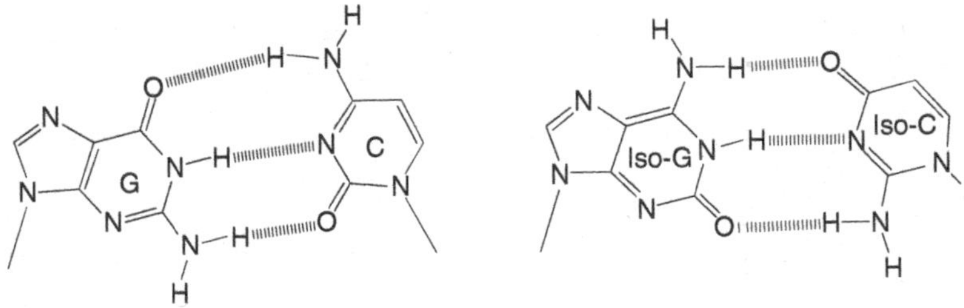

Figure 9.4 H-Bonding of G·C and isoG·isoC base pair.

stability of duplex DNA. Among the modifications reported for duplex stabilization, the alkynyl group at the 5-position of pyrimidine analogs[39–46] and at the 7-position of 7-deazapurine analogs have attracted considerable attention.[47] It is believed that the hydrophobic character of the alkynyl group and the increased polarizability of the nucleobase contribute to this improved DNA duplex stability. Chemical modifications at the 5-position of pyrimidines or the 7-position of 7-deazapurines are directed into the major groove of DNA and are well accommodated without significant steric hindrance. By contrast, the introduction of the same alkynyl group at the 8-position of 7-deazapurine lowers the stability of DNA, presumably due to steric constraints.[46,48] It is very important to note here that the 7-deaza0dG and 7-deaza-dA analogs are known destabilizing agents of ds DNA.[O,P]

Several other reported base modifications probe important biochemical processes or have useful chemotherapeutic properties. Isoguanosine (isoG) and isocytidine (isoC) are functional group transposition analogs of guanosine and cytidine (Figure 9.4) and participate in a form of base pairing analogous to Watson and Crick, except for a different pattern of hydrogen-bonding donor–acceptor pairs. More than 40 years ago, Rich proposed that the isoC–isoG base-pairing system could offer some advantages in nucleic acids chemistry because of its distinct hydrogen-bonding pattern that is different from the natural A–T (U) and G–C base pair.[49]

Benner and coworkers demonstrated the site-specific incorporation of isoGTP into oligonucleotides using deoxy-isoC containing DNA templates and T7 RNA polymerase.[50,51] Shortly after, Tor and Dervan reported the efficient incorporation of isoGTP opposite of C^5-methyl-deoxy-isoC in a DNA template by the T7 RNA polymerase.[52] In addition, they demonstrated the site-specific incorporation of N^6-(6-aminohexyl)-isoGTP into the RNA transcript by T7 RNA polymerase. The primary amine of N^6-(6-aminohexyl)-isoG can react with activated carboxylic acids leading to the functionalization of RNA in a site-specific manner and may be used for detection of nucleic acids. Separately, a group of researchers studied the wobble base pairing of 2, 6-diaminopurine riboside (DAP) and 5-methyl-isoC to understand the importance of G–U wobble base pairs, which are present in a wide variety of RNAs.[53,54] Recent theoretical calculations and thermodynamic data suggest

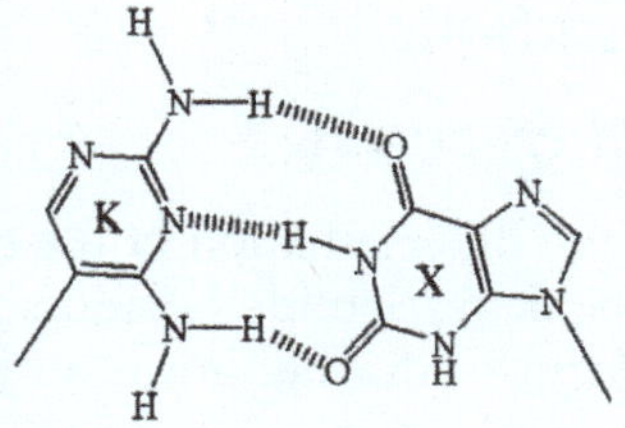

Figure 9.5 H-Bonding of diaminopyrimidine (K) and xanthosine (X).

that the isoC–isoG base pair is comparable in stability to the C–G base pair.[55] Although outside the scope of this review, Benner has expanded the genetic alphabet from 4 to 6 'letters' with a new Watson-Crick base pair of diaminopyrimidine C-nucleotide (K) with xanthosine (X) (Figure 9.5[Q,R]).

The 7-deazapurines, such as 7-deaza-adenosine (7-deaza-A) and 7-deaza-guanosine (7-deaza-G), are another type of base-modified nucleoside[56–60] and nucleotide.[61] The substitution of the nitrogen atom with a carbon at the 7-position (Figure 9.6) increases the stability of the glycosyl bond of these analogs compared to their parent nucleosides. In addition, on incorporation into DNA, do lower intrastrand (ss DNA) and Tm (ds DNA) stability via reduction of base stacking Hoogsteen interactions. Reduced intrastrand stability is important in electrophoretic mobility analysis of DNA to avoid/minimize compression artifacts.

It is believed that the lack of the N^7-nitrogen in 7-deaza-guanine can minimize or prevent unusual secondary structures (like tetraplexes) of oligonucleotides containing high GC content. In addition, these analogs can be modified further at the C^7 position to attach a wide variety of reporter groups for the detection of nucleic acids. The 8-halopurine and 5–halopyrimidine analogs were synthesized by direct halogenation methods and are photolabile. They are widely used in various nucleic acid-based technologies such as sequencing by mass spectrometry, detection of oligonucleotides using antibodies, and cross-linking studies probing the structure of protein–DNA complexes. Various azido-, halo-, and thioanalogs of purines and pyrimidines have also played a role as chemotherapeutic agents and/or in detection methods.

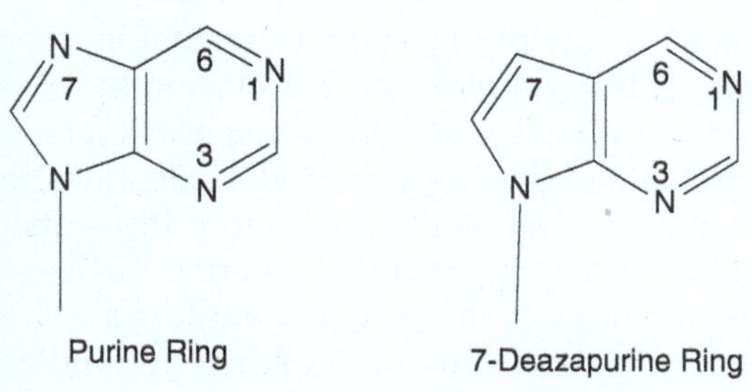

Purine Ring

7-Deazapurine Ring

Figure 9.6 Graphical representation of purine and 7-deazapurine ring systems.

9.4 Biological applications

9.4.1 Sequencing and genotyping

Genotyping is defined as the determination of the nucleotide sequence in a target nucleic acid at a specific nucleotide position. Sequencing is the determination of the total sequence information of an extended portion of the target nucleic acid.

The incorporation of 7-methyl-dGTP in a primer extension reaction, using Klenow fragment DNA polymerase, has been used to determine the genotype of a sample in which a T to C mutation was potentially present. The product DNA strands cleave on treatment with piperidine only if the methy-dG analog is present, inferring the presence of the dC mutation in the target. The identity of the cleaved DNA fragments is determined using MALDI mass spectrometry.[62] A similar approach has been developed for genotyping using a method known as incorporation and complete chemical cleavage. In this case, four modified nucleotide triphosphate analogs may be incorporated in separate PCRs. These analogs are then treated with oxidizers and bases to give specific cleavage products. These products can be separated and detected by standard methods, such as electrophoresis or mass spectroscopy. The analogs were 7-deaza-7-nitro-dATP; 7-deaza-7-nitro-dGTP; 5-hydroxy-dCTP; and 5-hydroxy-dUTP.[63,64]

Another method to improve peak shape in electrophoresis-based sequencing, in which 7-deaza-dITP is included in place of dGTP and N^4-Methyl dCTP for dCTP, has been described.[65] Jensen and co-workers reported the alleviation of band compression in sequencing by the use of 7-deaza-dATP analogous to the use of 7-deaza dGTP discussed previously, and the use of the two analogs together.[66] McDougall and colleagues have reported on the synthesis of 30 analogs of 2′-deoxyguanosine triphosphate; two of these, 7-ethyl-7-deaza-dGTP and 7-hydroxymethyl-7-deaza-dGTP, met the need of superior performance in sequencing reactions.[67] Further combinations of analogs have been examined in sequencing reactions, including the use of 7-deaza-dITP and 4-methyl-dCTP.[65]

The incorporation of 5-methyl-dCTP has been used to confer nuclease stability on randomly extended and labeled dsDNA fragments, which in turn can be used to generate sequence information. The frequently cutting restriction enzyme used cannot digest the DNA that incorporated the modified nucleotide, so all strands are truncated at fixed locations; this can then be separated by electrophoresis and the sequence inferred.[68] Koster et al. have shown the potential of using modified nucleotides in sequencing of DNA using mass spectrometry as the read-out of the sequence.[69–71]

Using chain-terminating and chain-elongating nucleotide triphosphate analogs, various techniques have been explored. These analogs — 4-thio-UTP, triphosphates of various 5-substituted pyrimidines, 8-substituted purines, and 7-deazapurines — confer mass differences on the extended and terminated DNA products. The MALDI or ES mass spectra of these products can be determined and the identity of the terminal base can be inferred. Rhodamine labeled ribonucleotide analogs have been reported for sequencing applications.[5,72] The fluorophores were attached to the 7-position of 7-deazapurines and to the 5-positions of pyrimidines using standard chemistries, based on the use of previously described dye-labeled deoxy analogs.[73]

9.4.2 Labeling

Ward and coworkers introduced the pivotal 5-substituted pyrimidine analogs, which allowed for the development of several successful DNA labeling systems.[74,75] Labeling systems developed since then will be covered elsewhere in this volume, although a set of base-modified analogs will be described in this section.

9.4.2.1 Immunofluorescence

The use of 5-bromo deoxyuridine triphosphate (BrdUTP) has revolutionized the way in which studies of cellular processes involving nucleic acid synthesis are undertaken. In general, this modified nucleoside triphosphate is administered to permeabilized cells, and the incorporation of the analog into DNA can be detected by treatment with antibodies specific to the modified base, after appropriate fixing of the cells.[76,77] For example, it has been reported that apoptotic cells feature numerous strand breaks. These breaks can be detected by incorporation of BrdUTP using terminal transferase, followed by treatment with a fluoresceinated anti-BrdU monoclonal antibody.[78] Compared to conventional labeling methods, when this technique is used enhanced detection of these breaks has been reported. Additionally, an alkaline phosphatase conjugated anti-BrdU antibody and colorimetric detection have been used to measure the reverse transcriptase (RT) activity of the HIV RT.[79–81] Similar methods have been applied to 5-Iodo-dUTP,[82] 5-fluoro-dU, and 5-nitro uracil.[83]

9.4.2.2 Photoaffinity labeling

The photoactive analog 8-azidoadenine-2′-deoxyriboside triphosphate (8-azido-dATP) has been used extensively to cross link DNA to proteins involved in nucleic acid processing. This property has been used to probe sites of contact between DNA and enzymes and other factors. The chemical synthesis and utility of the purine analogs has been described by Czarnecki and co-workers.[84] The analog is a substrate for several polymerases, which allows for its easy incorporation by nick translation.[85] Terminal transferase has been labeled using this substrate, thus allowing for identification of amino acid side chains involved in the enzymatic activity.[86,87] Other targets for labeling have included Klenow fragment,[88] tet repressor,[89] RNA polymerase subunits[90] and adenyl cyclase,[91] and have been reviewed previously by Meffert and colleagues.[92]

A photoactivatable and fluorescent analog that can be used as a probe of the nucleotide-binding site of kinases or cyclases has been described by

Sarfati.[93] The formation of an intramolecular cystine disulfide bond has been proposed as the mechanism for the photoinactivation of phosphoenolpyruvate carboxykinase when treated with 8-azido-GTP. However, the level of labeling by the analog was much less than the level of inactivation. The reactive intermediate formed during photolysis may have induced the intramolecular disulfide formation.[94]

Other forms of affinity labels have been described, including the use of 6-chloro-purine riboside 5′-triphosphate, as a way to modify inactivate adenylate cyclase irreversibly. The 6-thio analog of this compound also inactivated the enzyme, but was reversible by the addition of DTT. This implied the presence of an SH group in the purine amino-binding area of the enzyme active site.[95]

9.4.2.3 Thio analogs

Bullard et al. have shown that 4-thiouridine triphosphate can substitute for unmodified uridines in RNA using T7 RNA polymerase.[T] The thio group can serve as a point of conjugation to 5-Iodoacetamido-1,10-phenanthroline, which in turn can serve as a site-specific cleavage site for the nucleic acid, when contacted with Cu^{2+} in a reducing environment. This kind of modification has been used in footprinting protection assays to establish structural elements of tRNA. Eshaghpour et al. have demonstrated a similar alkylation reaction in which a variety of chemical labels are attached to the α-haloacetamido derivative used.[96] The analogous 4-thio-deoxyuridine has been used as a photoaffinity label when incorporated into nucleic acids. This has been used to label and identify the subunit of the HIV-1 RT that is in contact with the extending chain during polymerization.[97]

9.4.2.4 Spin labels

Toppin et al. have described the incorporation of spin-labeled nucleoside triphosphate analogs into DNA. These have the potential to be reporter molecules on the physiochemical environment of nucleic acids, as detected by EPR experiments.[98] C-nucleoside derived nucleoside triphosphate analogs have been prepared.[99,100] Many of the same functionality of side chains, linkers, and labels can be incorporated in these analogs. They exhibit a range of properties that are expected based on the known similarities and differences between these analogs and their naturally occurring counterparts.

9.4.3 Nuclease resistance

The incorporation of 5-methyl-dCTP in PCR has been used to protect restriction sites from degradation using enzymes sensitive to the presence of the 5-methyl group. This has allowed the generation of fragments that replicate those from a partial restriction enzyme digest of unmodified DNA, without the need for optimized enzyme digestion reactions. When 5′-labeled primers are used, the products can be easily visualized after electrophoretic separation.[101] This concept has been extended by Padgett and Sorge[102] to generate seamless junctions in ligation reactions, using the enzyme eam 1104I, which is sensitive to 5-methyl-dCTP incorporation and cuts several bases away from its recognition site. The incorporation of 5-methyl-dCTP into DNA from an RNA template has also been reported.[103]

9.4.4 PCR

Several BM-NTPs have been used in nucleic acid amplification systems. The most common form of this is the use of deoxyuridine triphosphate (dUTP) in PCR systems, which serves as a method to prevent carry-over contamination from previous PCRs. The incorporation of dUTP by the commonly used eubacterial thermostable and thermoactive polymerase enzymes generates PCR product that has complete or partial substitution of dU for thymidine, depending on whether dUTP is used alone or in a blend with TTP. In cases in which such a PCR product is transferred into a new PCR (carry-over contamination), which would otherwise give rise to a false positive result, the enzyme uracil-N-glycosylase (UNG, AmpERASE™) is used to excise the uracil base specifically, leaving an abasic site. This labile abasic site is subject to β-elimination and strand cleavage. In this way, any and all strands of DNA derived from a previous PCR can be treated with the enzyme before any thermocycling has been undertaken; after this, it cannot serve as a template for the generation of more PCR product.

Although dUTP is not normally found in nature, it can be easily made through the deamination of dCTP. The role of the UNG class of enzymes is to remove any dU bases that may have formed in DNA by the deamination of dC or via incorporation of dUTP in a host deficient in, or lacking, dUTPase activity. This mutation would otherwise result in a C–T transition mutation. The abasic site produced through the loss of the uracil base is then a substrate for other enzymes (initiated by "AP endonucleases") of the excision/repair complex to remove and repair this type mutagenic event.

In an alternate system, biotinylated dUTP can be incorporated into the DNA produced in one PCR. This labeled DNA can be treated with dnase 1–streptavidin complex, which delivers the nuclease to the amplified DNA and degrades it, rendering it unamplifiable in any future PCR. Because DNA from the sample is not biotinylated, it is not a substrate for the conjugate. Prevention of contaminant amplification and target amplification were carried out in the same vessel as part of a single procedure.[104]

Substitution of dCTP with 5-methyl-dCTP in PCR results in highly stable amplicons. This enhanced stability has been found to be inhibitory to PCR in certain systems; however, this inhibition can be overcome by increasing the denaturation temperature of the PCR or by including the destabilizing dITP analog.[27] In contrast, the structural analog, N^4-methyl-dCTP, is somewhat destabilizing when incorporated into double-stranded DNA. This has been used to advantage in sequencing applications, in which electrophoresis artifacts such as band compression and distension are reduced through incorporation of this analog as a means to improve denaturation of the sample.[105]

The most prevalent base-modified nucleotide triphosphate in use in biochemical applications is most likely 7-deaza-dGTP. This modification was introduced by Seela and co-workers,[61] Fernandez–Rachubinski et al.,[106] and Dierick[107] as a means to reduce the secondary structure of DNA with high G:C content. Compression in electrophoretic mobility of bands in sequencing gels was relieved by the inclusion of this modified triphosphate. There is evidence that efficient incorporation and generation of PCR product with these helix destabilizing nucleotide analogs both permits and requires modification of the PCR temperature profile. dI can be incorporated at 100% if T anneal/extend in lowered by the requisite extent. The inclusion of 7-deaza-dGTP confers considerable stability of the DNA toward a majority of restriction enzymes.[108] This analog was used successfully early in the development of PCR to amplify a target with 74% G:C content[109] and has been applied to the detection of fragile X syndrome.[110]

One problem induced by the incorporation of these 7-deaza analogs into ds DNA is that conventional staining and visualization of DNA by ethidium bromide or syBR Green I is extremely insensitive relative to normal DNA.[O] This post-electrophoretic detection issue has been ameliorated by the use of silver staining.[111]

Siegert et al. have used MALDI-TOF mass spectrometry for the fast and accurate detection of short PCR products.[112] In this case, the incorporation of 7-deaza purine analogs reduced fragmentation in the mass spectrometer, thus resulting in increased signal intensities and mass resolution. This method is suggested as a way to detect and analyze PCR products in diagnostic applications.

9.4.4.1 Electrochemical

Based on the differential redox activity of the analogs over the normal bases, 7-deazapurine nucleotide triphosphates have been incorporated into DNA in order to detect different sequences in PCR products.[113] Up to three different modified nucleotide triphosphates have been incorporated in one PCR. Kuwahara et al. reported the separate and simultaneous incorporation of the triphosphates of 5-(3-aminopropyl)dUTP, 5-((6-aminohexyl)carbamoylmethyl)dCTP and 2-amino-dATP, or N^6-Methyl-dATP.[114] Side chains on bases, which have the potential of exhibiting catalytic properties, have been incorporated into DNA again using Taq polymerase-based PCR. 5-Imidazolyl dUTPs and 3-(aminopropynyl)-7-deaza-dATP were incorporated into synthetic DNA and were refractory to a restriction enzyme.[115]

Researchers have incorporated 2-aminoadenine (or 2, 6-diaminopurine deoxyriboside triphosphate, DAPTP) into synthetic DNA chemically and enzymatically.[116] The extra hydrogen bond of this compound allows for stronger and potentially more specific hybridization of 2-aminoadenine containing probes to their targets. This analog again showed some inhibition of restriction enzymes. The inclusion of DAP into oligos has been shown to increase thermodynamic stability by around 1.5° per substitution.[29]

The future is limitless. Sophisticated chemistry (by sophisticated chemists) have provided and will continue to provide important molecular structures and mechanistic insights for nucleic acid chemistry and biochemistry as well as important tools for improving biotechnology applications to the life sciences.

Acknowledgments

We would like to acknowledge Dr. David Gelfand, for his careful review of this review, and for his constant support and guidance.

References

1. Auer, T., Sninsky, J.J., Gelfand, D.H., and Myers, T.W. Selective amplification of RNA utilizing the nucleotide analog dITP and *Thermus thermophilus* DNA polymerase. *Nucleic Acids Res.*, 1996, 24, 5021–5.
2. Perry, R.P. and Kelly, D.E. Existence of methylated messenger RNA in mouse L cells. *Cell*, 1974, 1, 37–42.
3. Furuichi, Y. Methylation-coupled transcription by virus-associated transcriptase of cytoplasmic polyhedrosis virus containing double-stranded RNA. *Nucleic Acids Res.*, 1974, 1, 809–822.
4. Hotchkiss, R.D. The quantitative separation of purines, pyrimidines and nucleosides by paper chromatography. *J. Bio. Chem.*, 1948, 175, 315–332.
5. Cheng, X. Structure and function of DNA methyltransferases. *Annu. Rev. Biophys. Biomol. Structr.*, 1995, 24, 293–318.
6. Kaito, C., Kai, M., Higo, T., Takayama, E., Fukamachi, H., Sekimizu, K., and Shiokawa, K. Activation of the maternally preset program of apoptosis by microinjection of 5-aza-2′-deoxycytidine and 5-methyl-2′-deoxycytidine-5′-triphosphate in Xenopus laevis embryos. *Dev. Growth Differentiation*, 2001, 43, 383–390.
7. Holliday, R. and Ho, T. Gene silencing and endogenous DNA methylation in mammalian cells. *Mutation Res.*, 1998, 400, 361–368.
8. Nyce, J. Gene silencing in mammalian cells by direct incorporation of electroporated 5-methyl-2′-deoxycytidine 5′-triphosphate. *Somatic Cell Molecular Genet.*, 1991, 17, 543–550.
9. Holliday, R. and Ho, T. Gene silencing in mammalian cells by uptake of 5-methyl deoxycytidine-5′-triphosphate. *Somatic Cell Molecular Genet.*, 1991, 17, 537–542.
10. Holliday, R. and Ho, T. Evidence for gene silencing by DNA methylation in normal human diploid fibroblasts. *Somatic Cell Molecular Genet.*, 1995, 21, 215–218.
11. Paulin, R.P., Ho, T., Balzer, H.J., and Holliday, R. Gene silencing by DNA methylation and dual inheritance in Chinese hamster ovary cells. *Genetics*, 1998, 149, 1081–1088.
12. Chen, J., Maxson, R., and Jones, P.A. Direct induction of DNA hypermethylation in sea urchin embryos by microinjection of 5-methyl dCTP stimulates early histone gene expression and leads to developmental arrest. *Dev. Biol.*, 1993, 155, 75–86.

13. Watt, F. and Molloy, P.L. Cytosine methylation prevents binding to DNA of a HeLa cell transcription factor required for optimal expression of the adenovirus major late promoter. *Genes Dev.*, 1988, 2, 1136–1142.
14. Vandeyar, M.A., Weiner, M.P., Hutton, C.J., and Batt, C.A. A simple and rapid method for the selection of oligodeoxynucleotide-directed mutants. *Gene*, 1988, 65, 129–133.
15. Holliday, R. and Ho, T. DNA methylation and epigenetic inheritance. *Methods*, 2002, 27, 179–183.
16. Keshet, I., Yisraeli, J., and Cedar, H. Effect of regional DNA methylation on gene expression. *Proc. Natl. Acad. Sci. USA*, 1985, 82, 2560–2564.
17. Dunn, D.B. The isolation of 1-methyladenylic acid and 7-methylguanylic acid from ribonucleic acid. *Biochem. J.*, 1963, 86, 14–15.
18. Igo–Kemenes TaZ, H.G. Involvement of 1-methyladenosine and 7-methylguanosine in the three-dimensional structure of (yeast) tRNA. *Eur. J. Biochem.*, 1971, 18, 292–298.
19. Rainen, L. and Stollar, B.D. Antibodies distinguishing between intact and alkali-hydrolyzed 7-methylguanosine. *Nucleic Acids Res.*, 1978, 5, 3877–3889.
20. Perry, R.P. and Scherrer, K. The methylated constituents of globin mRNA. *FEBS Lett.*, 1975, 57, 73–78.
21. Both, G.W., Banerjee, A.K., and Shatkin, A.J. Methylation-dependent translation of viral messenger RNAs *in vitro*. *Proc. Natl. Acad. Sci. USA*, 1975, 72, 1189–1193.
22. Muthukrishnan, S., Both, G.W., Furuichi, Y., and Shatkin, A.J. 5′-Terminal 7-methylguanosine in eukaryotic mRNA is required for translation. *Nature*, 1975, 255, 33–37.
23. Furuichi, Y., Morgan, M., Muthukrishnan, S., and Shatkin, A.J. Reovirus messenger RNA contains a methylated, blocked 5′-terminal structure: m-7G(5′)ppp(5′)G-MpCp. *Proc. Natl. Acad. Sci. USA*, 1975, 72, 362–366.
24. Engel, J.D. and von Hippel, P.H. Effects of methylation on the stability of nucleic acid conformations. Studies at the polymer level. *J. Biol. Chem.*, 1978, 253, 927–934.
25. Mace, D.C. N6-methyldeoxyadenosine 5′-triphosphate as a probe of the fidelity mechanisms of bacteriophage T4 DNA polymerase. *J. Biol. Chem.*, 1984, 259, 3616–3619.
26. Engel, J.D. and von Hippel, P.H. D(M6ATP) as a probe of the fidelity of base incorporation into polynucleotides by *Escherichia coli* DNA polymerase I. *J. Biol. Chem.*, 1978, 253, 935–939.
27. Wong, K.K. and McClelland, M. PCR with 5-methyl-dCTP replacing dCTP. *Nucleic Acids Res.*, 1991, 19, 1081–1085.
28. Trapane, T.L., Christopherson, M.S., Roby, C.D., Ts'o, P.O.P., and Wang, D. DNA triple helices with C-nucleosides (deoxypseudouridine) in the second strand. *J. Am. Chem. Soc.*, 1994, 116, 8412–8413.
29. Hoheisel, J.D. and Lehrach, H. Quantitative measurements on the duplex stability of 2,6-diaminopurine and 5-chloro-uracil nucleotides using enzymically synthesized oligomers. *FEBS Lett.*, 1990, 274, 103–106.
30. Lamm, G.M., Blencowe, B.J., Sproat, B.S., Iribarren, A.M., Ryder, U., and Lamond, A.I. Antisense probes containing 2-aminoadenosine allow efficient depletion of U5 snRNP from HeLa splicing extracts. *Nucleic Acids Res.*, 1991, 19, 3193–3198.

31. Bailly, C. and Waring, M.J. The use of diaminopurine to investigate structural properties of nucleic acids and molecular recognition between ligands and DNA. *Nucleic Acids Res.*, 1998, 26, 4309–4314.
32. Seela, F. and Zulauf, M. 7-Deaza-adenine-DNA: bulky 7-iodo substituents or hydrophobic 7-hexynyl chains are well accommodated in the major groove of oligonucleotide duplexes. *Chem. — Eur. J.*, 1998, 4, 1781–1790.
33. Seela, F. and Zulauf, M. Synthesis of oligodeoxyribonucleotides containing pyrazolo[3,4-d]pyrimidines: The influence of 7-substituted 8-aza-7-deazaadenines on the duplex structure and stability. *J. Chem. Soc., Perkin Trans.*, 1999, 1, 479–488.
34. Seela, F. and Becher, G. Oligonucleotides containing pyrazolo[3,4-d]pyrimidines: the influence of 7-substituted 8-aza-7-deaza-2′-deoxyguanosines on the duplex structure and stability. *Helvetica Chim. Acta*, 1999, 82, 1640–1655.
35. Seela, F., Muenster, I., Loechner, U., and Rosemeyer, H. 8-Azaadenosine and its 2′-deoxyribonucleoside. Synthesis and oligonucleotide base-pair stability. *Helvetica Chim. Acta*, 1998, 81, 1139–1155.
36. He, J. and Seela, F. 8-Aza-7-deazapurine-pyrimidine base pairs: the contribution of 2- and 7-substituents to the stability of duplex DNA. *Tetrahedron*, 2002, 58, 4535–4542.
37. Becher, G., He, J., and Seela, F. Major-groove-halogenated DNA: the effects of bromo and iodo substituents replacing H- C(7) of 8-aza-7-deazapurine-2,6-diamine or H - C(5) of uracil residues. *Helvetica Chim. Acta*, 2001, 84, 1048–1065.
38. Seela, F. and Becher, G. Pyrazolo[3,4-d]pyrimidine nucleic acids: adjustment of dA-dT to dG-dC base pair stability. *Nucleic Acids Res.*, 2001, 29, 2069–2078.
39. Sagi, J., Szemzo, A., Ebinger, K., Szabolcs, A., Sagi, G., Ruff, E., and Otvos L. Base-modified oligodeoxynucleotides. I. Effect of 5-alkyl, 5-(1-alkenyl) and 5-(1-alkynyl) substitution of the pyrimidines on duplex stability and hydrophobicity. *Tetrahedron Lett.*, 1993, 34, 2191–2194.
40. Barnes, T.W., III and Turner, D.H. Long-range cooperativity due to c5-propynylation of oligopyrimidines enhances specific recognition by uridine of ribo-adenosine over ribo-guanosine. *J. Am. Chem. Soc.*, 2001, 123, 9186–9187.
41. Barnes, T.W., III and Turner, D.H. C5-(1-Propynyl)-2′-deoxy-pyrimidines enhance mismatch penalties of DNA:RNA duplex formation. *Biochemistry*, 2001, 40, 12738–12745.
42. Wagner, R.W., Matteucci, M.D., Lewis, J.G., Gutierrez, A.J., Moulds, C., and Froehler, B.C. Antisense gene inhibition by oligonucleotides containing C-5 propyne pyrimidines. *Science*, 1993, 260, 1510–1513.
43. Froehler, B.C., Wadwani, S., Terhorst, T.J., and Gerrard, S.R. Oligodeoxyribonucleotides containing C–5 propyne analogs of 2′-deoxyuridine and 2′-deoxycytidine. *Tetrahedron Lett.*, 1992, 33, 5307–5310.
44. Gutierrez, A.J., Matteucci, M.D., Grant, D., Matsumura, S., Wagner, R.W., and Froehler, B.C. Antisense gene inhibition by C-5-substituted deoxyuridine-containing oligodeoxynucleotides. *Biochemistry*, 1997, 36, 743–748.
45. Ahmadian, M., Zhang, P., and Bergstrom, D.E. A comparative study of the thermal stability of oligodeoxyribonucleotides containing 5-substituted 2′-deoxyuridines. *Nucleic Acids Res.*, 1998, 26, 3127–3135.
46. Graham, D., Parkinson, J.A., and Brown, T. DNA duplexes stabilized by modified monomer residues: synthesis and stability. *J. Chem. Soc., Perkin Trans.* 1998, 1, 1131–1138.

47. Buhr, C.A., Wagner, R.W., Grant, D., and Froehler, B.C. Oligodeoxynucleotides containing C-7 propyne analogs of 7-deaza-2′-deoxyguanosine and 7-deaza-2′-deoxyadenosine. *Nucleic Acids Res.*, 1996, 24, 2974–2980.
48. Catalanotti, B., Galeone, A., Gomez–Paloma, L., Mayol, L., and Pepe, A. 2′-Deoxy-8-(propyn-1-yl)adenosine-containing oligonucleotides: effects on stability of duplex and quadruplex structures. *Bioorg. Med. Chem. Lett.*, 2000, 10, 2005–2009.
49. Rich, A. On the problems of evolution, and biochemical information transfer. Kasha, M. and Pullman, B., Eds., In: *Horizons in Biochemistry*. Academic Press: New York, 1962, 103–126.
50. Switzer, C., Moroney, S.E., and Benner, S.A. Enzymatic incorporation of a new base pair into DNA and RNA. *J. Am. Chem. Soc.*, 1989, 111, 8322–8323.
51. Switzer, C.Y., Moroney, S.E., and Benner, S.A. Enzymic recognition of the base pair between isocytidine and isoguanosine. *Biochemistry*, 1993, 32, 10489–10496.
52. Tor, Y. and Dervan, P.B. Site-specific enzymic incorporation of an unnatural base, N6-(6-aminohexyl)isoguanosine, into RNA. *J. Am. Chem. Soc.*, 1993, 115, 4461–4467.
53. Strobel, S.A., Cech, T.R., Usman, N., and Beigelman, L. The 2,6-diaminopurine riboside.5-methylisocytidine wobble base pair: an isoenergetic substitution for the study of G.U pairs in RNA. *Biochemistry*, 1994, 33, 13824–13835.
54. Musier–Forsyth, K., Shi, J.P., Henderson, B., Bald, R., Fuerste, J.P., Erdmann, V.A., and Schimmel P. Base-analog-induced aminoacylation of an RNA helix by a tRNA synthetase. *J. Am. Chem. Soc.*, 1995, 117, 7253–7254.
55. Roberts, C., Bandaru, R., and Switzer, C. Theoretical and experimental study of isoguanine and isocytosine: base pairing in an expanded genetic system. *J. Am. Chem. Soc.*, 1997, 119, 4640–4649.
56. Tolman, R.L., Robins, R.K., and Townsend, L.B. Pyrrolopyrimidine nucleosides. III. Total synthesis of toyocamycin, sangivamycin, tubercidin, and related derivatives. *J. Am. Chem. Soc.*, 1969, 91, 2102–2108.
57. Ramasamy, K., Imamura, N., Robins, R.K., and Revankar, G.R. A facile synthesis of tubercidin and related 7-deazapurine nucleosides via the stereospecific sodium salt glycosylation procedure. *Tetrahedron Lett.*, 1987, 28, 5107–5110.
58. Ramasamy, K., Imamura, N., Robins, R.K., and Revankar, G.R. Facile and improved synthesis of tubercidin and certain related pyrrolo[2,3-d]pyrimidine nucleosides by the stereospecific sodium salt glycosylation procedure. *J. Heterocyclic Chem.*, 1988, 25, 1893–1898.
59. Seela, F., Rosemeyer, H., Biesewig, A., and Juergens, T. 7-Carbapurine ribofuranosides: synthesis by solid-liquid phase-transfer glycosylation and nitrogen-15 NMR spectra. *Nucleosides Nucleotides*, 1988, 7, 581–584.
60. Townsend, L.B., Tolman, R.L., Robins, R.K., and Milne, G.H. The synthesis of 2-amino-7-(b-D-ribofuranosyl)pyrrolo[2,3-d]pyrimidin-4-one (7-deazaguanosine), a nucleoside Q and Q* analog. *J. Heterocyclic Chem.*, 1976, 13, 1363–1364.
61. Seela, F., Driller, H., Kehne, A., and Kaiser, K. Self-complementary oligomers containing 7-deaza-2′-deoxyguanosine or 2′-deoxytubercidin. *Chemica Scripta*, 1986, 26, 173–178.
62. Stanton, V.P., Jr., Wolfe, J.L., Kawate, T., and Verdine, G. Nucleic acid analysis based on enzymic incorporation of modified nucleotides and chemical cleavage. *PCT. Int. Appl. WO 2000018967*, 290 pp.

63. Wolfe, J.L., Kawate, T., Sarracino, D.A., Zillmann, M., Olson, J., Stanton, V.P., Jr., and Verdine, G.L. A genotyping strategy based on incorporation and cleavage of chemically modified nucleotides. *Proc. Natl. Acad. Sci. USA*, 2002, 99, 11073–11078.
64. Wolfe, J.L., Kawate, T., Allerson, C.R., and Stanton, V.P. Genotyping methods for detection of single nucleotide polymorphisms using base-modified oligonucleotides in nucleic acid amplification. *US Pat Appl. Publ. 2003165880*, 30 pp.
65. Reeve, M.A., Robinson, P.S., Franklin, H., McArdle, B., and Fuller, C. An improved method for dye-terminator sequencing of DNA using 7-deazapurine triphosphates and 4-methyl dCTP. *PCT Int. Appl. WO9720949*, 13 pp.
66. Jensen, M.A., Zagursky, R.J., Trainor, G.L., Cocuzza, A.J., Lee, A., and Chen, E.Y. Improvements in the chain-termination method of DNA sequencing through the use of 7-deaza-2′-deoxyadenosine. *DNA Seq.*, 1991, 1, 233–239.
67. McDougall, M.G., Sun, L., Livshin, I., Hosta, L.P., McArdle, B.F., Samols, S.B., Fuller, C.W., and Kumar, S. Analogs of guanine nucleoside triphosphates for sequencing applications. *Nucleosides Nucleotides Nucleic Acids*, 2001, 20, 501–506.
68. Sorge, J.A. and Blinderman, L.A. ExoMeth sequencing of DNA: eliminating the need for subcloning and oligonucleotide primers. *Proc. Natl. Acad. Sci. USA*, 1989, 86, 9208–9212.
69. Koster, H. Sanger DNA sequencing by mass spectrometry using base-specific chain termination and elongation, primer modification, nested fragments, reversible immobilization, and tag-specific probes. *US Pat 5547835*, 58 pp.
70. Koster, H. DNA sequencing by mass spectrometry. *US Pat 6194144*, 95 pp.
71. Kwon, Y.-S., Tang, K., Cantor, C.R., Koster, H., and Kang, C. DNA sequencing and genotyping by transcriptional synthesis of chain-terminated RNA ladders and MALDI-TOF mass spectrometry. *Nucleic Acids Res.*, 2001, 29, e 11/1–e 11/6.
72. Fisher, P.V., Vatta, P., and Khan, S.H. Dye-labeled ribonucleotide triphosphates for use in DNA sequencing and detection of mutations or 5-methylcytosine in DNA. *PCT Int. Appl. WO2003000841*, 96 pp.
73. Hobbs, F.W., Jr. and Cocuzza, A.J. Preparation of (aminoalkynyl)nucleotides as intermediates for fluorescent chain terminators for DNA sequencing. Eur. Patent Appl. 251786, 40 pp.
74. Ward, D.C., Waldrop, A.A., III, and Langer, P.R. Modified nucleotides and their use. *Eur. Pat. Appl. 63879*, 70 pp.
75. Langer, P.R., Waldrop, A.A., and Ward, D.C. Enzymatic synthesis of biotin-labeled polynucleotides: novel nucleic acid affinity probes. *Proc. Natl. Acad. Sci. USA*, 1981, 78, 6633–6637.
76. Thiry, M. Ultrastructural detection of DNA within the nucleolus by sensitive molecular immunocytochemistry. *Exp. Cell Res.*, 1992, 200, 135–144.
77. Cmarko, D., Verschure, P.J., Rothblum, L.I., Hernandez–Verdun, D., Amalric, F., van Driel, R., and Fakan, S. Ultrastructural analysis of nucleolar transcription in cells microinjected with 5-bromo-UTP. *Histochem. Cell Biol.*, 2000, 113, 181–187.
78. Li, X. and Darzynkiewicz, Z. Labeling DNA strand breaks with BrdUTP. Detection of apoptosis and cell proliferation. *Cell Proliferation*, 1995, 28, 571–579.
79. Ekstrand, D.H., Awad, R.J., Kallander, C.F., and Gronowitz, J.S. A sensitive assay for the quantification of reverse transcriptase activity based on the use of carrier-bound template and non-radioactive-product detection, with special reference to human-immunodeficiency-virus isolation. *Biotechnol. Appl. Biochem.*, 1996, 23 (Pt 2), 95–105.

80. Porstmann, T., Meissner, K., and Ruedinger, W. Solid-phase immunoassay and kit for reverse transcriptase. Ger. (East) Pat 289603, 4 pp.
81. Rytting, A.S., Akerblom, L., Gronowitz, J.S., and Kallander, C.F. Colorimetric capture assay for human-immunodeficiency-virus-I reverse transcriptase activity. *Biotechnol. Appl.Biochem.*, 1999, 29 (Pt 3), 241–250.
82. Vanderlaan, M., Watkins, B., Thomas, C., Dolbeare, F., and Stanker, L. Improved high-affinity monoclonal antibody to iododeoxyuridine. *Cytometry*, 1986, 7, 499–507.
83. Miller, M.R., Heyneman, C.., Walker, S., and Ulrich, R.G. Interaction of monoclonal antibodies directed against bromodeoxyuridine with pyrimidine bases, nucleosides, and DNA. *J. Immunol.*, 1986, 136, 1791–1795.
84. Czarnecki, J., Geahlen, R., and Haley, B. Synthesis and use of azido photoaffinity analogs of adenine and guanine nucleotides. *Methods Enzymol.*, 1979, 56, 642–653.
85. Meffert, R. and Dose, K. UV-induced cross-linking of proteins to plasmid pBR322 containing 8-azidoadenine 2′-deoxyribonucleotides. *FEBS Lett.*, 1988, 239, 190–194.
86. Evans, R.K., Beach, C.M., and Coleman, M.S. Photoaffinity labeling of terminal deoxynucleotidyl transferase. 2. Identification of peptides in the nucleotide binding domain. *Biochemistry*, 1989, 28, 713–720.
87. Evans, R.K. and Coleman, M.S. Photoaffinity labeling of terminal deoxynucleotidyl transferase. 1. Active site directed interactions with 8-azido-2′-deoxyadenosine 5′-triphosphate. *Biochemistry*, 1989, 28, 707–12.
88. Rush, J. and Konigsberg, W.H. Photoaffinity labeling of the Klenow fragment with 8-azido-dATP. *J. Bio. Chem.*, 1990, 265, 4821–4827.
89. Meffert, R., Rathgeber, G., Schaefer, H.J., and Dose, K. UV-induced cross-linking of tet repressor to DNA containing tet operator sequences and 8-azidoadenines. *Nucleic Acids Res.*, 1990, 18, 6633–6636.
90. Asano, Y., Mizumoto, K., Maruyama, T., and Ishihama, A. Photoaffinity labeling of influenza virus RNA polymerase PB1 subunit with 8-azido GTP. *J. Biochem.* (Tokyo), 1995, 117, 677–682.
91. Castets, F., Baillat, G., Mirzoeva, S., Mabrouk, K., Garin, J., d'Alayer, J., and Monneron, A. A brain synaptosomal adenylyl cyclase of high specific activity is photolabeled with azido-ATP. *Biochemistry*, 1994, 33, 5063–5069.
92. Meffert, R., Dose, K., Rathgeber, G., and Schaefer, H.-J. Ultraviolet crosslinking of DNA-protein complexes via 8-azidoadenine. *Methods Molecular Biol.* (Totowa, NJ), 1994, 30, 237–250.
93. Sarfati, R.S., Namane, A., Munier, H., and Barzu, O. New fluorescent and photoactivable analogs acting on nucleotide binding enzymes. *Tetrahedron Lett.*, 1991, 32, 4699–4702.
94. Lewis, C.T., Haley, B.E., and Carlson, G.M. Formation of an intramolecular cystine disulfide during the reaction of 8-azidoguanosine 5′-triphosphate with cytosolic phosphoenolpyruvate carboxykinase (GTP) causes inactivation without photolabeling. *Biochemistry*, 1989, 28, 9248–9255.
95. Skurat, A.V., Yurkova, M.S., Baranova, L.A., Gulyaev, N.N., Bulargina, T.V., and Severin, E.S. Evidence for the existence of a sulfhydryl group in the adenylate cyclase active site. *Biochem. Int.*, 1985, 10, 451–461.
96. Eshaghpour, H., Soell, D., and Crothers, D.M. Specific chemical labeling of DNA fragments. *Nucleic Acids Res.*, 1979, 7, 1485–1495.

97. Sheng, N. and Dennis, D. Active site labeling of HIV-1 reverse transcriptase. *Biochemistry*, 1993, 32, 4938–4942.
98. Toppin, C.R., Pauly, G.T., Devanesan, P., Kryak, D., and Bobst, A.M. Three novel spin-labeled substrates for enzymic incorporation into nucleic acid lattices. *Helvetica Chimica Acta*, 1986, 69, 345–349.
99. Muehlegger, K., Eltz, H., and Seela, F. Preparation of C-nucleoside derivatives for use in DNA sequencing, in situ hybridization, detection and marking of nucleic acids, and oligonucleotide chemical synthesis. *Ger. Offen. 19509038*, 10 pp.
100. Muhlegger, K., Vender Eltz, H., Seela, F., and Rosemeyer, H. C-nucleoside derivatives and their use in the detection of nucleic acids. *US Pat 6174998*, 13 pp.
101. Wong, K.-K., Markillie, L.M., and Saffer, J.D. A novel method for producing partial restriction digestion of DNA fragments by PCR with 5-methyl-CTP. *Nucleic Acids Res.*, 1997, 25, 4169–4171.
102. Padgett, K.A. and Sorge, J.A. Creating seamless junctions independent of restriction sites in PCR cloning. *Gene*, 1996, 168, 31–35.
103. Huse, W.D. and Hansen, C.J. Reverse transcription and double-stranded DNA using a deoxyribonucleotide triphosphate analog that confers nuclease resistance. *US Pat 5681726*, 13 pp.
104. Ullman, E.F., Chen, Y., and Rose, S.J. Method for preventing amplification of nucleic acid contaminants in amplification mixtures. *PCT Int. Appl. WO9602672*, 50 pp.
105. Pless, R.C. N4-methyl-2'-deoxycytidine 5'-triphosphate and its use in polymerase-catalyzed nucleic acid syntheses. *PCT Int. Appl. WO9405684*, 40 pp.
106. Fernandez–Rachubinski, F., Eng, B., Murray, W.W., Blajchman, M.A., and Rachubinski, R.A. Incorporation of 7-deaza dGTP during the amplification step in the polymerase chain reaction procedure improves subsequent DNA sequencing. *DNA Seq.*, 1990, 1, 137–140.
107. Dierick, H., Stul, M., De Kelver, W., Marynen, P., and Cassiman, J.J. Incorporation of dITP or 7-deaza dGTP during PCR improves sequencing of the product. *Nucleic Acids Res.*, 1993, 21, 4427–4428.
108. Seela, F. and Roling, A. 7-Deazapurine containing DNA: efficiency of c7GdTP, c7AdTP and c7IdTP incorporation during PCR-amplification and protection from endodeoxyribonuclease hydrolysis. *Nucleic Acids Res.*, 1992, 20, 55–61.
109. McConlogue, L., Brow, M.A., and Innis, M.A. Structure-independent DNA amplification by PCR using 7-deaza-2'-deoxyguanosine. *Nucleic Acids Res.*, 1988, 16, 9869.
110. Pergolizzi, R.G., Erster, S.H., and Brown, T.W. Detection of GC-rich DNA characteristic of Fragile X syndrome by PCR. *PCT Int. Appl. WO9315225*, 38 pp.
111. Weiss, J., Zucht, H.D., and Forssmann, W.G. Amplification of gene fragments with very high G/C content: c7dGTP and the problem of visualizing the amplification products. *PCR Methods Appl.* 1994, 4, 124–125.
112. Siegert, C.W., Jacob, A., and Koester, H. Matrix-assisted laser desorption/ionization time-of-flight mass spectrometry for the detection of polymerase chain reaction products containing 7-deazapurine moieties. *Anal. Biochem.*, 1996, 243, 55–65.
113. Yang, I.V., Ropp, P.A., and Thorp, H.H. Toward electrochemical resolution of two genes on one electrode: using 7-deaza analogues of guanine and adenine to prepare PCR products with differential redox activity. *Anal. Chem.*, 2002, 74, 347–354.

114. Kuwahara, M., Hososhima, S.-I., Takahata, Y., Kitagata, R., Shoji, A., Hanawa, K., Ozaki, A.N., Ozaki, H., and Sawai, H. Simultaneous incorporation of three different modified nucleotides during PCR. *Nucleic Acids Res. Suppl.*, 2003, 3, 37–38.
115. Gourlain, T., Sidorov, A., Mignet, N., Thorpe, S.J., Lee, S.E., Grasby, J.A., and Williams, D.M. Enhancing the catalytic repertoire of nucleic acids. II. Simultaneous incorporation of amino and imidazolyl functionalities by two modified triphosphates during PCR. *Nucleic Acids Res.*, 2001, 29, 1898–1905.
116. Chollet, A. and Kawashima, E. DNA containing the base analogue 2-aminoadenine, preparation, use as hybridization probes and cleavage by restriction endonucleases. *Nucleic Acids Res.*, 1988, 16, 305–317.

Added reference list

A. Hartmann, K.U. and C. Heidelberger (1961). Fluorinated pyrimidines. XIII. Inhibition of thymidylate synthetase. *J. Biol. Chem.*, 236: 3006–3013.
B. Newman, E.M. and D.V. Santi (1982). Metabolism and mechanism of action of 5-fluorodeoxycytidine. *Proc. Natl. Acad. Sci. U.S.A.*, 79(21): 6419–6423.
C. Wong, D.M., Ph.H. Weinstock, et al. (1991). Branch capture reactions: displacers derived from asymmetric PCR. *Nucleic Acids res.*, 19(9): 2251–2259.
D. Purmal, A.A., Y.W. Kow, et al. (1994). 5-Hydroxypyrimidine deoxynucleoside triphosphates are more efficiently incorporated into DNA by exonuclease-free Klenow fragment than 8-oxopurine deoxynucleoside triphosphates. *Nucleic Acids res.*, 22(19): 3930–3935.
E. Bruskov, V.I. and O.V. Kuklina (1988). 8-Oxy-GTP displays substrate properties of UTP in polynucleotide synthesis catalyzed by Escherichia coli RNA-polymerase on a poly[d(A-T)].poly[d(A-T)] template. *Molekulyarnaya Biologiya (Moscow)*, 22(3): 726–730.
F. Bruskov, V.I., V.V. Kuryavyi, et al. (1989). Ambiguous substrate-coding properties of 8-oxy-GTP in the reaction of oligonucleotide synthesis catalyzed by Escherichia coli RNA polymerase on DNA template of the DDIII mutant of T7 phage. *Doklady Akademii Nauk SSSR*, 307(1): 243–246, 2 plates [*Biochem.*].
G. Kapuler, A.M., D.C. Ward, et al. (1969). Utilization of substrate analogs by mengovirus-induced RNA polymerase. *Virology*, 37(4): 701–706.
H. Miller, H., R. Prasad, et al. (2000). 8-OxodGTP Incorporation by DNA Polymerase b Is Modified by Active-Site Residue Asn279. *Biochemistry*, 39(5): 1029–1033.
I. Kamath-Loeb, A.S., A. Hizi, et al. (1997). Incorporation of the guanosine triphosphate analogs 8-oxo-dGTP and 8-NH2-dGTP by reverse transcriptases and mammalian DNA polymerases. *J. Biol. Chem.*, 272(9): 5892–5898.
J. Bebenek, K., J.C. Boyer, et al. (1999). The base substitution fidelity of HIV-1 reverse transcriptase on DNA and RNA templates probed with 8-oxo-deoxyguanosine triphosphate. *Mutat. Res.*, 429(2): 149–158.
K. Pavlov, Y.I., D.T. minnick, et al. (1994). DNA Replication Fidelity with 8-oxodeoxyguanosine triphosphate. *Biochemistry*, 33(15): 4695–6701.
L. Kai, T., R. Matsunaga, et al. (2002). An oxidized nucleotide affects DNA replication through activation of protein kinases in Xenopus egg lysates. *Nucleic Acids Res.*, 30(2): 569–573.

M. Kamiya, H. and H. Kasai (1999). Preparation of 8-hydroxy-dGTP and 2-hydroxy-dATP by a phosphate transfer reaction by nucleoside-diphosphate kinase. *Nucleosides Nucleotides*, 18(3)L 307–310.

N. Bialkowski, K. and K.S. Kasprzak (1998). A novel assay of 8-oxo-2′-deoxyguanosine 5′-triphosphate pyrophosphohydrolase (8-oxo-dGTPase) activity in cultured cells and its use for evaluation of cadmium(II) inhibition of this activity. *Nucleic Acids Res.*, 26(13): 3194–3201.

O. Innis, M.A. (1990). PCR with 7-deaza-2′-deoxyguanosine triphosphate. *PCR Protocols: A Guide to Methods and Applications.* M.A. Innis, Gelfand, D.H., Sninsky, J.J., and White, T.J., Eds., Academic Press, 54–59.

P. McConlogue, L., M.A.D. Brow, et al. (1988). Structure-independent DNA amplification by PCR using 7-deaza-2′-deoxyguanosine. *Nucleic Acids Res.*, 16(20): 9869.

Q. Piccirilli, J.A., T. Krauch, et al. (1990). Enzymatic incorporation of a new base pair into DNA and RNA extends the genetic alphabet. *Nature,* 343(6253): 33–37.

R. Horlacher, J., M. Hottiger, et al. (1995). Recognition by viral and cellular DNA polymerases of nucleosides bearing bases with nonstandard hydrogen bonding patterns. *Proc. Natl. Acad. Sci. U.S.A.*, 92(14): 6329–6333.

S. Sasaki, N., M. Izawa, et al. (1998). Transcriptional sequencing: A method for DNA sequencing using RNA polymerase. *Proc. Natl. Acad. Sci. U.S.A.*, 95(7): 3455–3460.

T. Bullard, J.M., M.A. van Waes, et al. (1995). Regions of 23 S ribosomal RNA proximal to transfer RNA gound at the P and E sites. *J. Mol. Biol.*, 252(5): 572–582.

chapter ten

Nucleoside triphosphate analogs for nonradioactive labeling of nucleic acids

Glenn H. McGall
Affymetrix, 3380 Central Expressway, Santa Clara, CA

Contents

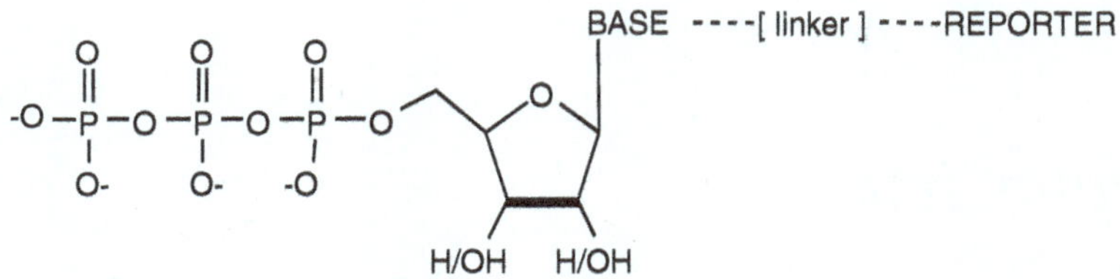

Figure 10.1 General structure of nucleoside triphosphate analogs for nonradioactive labeling.

10.1 Introduction

Labeled nucleoside triphosphate analogs (LNTPs) are used widely for *in-vitro* enzymatic incorporation of detectable reporter molecules or "tags" into DNA and RNA sequences derived from biological samples. The majority of analogs used for this purpose are based on the general structure shown in Figure 10.1, in which a detectable group — a luminescent molecule or a hapten to allow the attachment of a secondary signal-generating moiety — is attached via a linker to one of several modifiable positions on the heterocyclic nucleobase. A variety of analogs in which a label is attached to a position on the phosphate or sugar ring, or in which the sugar ring has been replaced with a noncanonical structure, have also been prepared for certain special applications.

Nonradioactive labels have become an increasingly popular choice over traditional radioisotope labels for most nucleic acid detection applications. In addition to being preferred for safety reasons, nonradioactive labels have the advantages of better stability, ease of use, and versatility in providing the ability to choose, from a variety of detection methods, the one best suited to a particular application or analysis platform. Optical detection technologies such as fluorescence microscopy have advanced to the point at which comparable, and in some cases better, sensitivity can be achieved relative to radioactive labeling. The use of labels such as biotin or other high-affinity ligands allows secondary signal generators, such as enzyme conjugates, to be bound immunochemically to achieve sensivities surpassing those of single flourescent molecules or radioisotopes.

Optical detection methods also provide intrinsically higher spatial resolution in comparsion with isotope detection using emulsion overlays or phosphor films, and this is a primary reason why nonradioactive detection is preferred for *in situ* hybridization and sample labeling for high-density probe arrays. Labeling with luminescent tags also provides the ability to use multiple spectrally resolvable labels to differentiate *in-situ* hybridization target sequences, sequencing fragments, and primer extension products for single nucleotide polymorphism (SNP) analysis.

Nonradioactive LNTPs can be incorporated into nucleic acid molecules using appropriate DNA or RNA polymerases; standard procedures are available for internal labeling of DNA sequences via

- Nick translation[1,2]
- Random priming[3,4]
- Specific primer extension[5]
- PCR amplification[6–8]
- Reverse transcription[9,10]
- 3′-end labeling with terminal transferase,[11–13] or Klenow DNA polymerase I[14,15]

Similarly, RNA molecules are labeled internally via *in vitro* transcription,[1,16–18] 3′-end labeled using terminal transferase,[19] or polyA polymerase.[20]

A number of chemical methods have been developed for directly labeling unmodified nucleic acids as an alternative to labeling with LNTPs. These include bisulfite-catalysed transamination of cytosine bases[21]; conversion of terminal phosphates to amine-reactive derivatives such as imidazolides and morpholidates[22,23]; and reductive amination of aldehyde groups generated via periodate oxidation of RNA 3′ termini[24–26] or by depurination of DNA.[27] Alkylating agents derived from aryl azides,[28] psoralen derivatives,[29,30] nitrogen mustards,[31] and platinum complexes[32–34] that react covalently with the heterocyclic bases are also commercially available for labeling applications. Reagents for specific alkylation of internal or terminal phosphate or phosphorothioate residues have also been developed.[35,36]

For certain applications, a chemical labeling approach may offer a more direct and cost-effective alternative to enzymatic labeling. Historically, however, the available reagents and methods for direct chemical labeling have had certain limitations that discouraged more widespread use. Native polynucleotides do not possess highly reactive functional groups, and the reagents used for chemical labeling must generally be very reactive or be used at high concentrations. This can lead to somewhat nonselective modification of the polynucleotides, variability in the extent of labeling, and potential toxicity and stability issues. The large excess of reagent typically used in chemical labeling methods usually requires an efficient clean-up step to minimize nonspecific background; this can lead to variation in recovery of the labeled polynucleotides, a drawback for applications requiring high-sensitivity detection.

By contrast, internal or terminal addition of labeled nucleotide analogs is specific to polynucleotide targets and typically requires millimolar or lower concentrations of the LNTP labeling reagent. When necessary, unincorporated nucleotides can be readily separated by solid phase extraction, which is now commercially available in convenient cartridge and 96-well plate formats. An additional consideration is that many chemical labeling methods result in modification of the nucleic acids at positions that may adversely affect normal Watson–Crick base pairing and/or duplex stability, and thereby the affinity and specificity of polynucleotides used in hybridization-based applications. Although LNTPs are not entirely benign in this regard, the position of attachment of the label is typically such that the impact of the modification on hybridization properties is minimal (Section 10.4.3).

Lastly, enzymatic labeling is most often preferred as a matter of convenience because nucleic acid samples are usually generated by polymerase-catalyzed extension, replication, or amplification reactions; the simultaneous incorporation of labels at this point in the process circumvents the need for subsequent labeling and clean-up steps. Thus, at present, enzymatic methods provide a more specific and convenient means of incorporating detectable labels into nucleic acids. However, as chemical labeling reagents and methods continue to improve, their use is likely to become more common for a number of general labeling applications.

10.2 Applications of labeled nucleotide analogs

Labeling with LNTPs has come to be a ubiquitous element in a broad range of research techniques used in genetics and molecular/cellular biology. Over the last decade, the utilization of LNTPs has experienced substantial growth due to international programs to sequence the genomes of humans and other organisms,[37–39] as well as the emergence of high-density DNA probe arrays as a standard tool for high-throughput screening and large-scale analysis of genetic variation and gene expression.[40–45] Exemplary applications of LNTPs include preparation of labeled probes for use in northern/southern blot, colony, or plaque hybridization,[46–50] and for a variety of *in-situ* hybridization techniques, including FISH and multicolor FISH or spectral karyotyping (SKY)[51–57]; comparative genomic hybridization (CGH)[58]; and fluorescence-based differential display.[59] They are also used for labeling DNA and RNA samples in hybridization-based sequence detection and analysis with DNA probe arrays.[8,10,18,60,61] Researchers in the area of cell molecular biology employ LNTPs in assays for DNA repair synthesis[62] and internucleosomal cleavage associated with apoptosis.[63] Biotin-LNTPs have been used for affinity labeling for the specific capture of polynucleotide sequences[64,65] and nucleic acid–protein complexes.[66]

Automated DNA sequencing by the Sanger "dideoxy" method is a particularly widespread commercial application in which fluorescent dideoxynucleotide chain terminators are a fundamental component.[67–71] In addition to primary sequencing, the dideoxy sequencing technology is also used extensively in SNP discovery and analysis[72,73] and serial analysis of gene expression (SAGE).[74] Fluorescent 2′-deoxy-NTP analogs have been employed in genomic sequence assembly[75,76] to label clone restriction digests for contig mapping of overlapping clones, as well as in a number of established genotyping technologies, such as microsatellite analysis[77] and single-stranded conformational polymorphism (SSCP) analysis.[78] LNTPs have also been utilized recently to introduce mass tags and affinity labels for mass spectrometry-based sequencing[79] and genotyping.[80]

In recent years, labeled deoxy- and dideoxynucleotide analogs have played an enabling role in the development and commercialization of arrayed single-base primer extension methods (SBE; also referred to as "minisequencing" or "digital genotyping") for SNP screening and mapping studies.[81–86] This method has proven particularly attractive for its simplicity, compatibility to various detection formats, and superior SNP discrimination relative to previous allele-specific hybridization methods. LNTPs continue to play an instrumental role in the ongoing development of a number of advanced technologies for large-scale sequencing and sequence analysis, based on sequencing by synthesis, exonucleolytic sequencing, and related strategies.[87–101]

Numerous practical guides are available that provide standardized protocols for enzymatic labeling of nucleic acid probes for blotting and *in-situ* hybridization[45–48]; DNA sequencing[102–106]; and sample preparation for DNA array-based analysis.[45,107] In addition, a great deal of practical information is available online from the various commercial providers of instrument platforms and reagents for genomic analysis.[108]

10.3 The synthesis of labeled nucleotide analogs

A number of NTP analogs have been prepared that possess modified heterocyclic bases exhibiting native fluorescence.[109–111] These have been utilized primarily as fluorescent probes for studying nucleotide-binding sites in proteins and enzymes.[109] Although some of these analogs can be recognized by polymerases and incorporated into DNA or RNA, their spectroscopic characteristics are generally not appropriate for the majority of labeling applications, in which high detection sensitivity is of primary importance. Most LNTPs for nucleic acid detection are of the type shown in Figure 10.1, which enables the incorporation of a wide range of luminescent labels and haptens. To achieve efficient labeling, LNTPs must be good substrates for DNA or RNA polymerases, so they are generally derived from ribo-, 2′/3′-deoxyribo- or 2′,3′-dideoxyribo-nucleotide analogs with a "canonical" purine or pyrimidine base (A, G, C, T, U) or some alternate heterocyclic base analog (see Appendix) at the β-C-1′ position. Enzymatic labeling efficiencies with LNTPs depend on the structure of the linker and labeling moieties, the position of their attachment, and the polymerase used (see Section 10.4).

LNTPs are prepared synthetically by one of a number of basic methods developed by researchers in the 1980s. These methods involve a sequence of manipulations to attach a linker, which has a terminal primary amine group, to the nucleobase of a nucleoside or nucleotide analog; conversion to a 5′-triphosphate (in the case of nucleoside intermediates); and, finally, coupling the label molecule as an amine-reactive derivative, typically an *N*-hydroxy-succinimidyl (NHS) ester. The first report of direct incorporation of an LNTP into nucleic acids involved C-5-biotinylaminoallyl-modified pyrimidine nucleotides.[1,112] The preparation of these compounds (shown in Figure 10.2) employed a mercuration and Pd(0)-catalysed alkenylation (Heck reaction) sequence demonstrated previously[113,114] to prepare the C-5 aminoallyl nucleotides; these were then reacted with biotin NHS ester. It is possible to introduce an entire linker–label construct as a prefabricated entity directly to the mercurated nucleotide[115,116]; however, this approach is rarely employed

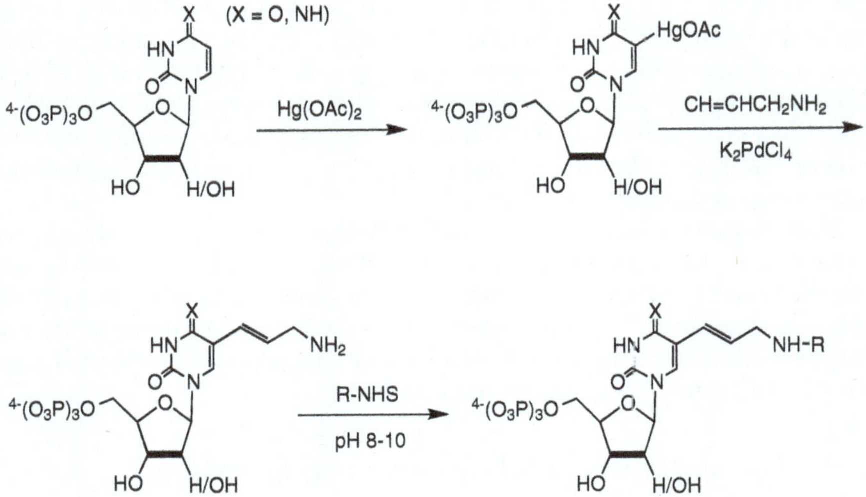

Figure 10.2 Synthesis of C-5 aminoallyl pyrimidine nucleotide analogs. (From Langer, P.R., Waldrop, A.A., and Ward, D.C. *Proc. Natl. Acad. Sci. USA*, 1981, 78, 6633–66371.)

because it is more efficient to prepare reagents from a common, commercially available aminoallylnucleotide intermediate.

Native purine nucleotides cannot be mercurated in this way, and alkylating the N-7 position of native purine bases is not a viable means of introducing a linker due to the hydrolytic instability this would cause. Ward and coworkers demonstrated that mercuration of the C-7 position of 7-deazapurine nucleotide analogs was also possible[117] and Pd(0)-catalysed coupling of alkenes to such intermediates has been reported.[118] However, this chemistry does not appear to have been useful for the preparation of 7-deazapurine-based labeling reagents.

Early examples of labeled purine nucleotide analogs were reported by Vincent et al.,[119,120] who prepared ATP derivatives bearing biotin- and 2,4-dinitrophenyl (DNP) moieties attached to the purine C-8 position via a C8-aminohexylamino linker. A reported method[121] was used to prepare aminohexylamino-ATP via the 8-bromo intermediate, and the haptens were coupled to the linker using available amine-reactive derivatives (Figure 10.3). These C8-modified dATP nucleotides appear to be substrates for terminal transferase in DNA end-labeling protocols.[119,120,122] However, incorporation of the analogs in primer extension and PCR reactions with T7 and *Taq* DNA polymerases is low,[123,124] and negligible with T7-RNA polymerase.[125] This appears to be attributable to the unfavorable steric effect of C8 substituents on nucleotide conformation (see Section 10.4.3.1); consequently, they have not seen extensive use as labeling reagents.

Shortly thereafter, however, labeled adenine nucleotides were prepared with aminoalkyl linkers on the exocyclic N-6 position[126] using similar nucleophilic substitution chemistry (Figure 10.3). Cytosine nucleotides labeled on

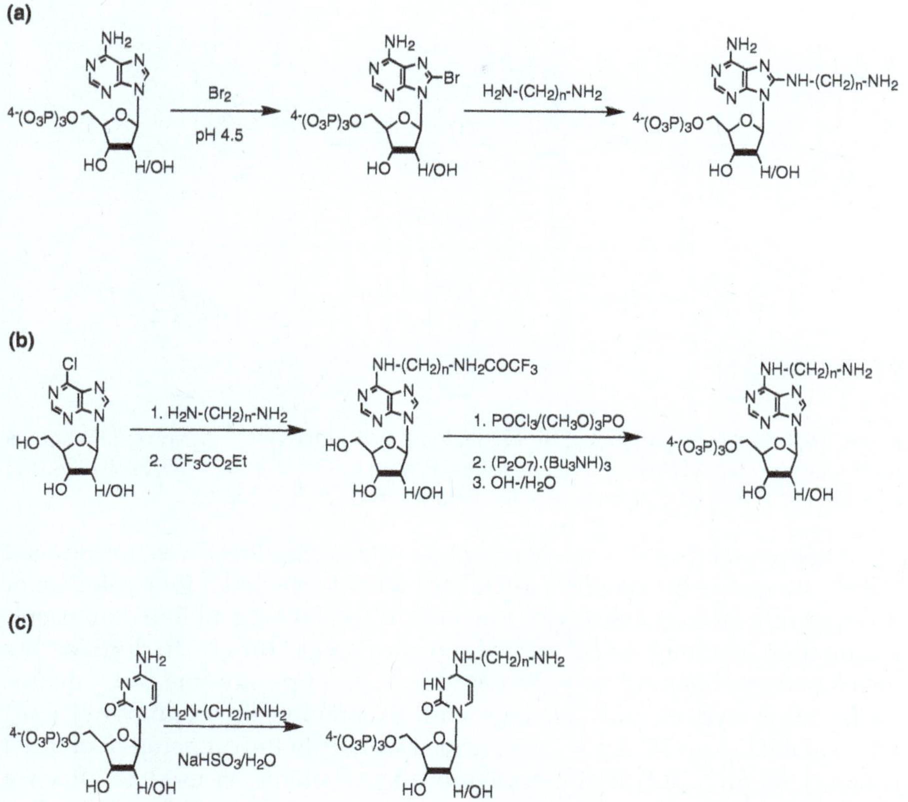

Figure 10.3 Scheme for the preparation adenine and cytosine nucleotide analogs with aminoalkyl linkers attached to the exocyclic nitrogen atoms. (From Vincent, C., Tchen, P., Cohen-Solal, M., and Kourilsky, P. *Nucl. Acids Res.*, 1982, 10, 6787–6797; Gebeyehu, G., Rao, P.Y., Soochan, P., Simms, D.A., and Klevan L. *Nucl. Acids Res.*, 1987, 15, 4513–4534; Gillam, I.C. and Tener, G.M. *Anal. Biochem.*, 1986, 157, 199–207.)

the exocyclic nitrogen atom[126,127] can also be prepared via activation and substitution at the 4-position and have proven to be very useful as nucleic acid-labeling reagents.[4]

In the mid-1980s, Sanger sequencing, using dideoxy-NTP terminators, had essentially replaced the Maxam–Gilbert method as the technology of choice, especially for the development of automated systems for the high-throughput sequencing needed to power the burgeoning genomics revolution. This provided the impetus for developing new labeling chemistries to enable a streamlined fluorescence-based readout. To meet this need, Smith et al. devised a four-color system based on the use of fluorescently labeled primers.[128] Using the unique fluorescent signature imparted by four spectroscopically resolvable dyes allowed differentiation of the extension fragments corresponding to each of the four termination reactions in a single electrophoresis lane.

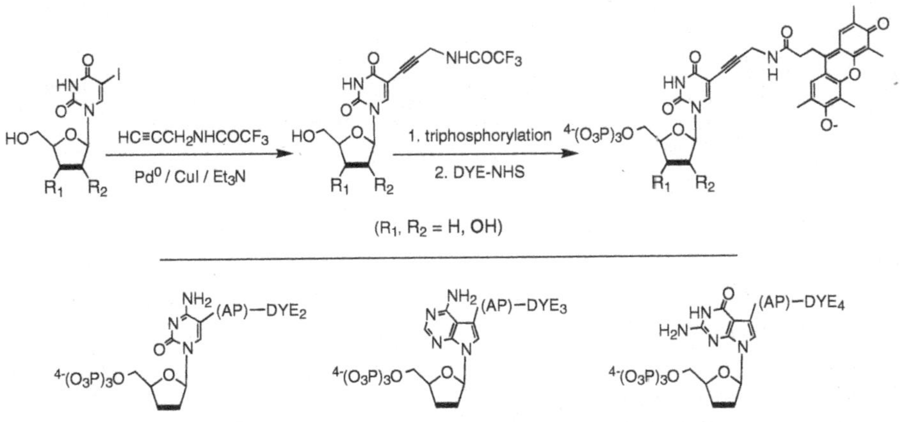

Figure 10.4 Synthesis of labeled dUTP analogs, and dideoxy-C, -A, and GTP analogs with aminopropynyl linkers. (From Hobbs, F.W., Jr. *J. Org. Chem.*, 1989, 54, 3420–3422 and Hobbs, F.W. and Cocuzza, A.J. US Patent 1991, US 5,047,519.)

Another landmark improvement to the technology was introduced shortly thereafter by scientists at DuPont who developed a four-color set of fluorescently labeled dideoxy terminators, thus allowing all four extension/termination reactions to be carried out in a single tube.[5] The terminators developed for this purpose were based on a set of pyrimidine and 7-deazapurine dideoxynucleotide analogs with a common aminopropynyl (AP) linker at the C-5 and C-7 positions, respectively.[129] In the preparation of these reagents (Figure 10.4), Pd(0)-catalysed alkynylation[130] is used to attach a protected aminopropynyl linker to the appropriate C-5 or C-7 iodo-nucleoside,[131] which is then phosphorylated, deprotected to reveal the primary amine, and then conjugated to the desired fluorescent dye.[132]

Many improvements have since been made to this basic set of reagents for DNA sequencing, including optimization of the linker and the fluorescent dyes to improve the enzymatic incorporation efficiencies; sensitivity; spectral resolution; and electrophoretic mobility characteristics. Kumar[71] contains a review on this area. However, the basic nucleobase and propynyl linker chemistry remains an industry standard for this application. It is also used to prepare a wide range of 2′-deoxy and ribo LNTPs currently used in a variety of other labeling applications (see Appendix).

Reagents based on the aminoallyl-pyrimidine and aminoalkynyl-pyrimidine/deazapurine nucleotides are incorporated efficiently by most polymerases and have seen the most extensive use for nucleic acid labeling. The N-6 and N-4 modified adenine and cytosine nucleotides also appear to be efficient polymerase substrates[126,127] and are offered commercially in addition to the aminoallyl-and aminopropynyl-based reagents. Numerous derivatives of these nucleotides modified with a variety of ligands, haptens, and luminescent labels are available from commercial sources as catalog or custom products.[108] Researchers can also prepare custom LNTP reagents directly from the unconjugated amino-linker modified nucleotide precursors.[133,134]

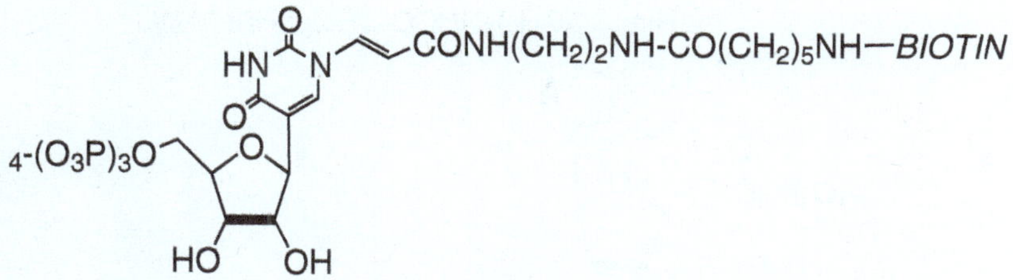

Figure 10.5 RNA labeling reagent based on pseudo-UTP. (From McGall, G. and Barone, A.D. US Pat. Appl. Publ. 2003, US2003180757 and Barone, A., Chen, C., Combs, D., Li, H., and McGall, G. Submitted for publication.)

The effectiveness of the 7-deazapurine nucleotide analogs as polymerase substrates and labeling reagents indicates that at least some departure from the four native nucleobase structures can be tolerated by the enzymes, with minimal impact on the hybridization characteristics of the polynucleotides in which they are incorporated. This has encouraged numerous explorations into LNTP analogs containing alternative noncanonical heterocyclic bases. Numerous examples can be found in the Appendix at the end of this chapter. Many of these reagents may seem to be mere equivalents and therefore redundant relative to the reagents described earlier. However, they appear to be of interest to reagent suppliers whose goal is to provide proprietary labeling reagents to support their specific analysis platform or to offer customers lower cost alternatives to existing commercial products.

One example is the pseudouridine-based analogs,[135–139] one of which is currently used for RNA labeling in array-based gene expression analysis products (Figure 10.5). Another reagent (Figure 10.6), used in array-based genotyping assays, lacks a nucleobase entirely, yet provides efficient, uniform 3′-end labeling of DNA with terminal deoxynucleotidyl transferase,[136] which is undoubtedly a consequence of the low nucleobase specificity of this non-template-directed polymerase.[140–143]

In at least some instances, it seems that alternate nucleobase structures may also offer specific improvements over the standard pyrimidine and purine/7-deazapurine LNTPs. For example, LNTPs with 8-aza-7-deazapu-

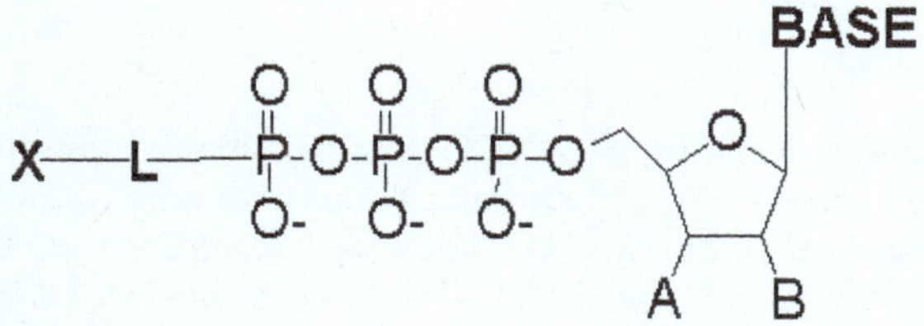

Figure 10.6 Abasic nucleotide analog for DNA 3′-end labeling. (From McGall, G. and Barone, A.D. US Pat. Appl. Publ. 2003, US2003180757; and Barone, A., Chen, C., Combs, D., Li, H., and McGall, G. Submitted for publication.)

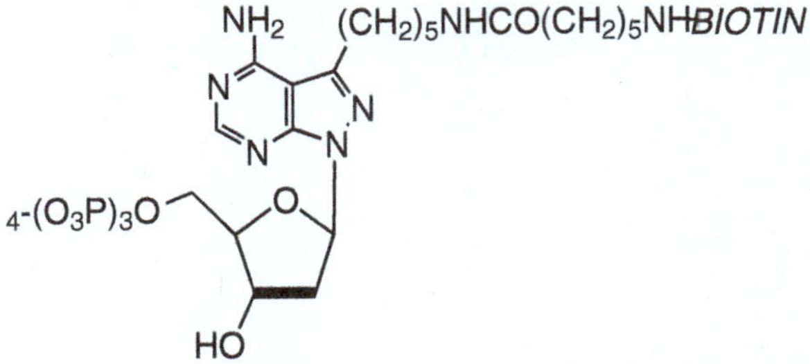

Figure 10.7 Pyrazolo[3,4-d]pyrimidine-based biotin-ATP analog. (From Petrie, C.R., Adams, A.D., Stamm, M., Van Ness, J., Watanabe, S.M., and Meyer, R.B., Jr. *Bioconjugate Chem.*, 1991, 2, 441–446.)

rine (pyrazolo[3,4-d]pyrimidine or "PP") bases with various linkers, including C-7 AP linkers (Figure 10.7), appear to be efficient polymerase substrates[136,144–146]; evidence indicates that the "PP" base analogs have duplex stabilizing effects that may be advantageous in hybridization-based applications.[147,148] Another stated rationale has been to utilize heterocycles, which behave as so-called "degenerate" or "universal" bases, with the aim of achieving uniform, sequence-independent labeling.[149] For example, the bicyclic 6H,8H-3-4dihydropyrimido[4,5-c]-[1,2]-oxazin-7-one derivatives shown in Figure 10.8 can be incorporated into DNA by Klenow DNA polymerase I, as an analog of TTP as well as dCTP.[150–152]

Finally, 5-bromo-uracil (BrU) deserves mention because it has also been used as an affinity label for nucleic acid detection in a number of *in-vivo* labelings in applications such as DNA/RNA synthesis and cell proliferation assays.[153] BrU nucleotides are efficiently incorporated by DNA and RNA polymerases and detectable with labeled anti-Br-U antibody conjugates.

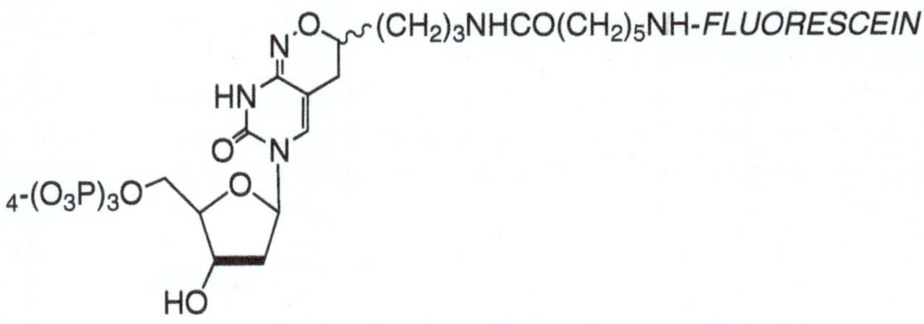

Figure 10.8 Bicyclic pyrimidine dCTP/TTP analog. (From Cummins, W.J., Smith, C.L., Nairne, R.J.D., Pickering, L., Simmonds, A.C., Hamilton, A.L., and Briggs, M.S.J. In: Loakes, D., Ed. *Modified Nucleosides: Synthesis and Applications*. Trivandrum, Research Signpost, 2002, 111–131; Brown, D., Hamilton, A., Loakes, D., Simmonds, A.C., and Smith, C. PCT Int. Appl. 1997, WO9728177; Cummins, W.J., Hamilton, A.L., Smith, C.L., and Briggs, M.S.J. *Nucleosides Nucleotides*, 2001, 20, 1049–1051; and Kumar, S., McDougall, M., Nampalli, S., Neagu, C., Loakes, D., and Brown, D. PCT Int. Appl. 2002, WO02002584.)

However, BrU has seen only limited use in hybridization-based detection applications such as blots, probe arrays, and the like.[154–156] This is presumably due to the relatively low affinity and/or specificity of the labeled anti-Br-U conjugates available for secondary detection, which do not, for example, recognize Br-U in double-stranded DNA or RNA hybrids.

10.4 Factors affecting labeled nucleotide incorporation efficiency

10.4.1 General considerations

As mentioned previously, labeling efficiencies with various LNTPs depend on the structural characteristics of the nucleotide-linker construct and labeling moiety. In general, the non-natural labeled NTP analogs may slow the rate of the polymerase extension reactions because they normally have higher K_m values than their native counterparts and act as competitive inhibitors. To achieve appreciable labeling, NTP analogs must compete with unlabeled nucleotides and must be present at a relatively high concentration to do so. The nature of the various interactions among the polymerase, LNTP, and the growing polynucleotide extension products is still largely obscure; polymerases vary significantly in their ability to discriminate against different types of label modifications. Undoubtedly, a variety of physicochemical characteristics of the analog, such as steric, electrostatic, and stereoelectronic effects; dipole moment; H-bonding; sugar ring conformation; etc., play a role in the recognition of the LNTP as a substrate by the polymerase and its ability to form a productive ternary enzyme–template–nucleotide complex. Because internal labeling of polynucleotides is typically the desired outcome using LNTPs, these characteristics will also influence the extent to which the modified nucleotide perturbs the resulting polynucleotide secondary structure and allows procession of the polymerase and continued synthesis of the product beyond the incorporation site.

Consequently, disruption of duplex structure by incorporation of modified nucleotides can result in premature chain termination events at or downstream of the incorporation site.[157] When double-stranded DNA is labeled via PCR, the accumulation of modified nucleotides in the template strands can exert a further inhibitory effect on polymerase processivity via secondary duplex structure perturbations.[157] These factors have the combined practical effect of reducing the processivity of the polymerase and inhibiting polynucleotide yield and average chain length.[17,157–162] Thus, attempting to maximize label density by increasing the LNTP concentration can be counterproductive. In practice, the ratio of the LNTP to its unlabeled counterpart is adjusted, typically within the range of 0.2 to 5.0, to achieve an optimum balance of product yield and label incorporation.[45–58,86,107]

Optimization of the labeling density is an important consideration in the preparation of labeled probes and targets for hybridization probe-based assays. It is frequently observed that an increase in the number of labels does not necessarily increase probe sensitivity, and may even decrease it, due to a number of factors. Overlabeling can reduce assay sensitivity by destabilization of the double-stranded complex[157,161–164]; in the case of biotinylated or other hapten-labeled polynucleotides, binding of the relatively bulky streptavidin or antibody molecules to additional labels becomes more sterically inhibited as the labels become too closely spaced.[17,165–167] In the case of fluorescent reporters, overlabeling can lead to fluorescence quenching[163] and increased nonspecific binding.[168]

In general, optimum performance in hybridization assays is usually achieved when label densities are in the range of 1 to 10% of bases substituted.[17,161–163,169] In contrast, a label density of 100%, using dye-labeled dNTP analogs, is required for one proposed exonucleotytic single-molecule sequencing scheme.[94–98] Amazingly, it appears that this can be achieved using available polymerases and LNTPs,[170–174] although dealing with the physical characteristics of such heavily labeled polynucleotides is bound to present some interesting challenges.

10.4.2 *Influence of the nucleobase*

Although detailed systematic studies have not been undertaken. However, as one would expect, it seems clear empirically that the optimum nucleotide structures for efficient incorporation — at least in the case of naturally occurring DNA and RNA polymerases — generally have characteristics that most closely resemble the canonical pyrimidine and purine nucleotides. With respect to the nucleobase, linker-modified pyrimidine and purine bases meet these requirements, of course, as long as the linker is attached to a favorable, noninterfering position (see Section 10.3 and Section 10.4.3.1). Reagents derived from 7-deazapurine analogs of ATP and GTP clearly meet these requirements as well. Together, the previously mentioned analogs constitute the majority of LNTPs currently available and in use. Numerous analogs based on other noncanonical bases (such as those mentioned in Section 10.3 and listed in the Appendix) have also been prepared; some appear to be reasonably efficient substrates and potentially useful labeling reagents. For many of these, however, only limited enzymatic incorporation data are available. The structure of the nucleobase is less critical when labeling with the non-template-directed polymerase TdT, which, not surprisingly, shows a relatively low discrimination against base modifications.[136,141–143]

10.4.3 *Influence of linker attachment site*

10.4.3.1 *Linkers attached to the nucleobase*

The position of the linker attachment is clearly a critical factor affecting labeling efficiency. From the point of view of B-form duplex DNA structure, the 5-position of pyrimidines and the 7-position of purines are the two base positions most accessible via the major groove.[175] The same is true in A-form duplexes (e.g., RNA–DNA hybrids), although the groove is narrower in this case. One can reasonably conclude that substitution at these positions, such as the attachment of a linker, should be least likely to perturb the normal duplex secondary structure and be most easily tolerated as a site for the introduction of a label.[1,112,176]

The destabilizing influence of substituents at the pyrimidine 6 position and at the purine 2 and 8 positions[110,177–183] indicates that these would not be optimal sites for the attachment of linkers. Steric accessibility would seem to explain the higher incorporation efficiencies generally observed for LNTPs based on 5-modified pyrimidines and 7-modified 7-deazapurine or 8-aza-7-deazapurine nucleotides (N7-modified purine nucleotides are hydrolytically unstable) compared to the low efficiencies observed with purine-C8 linker modifications.[124,125]

Analogs with alkyl linkers attached to the exocyclic N-6 position of adenine and N-4 of cytosine also appear to be efficient substrates of DNA and RNA polymerase[4,126,127,170]; this, too, can be rationalized on the basis of the steric accessibility of these sites via the major groove.[175] Even though they are involved in Watson–Crick hydrogen bonding, these positions appear to be able to tolerate a variety of alkyl linkers with only minor destabilization of the duplex structure.[127,184,185] In 3′-end labeling reactions using terminal transferase, N-biotinyl-AH-dCTP and dATP analogs are as efficient substrates for TdT as the C5-pyrimidine and C7-deaza modified nucleotides.[124,186] However, in comparison with similar C5-AA-modified dCTPs, the N4-aminoalkyl-modified dCTPs are generally not as efficiently incorporated in primer extension[170] and PCR labeling (Table 10.1).[124]

As noted previously, a variety of LNTP analogs have been reported in which linkers are attached to other non-canonical nucleobases, and one expects their effectiveness as polymerase substrates to be governed by the same geometric and conformational factors. Assuming that they adopt configurations analogous to their native counterparts, these analogs are invariably designed with linkers that occupy positions that are "sterically equivalent" to those which are optimal in pyrimidine and purine/7-deazapurine LNTPs.

10.4.3.2 Linkers attached to the furanose ring

Unlabeled 2′- and 3′-amino nucleotides are potent synthesis terminators of DNA polymerases; this has motivated explorations into the utility of nucleotide analogs in which linkers and labels have been attached to these positions, typically via amide or ester linkages (see Appendix, Figure 10.9). In general, however, most polymerases do not incorporate such compounds efficiently, even as chain terminators, presumably due to steric interference from the bulky substituent in the enzyme active site.[187] Incorporation of N-labeled 3′-amino nucleotide terminators by DNA pol-I,[188] TdT,[137,189] reverse transcriptase, and RNA polymerase[190] is generally poor; it usually requires high concentrations of the analog and often complete replacement of the native NTP. By using polymerases and conditions optimized for dideoxy sequencing, N-labeled 3′-amino 2′,3′-dideoxy-NTP analogs have performed

Table 10.1 Relative Incorporation Efficiency (Dyes/Amplicon)[a,b] of dNTPs Labeled with Fluorescein at Various Modification Sites in PCR Reactions

	Nucleotide analog				
	FL-dATP			FL-dCTP	
Polymerase	C7-deaza[c]	N6[d]	C8[e]	N4[f]	C5[g]
AmpliTaq	26	1.8	1.5	15	11
DeEPVent-	28	4.4	1.5	4.7	46
PFu-	7.8	2.0	1.3	1.1	11
rTth	38	1.9	1.3	—	15
Tfl	17	5.1	1.3	0	4.1
TLi	1.3	1.3	1.4	5.6	0
AmpliTaq-FS	23	2.1	1.5	11	24

[a] Target: HIV HBX2 plasmid insert
[b] Experiments run in triplicate
[c] Deazaadenine C7-12 atom AP linker (source: NEN)
[d] Adenine N6-12 atom alkyl linker (source: NEN)
[e] Adenine C8-15 atom alkylamino linker (source: Roche)
[f] Cytosine N4-14 atom alkyl linker (source: NEN)
[g] Cytosine C5-12 atom AP linker (source: NEN)
Source: McGall, G.H. and Barone, A.D. Unpublished data.

Figure 10.9 Nucleotide analogs labeled on the furanose ring.

as sequencing terminators with reasonable efficacy,[191] but little interest has been shown in exploiting such reagents for this application.

Labels linked to the 3′-oxygen via a cleavable ester or ether bonds have been of interest as "reversible" chain terminators for so-called base addition sequencing or sequencing-by-synthesis schemes.[87,88] In this scheme, single-base extension of sequencing primers is carried out with sequential rounds of reversible terminator additions. Following each round, the identity of the terminator incorporated in each particular target sequence is determined by the wavelength of the reporter dye, whereupon the reporter is released by chemical, photochemical, or enzymatic means, thereby unmasking the 3′-OH to reinitiate polymerization for the next round of terminator NTP addition.

Unfortunately, the generally poor substrate characteristics of these analogs; interference from unlabeled nucleotide impurities[192]; and a tendency for the catalytic editing function of DNA polymerases to remove 3′ modifications prematurely[193] have confounded previous attempts to implement this scheme.[88] Nevertheless, interest in enabling the approach or variations of it for single-molecule sequencing methods has recently been revived.[89–93] One promising proposal employs nucleotides labeled on the nucleobase via cleavable linkers, and small, removable protecting groups attached the 3′- or 4′-positions[89,90] to block chain extension.

Other proposed methods directly monitor the sequential incorporation of nonterminating LNTPs into the growing polynucleotide as a function of time using optical techniques with high spatial and temporal resolution.[91,93] A variation of this latter method would monitor the sequential release of flourescent labels attached to the γ-phosphate of otherwise unmodified NTPs.[99–101] It remains to be established whether these variations will lead to practicable solutions to problems that have thus far hindered the exploitation of previous sequencing-by-synthesis schemes.

A thymine 2′,3′-dideoxynucleotide analog with a labeled linker extending from the β C1′ position has been reported for putative use as sequencing terminator,[194] but its characteristics as a polymerase substrate were not described. In view of the substrate properties of simple C′-methyl-substituted NTP analogs,[195,196] it might also be possible to attach linkers to the 2′ (ara), 4′, and 5′ positions on the furanose ring to generate analogs for labeling, if there were some motivation for doing so.

10.4.4 Influence of linker structure

The structure of the linker has also been recognized as an important factor influencing the substrate properties of LNTPs and their relative incorporation efficiencies. An early comparison of biotinylated N6-linker-dATP analogs[126] illustrates that overall linker length and internal structure are relevant factors (Table 10.2). Similar observations have been reported for C5 linker-modified dUTP analogs.[2,161,197] Presumably, longer linkers can reduce inhibition of incorporation rates due to unfavorable steric interactions with the label. From the available data, however, it would appear that little further improvement occurs beyond a length of about 8 to 10 atoms, and very long linkers can actually result in decreased incorporation levels.[126,170,171,198] In applications using secondary detection with labeled streptavidin or antibody conjugates, the use of longer linkers can allow more efficient binding of the large secondary labeling molecules.[199,200] In this case, the best linker length in terms of assay sensitivity would strike an optimum balance between incorporation and binding efficiency.

The geometry and flexibility of the linker can exert profound effects, as illustrated by comparisons of incorporation levels for alkyl-, alkenyl-, and alkynyl-modified C5-dUTP and C7-7-deazaATP analogs in chain extension reactions with *Taq* polymerase[201–204] (Figure 10.10). For example, the *cis*-alkenyl and flexible saturated alkyl linkers are essentially devoid of substrate activity. Studies based on oligonucleotide hybridization[184,205] indicate that a saturated carboxypropyl linker should not significantly destabilize the

Table 10.2 Relative Incorporation Levels of Biotinylated N6-Linker-dATP Analogs by Nick Translation

Linker	% Incorporation
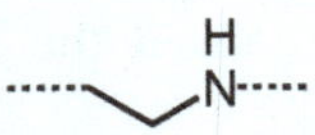	26
	56
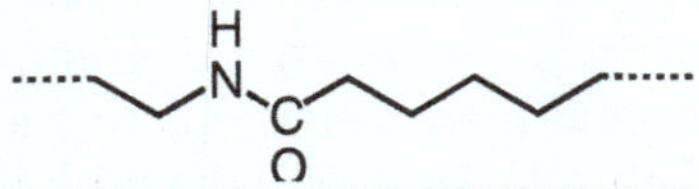	16
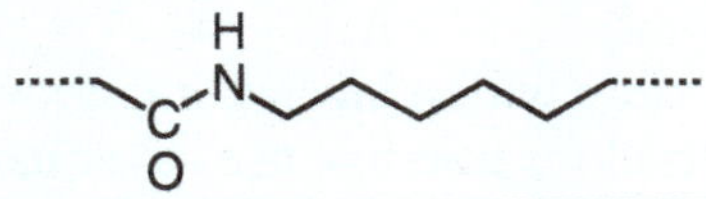	7
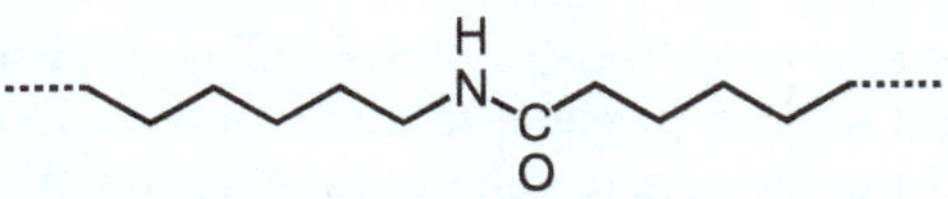	36
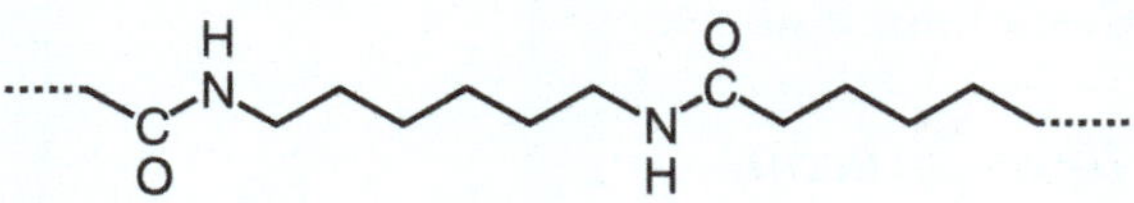	7

Source: Gebeyehu, G., Rao, P.Y., Soochan, P., Simms, D.A., and Klevan, L. *Nucl. Acids Res.*, 1987, 15, 4513–4534.

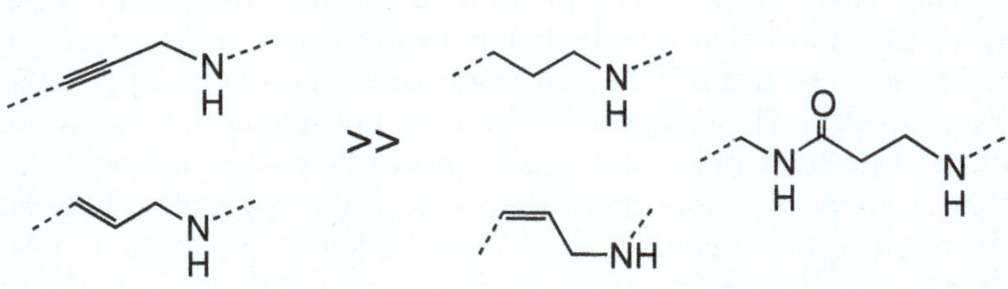

Figure 10.10 Influence of various linker structures on the relative incorporation efficiency of C5-modified dUTP analogs by *Taq* polymerase.

resulting duplex. Therefore, these observations suggest that the *cis*-alkenyl and flexible saturated alkyl linkers allow unfavorable contacts within the enzyme–substrate complex, and the more rigid, linear propynyl and *trans*-alkenyl linkers remain sterically unencumbered. Another thermophilic DNA polymerase (KOD dash) reportedly tolerates saturated C5-alkyl linkers to a significantly greater extent, although a *trans*-alkenyl linker was still preferred.[206] A similar preference for an alkenyl linker structure over a saturated alkyl linker has also been observed in the incorporation of N1-biotinylated *pseudo*UTP analogs by RNA polymerase.[139]

Another favorable aspect of the alkynyl linkers is that these substituents exert a significant stabilizing effect on the DNA duplex,[207–209] due to a combination of hydrophobicity and increased polarizability of the nucleobase. This effect may also contribute to the somewhat higher incorporation levels observed for LNTPs with these linkers.

Most LNTP linkers are electrostatically neutral, although a variety of fluoresecent dideoxynucleotide terminators with charge-modified linkers has been prepared for DNA sequencing (see Appendix and Kumar[71]); the purpose is to improve the electrophoretic characterisitics of sequencing fragments to achieve cleaner, longer reads. This aspect of linker structure will likely influence incorporation efficiencies significantly, but this area has not been well studied.

10.4.5 *Influence of label and polymerase*

10.4.5.1 *General labeling*

A wide range of labeling groups has been directly incorporated into nucleic acids via LNTPs (see Appendix). Certainly the size, shape, charge, and hydrophobicity of the labeling molecule will be important factors affecting substrate characteristics of the LNTP, and some promising efforts have been made to clarify the optimal characteristics of the label and the polymerases used for incorporating labels in primer extension and amplification reactions.[172,210] Outside the area of dideoxy sequencing, however, few detailed studies have been carried out. Based on the available information, one broad conclusion seems valid: when all other factors are approximately equal, the smaller, less obtrusive labels are incorporated more efficiently than larger molecules. For example, the relative incorporation of unlabeled aminoallyl or aminopropynyl-NTPs > biotinyl-NTPs > fluorescent dye-NTPs, although the many different fluorescent dyes evaluated vary significantly.[172,210]

The higher labeling efficiencies achievable with small molecule tags is one reason for the persistent popularity of the original indirect affinity-labeling approach used by Ward and coworkers,[1] is wherein a streptavidin-dye or -enzyme conjugate is bound to the polynucleotide products that have been prelabeled with biotinylated LNTPs. This approach also affords a great deal of flexibility by allowing a wide variety of signal-generating groups to be introduced after incorporation of the biotin ligand. Haptens such as biotin, digoxigenin, 2,4-dinitrophenyl, and fluorescein are also frequently used in concert with labeled conjugates of their corresponding antibodies. The combined use of several different hapten labels has enabled simultaneous detection of multiple target sequences in multicolor FISH protocols.[51–57]

The even higher incorporation efficiencies that can be attained using unlabeled amine-modified NTPs have also motivated the use of a chemical postincorporation labeling strategy. In this case, dyes and other labeling groups are coupled to the amine-modified polynucleotide using appropriate reactive derivatives such as NHS esters. This circumvents the negative effect that these dyes, in a direct incorporation format, would otherwise have on labeling efficiency and product yields. This approach has been used for generating labeled cDNA[212–216] and RNA.[17,158,217] In addition to primary amines, other modifiable functional groups, such as thiol,[218–221] maleimide,[221] aminooxyl,[222] and carbonyl[223] have been incorporated into nucleic acids via modified NTP analogs. The chemical postlabeling strategy offers flexibility and the ability to achieve higher labeling densities; however, it is less convenient than direct incorporation and complete labeling of the functional groups is not always guaranteed.

Although most of the DNA polymerases traditionally used for labeling DNA are A-family polymerases (Klenow DNA pol I, T4, T7, and *Taq* DNA polymerases), recent reports indicate that, in some cases, B-family archaeon polymerases such as Tli/Vent, Pfu, and Tgo (exo-) may provide superior incorporation of dye-labeled dNTPs in simple primer extension reactions.[171,172,198] Improved incorporation efficiencies can also be achieved via genetic modification, and mutant forms of the A-family *Taq*[224,377] and bacteriophage T4 DNA polymerases[97,225] have been identified that provide substantially improved utilization of LNTPs.

Essentially, complete labeling of a single-strand primer extension product in the sole presence of dye-labeled NTPs has been achieved with Vent and Tgo (exo-).[171,172] Complete labeling with dye-labeled pyrimidine dNTPs has also been accomplished in single-round primer extension reactions using wildtype Klenow fragment DNA pol I.[173,174] The purpose of achieving such high labeling levels is to generate target sequences for an exonuclease-based sequencing strategy.[94–98] For cDNA labeling by reverse transcription, FluoroScript (InVitrogen), an avian reverse transcriptase with reduced RNAse H activity, has been shown to incorporate fluorescently labeled dUTPs more efficiently than the AMV and MMLV enzymes.[226,227] The incorporation of LNTPs into RNA via *in vitro* transcription has not been optimized as extensively as DNA labeling reactions. However, a comparison of three commonly used RNA polymerases indicates that incorporation levels for dye-labeled UTP analogs with C5-AP linkers are in the following order: T7 > T3 > SP6; all three RNA polymerases showed a strong preference for incorporating fluorescein-UTP over cyanine-UTP.[124] In addition to C5-labeled pyrimidine nucleotides, T7 RNA polymerase also efficiently incorporates LNTPs derived from the C-nucleosides *pseudo*uridine and *pseudoiso*cytidine.[136–139] Labeled ATP analogs are generally not well tolerated by the enzyme.[136]

10.4.5.2 Dideoxy DNA sequencing

Since the development of the first four-color set of fluorescently labeled dideoxynucleotide terminators,[5] the study of dye-NTP incorporation by various DNA polymerases has been pursued extensively in order to achieve optimum performance on commercial sequencing systems (see Kumar[71] for a review). Here the goal is not only uniform incorporation levels but also uniform spectroscopic characteristics (sensitivity and resolution) and electrophoretic behavior are also key considerations in the selection and optimization of dyes and linkers. The polymerases used most commonly in DNA sequencing are T7 and *Taq* DNA polymerases; mutant forms that exhibit minimal discrimination against dideoxynucleotides (e.g., Sequenase, Thermo-Sequenase, and AmpliTaq FS)[228–230] have been developed. The thermostable polymerases are currently preferred because they enable cycle sequencing for linear signal amplification.[231]

Fluorescein-type dye-labeled ddNTPs were initially developed for sequencing with T7 DNA polymerases,[5,232] but fluorescein-labeled pyrimidine ddNTPs are less efficiently incorporated by *Taq* polymerase, as are terminators labeled with BODIPY and carbocyanine dyes.[232,233] The zwitterionic rhodamine family of dyes is incorporated most efficiently and uniformly by *Taq* polymerase, and mechanistic studies indicate that fluorescein dyes attached to a linker on the pyrimidine C5 position interfere with a conformational change step that takes place after the nucleotide binds to the polymerase. The enzyme seems to be less discriminating between rhodamine and flourescein dyes when attached to the C7 of 7-deazapurine NTPs.[234] The current "standard" four-color dye terminators are based on a set of 4,7-dichloro-substituted rhodamine dyes[235]; this has been recently supplemented by inclusion of one or more rhodamine-flourescein FRET dyes,[235–237] which provide improved sensitivities and read lengths with the current instrumentation platforms. An improved mutant T7 polymerase, Thermo Sequenase II, which affords more uniform incorporation of rhodamine-ddNTPs, is now available.

Nevertheless, the rhodamine-based terminators have certain drawbacks that fluorescein-based terminators apparently do not,[234] so there continues to be an interest in studying other thermostable polymerases that may incorporate the latter more efficiently and uniformly. Toward this end, directed mutagenesis of *Taq* polymerase has shown significant potential for improvement.[224] Mutants of the B-family archeon polymerase Vent,[238] which show a preference for "acyclo"-labeled terminators,[239] have been generated (Figure 10.11). These appear to be less discriminating with respect to the dye, incor-

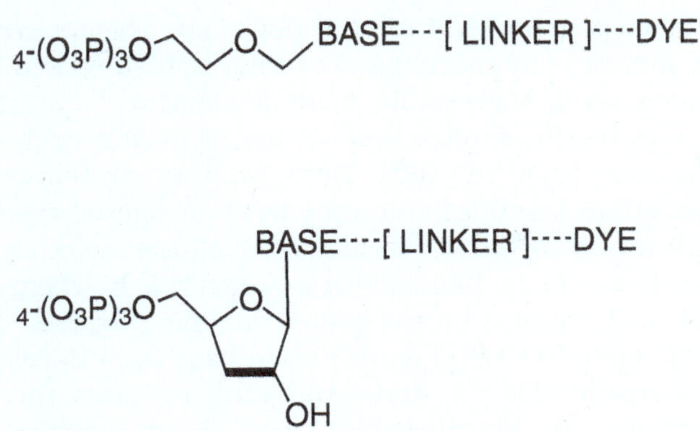

Figure 10.11 Sequencing dye terminators based on "acyclo" and 3′-deoxy NTP analogs. (From Trainor, G.L. US Patent, 1996, US 5,558,991; Sasaki, N., et al., *Proc. Natl. Acad. Sci. U.S.A.*, 1998, 95, 5455–5460.)

porating various rhodamine, fluorescein, carbocyanine, and BODIPY dye-labeled acyclo-NTPs with similar efficiency.[240,241] Other mutant forms of B-family polymerases, such as Pfu,[242] KOD dash,[206] and JDF-3,[243] have also been considered as potentially improved sequencing enzymes. For an informative overview of DNA polymerases in sequencing and other areas of biotechnology, see Hamilton et al.[244]

A modified sequencing methodology, known as "transcriptional sequencing," has also been developed in order to mitigate the drawbacks associated with existing cycle-sequencing protocols.[245,246] This method employs dye-labeled 3′-deoxy NTP terminators based on AP linker chemistry (Figure 10.11); these are utilized by a mutant T7 RNA polymerase to generate a cRNA sequencing ladder. The apparent advantages include uniform terminator incorporation, rapid isothermal signal amplification, and longer read lengths without the need to remove unreacted terminators. Rhodamine dyes have been used in the terminators for transcriptional sequencing, although it is not reported whether fluorescein or other dye families were evaluated in this system.

In another proposed sequencing method, direct PCR sequencing (DPCRS), a mutant DNA polymerase is employed to incorporate labeled ribonucleotide analogs randomly in a PCR reaction.[247,248] The incorporation of labeled ribonucleotides does not result in chain termination, thus allowing geometric amplification to be achieved. The amplified sequences are then subjected to chemical or enzymatic hydrolysis at the ribo-NTP incorporation sites to generate 3′-end labeled sequencing fragments. As yet, the performance characteristics of this method have not been reported; nevertheless, this approach would also be potentially useful as a sample preparation method for a number of DNA array applications, which require randomly fragmented and uniformly labeled DNA targets.

10.5 Cleavable linkers

Researchers have prepared and utilized nucleotide analogs with cleavable linkers for a number of applications. For example, biotinylated LNTPs with linkers containing a thiol cleavable disulfide element (Figure 10.12a) have been used to enable the capture and release of nucleic acids and nucleic acid–protein complexes to study DNA binding proteins.[64–66] Also, a four-color set of thiolcleavable fluorescent dNTP analogs was used in a recent demonstration of an *in-situ* sequencing-by-synthesis methodology[92] with single-molecule sensitivity. Linkers that are cleavable by chemical or enzymatic hydrolysis have also been explored for this purpose.[87–92] Similarly, photocleavable biotin-dNTPs (Figure 10.12b) have been developed for the isolation of biomolecules[249,250] and sequencing by mass spectrometry[310]; flourescent photocleavable nucleotides have been prepared for use in sequencing-by-synthesis schemes.[88,90] Compared with chemically cleavable linkers, photocleavable linkers offer a more convenient reagent-free process for releasing labels after incorporation into polynucleotides (Figure 10.12).

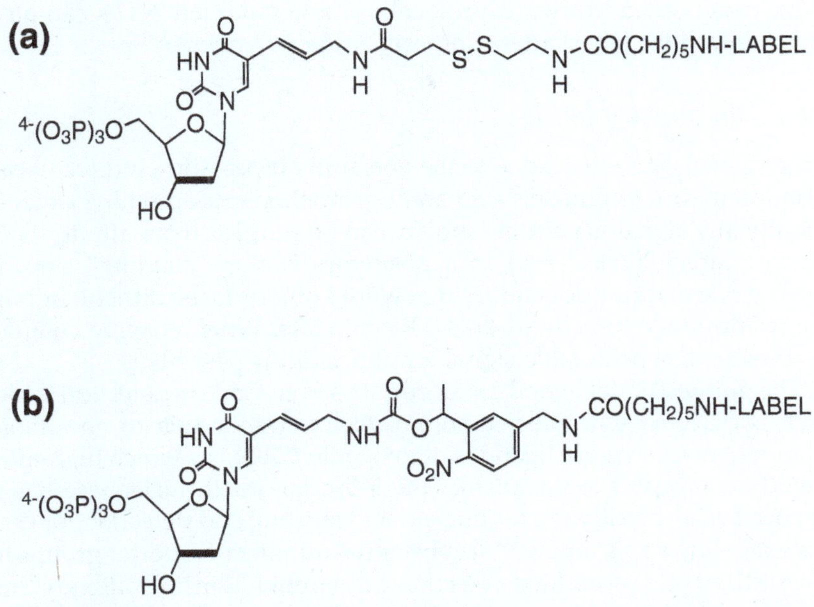

Figure 10.12 Examples of nucleotide analogs with (a) chemically cleavable and (b) photochemically cleavable linkers.

10.6 Labels

10.6.1 Primary labels

A large number of "primary" reporter groups have been conjugated to NTP analogs for direct detection of nucleic acids. The selection of primary reporter groups is based on the detection requirements of the particular genomic research or diagnostic application/instrumentation platform with which it is to be used. Reviews have recently been published on the general area of nucleic acid detection modalities.[71,251,252] The Appendix provides a reasonably comprehensive list of LNTP analogs for which published accounts describe preparation of the reagents. The most common primary reporter groups are fluorescent dye molecules and "affinity tags" — ligands that allow subsequent recognition and binding of a secondary signal-generating entity.

The most intensive use of primary fluorescent dye labels has been for the preparation of dideoxynucleotide analogs for use as DNA sequencing terminators. Numerous other labels have been used as primary reporters, including photocrosslinkers, redox labels, nitroxyl spin labels, phosphors, mass tags, etc. Commercial providers[108] currently have a significant number of LNTPs available whose preparation was not been specifically described in the scientific or patent literature, but which clearly employed the general methods described in Section 10.3. As pointed out in that section, the availability of basic aminoallyl- and aminopropynyl-modified nucleotide analogs enables researchers to prepare LNTPs with virtually any desired reporter group, using straightforward protocols; amine-modified NTPs can also be used to introduce labels after synthesis of the polynucleotide.

10.6.2 Secondary labels

As mentioned previously, due to the versatility it provides, indirect labeling is commonly used in nucleic acid analysis methods based on hybridization. Virtually any signal-generating moiety can be coupled to an affinity-labeled probe or target nucleic acid in a posthybridization "staining" step, thus allowing one to introduce entities that would otherwise be difficult or impossible to introduce directly as an NTP conjugate. When enzyme conjugates are employed, considerable signal amplification is possible.

The quintessential ligand for affinity-based secondary detection is biotin, which is detected with labeled conjugates of streptavidin or an antibiotin antibody. Another useful ligand is digoxigenin (DIG), for which high-affinity antibody conjugates are available; the 5-bromo-uracil nucleobase has also been used as an affinity tag for nucleic acid labeling and detection for certain *in-vivo* labeling applications.[153] A substantial number of reporter groups have been utilized for secondary detection via streptavidin or antibody conjugates; Table 10.3 provides a representative list.

10.7 Summary and future directions

The reagents and methods available for nonenzymatic chemical labeling of nucleic acids will continue to improve and find increasing use as alternatives in many of the applications in which LNTPs are currently employed. However, nonradioactive enzymatic labeling with LNTP analogs will likely continue to play a predominant role in genomic analysis applications for some time. This seems probable because LNTPs have provided the foundation on which a number of established technologies for genetic analysis have been built; they currently require these reagents as an essential element (e.g., DNA sequencing with fluorescent terminators and fluorescent digital genotyping with arrayed primer extension). Furthermore, existing assays and instrument platforms for a number of applications are rapidly evolving towards a high level of standardization — especially those under consideration for regulated diagnostic or clinical use (e.g., DNA microarrays).

Technologies that utilize LNTPs for labeling in a research and development setting will likely continue to use them as they transfer into and become established in more regulated laboratory environments. Another consideration is that enzymatic labeling is usually the most cost-effective labeling strategy for analytical assays because polynucleotide analytes are commonly prepared by polymerase-catalyzed extension, replication, or amplification reactions. Simultaneously incorporating labels during these steps is more streamlined and efficient than methods that require additional chemical labeling and clean-up steps.

Table 10.3 Secondary Signal-Generating Streptavidin or Antibody Conjugates

Reporter type	Examples	Ref.
Fluorescent dye	Fluorescein	253, 254
	IRD40	255
Fluorescent lanthanide complex	Eu(III)-$(BiPy)_3$	256, 257
Fluorescent protein	Phycoerythrin	258
Fluorescent nanoparticle	Organic dye-impregnated latex particle	259
	Inorganic semiconductor crystal ("quantum-dots")	260
Electroluminescent complex	$Ru(BiPy)_2$	261
Chemiluminophore	Aequorin	262–264
Colloidal metal nanoparticles	Au (absorption)	265
	Au (electron microscopy)	266–268
	Se, Au (resonance light scattering)	269–271
Inorganic phosphorescent particles	Down-converting	272
	Up-converting	273
Magnetic microparticles	Fe_2O_3-impregnated latex	274,378
Enzymes		
Peroxidase	Colorimetric	165, *200*, 253
	Chemiluminescence (luminol)	275, 276
	Tyramide signal amplification	277, 278
Alkaline phosphatase	Colorimetric (NBT/BCIP; HNPP/HBNP)	2, 14, 17, 81, 126
	Fluorescence (Attophos)	279, 280
	Fluorescence (ELF)	281
	Chemiluminescence (dioxetane)	282, 283
	Chemiluminescence (acridinium)	284
	Chemiluminescence (luciferin)	285
	Colorimetric	286
Urease	Chemiluminescence (coupled luciferin/ase)	287
G6PDH		
β-Galactosidase	Fluorescence (coumarin phosphate)	288

In terms of new developments and applications, the field is likely to see increased utilization of novel polymerases, fluorophores (e.g., quantum dots) and other labels in combination with improved detection strategies. Fluorescent LNTPs are playing a key role in a number of new single-molecule sequencing technologies based on sequencing by synthesis, exonucleolytic sequencing, and related strategies currently under development.[86–101] These technologies generally utilize highly sensitive fluorescence microscopic techniques, with high spatial and temporal resolution, to monitor the enzymatic addition of nucleotides to, or their removal from, single polynucleotide fragments in real time. Several proposed methods, for example, would utilize

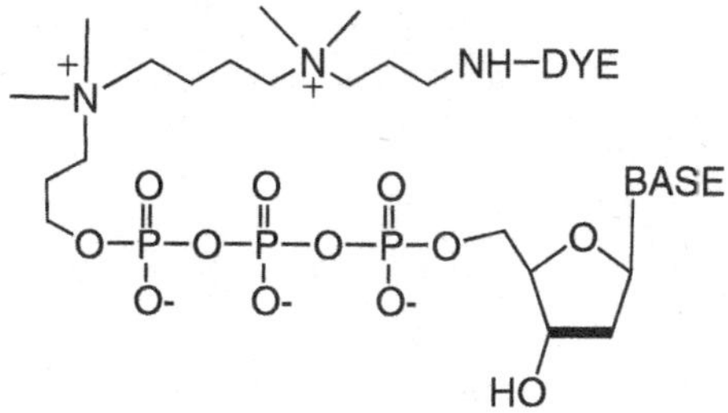

Figure 10.13 Phosphate-labeled nucleotide analog for use in a proposed single-molecule sequencing method. (From Williams, J.G.K. PCT Int. Pat. Appl. 2000, WO00036152; and Williams, J.G.K., Bashford, G.R., Chen, J., Draney, D., Narayanan, N., Reynolds, B.L., and Sheaff, P. PCT Int. Appl. 2001, WO01094609.)

NTPs labeled with fluorescent dyes attached to the γ-phosphate (e.g., Figure 10.13) and detect the insertion or release of the fluorophore from the polymerase active site[99–101] during chain extension.

These advanced technologies promise to provide a lower-cost, higher-throughput alternative to conventional methods for large-scale sequencing. Significant challenges must be overcome in order to implement these novel technologies routinely, and there will certainly be many innovative developments in this area and will likely take place in the near future.

References

1. Langer, P.R., Waldrop, A.A., and Ward, D.C. Enzymatic synthesis of biotin-labeled polynucleotides: novel nucleic acid affinity probes. *Proc. Natl. Acad. Sci. USA*, 1981, 78, 6633–6637.
2. Leary, J.J., Brigati, D.J., and Ward, D.C. Rapid and sensitive colorimetric method for visualizating biotin-labeled DNA probes hybridized to DNA or RNA immobilized on nitrocellulose: bio-blots. *Proc. Natl. Acad. Sci. USA*, 1983, 80, 4045–4049.
3. Feinberg, A.P. and Vogelstein, B. A technique for radiolabeling DNA restriction endonuclease fragments to high specific activity. *Anal. Biochem.*, 1984, 137, 266–267.
4. Klevan, L. and Gebeyehu, G. Biotinylated nucleotides for labeling and detecting DNA. *Methods Enzymol.*, 1990, 184, 560–561.
5. Prober, J.M., Trainor, G.L., Dam, R.J., Hobbs, F.W., Robertson, C.W., Zagursky, R.J., Cocuzza, A.J., Jensen, M.A., and Baumeister, K. A system for rapid DNA sequencing with fluorescent chain-terminating dideoxynucleotides *Science*, 1987, 238, 336–341.
6. Lo, Y.-M., Mehal, W.Z., and Fleming, K.A. Rapid production of vector-free biotinylated probes using the polymerase chain reaction. *Nucl. Acids Res.*, 1988, 16, 8719.
7. Dennis, L.Y.-M., Mehal, W.Z., and Fleming, K.A. Nonradioactive detection of PCR products. In: Innis, M.A., Gelfand, D.H., Sninsky, J.J., and White, T.J., Eds. *PCR Protocols: A Guide to Methods and Applications*. San Diego: Academic Press, 1990, 113–118.

8. Cronin, M.T., Fucini, R.V., Kim, S.M., Masino, R.S., Wespi, R.M., and Miyada, C.G. Cystic fibrosis mutation detection by hybridization to light-generated DNA probe arrays. *Hum. Mutation*, 1996, 7, 244–255.
9. Mogensen, J., Kolrava, S., Hindkjaer, J., Petersen, S., Koch, J., Nygard, M., Jensen, T., Gerersen, N., Junker, S., and Bolund, L. Nonradioactive, sequence-specific detections of RNA *in situ* by primed *in-situ* labeling (PRINS). *Exp. Cell Res.*, 1991, 19, 92–98.
10. Schena, M., Shalon, D., Davis, R.W., and Brown, P.O. Quantitative monitoring of gene expression patterns with a complementary DNA microarray. *Science*, 1995, 27, 467–470.
11. Brakel, C.L. and Engelhardt, D.L. In: Kingsbury, D.T. and Falkow, S., Eds. *Rapid Detection and Identification of Infectious Agents*. New York: Academic Press, 1985, 235–243.
12. Riley, L.K., Marshall, M.E., and Coleman, M.S. A method for biotinylating oligonucleotide probes for use in molecular hybridizations. *DNA*, 1986 5, 333–337.
13. Schmitz, G.G., Walter, T., Seibl, R., and Kessler, C. Nonradioactive labeling of oligonucleotides *in vitro* with the hapten digoxigenin by tailing with terminal transferase. *Anal. Biochem.*, 1991, 192, 222–231.
14. Murasugi, A. and Wallace, R.B. Biotin-labeled oligonucleotides: enzymatic synthesis and use as hybridization probes. *DNA*, 1984, 3, 269–277.
15. Inazuka, M., Tahira, T., and Hayashi, K. One-tube post-PCR fluorescent labeling of DNA fragments. *Genome Res.*, 1996, 6, 551–557.
16. McCracken, S. Preparation of labeled RNA transcripts using SP6 RNA polymerase. *Focus* (BRL/Life Technol), 1985, 7.2, 5–8.
17. Folsom, V., Hunkeler, M., Haces, A., and Harding, J.D. Detection of DNA targets with biotinylated and fluoresceinated RNA probes. *Anal. Biochem.*, 1989, 182, 309–314.
18. Lockhart, D.J., Dong, H., Byrne, M.C., Follettie, M.T., Gallo, M.V., Chee, M.S., Mittmann, M., Wang, C., Kobayashi, M., Horton, H., and Brown, E.L. Expression monitoring by hybridization to high-density oligonucleotide arrays. *Nat. Biotechnol.*, 1996, 14, 1675–1680.
19. Rosenmyer, V., Laubrock, A., and Seibl, R. Nonradioactive 3′-end-labeling of RNA molecules of different lengths by terminal deoxynucleotidyltransferase. *Anal. Biochem.*, 1995, 224, 446–449.
20. Martin, G. and Keller, W. Tailing and 3′-end labeling with yeast poly(A) polymerase and various nucleotides. RNA 1998, 4, 226–230.
21. Draper, D.E. and Gold, L. A method for linking fluorescent labels to polynucleotides: application to studies of ribosome-ribonucleic acid interactions. Biochemistry, 1980, 19, 1774–1781.
22. Chu, B.C.F. Wahl, G.M., and Orgel, L.E. Derivatization of unprotected polynucleotides. *Nucl. Acids Res*, 1986, 14, 5591–5603.
23. Ghosh, S.S., Kao, P.M., and Kwoh, D.Y. Synthesis of 5′-oligonucleotide hydrazide derivatives and their use in preparation of oligonucleotide-enzyme hybridization probes. *Anal. Biochem.*, 1988, 178, 43–51.
24. Hansske, F. and Cramer, F. Modification of the 3′ terminus of tRNA by periodate oxidation and subsequent reaction with hydrazides. *Methods Enzymol.*, 1979, 59, 172–181.
25. Odom, O.W., Jr., Robbins, D.J., Lynch, J., Dottavio–Martin, D., Kramer, G., and Hardesty, B. Distances between 3′ ends of ribosomal ribonucleic acids reassembled into *Escherichia coli* ribosomes. *Biochemistry*, 1980, 19, 5947–5954.

26. Bauman, J.G.J., Wiegant, J., and van Duijn, P. Cytochemical hybridization with fluorochrome-labeled RNA. I. Development of a method using nucleic acids bound to agarose beads as a model. *J. Histochem. Cytochem.*, 1981, 29, 227–237.
27. Proudnikov, D. and Mirzabekov, A. Chemical methods of DNA and RNA fluorescent labeling. *Nucl. Acids Res.*, 1996, 24, 4535–4542.
28. Forster, A.C., McInnes, J.L., Skingle, D.C., and Symons, R.H. Nonradioactive hybridization probes prepared by the chemical labelling of DNA and RNA with a novel reagent, photobiotin. *Nucl. Acids Res.*, 1985, 13, 745–761.
29. Sheldon, E.L., Kellogg, D.E., Watson, R.E., Levinson, C.H., and Erlich, H.A. Use of nonisotopic M13 probes for genetic analysis: application to HLA class II loci. *Proc. Natl. Acad. Sci. USA*, 1986, 83, 9085–9089.
30. Albarella, J.P., Minegar, R.L., Patterson, W.L., Dattagupta, N., and Carlson, E. Monoadduct forming photochemical reagents for labeling nucleic acids for hybridization. *Nucl. Acids Res.*, 1989, 17, 4293–4308.
31. Slattum, P.S., Loomis, A.G., Machnik, K.J., Watt, M.-A., Duzeski, J.L., Budker, V.G., Wolff, J.A., and Hagstrom, J.E. Efficient *in vitro* and *in vivo* expression of covalently modified plasmid DNA. *Molec. Therapeutics*, 2003, 8, 255–263.
32. van Belkum, A., Linkels, E., Jelsma, T., Houthoff, H.J., van den Berg, F., and Quint, W. Application of a new, universal DNA labeling system in the PCR mediated diagnoses of *Chlamydia trachomatis* and human papillomavirus type 16 infection in cervical smears. *J. Virol. Methods*, 1993, 45, 189–200.
33. van Gijlswijk, R.P.M., Talmane, G., Janssen, P.J.A., Snoeijers, S.S., Killian, J., Tanke, H.J., and Heetebrij, R.J. Universal linkage system: versatile nucleic acid labeling technique. *Expert Rev. Molec. Diagnost.*, 2001, 1, 81–91.
34. Gupta, V., Cherkassky, A., Chatis, P., Joseph, R., Johnson, A.L., Broadbent, J., Erickson, T., and DiMeo, J. Directly labeled mRNA produces highly precise and unbiased differential gene expression data. *Nucl. Acids Res.*, 2003, 31, E13.
35. Czworkowski, J. and Hardesty, B. Fluorescence study of the topology of messenger RNA bound to the 30S ribosomal subunit of *Escherichia coli*. *Biochemistry*, 1991, 30, 4821–4830.
36. Laayoun, A., Kotera, M., Sothier, I., Trevisiol, E., Bernal–Mendez, E., Bourget, C., Menou, L., Lhomme, J., and Troesch, A. Aryldiazomethanes for universal labeling of nucleic acids and analysis on DNA chips. *Bioconj. Chem.*, 2003, 14, 1298–1306.
37. Venter, J.C., Adams, M.D., Sutton, G.G., Kerlavage, A.R., Smith, H.O., and Hunkapiller, M. Shotgun sequencing of the human genome. *Science*, 1998, 280, 1540–1542.
38. Tsui, L.C. and Scherer, S.W. The human genome project. In: Sensen, C.W., Ed. *Essentials of Genomics and Bioinformatics*. Weinheim: Wiley–VCH, 2002, 41–60.
39. Pühler, A., Jording, D., Kalinowski, J., Buttgereit, D., Renkawitz–Pohl, R., Altschmied, L., Danchin, A.E., Feldmann, H., Klenk, H.-P., and Kröger, M. Genome projects of model organisms. In: Sensen, C.W., Ed. *Essentials of Genomics and Bioinformatics*. Weinheim: Wiley–VCH, 2002, 5–40.
40. Various authors. In: Phimister, B., Ed. The chipping forecast. *Nat. Genet.*, 1999, 21(Suppl).
41. Young, R.A. Biomedical discovery with DNA arrays. *Cell*, 2000, 102, 9–15.
42. Lockhart, D.J. and Winzeler, E.A. Genomics gene expression and DNA arrays. *Nature*, 2000, 405, 827–836.

43. Patil, N. et al. Blocks of limited haplotype diversity revealed by high-resolution scanning of human chromosome 21. *Science*, 2001, 294, 1719–1723.
44. Cawley, S., Drenkow, J., Bekiranov, S., Strausberg, R.L., Fodor, S.P.A., and Gingeras, T.R. Large-scale transcriptional activity in chromosomes 21 and 22. *Science*, 2002, 296, 916–919.
45. Schena, M. *Microarray Analysis*. New York: John Wiley & Sons, 2002.
46. Mundy, C.R., Cunningham, M.W., and Read, C.A. Nucleic acid labeling and detection. In Brown, T.A., Ed. *Essential Molecular Biology*, 2nd ed., Oxford: Oxford University Press, 2001, 63–107.
47. Various authors. In: Rapley, R., Ed. *The Nucleic Acid Protocols Handbook*. Totowa: Humana Press, 2000, 117–173.
48. Kessler, C. Nonradioactive labelling methods for nucleic acids. In: Kricka, L.J., Ed. Nonisotopic probing blotting and sequencing. New York: Academic Press, 1995, 41–109.
49. Keller, G.H. Nonradioactive labeling procedures. In: Keller, G.H. and Manak, M.M., Eds. *DNA Probes*, 2nd ed., New York: Stockton Press, 1993, 173–198.
50. Tijssen, P., Ed. *Laboratory Techniques in Biochemistry and Molecular Biology*. Vol. 24: *Hybridization with Nucleic Acid Probes*. New York: Elsevier, 1993.
51. Nederloff, P.M., Robinson, D., Abuknesha, R., Wiegant, J., Hopman, A.S.H.N., Tanke, H.J., and Raap, A.K. Three-color fluorescence *in-situ* hybridization for the simultaneous detection of multiple nucleic acid sequences. *Cytometry*, 1989, 10, 20–27.
52. Reid, T., Baldini, A.R., Baldini, T.C., and Ward, D.C. Simultaneous visualization of seven different DNA probes by *in-situ* hybridization using combinatorial fluorescence and digital imaging microscopy. *Proc. Natl. Acad. Sci. USA*, 1992, 89, 1388–1392.
53. Schröck, E., du Manoir, S., Veldman, T., Schoell, B., Wienberg, J., Ferguson–Smith, M.A., Ning, Y., Ledbetter, D.H., Bar–Am, I., Soenkensen, D., Garini, Y., and Ried, T. Multicolor spectral karyotyping of human chromosomes. *Science*, 1996, 273, 494–497.
54. Speicher, M.R., Gwyn Ballard, S., and Ward, D.C. Karyotyping human chromosomes by combinatorial multi-fluor FISH. *Nat. Genet*., 1996, 12, 368–375.
55. Tanke, H.J., Wiegant, J., van Gijlswijk, R.P., Bezrookove, V., Pattenier, H., Heetebrij, R.J., Talmane, G., Raap, A.K., and Vrolijk, J. New strategy for multicolour fluorescence *in-situ* hybridisation: COBRA: COmbined Binary RAtio labelling. *Eur. J. Hum. Genet*., 1999, 7, 2–11.
56. Henegariu, O., Heerema, N.A., Bray–Ward, P., and Ward, D.C. Color-changing karyotyping: an alternative to M-FISHSKY. *BioTechniques*, 1999, 23, 263–264.
57. Levsky, J.M. and Singer, R.H. Fluorescence *in-situ* hybridization: past, present and future. *J. Cell Sci*., 2003, 116, 2833–2838.
58. Kallioniemi, A., Kallioniemi, O.P., Sudar, D., Rutovitz, D., Gray, J.W., Waldman, F., and Pinkel, D. Comparative genomic hybridization for molecular cytogenetic analysis of solid tumors. *Science*, 1992, 258, 818–821.
59. Reinhardt, B., Frank, U., Gellner, K., Lohmann, J.U., and Bosch, T.C.G. High-resolution, fluorescence-based differential display on a DNA sequencer followed by band excision. *Bio.Tech*., 1999, 27, 268–271.
60. Kozal, M.J., Shah, N., Shen, N., Yang, R., Fucini, R., Merigan, T.C., Richman, D.D., Morris, M.S., Hubbell, E., Chee, M.S., and Gingeras, T.R. Extensive polymorphisms observed in HIV-1 clade B protease gene using high-density oligonucleotide arrays. *Nat. Med*., 1996, 2, 753–759.

61. Wang, D.G. et al. Large-scale identification mapping and genotyping of single-nucleotide polymorphisms in the human genome. *Science*, 1998, 280, 1077–1082.
62. Lee, C.H., Khalifah, S., and Chin, K.-V. Nonradioactive detection of DNA repair synthesis. *Molec. Biotechnol.*, 1999, 11, 263–266.
63. Wu, C.F., Bishopric, N.H., and Pratt, R.E. Nonradioactive method for determination of internucleosomal cleavage associated with apoptosis. *Bio. Techn.*, 1997, 23, 840–843.
64. Shimkus, M., Levy, J., and Herman, T. A chemically cleavable biotinylated nucleotide: usefulness in the recovery of protein–DNA complexes from avidin affinty columns. *Proc. Natl. Acad. Sci. USA*, 1985, 82, 2593–2597.
65. Dawson, B.A., Herman, T., and Lough, J. Affinty isolation of transcriptionally active murine erythroleukemia cell DNA using a cleavable biotinylated nucleotide analog. *J. Biol. Chem.*, 1989, 264, 12830–12837.
66. Persinger, J. and Bartholomew, B. Site-directed DNA photoaffinity labeling of RNA polymerase III transcription complexes. *Methods Molec. Biol.*, 2001, 148 (DNA–protein interactions 2nd ed.), 363–381.
67. Hunkapillar, T., Kaiser, R.J., Koop, B.F., and Hood, L. Large scale and automated DNA sequence determination. *Science*, 1991, 254, 59–67.
68. Efcavitch, J.W. Electrophoresis-based fluorescent dideoxy-terminator sequencing. Novartis Foundation Symposium, 2000, 229 (From Genome to Therapy: Integrating New Technologies with Drug Development), 5–13.
69. Middendorf, L.R., Humphrey, P.G., Narayanan, N., and Roemer, S.C. Sequencing technology. In: Sensen, C.W., Ed. *Biotechnology*, Vol 5b, 2nd ed. Weinheim: Wiley–VCH, 2001, 183–226.
70. Marziali, A. and Akeson, M. New DNA sequencing methods. *Annu. Rev. Biomed. Eng.*, 2001, 3, 195–223.
71. Kumar, S. Fluorescent dye nucleotide conjugates for DNA sequencing. In: Loakes, D., Ed. *Modified Nucleosides: Synthesis and Applications*. Trivandrum, India: Research Signpost, 2002, 87–110.
72. Taillon–Miller, P., Piernot, E.E., and Kwok, P.-Y. Efficient approach to unique single-nucleotide polymorphism discovery. *Genome Res.*, 1999, 9, 499–505.
73. Kwok, P.-Y., Carlson, C., Yager, T.D., Ankener, W., and Nickerson, D.A. Comparative analysis of human DNA variations by fluorescence-based sequencing of PCR products. *Genomics*, 1994, 23, 138–144.
74. Velculescu, V.E., Zhang, L., Vogelstein, B., and Kinzler, K.W. Serial analysis of gene expression. *Science*, 1995, 270, 484–487.
75. Gillett, W., Hanks, L., Wong, G.K.-S., Yu, J., Lim, R., and Olson, M.V. Assembly of high-resolution restriction maps based on multiple complete digests of a redundant set of overlapping clones. *Genomics*, 1996, 33, 389–408.
76. Gregory, S.G., Howell, G.R., and Bentley, D.R. Genome mapping by fluorescent fingerprinting. *Genome Res.*, 1997, 7, 1162–1168.
77. Kondo, H., Tahira, H., Hayashi, K., Oshima, K., and Hayashi, K. Microsatellite genotyping of post-PCR fluorescently labeled markers. *BioTechniques*, 2000, 29, 868–872.
78. Inazuka, M., Wenz, H.-M., Sakabe, M., Tahira, T., and Hayashi, K.A. Streamlined mutation detection system: multicolor post-PCR fluorescence labeling and single-strand conformational polymorphism analysis by capillary electrophoresis. *Genome Res.*, 1997, 7, 1094–1103.

79. Guo, B. Mass spectrometry in DNA analysis. *Anal. Chem.* (review), 1999, 71, 333R–337R.
80. Kim, S., Ruparel, H.D., Gilliam, T.C., and Ju, J. Innovation: digital genotyping using molecular affinity and mass spectrometry. *Nat. Rev. Genet.*, 2003, 4, 1001–1008.
81. Harju, L., Weber, T., Alexandrova, L., Lukin, M., Ranki, M., and Jalanko, A. Colorimetric solid-phase minisequencing assay illustrated by detection of alpha 1-antitrypsin Z mutation. *Clin. Chem.*, 1993, 39, 2282–2287.
82. Shumaker, J.M., Metspalu, A., and Caskey, C.T. Mutation detection by solid phase primer extension. *Hum. Mutation*, 1995, 7, 346–354.
83. Pastinen, T., Kurg, A., Metspalu, A., Peltonen, L., and Syvanen, A.-C. Minisequencing: a specific tool for DNA analysis and diagnostics on oligonucleotide arrays. *Genome Res.*, 1997, 7, 606–614.
84. Fan, J.-B., Chen, X., Halushka, M.K., Berno, A., Huang, X., Ryder, T., Lipshutz, R.J., Lockhart, D.J., and Chakravarti, A. Parallel genotyping of human SNPs using generic high-density oligonucleotide tag arrays. *Genome Res.*, 2000, 10, 853–860.
85. Pastinen, T., Raitio, M., Lindroos, K., Tainola, P., Peltonen, L., and Syvänen, A.-C. A system for specific, high-throughput genotyping by allele-specific primer extension on microarrays. *Genome Res.*, 2000, 10, 1031–1042.
86. Zhu, J., Shendure, J., Mitra, R.D., and Church, G.M. Single molecule profiling of alternative pre-mRNA splicing. *Science*, 2003, 301, 836–839.
87. Canard, B. and Sarfati, S.R. DNA polymerase fluorescent substrates with reversible 3′-tags. *Gene*, 1994, 148, 1–6.
88. Welch, M.B., Martinez, C.I., Zhang, A.J., Jin, S., Gibbs, R., and Burgess, K. Syntheses of nucleosides designed for combinatorial DNA sequencing. *Chem. Eur. J.*, 1999, 5, 951–960.
89. Odedra, R., Simmonds, A., and Pickering, L. Nucleotide analogs comprising a reporter moiety and a polymerase-blocking moiety. PCT Int. Appl. 2001, WO0192284.
90. Li, Z., Bai, X., Ruparel, H., Kim, S., Turro, N.J., and Ju, J. A photocleavable fluorescent nucleotide for DNA sequencing and analysis. *Proc. Natl. Acad. Sci. USA*, 2003, 100, 414–419.
91. Braslavsky, I., Hebert, B., Kartalov, E., and Quake, S.R. Sequence information can be obtained from single DNA molecules. *Proc. Natl. Acad. Sci. USA*, 2003, 100, 3960–3964.
92. Mitra, R.D., Shendure, J., Olejnik, J., Krzymanska–Olejnik, E., and Church, G.M. Fluorescent *in-situ* sequencing on polymerase colonies. *Anal. Biochem.*, 2003, 320, 55–65.
93. Levene, M.J., Korlach, J., Turner, S., Foquet, M., Craighead, H.G., and Webb, W.W. Zero-mode waveguides for single-molecule analysis at high concentrations. *Science*, 2003, 299, 682–686.
94. Jett, J.H., Keller, R.A., Martin, J.C., Marrone, B.L., Moyzis, R.K., Ratliff, R.L., Seitzinger, N.K., Shera, E.B., and Stewart, C.C. High-speed DNA sequencing: an approach based upon fluorescence detection of single molecules. *J. Biomol. Struct. Dyn.*, 1989, 7, 301–309.
95. Eigen, M. and Rigler, R. Sorting single molecules: application to diagnostics and evolutionary biotechnology. *Proc. Natl. Acad. Sci. USA*, 1994, 91, 5740–5747.

96. Various authors. In: Rigler, R. and Seela, F., Eds. *J. Biotechnol.*, 2001, 86 (DNA sequencing at the single molecule level), 161–308.
97. Werner, J.H., Cai, H., Jett, J.H., Reha–Krantz, L., Keller, R.A., and Goodwin, P.M. Progress towards single-molecule DNA sequencing: a one color demonstration. *J. Biotechnol.*, 2003, 102, 1–14.
98. Harding, J.D. and Keller, R.A. Single-molecule detection as an approach to rapid DNA sequencing. *TIBTECH*, 1992, 10, 55–57.
99. Williams, J.G.K. Nucleic acid sequencing of single molecules by polymerase extension releasing labeled diphosphates. PCT Int. Pat. Appl. 2000, WO00036152.
100. Williams, J.G.K., Bashford, G.R., Chen, J., Draney, D., Narayanan, N., Reynolds, B.L., and Sheaff, P. Charge-switch nucleotides for use in nucleic acid sequencing. PCT Int. Appl. 2001, WO01094609.
101. Hardin, S.H., Briggs, J.M., Tu, S.-C., Gao, X., and Willson, R. Real-time sequence determination using labeled polymerizing enzymes and/or labeled monomers. PCT Int. Appl. 2002, WO0200468.
102. Graham, C.A. and Hill, A.J., Eds. *DNA Sequencing Protocols*. 2nd ed. Totowa: Humana Press, 2001.
103. Various authors. In: Rapley, R., Ed. *The Nucleic Acid Protocols Handbook*. Totowa: Humana Press, 2000, 481–565.
104. Alphey, L. *DNA Sequencing*. New York: Springer–Verlag, 1997.
105. Ansorge, W., Voss, H., and Zimmermann, J., Eds. *DNA Sequencing Strategies: Automated and Advanced Approaches*. New York: Wiley, 1997.
106. Adams, M.D., Venter, J.C., and Fields, C., Eds. *Automated DNA Sequencing and Analysis*. London: Elsevier, 1994.
107. Bittner, M. et al. In: Bowtell, D. and Sambrook, J., Eds. *DNA Microarrays*. Cold Spring Harbor: CSHL Press, 2002, 173–239.
108. Stull, D. A feast of fluorescence. *Scientist*, 2001, 15, 20 (recent list of commercial suppliers of labeled NTPs).
109. Cremo, C.R. Fluorescent nucleotides: synthesis and characterization. *Methods Enzymol.*, 2003, 360 (Biophotonics, Part A), 128–177 (recent review).
110. Seela, F., Zulauf, M., Sauer, M., and Deimel, M. 7-Substituted 7-deaza-2′-deoxyadenosines and 8-aza-7-deaza-2′-deoxyadenosines: fluorescence of DNA-base analogs induced by the 7-alkynyl side chain. *Helv. Chim. Acta*, 2000, 83, 910–927.
111. Conrad, M.J. Applications of fluorescent N-nucleosides and fluorescent structural analogs of N-nucleosides. PCT Int. Appl. 1993, WO9316094.
112. Ward, D.C., Waldrop, A.A., and Langer, P.R. Modified nucleotides and methods of preparing and using same. US Patent, 1987, US 4,711,955.
113. Ruth, J.L. and Bergstrom, D.E. C-5 substituted pyrimidine nucleosides. 1. Synthesis of C-5 allyl, propyl, and propenyl uracil and cytosine nucleosides via organopalladium intermediates. *J. Org. Chem.*, 1978, 43, 2870–2876.
114. Bigge, C.F., Kalaritis, P., Deck, J.R., and Merte, M.P. Palladium-catalyzed coupling reactions of uracil nucleosides and nucleotides. *J. Am. Chem. Soc.*, 1980, 102, 2033–2038.
115. Sarfati, S.R., Pochet, S., Neumann, J.M., and Igolen, J. Synthesis of uridine and 2′-deoxyuridine mono- and tri-phosphates alkylated int position 5 by glycosides of a-D-mannose and N-acetyl-b-dglucoseamine: DNA and RNA monomers with tethered lectin targets. *J. Chem. Soc. Perkin Trans.*, 1, 1990, 1065–1070.

116. Nelson, P.A., Bahl, C., and Gibbons, I. A convenient preparation of a C-5 biotinylated dUTP derivative. *Nucleosides Nucleotides*, 1986, 5, 233–241.
117. Dale, R.M.K., Livingston, D.C., and Ward, D.C. The synthesis and enzymatic polymerization of nucleotides containing mercury: potential tools for nucleic acid sequencing and structural analysis. *Proc. Natl. Acad. Sci.*, 1973, 70, 2238.
118. Bergstrom, D.E., Brattesani, A.J., Ogawa, M.K., and Schweickert, M.J. Pyrrolo[2,3-d]pyrimidine nucleoside antibiotic analogs. Synthesis via organopalladium intermediates derived from 5-mercuritubercidin. *J. Org. Chem.*, 1981, 46, 1423–1431.
119. Vincent, C., Tchen, P., Cohen–Solal, M., and Kourilsky, P. Synthesis of 8-(2-4 dinitrophenyl 2-6 aminohexyl) amino-adenosine 5′ triphosphate: biological properties and potential uses. *Nucl. Acids Res.*, 1982, 10, 6787–6797.
120. Kourilsky, P., Vincent, C., and Tchen, P. DNA fragments marked at least at one of the ends thereof by modified ribonucleotides recognizable by related molecules and method for analysing such fragments. PCT Int. Appl. 1983, WO8302277.
121. Lee, C.-Y., Lazarus, L.H., Kabakoff, D.S., Laver, M.B., Russell, P.J., and Kaplan, N.O. Purification of kinases by general ligand chromatography. *Arch. Biochem. Biophys.*, 1977, 178, 8–18.
122. Dattagupta, N. and Kocher, J. Novel fluorescent label. Eur. Pat. Appl. 1997, EP0527433.
123. Steffens, D.L., Jang, G.Y., Sutter, S.L., Brumbaugh, J.A., Middendorf, L.R., Muhlegger, K., Mardis, E.R., Weinstock, L.A., and Wilson, R.K. An infrared fluorescent dATP for labeling DNA. *Genome Res.*, 1995, 5, 393–399.
124. McGall, G.H. and Barone, A.D. Unpublished data.
125. Ito, Y. Modified nucleic acids for *in vitro* selection. *Nucl. Acids Symp. Series*, 1997, 37, 259–260.
126. Gebeyehu, G., Rao, P.Y., Soochan, P., Simms, D.A., and Klevan, L. Novel biotinylated nucleotide-analogs for labeling and colorimetric detection of DNA. *Nucl. Acids Res.*, 1987, 15, 4513–4534.
127. Gillam, I.C. and Tener, G.M. N4-(6-aminohexyl)cytidine and -deoxycytidine can be used to label DNA. *Anal. Biochem.*, 1986, 157, 199–207.
128. Smith, L.M., Sander, J.Z., Kaiser, R.J., Hughes, P., Dodd, C., Connell, C.R., Heiner, C., Kent, S.B., and Hood, L.E. Fluorescence detection in automated sequence analysis. *Nature*, 1986, 321, 674–679.
129. Hobbs, F.W., Jr. Palladium-catalyzed synthesis of alkynylamino nucleosides. A universal linker for nucleic acids. *J. Org. Chem.*, 1989, 54, 3420–3422.
130. Robins, M.J. and Barr, P.J. Nucleic acid related compounds. 39. Efficient conversion of 5-iodo to 5-alkynyl and derived 5-substituted uracil bases and nucleosides. *J. Org. Chem.*, 1983, 48, 1854–1862.
131. Coccuza, A.J. Total synthesis of 7-iodo-2′,3′-dideoxy-7-dezapurine nucleosides, key intermediates in the preparation of reagents for the automated sequencing of DNA. *Tetrahedron Lett.*, 1988, 29, 4061–4064.
132. Hobbs, F.W. and Cocuzza, A.J. Alkynylamino-nucleotides. US Patent, 1991, US 5,047,519.
133. Nimmakayalu, M., Henegariu, O., Ward, D.C., and Bray–Ward, P. Simple method for preparation of fluorhapten-labeled dUTP. *BioTechniques*, 2000, 28, 518–522.
134. Henegariu, O., Bray–Ward, P., and Ward, D.C. Custom fluorescent nucleotide synthesis as an alternative method for nucleic acid labeling. *Nat. Biotechnol.*, 2000, 18, 345–348.

135. Muhlegger, K., Eltz, H., and Seela, F. C-Nucleosides derivatives and their use in the detection of nucleic acids. PCT Int. Appl. 1996, WO9628640.
136. Barone, A.D., Chen, C., McGall, G.H., Rafii, K., Buzby, P.R., and Dimeo, J.J. Novel nucleoside triphosphate analogs for the enzymatic labeling of nucleic acids. *Nucleosides Nucleotides*, 2001, 20, 525–532.
137. McGall, G. and Barone, A.D. Nucleic acid labeling compounds. US Pat. Appl. Publ. 2002, US20020165372.
138. McGall, G. and Barone, A.D. Heterocyclic nucleoside derivatives for labeling of nucleic acids. US Pat. Appl. Publ. 2003, US20031807 57.
139. Barone, A., Chen, C., Combs, D., Li, H., and McGall, G. Novel nucleotide analogs for nucleic acid labeling. Submitted.
140. Matsuzake, H., et al. Parallel genotyping of over 10,000 SNPs using a one-primer assay on a high-density oligonucleotide array. *Genome Res.*, 2004; 14, 414-425.
141. Khandazhinskaya, A.L., Jasko, M.V., Shirokov, E.A., and Kukhanova, M.K. New substrates of terminal deoxynucleotidyl transferase: synthesis and biological evaluation. *Collect Symp. Ser.*, 2002, 5(chemistry), 344–347.
142. Delarus, M., Boulé, J.B., Lescar, J., Expert–Bezançon, N., Jourdan, N., Sukumar, N., Rougeon, F., and Papanicolaou, C. Crystal structures of a template-independent DNA polymerase: murine terminal deoxynucleotidyl transferase. *EMBO J.*, 2002, 21, 427–439.
143. Kreyevsky, A.A., Victorova, L.S., Arzumanov, A.A., and Jasko, M.V. Terminal deoxynucleotidyl transferase: catalysis of DNA (oligodeoxynucleotide) phosphorylation. *Pharmacol. Therapeut.*, 2000, 85, 165–173.
144. Sugimoto, S. and Sato, T. Labeled nucleotides and polynucleotides. Japanese Patent, 1986, Sho 61–109717.
145. Petrie, C.R., Adams, A.D., Stamm, M., Van Ness, J., Watanabe, S.M., and Meyer, R.B., Jr. A novel biotinylated adenylate analog derived from pyrazolo[3,4-d]pyrimidine for labeling DNA probes. *Bioconjugate Chem.*, 1991, 2, 441–446.
146. Seela, F., Ramzaeva, N., and Becher, G. 7-Deazapurine DNA: oligonucleotides containing 7-substituted 7-deaza -2′-deoxyguanosine and 8-aza-7-deaza-2′-deoxyguanosine. *Collect Czech. Chem. Commun.*, 1996, 61 (special issue), S258–S261.
147. Kutyavin, I.V., Lokhov, S.G., Afonina, I.A., Dempcy, R., Gall, A.A., Gorn, V.V., Lukhtanov, E., Metcalf, M., Mills, A., Reed, M.W., Sanders, S., Shishkina, I., and Vermeulen, N.M.J. Reduced aggregation and improved specificity of G-rich oligodeoxyribonucleotides containing pyrazolo[3,4-d]pyrimidine guanine bases. *Nucl. Acids Res.*, 2002, 30, 4952–4959.
148. He, J., Becher, G., and Seela, F. Pyrazolo[3,4-d]pyrimidine nucleic acids: adjustment of the dA-dT to the dG-dC base pair stability. *Nucleosides Nucleotides*, 2003, 22, 573–576.
149. Cummins, W.J., Smith, C.L., Nairne, R.J.D., Pickering, L., Simmonds, A.C., Hamilton, A.L., and Briggs, M.S.J. Synthesis of novel modified five membered and bicyclic heterocyclic nucleotide analogs and their use in labelling and detection. In: Loakes, D., Ed. *Modified Nucleosides: Synthesis and Applications*. Trivandrum, Research Signpost, 2002, 111–131.
150. Brown, D., Hamilton, A., Loakes, D., Simmonds, A.C., and Smith, C. Nucleoside analogs. PCT Int. Appl. 1997, WO9728177.

151. Cummins, W.J., Hamilton, A.L., Smith, C.L., and Briggs, M.S.J. Synthesis and study of the fluorescein conjugate of the nucleotide dPTP. *Nucleosides Nucleotides*, 2001, 20, 1049–1051.
152. Kumar, S., McDougall, M., Nampalli, S., Neagu, C., Loakes, D., and Brown, D. Preparation of nucleotide analogs and their enzymic incorporation into oligodeoxyribonucleotide duplexes. PCT Int. Appl. 2002, WO02002584.
153. Dolbeare, F. Bromodeoxyuridine: a diagnostic tool in biology and medicine. Part III. *Histochem. J.*, 1996, 28, 531–575.
154. Sakamoto, H., Traincard, F., Vo Quang, T., Ternynck, T., Guesdon, J.L., and Avrameas, S. 5-Bromodeoxyuridine *in-vivo* labeling of M13 DNA, and its use as a nonradioactive probe for hybridization experiments. *Molec. Cell Probes*, 1987, 1, 109–120.
155. Tabibzadeh, S., Bhat, U.G., and Sun, X. Generation of nonradioactive bromodeoxyuridine labeled DNA probes by polymerase chain reaction. *Nucl. Acids Res.*, 1991, 19, 2783.
156. Darzynkiewicz, Z., Li, X., and Traganos, F. Methods for labeling DNA ends with halogenated nucleotides and detecting same with antibodies. PCT Int. Appl. 1997, WO9708345.
157. Zhu, Z. and Waggoner, A.S. Molecular mechanism controlling the incorporation of fluorescent nucleotides into DNA by PCR. *Cytometry*, 1997, 28, 206–21.
158. Luehrsen, K.R. and Baum, M.P. *In-vitro* synthesis of biotinylated RNA probes from A-T rich templates: problems and solutions. *BioTechniques*, 1987, 5, 660–670.
159. Heer, H.E., Keyszer, G.M., Gay, R.E., and Gay, S. Inhibition of RNA polymerase by digoxigenin-labeled UTP. *BioTechniques*, 1994, 16, 54–55.
160. Hiyoshi, M. and Hosoi, S. Assay of DNA denaturation by polymerase chain reaction-driven fluorescent label incorporation and fluorescence resonance energy transfer. *Anal. Biochem.*, 1994, 221, 306–311.
161. Zhu, Z., Chao, J., Yu, H., and Waggoner, A.S. Directly labeled DNA probes using fluorescent nucleotides with different length linkers. *Nucl. Acids Res.*, 1994, 22, 3418–3422.
162. Yu, H., Chao, J., Patek, D., Mujumdar, R., Mujumdar, S., and Waggoner, A.S. Cyanine dye dUTP analogs for enzymatic labeling of DNA probes. *Nucl. Acids Res.*, 1994, 22, 3226–3232.
163. Randolph, J.B. and Waggoner, A.S. Stability, specificity and fluorescence brightness of multiply labeled fluorescent DNA probes. *Nucl. Acids Res.*, 1997, 25, 2923–2929.
164. Haralambidis, J., Chai, M., and Tregear, G.W. Preparation of base-modified nucleosides suitable for non-radioactive label attachment and their incorporation into synthetic oligodeoxyribonucleotides. *Nucl. Acids Res.*, 1987, 15, 4857–4876.
165. Cook, A.F., Vuocolo, E., and Brakel, C.L. Synthesis and hybridization of a series of biotinylated nucleotides. *Nucl. Acids Res.*, 1988, 16, 4077–4095.
166. Holtke, H.-J., Seibl, R., Burg, J., Meuhlegger, K., and Kessler, C. Nonradioactive detection of nucleic acids. II. Optimization of the digoxignenin system. *Biol. Chem. Hoppe–Seyler*, 1990 371, 929–938.
167. Chollet, A. and Kawashima, E.H. Biotin-labeled synthetic oligodeoxyribonucleotides: chemical synthesis and uses as hybridization probes. *Nucl. Acids Res.*, 1985, 13, 1529–1541.

168. Wallner, G., Amann, R., and Beisker, W. Optimizing fluorescent *in situ* hybridization with rRNA-targeted oligonucleotide probes for flow cytometric identification of microorganisms. *Cytometry*, 1993, 14, 136–143.
169. Dorris, D.R., Ramakrishnan, R., Trakas, D., Dudzik, F., Belval, R., Zhao, C., Nguyen, A., Domanus, M., and Mazumder, A. A highly reproducible, linear, and automated sample preparation method for DNA microarrays. *Genome Res.*, 2002, 12, 976–984.
170. Harding, J.D., Gebeyehu, G., Laskin, R., and Haces, A. Synthesis of fluorescence-labeled nucleic acids. PCT Int. Appl. 1993, WO9319206.
171. Augustin, M.A., Ankenbauer, W., and Angerer, B. Progress towards single-molecule sequencing: enzymatic synthesis of nucleotide-specifically labeled DNA. *J. Biotechnol.*, 2001, 86, 289–301.
172. Tasara, T., Angerer, B., Damond, M., Winter, H., Dörhöfer, S., Hübscher, U., and Amacker, M. Incorporation of reporter molecule-labeled nucleotides by DNA polymerases. II. High-density labeling of natural DNA. *Nucl. Acids Res.*, 2003, 31, 2636–2646.
173. Brakmann, S. and Nieckchen, P. Large fragment of *E. coli* DNA polymerase I can synthesize DNA exclusively from fluorescently labeled nucleotides. *ChemBioChem*, 2001, 2, 773–777.
174. Brakmann, S. and Löbermann, S. High-density labeling of DNA: preparation and characterization of the target material for single-molecule sequencing. *Angew Chem. Int. Ed.*, 2001, 40, 1427–1429.
175. Pullman, A. and Pullman, B. Molecular electrostatic potential of the nucleic acids. *Q. Rev. Biophys.*, 1981, 14, 289–380.
176. Seela, F. and Zulauf, M. Synthesis of 7-alkynylated 8-aza-7-deaza-2′-deoxyadenosines via the Pd-catalyzed cross-coupling reaction. *J. Chem. Soc. Perkin Trans.*, I 1998, 3233–3240.
177. Kanaya, E.N., Howard, F.B., Frazier, J., and Miles, H.T. Poly(2-amino-8-methyldeoxyadenylic acid): contrasting effects in deoxy- and ribopolynucleotides of 2-amino and 8-methyl substituents. *Biochemistry*, 1987, 26, 7159.
178. Birnbaum, G.I., Hruska, F.E., and Niemczura, W.P. A pyrimidine nucleoside constrained in the syn form. Structure and conformation of 6-methyl-2′-deoxyuridine. *J. Am. Chem. Soc.*, 1980, 102, 5586–5590.
179. Sugiyama, H., Kawai, K., Matsunaga, A., Fujimoto, K., Saito, I., Robinson, H., and Wang, A.H. Synthesis, structure and thermodynamic properties of 8-methylguanine-containing oligonucleotides: Z-DNA under physiological salt conditions. *Nucl. Acids. Res.*, 1996, 24, 1272–1278.
180. Dias, E., Battiste, J.L., and Williamson, J.R. Chemical probe for glycosidic conformation in telomeric DNAs. *J. Am. Chem. Soc.*, 1994, 116, 4479–4480.
181. Seela, F., Chen, Y., and Mittelbach, C. The effects of 7-deaza-8-methylguanine on DNA structure and stability. *Helv. Chim. Acta*, 1998, 81, 570–583.
182. Graham, D., Parkinson, J.A., and Brown, T. DNA duplexes stabilized by modified monomer residues: synthesis and stability. *J. Chem. Soc. Perkin Trans.*, I. 1998, 1131.
183. Eason, R.G., Burkhardt, D.M., Phillips, S.J., Smith, D.P., and David, S.S. Synthesis and characterization of 8-methoxy-2′-deoxyadenosine-containing oligonucleotides to probe the syn glycosidic conformation of 2′-deoxyadenosine within DNA. *Nucl. Acids Res.*, 1996, 24, 890–897.

184. Telser, J., Cruickshank, K.A., Morrison, L.E., and Netzel, T.L. Synthesis and characterization of DNA oligomers and duplexes containing covalently attached molecular labels: comparison of biotin, fluorescein, and pyrene labels by thermodynamic and optical spectroscopic measurements. *J. Am. Chem. Soc.*, 1989, 111, 6966–6976.
185. Miller, P.A. and Cushman, C.D. Selective modification of cytosines in oligodeoxyribonucleotides. *Bioconjugate Chem.*, 1992, 3, 74–79.
186. Flickinger, J.L., Gebeyehu, G., Buchman, G., Haces, A., and Rashtchian, A. Differential incorporation of biotinylated nucleotides by terminal deoxynucleotidyl transferase. *Nucl. Acids Res.*, 1992, 20, 2382.
187. Pelletier, H., Sawaya, M.R., Kumar, A., Wilson, S.H., and Kraut, J. Structures of ternary complexes of rat DNA polymerase β, a DNA template-primer, and ddCTP. *Science*, 1994, 264, 1891–1903.
188. Chidgeavadze, Z.G., Beabealashvilli, R., Kreyevsky, A.A., and Kukhanova, M.K. Nucleoside 5′-triphosphates with modified sugars as substrates for DNA polymerases. *Biochem. Biophys. Acta*, 1986, 868, 145–152.
189. Beabealashvilli, R., Scamrov, A.V., Kutateladze, T.V., Mazo, A.M., Krayevsky, A.A., and Kukhanova, M.K. Nucleoside 5′-triphosphates modified at sugar residues as substrates for calf thymus terminal deoxynucleotidyl transferase and for AMV reverse transcriptase. *Biochem. Biophys. Acta*, 1986, 868, 136–144.
190. Kutateladze, T., Beabealashvilli, R., Aleksandrova, L.A., Obukhov, A.G., and Kraevskii, A.A. Analogs of nucleoside triphosphates with a modified carbohydrate residue as substrates for RNA polymerase. *Molek. Biol.* (Moscow), 1986, 20, 267–276.
191. Faulstich, K., Herrlein, M.K., Brandtner, S., Konrad, R.E., Griesinger, C., and Engels, J.W. A new class of 3′-modified nucleotides as chain terminators for DNA sequencing. *Bioorg. Med. Chem. Lett.*, 1994, 4, 1975–1978.
192. Metzker, M.L., Raghavachari, R., Burgess, K., and Gibbs, R.A. Elimination of residual natural nucleotides from 3′-0-modified-dNTP syntheses by enzymatic mop-up. *BioTechniques*, 1998, 25, 814–817.
193. Canard, B., Cardona, B., and Sarfati, R.S. Catalytic editing properties of DNA polymerases. *Proc. Natl. Acad. Sci. USA*, 1995, 92, 10859–10863.
194. Sabesan, S. and Trainor, G.L. Dideoxyfructonucleosides and deoxyfructonucleotides. US Patent, 1994, 5, 276,143.
195. Tunitskaya, V.L., Rusakova, E.E., Padyukova, N.S., Ermolinsky, B.S., Chernyi, A.A., Kochetkov, S.N., Lysov, Y.P., and Mikhailov, S.N. Substrate properties of C′-methyl UTP derivatives in T7 RNA polymerase reactions. Evidence for N-type NTP conformation. *FEBS Lett.*, 1997, 400(3), 263–266.
196. Mikhailov, S.N., Padyukova, N.Sh., Lysov, Y.P., Savochkina, L.P., Chidgivadze, Z.G., and Bibilashvili, R.S. Substrate properties of C′-methylnucleoside and C′-methyl-2′-deoxynucleoside 5′-triphosphates in RNA and DNA synthesis reactions catalyzed by RNA and DNA polymerases. *Nucleosides Nucleotides*, 1991, 10, 339–343.
197. Muehlegger, K., Batz, H.-G., Bohm, S., Von der Eltz, H., Holtke, H.-J., and Kessler, C. Synthesis and use of new digoxigenein-labeled nucleotides in nonradioactive labeling and detection of nucleic acids. *Nucleosides Nucleotides*, 1989, 8, 1161–1163.

198. Livak, K.J., Hobbs, F.W., and Zagursky, R.J. Detection of single base differences using biotinylated nucleotides with very long linker arms. *Nucl. Acids Res.*, 1992, 20, 4831–4837.
199. Brigati, D.J., Myerson, D., Leary, J.J., Spalholz, B., Travis, S.Z., Fong, C.K.Y., Hsiung, G.D., and Ward, D.C. Detection of viral genomes in cultured cells and paraffin-embedded tissue sections using biotin-labeled hybridization probes. *Virology*, 1983, 126, 32–50.
200. McClean, J., Davison, A., Rao, M. Vaman, and Brown, T. Antibody-mediated detection and physical properties of oligonucleotides labelled with multiple internal and terminal 2,4-dinitrophenyl groups. *Biomed. Peptides Proteins Nucl. Acids*, 1996, 2, 1–6.
201. Pauly, G.T., Bobst, E.V., Bruckman, D., and Bobst, A.M. Template-dependent incorporation of spin-labeled thymidine analogs into viral DNA. *Helv. Chim. Acta*, 1989, 72, 110–116.
202. Godovikova, T.S., Kolpashchikov, D.M., Orlova, T.N., Richter, V.A., Ivanova, T.M., Grochovsky, S.L., Nasedkina, T.V., Victorova, L.S., and Poletaev, A.I. 5-[3-(E)-(4-Azido-2356-tetrafluorobenzamido)propenyl-1]-2′-deoxy- uridine-5′-triphosphate substitutes for thymidine-5′-triphosphate in the polymerase chain reaction. *Bioconjugate Chem.*, 1999, 10, 529–537.
203. Lee, S.E., Sidorov, A., Gourlain, T., Mignet, N., Thorpe, S.J., Brazier, J.A., Dickman, M.J., Hornby, D.P., Grasby, J.A., and Williams, D.M. Enhancing the catalytic repertoire of nucleic acids: a systematic study of linker length and rigidity. *Nucl. Acids Res.*, 2001, 29, 1565–1573.
204. Gourlain, T., Sidorov, A., Mignet, N., Thorpe, S.J., Lee, S.E., Grasby, J.A., and Williams, D.M. Enhancing the catalytic repertoire of nucleic acids II Simultaneous incorporation of amino and imidazolyl functionalities by two modified triphosphates during PCR. *Nucl. Acids Res.*, 2001, 29, 1898–1905.
205. Ruth, J.L. Oligodeoxynucleotides with reporter groups attached to the base. In: Eckstein, F., Ed, *Oligonucleotides and Analogs*. Oxford: IRL Press, 1991, 255–282.
206. Sawai, H., Ozaki–Nakamura, A., Mine, M., and Ozaki, H. Synthesis of new modified DNAs by hyperthermophilic DNA polymerase: substrate and template specificity of functionalized thymidine analogues bearing an sp3-hybridized carbon at the C5-position for several DNA polymerases. *Bioconjugate Chem.*, 2002, 13, 309–316.
207. Froehler, B.C., Wadwani, S., Terhorst, T.J., and Gerrard, S.R. Oligonucleotides containing C-5 propyne analogs of 2′-deoxyuridine and 2′-deoxycytidine. *Tetrahedron Lett.*, 1992, 33, 5307–5310.
208. Buhr, C.A., Wagner, R.W., Grant, D., and Froehler, B.C. Oligodeoxynucleotides containing C-7 propyne analogs of 7-deaza-2′- deoxyguanosine and 7-deaza-2′-deoxyadenosine. *Nucl. Acids Res.*, 1996, 24, 2974–2980.
209. He, J. and Seela, F. Propynyl groups in duplex DNA: stability of base pairs incorporating 7-substituted 8-aza-7-deazapurines or 5-substituted pyrimidines. *Nucl. Acids Res.*, 2002, 30, 5485–5496.
210. Giller, G., Tasara, T., Angerer, B., Mühlegger, K., Amacker, M., and Winter, H. Incorporation of reporter molecule-labeled nucleotides by DNA polymerases. I. Chemical synthesis of various reporter group-labeled 2′-deoxyribonucleoside-5′-triphosphates. *Nucl. Acids Res.*, 2003, 31, 2630–2635.

211. Obayashi, T., Masud, M.M., Ozaki, A.N., Ozaki, H., Kuwahara, M., and Sawai, H. Enzymatic synthesis of labeled DNA by PCR using new fluorescent thymidine nucleotide analogue and superthermophilic KOD dash DNA polymerase. *Bioorg. Med. Chem. Lett.*, 2002, 12, 1167–1170.
212. Forghani, B., Yu, G.J., and Hurst, J.W. Comparison of biotinylated DNA and RNA probes for rapid detection of varicella–zoster virus genome by *in-situ* hybridization. J. *Clin. Microbiol.*, 1991, 29, 583–591.
213. Richter, A., Schwager, C., Hentze, S., Ansorge, W., Hentze, M.W., and Muckenthaler, M. Comparison of fluorescent tag DNA labeling methods used for expression analysis by DNA microarrays. *BioTechniques*, 2002, 33, 620–628.
214. Yu, J., Othman, M.I., Farjo, R., Zareparsi, S., MacNee, S.P., Yoshida, S., and Swaroop, A. Evaluation and optimization of procedures for target labeling and hybridization of cDNA microarrays. *Molec. Vision*, 2002, 8, 130–137.
215. Xiang, C.C., Kozhich, O.A., Chen, M., Inman, J.M., Phan, Q.N., Chen, Y., and Brownstein, M.J. Amine-modified random primers to label probes for DNA microarrays. *Nat. Biotechnol.*, 2002, 20, 738–742.
216. Cox, W.G. and Singer, V.L. Fluorescent DNA hybridization probe preparation using amine modification and reactive dye coupling. *BioTechniques*, 2004, 36, 114–122.
217. 't Hoen, P.A.C., de Kort, F., van Ommen, G.J.B., and den Dunnen, J.T. Fluorescent labeling of cRNA for microarray applications. *Nucl. Acids Res.*, 2003, 31, e20.
218. Short JM. Modified nucleotides and methods useful for nucleic acid sequencing. PCT Int. Appl. 1999, WO9949082.
219. Sperling, J. and Medalia, O. Metal cluster containing nucleotides and nucleic acids and intermediates therefor. PCT Int. Appl. 2001, WO0120017.
220. Held, H.A., Roychowdhury, A., and Benner, S.A. C-5 Modified nucleosides: direct insertion of alkynyl-thio functionality in pyrimidines. *Nucleosides Nucleotides*, 2003, 22, 391–404.
221. Nampalli, S., McDougall, M.G., Lavrenov, K., Xiao, H., and Kumar, S. Utility of thiol-cross-linked fluorescent dye labeled terminators for DNA sequencing. *Bioconjugate Chem.*, 2002, 13, 468–473.
222. Trevisiol, E., Defranq, E., Lhomme, J., Laayoun, A., and Cros, P. Synthesis of nucleoside triphosphates that contain an aminooxy function for "post-amplification labeling." *Eur. J. Org. Chem.*, 2000, 211–217.
223. Trevisiol, E., Defranq, E., Lhomme, J., Laayoun, A., and Cros, P. Synthesis of methylketone containing nucleoside triphophates for RNA labeling. *Tetrahedron*, 2000, 56, 6501–6510.
224. Gelfand, D.H., Kalman, L.V., Myers, T.W., Reichert, F.L., and Sigua, C.L. Analogs of thermostable DNA polymerases with broadened substrate specificities for DNA sequencing with dye-labeled nucleotides. Eur. Pat. Appl. 1999, EP902035.
225. Goodman, M.F. and Reha–Krantz, L. Synthesis of fluorophore-labeled DNA. PCT Int. Appl. 1996, WO9739150.
226. Gruber, C. and Shih, P.-J. FluoroScript RT: a superior method to produce fluorescently labeled cDNA targets for microarray screening. *Focus* (Invitrogen), 2002, 24, 2–5.
227. Wrobel, G., Schlingemann, J., Hummerich, L., Kramer, J., Lichter, P., and Hahn, M. Optimization of high-density cDNA-microarray protocols by design of experiments. *Nucl. Acids Res.*, 2003, 31, e67.

228. Tabor, S. and Richardson, C.C. DNA sequence analysis with a modified bacteriophage T7 DNA polymerase. Effect of pyrophosphorolysis and metal ions. *J. Biol. Chem.*, 1990, 265, 8322–8328.
229. Tabor, S. and Richardson, C.C. A single residue in DNA Polymerases of the *Escherichia coli* DNA polymerase I family is critical for distinguishing between deoxy- and dideoxyribonucleotides. *Proc. Natl. Acad. Sci. USA*, 1995, 92, 6339–6343.
230. Reeve, M.A. and Fuller, C.W. A thermostable DNA polymerase for DNA sequencing. *Nature*, 1995, 376, 796–797.
231. Craxton, M. Linear amplification sequencing: a powerful method for sequencing DNA. *Methods* , 1991, 3, 20–26.
232. Lee, L.G., Connell, C.R., Woo, S.L., Cheng, R.D., McArdle, B.F., Fuller, C.W., Halloran, N.D., and Wilson, R.K. DNA sequencing with dye-labeled terminators and T7 DNA polymerase: effect of dyes and dNTPs on incorporation of dye-terminators and probability analysis of termination fragments. *Nucl. Acids Res.*, 1992, 20, 2471–2483.
233. Bergot, J.B., Chakerian, V., Connell, C.R., Eadie, J.S., Fung, S., Hershey, N.D., Lee, L.G., Menchen, S.M., and Woo, S.L. Spectrally resolvable rhodamine dyes for nucleic acid sequence determination. US Patent, 1994, 5, 366, 860.
234. Brandis, J.W. Dye structure affects *Taq* DNA polymerase terminator selectivity. *Nucl. Acids Res.*, 1999, 25, 4500–4504.
235. Rosenblum, B.B., Lee, L.G., Spurgeon, S.L. Khan, S.H., Menchen, S.M., Heiner, C.R., and Chen, S.M. New dye-labeled terminators for improved DNA sequencing patterns. *Nucl. Acids Res.*, 1997, 25, 4500–4504.
236. Nampalli, S., Khot, M., and Kumar, S. Fluorescence resonance energy transfer terminators for DNA sequencing. *Tetrahedron Lett.*, 2000, 41, 8867–8871.
237. Kumar, S., Nampalli, S., Finn, P.J., Rao, T.S., Chen, C.-Y., Flick, P.K., and Fuller, C.W. Synthesis and ET efficiency of FRET terminators derived from different linkers. *Nucleosides Nucleotides*, 2003, 22, 1595–1598.
238. Gardner, A.F. and Jack, W.E. Determinatnts of nucleotide sugar recognition in an archaeon DNA polymerase. *Nucl. Acids Res.*, 1999, 27, 2545–2553.
239. Trainor, G.L. DNA sequencing method using acyclonucleoside triphosphates. US Patent, 1996, US 5,558,991.
240. Gardner, A.F. and Jack, W.E. Acyclic and dideoxy terminator preferences denote divergent sugar recognition by archaeon and *Taq* DNA polymerases. *Nucl. Acids Res.*, 2002, 30, 605–613.
241. Gardner, A.F., Joyce, C.M., and Jack, W.E. Comparative kinetics of nucleotide analog incorporation by vent DNA polymerase. *J. Biol. Chem.*, 2004, 279, 11834–11842.
242. Evans, S.J., Fogg, M.J., Mamone, A., Davis, M., Pearl, L.H., and Connolly, B.A. Improving dideoxynucleotide-triphosphate utilisation by the hyper-thermophilic DNA polymerase from the archaeon *Pyrococcus furiosus*. *Nucl. Acids Res.*, 2000, 28, 1059–1066.
243. Arezi, B., Hansen, C.J., and Hogrefe, H.H. Efficient and high fidelity incorporation of dye-terminators by a novel archaeal DNA polymerase mutant. *J. Molec. Biol.*, 2002, 322, 719–729.
244. Hamilton, S.C., Farchaus, J.W., and Davis, M.C. DNA polymerases as engines for biotechnology. *BioTechniques*, 2001, 31, 370–383.

245. Sasaki, N., Izawa, M., Watahiki, M., Ozawa, K., Tanaka, T., Yoneda, Y., Matsuura, S., Carninci, P., Muramatsu, M., Okazaki, Y., and Hayashizaki, Y. Transcriptional sequencing: a method for DNA sequencing using RNA polymerase. *Proc. Natl. Acad. Sci. USA*, 1998, 95, 5455–5460.
246. Shibata, K. et al. RIKEN integrated sequence analysis (RISA) system-384-format sequencing pipeline with 384 multicapillary sequencer. *Genome Res.*, 2000, 10, 1757–1771.
247. Gelfand, D.H., Kalman, L.V., and Reichert, F.L. Thermostable DNA polymerases having reduced discrimination against ribo-NTPs. US Patent, 1999, US 5,939,292.
248. Fisher, P.V., Vatta, P., and Khan, S.H. Dye-labeled ribonucleotide triphosphates. PCT Int. Appl. 2003, WO0300841.
249. Olejnik, J., Krzymańska–Olejnik, E., and Rothschild, K.J. Photocleavable affinity tags for isolation and detection of biomolecules. *Methods Enzymol.*, 1998, 291, 135–154.
250. Olejnik, J., Sonar, S.M., Krzymanska–Olejnik, E., and Rothschild, K.J. Photocleavable biotin derivatives: a versatile approach for the isolation of biomolecules. *Proc. Natl. Acad. Sci. USA*, 1995, 92, 7590–7594.
251. Kricka, L.J. Stains labels and detections strategies for nucleic acids assays. *Ann. Clin. Biochem.*, 2002, 39, 114–129.
252. Van Ness, J. Present and future of tags and labels in high throughput genomic analysis. *Functional Genomics Ser.*, 2002, 1 (Genomic Technologies), 125–161.
253. Manuelidis, L., Langer–Safer, P.R., and Ward, D.C. High-resolution mapping of satellite DNA using biotin-labeled DNA probes. *J. Cell Biol.*, 1982, 95, 619–625.
254. Moens, P.B. and Pearlman, R.E. *Chromosoma*, 1989 , 98, 287–294.
255. Waddell, E., Wang, Y., Stryjewski, W., McWhorter, S., Henry, A.C., Evans, D., McCarley, R.L., and Soper, S.A. *Anal. Chem.*, 2000, 72, 5907–5917.
256. Turro, C., Fu, P.K.-L., and Bradley, P.M. In: Sigel, A. and Sigel, H., Eds. *The Lanthanides and Their Interrelations with Biosystems*. In: Vol. 40. *Metal Ions in Biological Systems*. Marcel Dekker: New York, 2002, 324–353.
257. Lopez, E., Chypre, C., Alpha, B., and Mathis, G. Europium(III) trisbipyridine cryptate label for time-resolved fluorescence detection of polymerase chain reaction products fixed on a solid support. *Clin. Chem.*, 1993, 39, 196–201.
258. Chee, M., Yang, R., Hubbell, E., Berno, A., Huang, X.C., Stern, D., Winkler, J., Lockhart, D.J., Morris, M.S., and Fodor, S.P.A. Accessing genetic information with high-density DNA arrays. *Science*, 1996, 274, 610–614.
259. Taylor, J.R., Fang, M.M., and Nie, S. Probing specific sequences on single dna molecules with bioconjugated fluorescent nanoparticles. *Anal. Chem.*, 2000, 72, 1979–1986.
260. Gerion, D., Chen, F., Kannan, B., Fu, A., Parak, W.J., Chen, D.J., Majumdar, A., and Alivisatos, A.P. Room-temperature single-nucleotide polymorphism and multiallele DNA detection using fluorescent nanocrystals and microarrays. *Anal. Chem.*, 2003, 75, 4766–4772.
261. Gudibande, S.R., Kenten, J.H., Link, J., Friedman, K., and Massey, R.J. Rapid nonseparation electrochemiluminescent DNA hybridization assays for PCR products using 3′-labeled oligonucleotide probes. *Molec. Cell. Probes*, 1992, 6, 495–503.

262. Stults, N.L., Stocks, N.F., Rivera, H., Gray, J., McCann, R.O., O'Kane, D., Cummings, R.D., Cormier, M.J., and Smith, D.F. Use of recombinant biotinylated aequorin in microtiter and membrane-based assays: purification of recombinant apoaequorin from *Escherichia coli*. *Biochemistry*, 1992, 31, 1433–1442.
263. Galvan, B. and Christopoulos, T.K. Bioluminescence hybridization assays using recombinant aequorin. Application to the detection of prostate-specific antigen mRNA. *Anal. Chem.*, 1996, 68, 3545–3550.
264. Laios, E., Ioannou, P.C., and Christopoulos, T.K. Enzyme-amplified aequorin-based bioluminometric hybridization assays. *Anal. Chem.*, 2001, 73, 689–692.
265. Tomlinson, S., Lyga, A., Huguenel, E., and Dattagupta, N. Detection of biotinylated nucleic acid hybrids by antibody-coated gold colloid. *Anal. Biochem.*, (1988), 171(1), 217–222.
266. Hutchison, N.J., Langer–Safer, P.R., Ward, D.C., and Hamkalo, B.A. *In situ* hybridization at the electron microscope level: hybrid detection by autoradiography and colloidal gold. *J. Cell Biol.*, 1982, 95, 609–618.
267. Hacker, G.W., Hauser–Kronberger, C., Zehbe, I., Su, H., Schiechl, A., Dietze, O., and Tubbs, R. *In-situ* localization of DNA and RNA sequences: super-sensitive *in-situ* hybridization using streptavidin-Nanogold™-silver staining: minireview, protocols and possible applications. *Cell Vision*, 1997, 4, 54–65.
268. Schofer, C., Weipoltshammer, K., Hauser–Kronberger, C., and Wachtler, F. High-resolution detection of nucleic acids at the electron microscopic level — review of *in-situ* hybridization technology, the use of gold, and catalysed reporter deposition. *Cell Vision*, 1997, 4, 443–454.
269. Stimpson, D.I., Hoijer, J.V., Hsieh, W., Jou, C., Gordon, J., Theriault, T., Gamble, R., and Baldeschwieler, J.D. Real-time detection of DNA hybridization and melting on oligonucleotide arrays using optical wave guides. *Proc. Natl. Acad. Sci. USA*, 1995, 92, 6379–6383.
270. Yguerabide, J. and Yguerabide, E.E. Resonance light scattering particles as ultrasensitive labels for detection of analytes in a wide range of applications. *J. Cell Biochem.*, Suppl, 2001, 37, 71–81.
271. Bao, P., Frutos, A.G., Greef, C., Lahiri, J., Muller, U., Peterson, T.C., Warden, L., and Xie, X. High-sensitivity detection of DNA hybridization on microarrays using resonance light scattering. *Anal. Chem.*, 2002, 74, 1792–1797.
272. Beverloo, H.B., Schadewijk, A. van, Zijlmans, H.J.M.A.A., and Tanke, H.J. Immunochemical detection of proteins and nucleic acids on filters using small luminescent inorganic crystals as marker. *Anal. Biochem.*, 1992, 203, 326–334.
273. Rijke, F.V.D., Zijlmans, H., Li, S., Vail, T., Raap, A.K., Niedbala, R.S., and Tanke, H.J. Up-converting phosphor reporters for nucleic acid microarrays. *Nat. Biotechnol.*, 2001, 19, 273–276.
274. Miller, M.M., Sheehan, P.E., Edelstein, R.L., Tamanaha, C.R., Zhong, L., Bounnack, S., Whitman, L.J., and Colton, R.J. A DNA sensor utilizing magnetic microbeads and magnetoelectronic detection. *J. Mag. Mag. Mater.*, 2001, 225, 138–44.
275. Mathews, J.A., Batki, A., Hynds, C., and Kricka, L.J. Enhanced chemiluminescent method for the detection of DNA dot-hybridization assays. *Anal. Biochem.*, 1985, 151, 205–209.
276. Stone, T. and Durrant, I. Enhanced chemiluminescence for the detection of membrane-bound nucleic acid sequences: advances of the Amersham system. *Genet. Anal.: Techn. Appl.*, 1991, 8, 230–237.

277. Wigle, D.A., Radakovic, N.N., Venance, S.L., and Pang, S.C. Enhanced chemiluminescence with catalyzed reporter deposition for increasing the sensitivity of Western blotting. *BioTechniques*, 1993, 14, 562–563.
278. Speel, E.J.M., Ramaekers, F.C.S., and Hopman, A.H.N. Sensitive multicolor fluorescence *in-situ* hybridization using catalyzed reporter deposition (CARD) amplification. *J. Histochem. Cytochem.*, 1997, 45, 1439–1446.
279. Cano, R.J., Klem, R.E., Palomares, J.C., and Torres, M.J. DNA hybridization assay using AttoPhos, a fluorescent substrate for alkaline phosphate. *BioTechniques*, 1992, 12, 264–269.
280. Maier, E., Crollius, H.R., and Lehrach, H. Hybridisation techniques on gridded high density DNA and *in-situ* colony filters based on fluorescence detection. *Nucl. Acids Res.*, 1994, 22, 3423–3424.
281. Paragas, V.B., Zhang, Y.-Z., Haugland, R.P., and Singer, V.L. The ELF-97 alkaline phosphatase substrate provides a bright photostable fluorescent signal amplification method for FISH. *J. Histochem. Cytochem.*, 1997, 45, 345–358.
282. Bronstein, I., Cate, R.I., Ramachandran, K.L., and Voyta, J.L. Chemiluminescent 12-dioxetane based enzyme sunstrates and their application in the detection of DNA. *Photochem. Photobiol.*, 1989, 49, 9–15.
283. Schaap, A.P., Akhavan, H., and Romano, L.J. Chemiluminescent substrates for alkaline phosphatase: application to ultrasensitive enzyme-linked immunoassays and DNA probes. *Clin. Chem.*, 1989, 35, 1863–1864.
284. Akhavan–Tafti, H., Lreddy, L.V., Siripurapu, S., Schoenfelner, B.A., Handley, R.S., and Schaap, A.P. Chemiluminescent detection of DNA in low- and medium-density arrays. *Clin. Chem.*, 1998. 44, 2065–2066.
285. Hauber, R., Miska, W., Schleinkofer, L., and Geiger, R. New sensitive radioactive-free bioluminescence-enhanced detection system in protein blotting and nucleic acid hybridization. *J. Biolum. Chemilum.*, 1989, 4, 367–372.
286. Dill, K., Chan, S.D.H., and Gibbs, T.W.J. Detection of plasmids using DNA and RNA probes and the light-addressable potentiometric sensor. *Biochem. Biophys. Methods*, 1997, 35, 197–202.
287. Belager, P., Terouamme, B., Boussioux, A.-M., and Nicolas, J.-C. Use of bioluminescence in nucleic acid hybridization reactions. *J. Biolum. Chemilum.*, 1989, 4, 302–309.
288. Nagata, Y., Yokota, H., Kosuda, O., Takemura, K., and Kikiuchi, T. Quantification of picogram levels of specific DNA immobilized in microtiter wells. *FEBS Lett.*, 1985, 183, 379–382.
289. Aleksandrova, L.A., Atrazhev, A.M., Kukhanova, M.K., and Kraevskii, A.A. Incorporation, into the DNA, chain of a fluorescent derivative of 2-deoxyuridylic acid during catalysis of synthesis by DNA-polymerase A. *Bioorg. Chem.*, 1987, 13, 643–647.
290. Toppin, C.R., Pauly, G.T., Devanesan, P., Kryak, D., and Bobst, A.M. Three novel spin-labeled substrates for enzymatic imcorporation into nucleic acid lattices. *Helv. Chim. Acta*, 1986, 69, 345–349.
291. Muehlegger, K., Huber, E., Von der Eltz, H., Rueger, R., and Kessler, C. Nonradioactive labeling and detection of nucleic acids IV synthesis and properties of digoxigenin-modified 2′-deoxyuridine-5′-triphosphates and a photoactivatable analog of digoxigenin (photodigoxigenin). *Biol. Chem. Hoppe–Seyler*, 1990, 371, 953–965.

292. Collasius, M. and Valet, G. New nucleoside tri:phosphoric acid ester derivatives used for preparation of polymerase chain reaction probes amplification of nucleic acid sequences and as chain terminators for sequences. Ger. Patent, 1992, DE 4119075.
293. Koecher, J., Springer, W., Kuckert, E., and Boecker, T. Sulpho-cumarin containing nucleotiden and their use in detection methods for nucleic acids. Eur. Pat. Appl. 1994, EP608737.
294. Evangelista, R.A., Liu, M.-S., Rampal, S., and Chen, F.-T.A. Characterization of fluorescent nucleoside triphosphates by capillary electrophoresis with laser-induced fluorescence detection, action of alkaline phosphatase and DNA polymerase. *Anal. Biochem.*, 1996, 235, 89–97.
295. Moller, U., Schubert, F., and Cech, D. Synthesis of methylene blue modified 2′-deoxyuridine-5′-triphosphate and its use in DNA labeling reactions. *Collect. Czech. Chem. Commun.*, 1996 61 S164–166.
296. van de Corput, M.P.C., Dirks, R.W., Wiegant, W.W., Wiegant, J., Mühlegger, K., and Raap, A.K. Oestradiol, a new hapten for detecting nucleic acid sequences by FISH. *Histochem. Cell Biol.*, 1997, 108, 359–364.
297. Stolowitz, M.S.l. and Kaiser, R.J. Novel boronic acid-containing nucleic acids and their use as diagnostic agents. PCT Int. Appl. 1998, WO9805672.
298. Wlassoff, W.A. and King, G.C. Ferrocene conjugates of dUTP for enzymatic redox labelling of DNA. *Nucl. Acids Res.*, 2002, 30, e581–587.
299. Alpha–Bazin, B., Bazin, H., Guellemer, S., Sauvaigo, S., and Mathis, G. Europium cryptate labeled deoxyuridine-triphosphate analog: synthesis and enzymatic incorporation. Nucleosides and nucleotides. *Nucl. Acids*, 2000, 19, 1463–1464.
300. Alpha–Bazin, B., Bazin, H., Boissey, L., and Mathis, G. Europium cryptate-tethered ribonucleotide for the labeling of RNA and its detection by time-resolved amplification of cryptate emission. *Anal. Biochem.*, 2000, 286, 17–25.
301. De Haas, R.R., van Gijlswijk, R.P., van der Tol, B., Veuskens, J., van Gijssel, H.E., Tijdens, R.B., Bonnet, J., Verwoerd, N.P., and Tanke, H.J. Phosphorescent platinum palladium coproporphyrins for time-resolved luminescence microscopy. *J. Histochem. Cytochem.*, 1999, 47, 183–196.
302. Sakthivel, K. and Barbas, C.F. Expanding the potential for binding and catalysis: highly functionalized dUTP derivatives that are substrates for thermostable DNA polymerases. *Bioconjugate Chem.*, 2001, 12, 396–405.
303. Bartholomew, B., Kassavetis, G.A., Braun, B.R., and Geiduschek, E.P. The subunit structure of *Saccharomyces cerevisiae* transcription factor IIIC probed with a novel photocrosslinking reagent. *EMBO J.*, 1990, 9, 2197–2205.
304. Persinger, J. and Bartholomew, B. Mapping the contacts of yeast TFIIIB and RNA polymerase III at various distances from the major groove of DNA by DNA photoaffinity labeling. *J. Biol. Chem.*, 1996, 271, 33039–33046.
305. Tate, J.J., Persinger, J., and Bartholomew, B. Survey of four different photoreactive moieties for DNA photoaffinity labeling of yeast RNA polymerase III transcription complexes. *Nucl. Acids Res.*, 1998, 26, 1421–1426.
306. Zofall, M. and Bartholomew, B. Two novel dATP analogs for DNA photoaffinity labeling. *Nucl. Acids Res.*, 2000, 28, 4382–4390.
307. Shimkus, M.L., Guaglianone, P., and Herman, T.M. Synthesis and characterization of biotin-labeled nucleotide analogs. *DNA*, 1986, 5, 247–255.
308. Herman, T. and Fenn, B.J. Chemically cleavable biotin-labeled nucleotide analogs. *Methods Enzymol.*, 1990, 184, 577–583.

309. Barstow, D.A., Garman, A.J., and Parker, J.R. Nonisotopic nucleic acid labeling method. PCT Int. Appl. 1992, WO9200989.
310. Bai, X., Kim, S., Li, Z., Turro, N.J., and Ju, J. Design and synthesis of a photocleavable biotinylated nucleotide for DNA analysis by mass spectrometry. *Nucl Acids Res.*, 2004, 32, 535–541.
311. Prober, J.M., Dam, R.J., Robertson, C.W., Hobbs, F.W., and Trainor, G.L. Method, system and reagents for DNA sequencing. Eur. Pat. Appl. 1987, EP0252683.
312. Metzger, M.L. and Gibbs, R.A.K. Alternative dye-labeled primers, ribonucleotides, deoxyribonucleotides, and dideoxyribonucleotides for automated analysis and homogeneous amplificationdetection assays. PCT Int. Appl. 1997, WO9700967
313. Menchen, S.M., Benson, S.C., Rosenblum, B.B., and Khan, S.H. Mobility-modified cyanine dyes. PCT Int. Appl. 2001, WO0149790.
314. Menchen, S.M., Benson, S.C., Lam, J.Y.L., Sun, D., Rosenblum, B.B., Khan, S.H., and Taing, M. Sulfonated diarylrhodamine dyes. PCT Int. Appl. 2002, WO0244416.
315. Finn, P.J., Sun, L., Nampalli, S., Xiao, H., Nelson, J.R., Mamone, J.A., Grossmann, G., Flick, P.K., Fuller, C.W., and Kumar, S. Synthesis and application of charge-modified dye-labeled dideoxynucleoside-5′-triphosphates to "direct-load" DNA sequencing. *Nucl. Acids Res.*, 2002, 30, 2877–2885.
316. Kumar, S., Flick, P.K., Nelson, J.R., Finn, P.J., Nampalli, S., and Bull, M. Charge-modified nucleic acid terminators. PCT Int. Appl. 2001, WO0119841.
317. Roemer, S.C., Johnson, C.M., Boveia, V.R., Buzby, P.R., DiMeo, J.J., Draney, D., Narayanan, N., and Olive, D.M. Novel fluorescent nonnucleoside triphosphates as terminators of enzyme-directed DNA synthesis. *Proc. SPIE*, 2001, 4264 (Genomics Proteomics Technol.), 1–8.
318. Uchida, T., Ishikawa, H., Koyama, K. Kurano, Y., and Uchida, Y. Phenytoin derivatives. Eur. Patent Appl. 1992, EP471338.
319. Kornher, J.S. and Livak, K.J. Mutation detection using nucleotide analogs that alter electrophoretic mobility. *Nucl. Acids Res.*, 1989, 17, 7779–7784.
320. Braman, J.C., Moores, J.C., Anderson, J.D., and Huang, H. Porphyrin labeling of polynucleotides. PCT Int. Appl. 1996, WO9611937.
321. Matjulic–Adamic, J., Daniher, A.T., Karpeisky, A., Haeberli, P., Sweedler, D., and Beigelman, L. Functionalized nucleoside 5′-triphosphates for *in vitro* selection of new catalytic ribonucleic acids. *Bioorg. Med. Chem. Lett.*, 2000, 10, 1299–1302.
322. Anne, A., Blanc, B., and Moiroux, J. Synthesis of the first ferrocenyl-labeled dideoxynucleotide and itsuse for 3′-redox end-labeling of 5′-modified single-stranded oligonucleotides. *Bioconjugate Chem.*, 2001, 12, 396–405.
323. Koster, H. DNA sequencing by mass spectrometry. US Patent, 2000, 5,547,835.
324. Koster, H. DNA sequencing by mass spectrometry via exonuclease degradation. US Patent, 2000, 6,074,823.
325. Held, H.A., Roychowdhury, A., and Benner, S.A. C-5 modified nucleosides: direct insertion of alkynyl–thio functionality in pyrimidines. *Nucleosides Nucleotides*, 2003, 22, 391–404.
326. Kumar, A., Tchen, P., Roullet, F., and Cohen, J. Nonradioactive labeling of synthetic oligonucleotide probes with terminal deoxynucleotidyl transferase. *Anal. Biochem.*, 1988, 169, 376–82.
327. Tyagi, S.C. Spin-labeled nucleotide substrates for DNA-dependent RNA polymerase from *Escherichia coli*. *J. Biol. Chem.*, 1991, 266, 17936–17940.

328. Muhlegger, K., Eltz, H., Seela, F., and Rosemeyer, H. C-nucleosides derivatives and their use in the detection of nucleic acids. US Patent, 2001, 6,174,998.
329. Muhlegger, K. and Von der Eltz, H. Heterocyclic compounds and their use for isolating nucleic acids. PCT Int. Appl. 1998, WO9811104
330. Muhlegger, K. and Von der Eltz, H. Heterocyclic compounds and their use for isolating nucleic acids. US Patent, 2002, US 6,403,786.
331. Smith, C., Cummins, W.J., and Nairne, R.J.D. Nucleoside analogs. PCT Int. Appl. 2001, WO0116150.
332. Brown, D., Hamilton, A., Loakes, D., Simmonds, A.C., and Smith, C. Nucleoside analogues. US Patent, 2001, 6,313,286.
333. Strobel, O.K., Bobst, E.V., and Bobst, A.M. Nick translation of l phage DNA with a deoxycytidine analog spin labeled in the 5-position. *Arch. Biochem. Biophys.*, 1989, 273, 597–601.
334. Strobel, O.K., Keyes, R.S., and Bobst, A.M. Base dynamics of local Z–DNA-conformations as detected by electron paramagnetic resonance with spin-labeled deoxycytidine analogues. *Biochemistry,* 1990, 29, 8522–8528.
335. Khan, S.H., Rosenblum, B.B., Zhen, W., and Menchen, S.M. Fluorescent nucleobase conjugates having anionic linkers. PCT Int. Appl. 2002, WO0230944.
336. Khan, S.H., Menchen, S.M., and Rosenblum, B.B. Propargylethoxyamino nucleotides. US Patent 1998, 5,821,356.
337. Khan, S.H. Menchen, S.M., and Rosenblum, B.B. Substituted propargylethoxyamido nucleosides, oligonucleotides and methods for using same. US Patent 1998, 5,770,716.
338. Khan, S.H., Rosenblum, B.B., Zhen, W., and Menchen, S.M. Nucleotide compounds including a rigid linker. US Patent 1999, 5,948,648.
339. Hyashizaki, Y., Ozawa, K., Fujio, K., and Tanaka, T. 3′-Deoxyribonucleotide derivatives PCT Int. Appl. 1999, WO9902543.
340. Lannutti, B.J., Persinger, J., and Bartholomew, B. Probing the protein–DNA contacts of a yeast RNA polymerase III transcription complex in a crude extract: solid phase synthesis of DNA photoaffinity probes containing a novel photoreactive deoxycytidine analog. *Biochemistry,* 1996, 35, 9821–9831.
341. Mishra, N.C., Khorshidi, H.S., Szweda, P., and George, J. Modified nucleotides for nucleic acid labeling. US Patent 1997, 5,684,142.
342. Stolze, K., Ulrcih, K., Klingel, S., Sagner, G., Wartbichler, R., and Engels, J.W. Synthesis of 3′-sugar- and base-modified nucleotides and their application as potent chain terminators in DNA sequencing. *Helv. Chim. Acta*, 1999, 1311–1323.
342. Brush, C.K. and Reimer, N.D. Nonsulfonated cyanine dyes for labeling nucleosides and nucleotides. PCT Int. Appl. 1998, WO9858942.
344. Hurskainen, P. Nucleic acid labelling employing lanthanide chelates. *J. Alloys Comp.*, 1995, 225, 489–491.
345. Guillou–Bonnici, F., Defranc, E., Hoang, A., Laayoun, A., Lhomme, J., and Trevisiol, E. Target nucleic acid sequence amplification method. PCT Int. Appl. 1998, WO9805766
346. McGall, G.H. and Barone, A.D. Nucleic acid labeling compounds. PCT Int. Appl. 2000, WO0006771.
347. Simmonds, A.C., Hamilton, A., Smith, C., Loakes, D., Brown, D., Hill, F., Kumar, S., Nampali, S., and McDougall, M. Base analogues. US Patent 2002, 6,444,682.

348. Simmonds, A.C., Hamilton, A., Smith, C., Loakes, D., Brown, D., Hill, F., Kumar, S., Nampali, S., and McDougall, M. Base analogues. US Patent 2003, 6,605,611.
349. Simmonds, A.C., Hamilton, A., Smith, C., Loakes, D., Brown, D., Hill, F., Kumar, S., Nampali, S., and McDougall, M. Base analogues. US Pat Appl Pub 2003, 20030060431.
350. Klevan, L., Gebeyehu, G., and Rao, P.Y. Nucleotide analogs for nucleic acid labeling and detection. US Patent 1989, 4,828,979.
351. Schoetzau, T., Klingel, S., Wartbichler, R., Koert, U., and Engels, J.W. Synthesis of a fluorescent derivative of 6-N-[N-(6-aminohexyl)carbamoyl)-2′,3′-dideoxyadenosine 5′-triphosphate for detection of nucleic acids. *J. Chem. Soc. Perkin Trans.*, 1, 2000, 1411–1415.
352. Flanagan, J.H., Khan, S.H., Menchen, S., Soper, S.A., and Hammer, R.P. Functionalized tricarbocyanine dyes as near-infrared fluorescent probes for biomolecules. *Bioconjugate Chem.*, 1997, 8, 751–756.
353. Seela, F., Feiling, E., Gross, J., Hillenkamp, F., Ramzaeva, N., Rosemeyer, H., and Zulauf, M. Fluorescent DNA: the development of 7-deazapurine nucleoside triphosphates applicable for sequencing at the single molecule level. *J. Biotechnol.* 2001, 86, 269–279.
354. Strobel, H., Dugué, L., Marlière, P., and Pochet, S. Synthesis and recognition by DNA polymerases of a reactive nucleoside, 1-(2-deoxy-β-D-erythro-pentofuranosyl)-imidazole-4-hydrazide. *Nucl. Acids Res.*, 2002, 30, 1869–1878.
355. Strobel, H., Dugué, L., Marlière, P., and Pochet, S. A parallel synthesis scheme for generating libraries of DNA polymerase substrates and inhibitors. *Chem. Biochem.*, 2002, 3, 1251–1256.
356. Kapmeyer, H., Lappi, D.A., and Kaplan, N.O. The synthesis of 8-(6-aminohexyl)-amino-GTP and -GDP and their application as ligands in affinity chromatography. *Anal. Biochem.*, 1979, 99, 189–199.
357. Hiratsuka, T. New ribose-modified fluorescent analogs of adenine and guanine nucleotides available as subtrates for various enzymes. *Biochem. Biophys. Acta*, 1983, 742, 496–508.
358. Sarfati, R.S., Kansal, V.K., Munier, H., Glaser, P., Gilles, A.M., Labruyere, E., Mock, M., Danchin, A., and Barzu, O. Binding of 3′-anthraniloyl-2′-deoxy-ATP to calmodulin-activated adenylate cyclase from *Bordetella pertussis* and *Bacillus anthracis*. *J. Biol. Chem.*, 1990, 265, 18902–18906.
359. Tsien, R.Y., Ross, P., Fahnestock, M., and Johnson, A.J. DNA sequencing. PCT Int. Appl. 1991, WO9106678.
360. Sarfati, S.R., Berthod, T., Guerreiro, C., and Canard, B. Synthesis of fluorescent derivatives of 3′-O-(6-aminohexanoyl)pyrimidine nucleosides 5′-triphosphates that act as DNA polymerase substrates reversibly tagged at C-3′. *J. Chem. Soc. Perkin Trans.*, 1 1995, 1163–1171.
361. Welch, M.B. and Burgess, K. Synthesis of fluorescent, photolabile 3′-O-protected nucleoside triphosphates for the base addition sequencing scheme. *Nucleosides Nucleotides*, 1999, 18, 197–201.
362. Kwiatkowski, M. Novel chain terminators, the use thereof for nucleic acid sequencing and synthesis and methods of their preparation. PCT Int. Appl. 1996, WO9623807A1.
363. Kwiatkowski, M. Compounds for protecting hydroxyls and methods for their use. PCT Int. Appl. 2001, WO0125247.

364. Kutateladze, T., Beabealashvilli, R., Azhayev, A., and Kreyevsky, A. 3′-deoxy-3′-aminonucleoside 5′-triphosphates — terminators of RNA synthesis, catalyzed by DNA-dependent RNA polymerase from *Escherichia coli*. *FEBS Lett.*, 1983, 153, 420–426.
365. Chidgeavadze, Z.G., Beabealashvilli, R., Atrazhev, A.M., Kukhanova, M.K., Azhayev, A.V., and Kreyevsky, A.A. 2′,3′-Dideoxy-3′ aminonucleoside 5′-triphosphates are the terminators of DNA synthesis catalyzed by DNA polymerases. *Nucl. Acids Res.*, 1984, 12, 1671–1686.
366. Chidgeavadze, Z.G., Beabealashvilli, R., Kreyevsky, A.A., and Kukhanova, M.K. Nucleoside 5′-triphosphates with modified sugars as substrates for DNA polymerases. *Biochem. Biophys. Acta*, 1986, 868, 145–152.
367. Herrlein, M.K., Konrad, R.E., Engels, J.W., Holletz, T., and Cech, D. 3′-Amino-modified nucleotide useful as potent chain terminators for current DNA sequencing methods. *Helv. Chim. Acta*, 1994, 77, 586–596.
368. Wojczewski, C., Faulstich, K., and Engels, J.W. Synthesis and application of 3′-amino-dye terminators for DNA sequencing. *Nucleosides Nucleotides*, 1997, 16, 751–754.
369. Cech, D., Schoetzau, T., Holletz, T., and Zehl, A. New approaches toward the synthesis of nonradioactively labelled nucleoside triphosphates as terminators for DNA sequencing. *Collect. Czech. Chem. Commun.*, 1996, 61, S297–S300.
370. Wojczewski, C. and Engels, J.W. Synthesis of 3′-S-(2-aminoethylthio)-3′-deoxythymidine-5′-triphosphates. *Synthesis*, 2000, 149–153.
371. Raines, K. and Gottlieb, P.A. Enzymatic incorporation of 2′-thio-CTP into the HDV ribozyme. *RNA*, 1998, 4, 340–345.
372. Sabesan, S. and Trainor, G.L. Dideoxyfructonucleosides and deoxyfructonucleotides. US Patent 1994, 5,276,143.
373. Hileman, R.E., Parkhurst, K.M., Gupta, N.K., and Parkhurst, L.J. Synthesis and characterization of conjugates formed between periodate-oxidized ribonucleotides and amine-containing fluorophores. *Bioconjugate Chem.*, 1994, 5, 436–444.
374. Marciacq, F., Sauvaigo, S., Issartel, J.-P., Mouret, J.-F., and Molko, D. Methods for making morpholino-nucleotides, and their use for analyzing and marking nucleic acid sequences. PCT Int. Appl. 2000, WO0050626.
375. Nairne, R.J.D., Pickering, L., and Smith, C.L. Synthesis of pyrrole carboxamide nucleotide triphosphates-putative labeled nucleotide analogues. *Tetrahedron Lett.*, 2002, 43, 2289–2291.
376. Schwartz, D.A. and Hogrefe, R.I. Triphosphate oligonucleotide modification reagents and uses thereof. US Patent 2004, US 6,686,461.
377. Ghadessy, F.J., Ramsay, N., Boudsocq, N., Loakes, D., Brown, A., Iwai, S., Vaisman, A., Woodgate, R., and Holliger, P. Generic expansion of the substrate spectrum of a DNA polymerase by directed evolution. *Nat. Biotechnol.*, 2004, 22, 755–759.
378. Schotter, J., Kamp, P.B., Becker, A., Pühler, A., Reiss, G., and Brückl, H. Comparison of a prototype magnetoresistive biosensor to standard fluorescent DNA detection. *Biosensors Bioelectron.*, 2004, 19(10), 1149–1156.
379. Yarbrough, L.R., Schlageck, J.G., and Baughman, M. Synthesis and properties of fluorescent nucleotide substrates for DNA-dependent RNA polymerases. *J. Biol. Chem.*, 1979, 254, 12069–12073.

380. Dhar, G. and Bhaduri, A. Synthesis and characterization of stacked and quenched uridine nucleotide fluorophores. *J. Biol. Chem.*, 1999, 274, 14568–14572.
381. Sastry, S. A fluorescence-based assay for transcription using a novel fluorescent GTP analog. *Biophys. Chem.*, 2001, 91, 191–208.
382. Wu, F.Y., Abdulwajid, A.W., and Solaiman, D. Synthesis and properties of adenosine-5′-triphosphoro-gamma-1-(5-sulfonic acid)naphthyl ethylamidate: a fluorescent nucleotide substrate for DNA-dependent RNA polymerase from *Escherichia coli. Arch. Biochem. Biophys.*, 1986, 246, 564–571.
383. Draganescu, A., Hodawadekar, S.C., Gee, K.R., and Brenner C. Fhit-nucleotide specificity probed with novel fluorescent and fluorogenic substrates. *J. Biol. Chem.*, 2000, 275, 4555–4560.
384. McEwen, D.P., Gee, K.R., Kang, H.C., and Neubig, R.S. Fluorescent BODIPY-GTP analogs: real-time measurement of nucleotide binding to G proteins. *Anal. Biochem.*, 2001, 292, 109–117.
385. Lawler, J.F., Jr. Reagents for monitoring nucleic acid amplification and methods of using same. PCT Int. Appl. 2003, WO03089670.
386. Nelson, J., Fuller, C., Sood, A., and Kumar S. Terninal phosphate-labeled nucelotides and methods of use. PCT Int. Appl. 2003, WO0020984.
387. Kumar, S. and Sood, A. Preparation of fluorescent dye-labeled nucleoside polyphosphates as substrates for nucleic acid polymerases. PCT Int. Appl. 2003, WO03020734.

Appendix: Labeled nucleotide analogs

A.10.1 Uracil nucleotide analogs

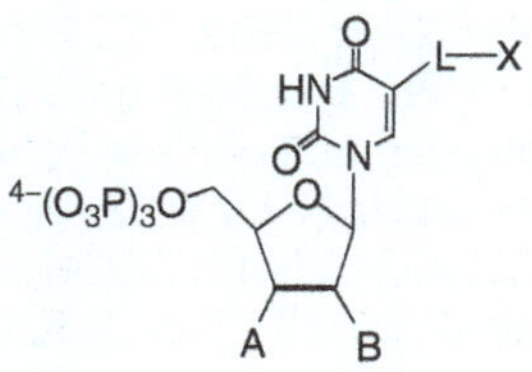

A	B	L	X	Ref.
		Amino based:		
OH	H,OH	-CH=CHCH$_2$NH-	H, biotin	1
OH	H	-CH=CHCH$_2$[NHCO(CH$_2$)$_5$]$_{0\text{-}2}$NH-	Biotin	184
OH	H	-CH=CHCH$_2$NHCO(CH$_2$)$_5$NHCOCHNH-	Dansyl	270
OH	H	-CH=CHCH$_2$[NHCO(CH$_2$)$_4$]$_{0,1}$NH-	Nitroxyl spin label	201, 290
OH	H	-CH=CHCH$_2$NH[CO(CH$_2$)$_5$]$_{0\text{-}3}$NHCO (CH$_2$)$_2$CO$_2$-	DIG	197, 291
H	H	-CH=CHCH$_2$NHCO(CH$_2$)$_5$NHCO(CH$_2$)$_2$-CO$_2$-	DIG	13
OH	H	-CH=CHCH$_2$NHCO(-*o*-C$_6$H$_4$)NH-	Tb chelate	292
OH	H	-CH=CHCH$_2$[NHCO(CH$_2$)$_5$]$_{0,1}$NH-	FAM, TMR, coumarin	170
OH	H	-CH=CHCH$_2$NH-	DNP	81, 134
OH	H	-CH=CHCH$_2$NHCO(CH$_2$)$_3$-	Sulfocoumarin dyes	122, 293
OH	H	-CH=CHCH$_2$NH[CO(CH$_2$)$_5$]$_{0\text{-}3}$NH-	CY5	157, 161, 162
OH	H	-CH=CHCH$_2$NH-	Methylene blue	295
OH	H	-CH=CHCONH(CH$_2$)$_6$NH-	Estradiol	296
OH	H	-CH=CHCH$_2$NHCO(CH$_2$)$_5$NHCO(CH$_2$)$_2$ CONH-	Arylboronate	297
OH	H	-CH=CHCH$_2$NHCO(CH$_2$)$_2$S-	S-(2-pyridyl)	218
OH	H	-CH=CHCH$_2$NHCO(CH$_2$)$_{0,1}$-	Ferrocene	296
OH	H, OH	-CH=CHCH$_2$NHCO(CH$_2$)$_5$NH-	Eu cryptate	299, 300
OH	H	-CH=CHCH$_2$NH-	Pd, Pt coproporphyrin	301
OH	H	-CH=CHCH$_2$NH-	H, acyl moieties	302
OH	H	-CH=CHCH$_2$(NHCOCH$_2$)$_{0\text{-}3}$NH-	Photocrosslinkers	303–306
OH	H	-CH=CHCH$_2$NHC(=NH)(CH$_2$)$_4$S-	H	219
OH	H	-CH=CHCH$_2$NHCO(CH$_2$)$_2$SS(CH$_2$)$_{2,5}$NH-	H, biotin	307
OH	H	-CH=CHCH$_2$NHCO(CH$_2$)$_2$SS(CH$_2$)$_2$ NHCO(CH$_2$)$_5$NH-	Biotin	308
OH	H,OH	-CH=CHCH$_2$NHCO(CH$_2$)$_5$NHCO(CH$_2$)$_2$ SS(CH$_2$)$_2$NH-	Biotin	309
OH	H	-CH=CHCH$_2$NHCO$_2$CH(CH$_3$)[(6-NO$_2$) C$_6$H$_3$-3-]-CH$_2$[NHCO(CH$_2$)$_5$]$_{0,1}$NH-	H, biotin, TMR, Cy5, BODIPY (photocleavable)	90, 249, 250

A	B	L	X	Ref.
		Aminoalkynyl based:		
H	H	-C≡CCH$_2$NH-	H, xanthene dyes	5, 132, 311
H	H	-C≡CHCH$_2$NH-	Rhodamine dyes	233
H	H	-C≡CHCH$_2$[O(CH$_2$)$_2$]$_{0,1}$NH-	Cl$_2$rhodamine, FRET dyes	235
H	H	-C≡CCH$_2$NH-	BODIPY dyes	312
H	H	-C≡CHCH$_2$[O(CH$_2$)$_2$]$_{0,1}$NH-	Carbocyanine dyes	313
H	H	-C≡CHCH$_2$[O(CH$_2$)$_2$]$_{0,1}$NH-	Diarylrhodamine dyes	314
H	H	-C≡CHCH$_2$[NHCO(CH$_2$)$_5$]$_{1\text{-}2}$-[NHCOCH(SO$_3^{(-)}$)CH$_2$]$_{1\text{-}3}$NH-	Fluor, rhod, dyes	221, 315
H	H	-C≡CHCH$_2$[NHCOCH((CH$_2$)$_5$N$^{(+)}$(CH$_3$)$_2$)]$_{1\text{-}6}$ NH-	Fluor, rhod, FRET dyes	316
- acyclo -		-C≡CCH$_2$NH-	H, TFA, var. fluors	239
- acyclo -		-C≡CCH$_2$NH-	"IRD800" cyanine dye	317
OH	H	-C≡CCH$_2$NHCO(CH$_2$)$_2$CONH(CH$_2$)$_6$-	Phenytoin	318
OH	H	-C≡CCH$_2$(OCH$_2$CH$_2$)$_6$[NHCO(CH$_2$)$_5$]$_2$NH-	Biotin	198, 319
OH	H	-C≡CCH$_2$NHCO(CH$_2$)$_2$SS(CH$_2$)$_3$-	Cy5 cycanine dye	92
OH	H	-C≡CCH$_2$NHCO(CH$_2$)$_5$NH-	Mn-porhpyrin	320
H	F, OMe	-C≡CCH$_2$NH-;	H	321
OH,H	H	-C≡CCH$_2$NHCO(CH$_2$)$_2$-	Ferrocenyl	322
OH	OH	-C≡CCH$_2$NHCOCH$_2$-	ONH$_2$	222
OH	OH	-C≡CCH$_2$NHCO(CH$_2$)$_4$-	COCH$_3$	223
H,OH	H	-C≡CCH$_2$NHCOCH$_2$NH-; -CCCH$_2$NHCO(CH$_2$)$_2$NH-	H	323, 324
OH	H	-C≡C(CH$_2$)$_{1,2}$S-	S-tBu	325
OH	OH	-C≡CCH$_2$NH-; -CCCH$_2$O(CH$_2$)$_2$NH-	H, TFA, rhodamine dyes	248
H	OH	-C≡CH(CH$_2$)$_{1,4}$NH -	H, TFA, rhodamine dye	245
H	H	-C≡CCH$_2$[NHCO(CH$_2$)$_5$]$_{0,2}$NH-	H, thiol-reactive grps	221
OH	H	-C≡CCH$_2$NH(COCH$_2$NH)$_3$COCH(CH$_3$) NHCO-(CH$_2$)$_5$NH-	H, TFA, FAM	89
		Miscellaneous linkers:		
OH	H	-CH=CHCOO(CH$_2$)$_6$NH-	Biotin	116
OH	H,OH	-CH=CHCONH(CH$_2$)$_6$NH-	Saccharide moieties	115
OH	H	-(CH$_2$)$_3$[NHCO(CH$_2$)$_4$]$_{0,1}$NH-	Nitroxyl spin label	201
H	F, OMe	-(CH$_2$)$_3$NH-	H	321
OH	H	-(CH$_2$)$_3$NH-	H, TFA, imidazolyl	203
OH	H	-(CH$_2$)$_3$NH-	H	206
OH	H	-CH$_2$NHCO(CH$_2$)$_2$NH-;	H, 4-azidobenzoyl	202
OH	H	-CH$_2$CONH(CH$_2$)$_6$[NHCO(CH$_2$)$_5$]$_{0,1}$NH-	H, TFA, FAM, biotin	206, 211
OH	H	-NHCO(CH$_2$)$_6$NH-	Biotin	326
OH	OH	-NH-	Nitroxyl spin label	327

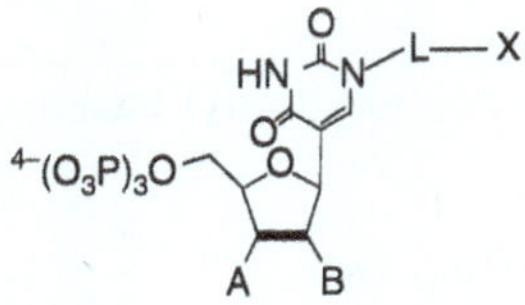

A	B	L	X	Ref.
OH	H	$-(CH_2)_2CONH(CH_2)_3NHCO(CH_2)_5NH-$	DIG, TMR	135, 328
OH	OH	$-(CH_2)_2CONH(CH_2)_2NHCO(CH_2)_5NH-$	H, TFA, fluor, biotin	136–139
OH	OH	$-C{=}CHCONH(CH_2)_2NHCO(CH_2)_5NH-$	Biotin	138, 139

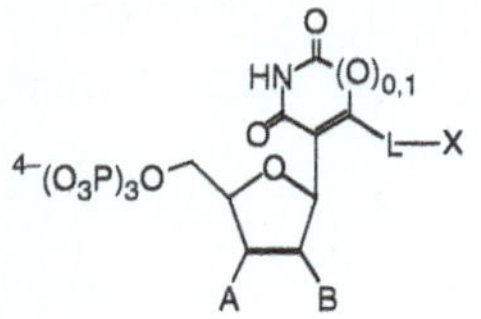

A	B	L	X	Ref.
OH	H	$-NH(CH_2)_5NH-$	FAM	329

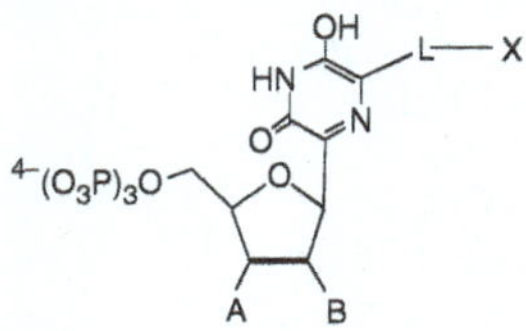

A	B	L	X	Ref.
OH	H	$-NH(CH_2)_2O(CH_2)_2O(CH_2)_2NH-$	TMR	329, 330

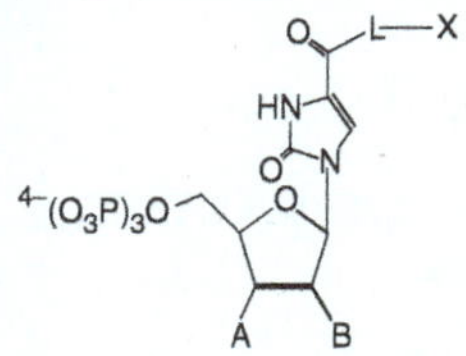

A	B	L	X	Ref.
OH	H	$-NH(CH_2)_4NHCO(CH_2)_5NH-$	FAM	331

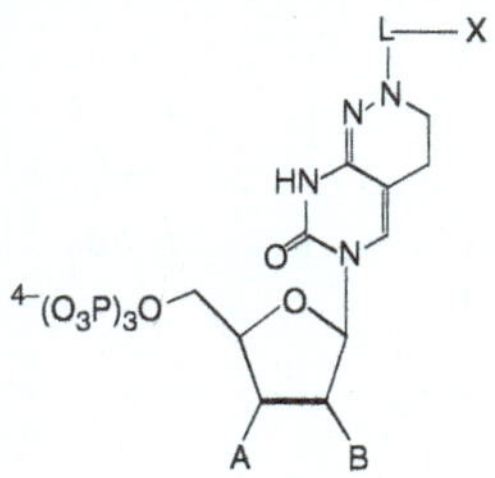

A	B	L	X	Ref.
OH	H	$-CO(CH_2)_5NH-$	FAM	332

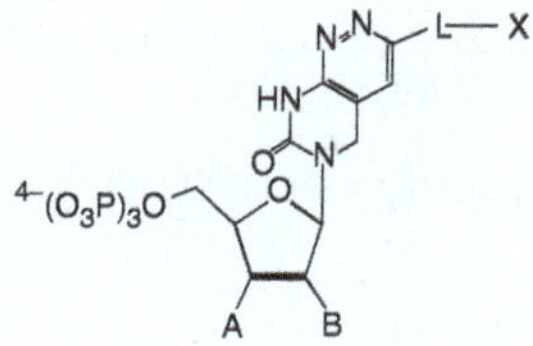

A	B	L	X	Ref.
OH	OH	$-CO(CH_2)_5NH-$	H, FAM, biotin	152

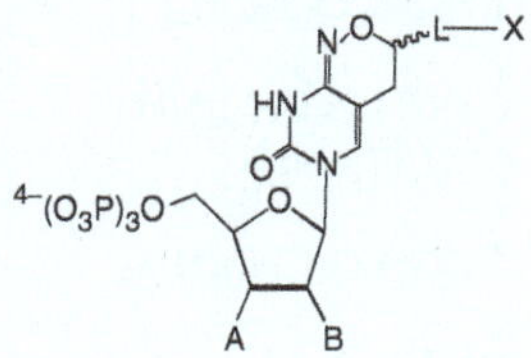

A	B	L	X	Ref.
OH	H	$-(CH_2)_3[NHCO(CH_2)_5]_{0,1}NH-$	H, FAM, DNP	331, 332

A.10.2 Cytidine nucleotide analogs

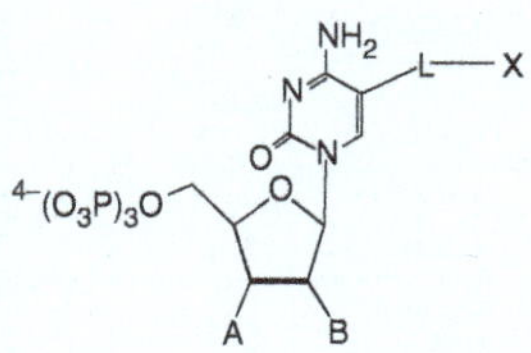

A	B	L	X	Ref.
OH	H, OH	$-CH{=}CHCH_2NH-$	H, biotin	112
OH	H	$-CH{=}CHCH_2NH[CO(CH_2)_4]_{0,1}NH-$	H, nitroxyl spin label	333, 334
OH	H	$-CH{=}CHCH_2NH-$	2,4-DNP	81
OH	H	$-C{\equiv}CCH_2(OCH_2CH_2)_6[NHCO(CH_2)_5]_2NH-$	Biotin	198, 319
OH	H	$-C{\equiv}CCH_2NH-$	H, TFA, xanthene dyes	132
OH	H	$-C{\equiv}CCH_2NHCO(CH_2)_2SS(CH_2)_3-$	Cy5 cycanine dye	92
H	H	$-C{\equiv}CCH_2NH-$	H, xanthene dyes	5, 132, 311
H	H	$-C{\equiv}CHCH_2NH-$	Rhodamine dyes	233
H	H	$-C{\equiv}CCH_2[O(CH_2)_2]_{0,1}NH-$	Cl_2rhodamine, FRET dyes	235
H	H	$-C{\equiv}CCH_2NH-$	BODIPY dyes	312
H	H	$-C{\equiv}CCH_2[O(CH_2)_2]_{0,1}NH-$	Carbocyanine dyes	313
H	H	$-C{\equiv}CCH_2[O(CH_2)_2]_{0,1}NH-$	Diarylrhodamine dyes	313
H	H	$-C{\equiv}CCH_2[NHCO(CH_2)_5]_{1\text{-}2}\text{-}[NHCOCH(SO_3^{(-)})CH_2]_{1\text{-}3}NH-$	Fluor, rhod, dyes	236, 315

A	B	L	X	Ref.
H	H	$-C{\equiv}CCH_2[NHCOCH((CH_2)_5N^{(+)}(CH_3)_2)]_{1-6}NH-$	Fluor, rhod, FRET dyes	316
H	H	$-C{\equiv}CCH_2O(CH_2)_2NHCOCH_2OPO_2^{(-)}O(CH_2)_2NH-$	FRET dye	335
- acyclo -		$-C{\equiv}CCH_2NH-$	H, TFA, var. fluors	239
- acyclo -		$-C{\equiv}CCH_2NH-$	"IRD800" cyanine dye	317
H	H	$-C{\equiv}CCH_2O(CH_2)_2NH-$	FAM, HEX, TET, TMR	336
H	H	$-C{\equiv}CHCH_2O(CH_2)_2NHCO(p\text{-}C_6H_4)_{0,1}CH_2NH-$	FAM, HEX, TET, TMR	337
F	H	$-C{\equiv}C(C_6H_4\text{-}4\text{-})_{0\text{-}2}C\ CCH_2O(CH_2)_2NH-$	Rhodamine dyes	338
H	OH	$-C{\equiv}C(CH_2)_4NH\text{-}CO(CH_2)_5NH-$	H, TFA, rhodamine dye	245, 339
OH	OH	$-C{\equiv}CCH_2NH-$; $-C\ CCH_2O(CH_2)_2NH-$	H, TFA, rhodamine dyes	248
OH	H	$-C{\equiv}CCH_2NHCO(CH_2)_5-$	ONH_2	376
OH	H	$-C{\equiv}CCH_2NHCO(1\text{-}C_6H_4\text{-}4)-$	CHO	376
OH	H	$-C{\equiv}CCH_2NHCO(3\text{-}C_5H_3N\text{-}5)_{0,1}-$	$NHNH_2$	376

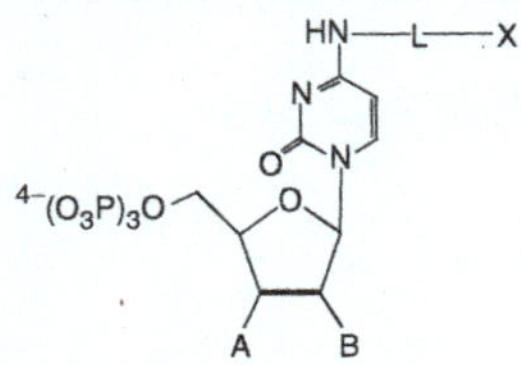

A	B	L	X	Ref.
OH	H, OH	$-(CH_2)_6NH-$	H, biotinyl	127
OH	H	$-(CH_2)_{2,6}[NHC(O)(CH_2)_5]_{0,1}NH-$	H, biotinyl	126, 127
OH	H	$-(CH_2)_6[NHC(O)(CH_2)_5]_{0,1}NH-$	FAM, TMR	170
OH	H	$-[(CH_2)_2O]_2(CH_2)_2NH-$	FAM, TMR	170
OH	H	$-(CH_2)_2NH-$	Aryl azide	340
OH	H	$-(CH_2)_2O(CH_2)_2NH-$	Biotin, dig, fluors	341
H	H	$-(CH_2)_{4/6}NH$ -	Phenoxazine dye	342
H,OH	H, OH	$-(CH_2)_6NHC(O)(CH_2)_2C(O)O-$	Indocarbocyanine dyes	343
OH	H	$-(CH_2)_6NHC(O)OCH(CH_3)[(6\text{-}NO_2)C_6H_3\text{-}3\text{-}]\text{-}CH_2NHC(O)(CH_2)_6NH-$	Biotin (photocleavable)	249, 250
OH	H	$-(CH_2)_6NHC(S)NH(4\text{-}C_6H_4)(CH_2)_2-$	Eu chelate	344
OH	OH	$-(CH_2)_6-$	ONH_2	345
OH	OH	$-(CH_2)_4-$	$COCH_3$	223

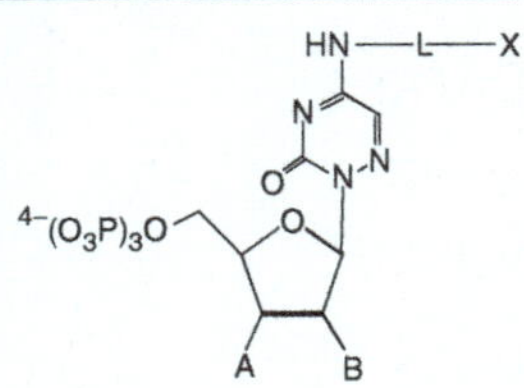

A	B	L	X	Ref.
OH	OH	$-(CH_2)_4NHCO(CH_2)_5NH-$	FAM, biotin	136, 137, 346

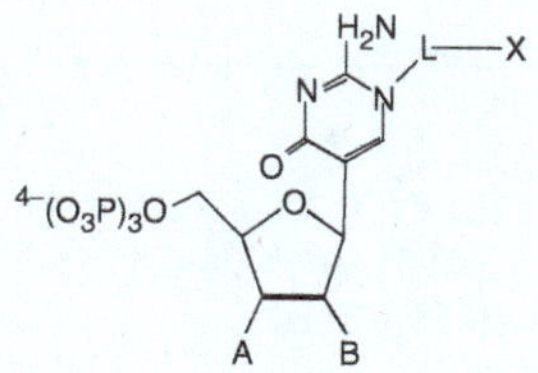

A	B	L	X	Ref.
OH	OH	-CH=CHCONH$(CH_2)_2$NH-CO$(CH_2)_5$NH-	FAM, biotin	138, 139

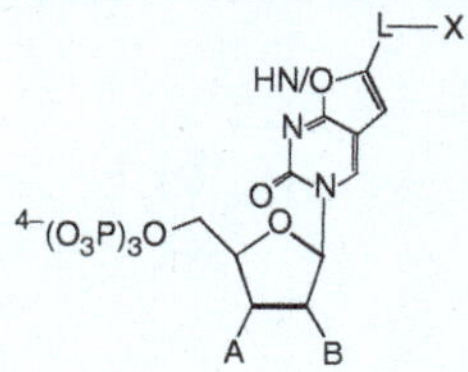

A	B	L	X	Ref.
OH	H	-CH_2NH[CO$(CH_2)_5$NH]$_{0,1}$-	H, TFA, FAM, Cy5	347–349

A.10.3 Adenine nucleotide analogs

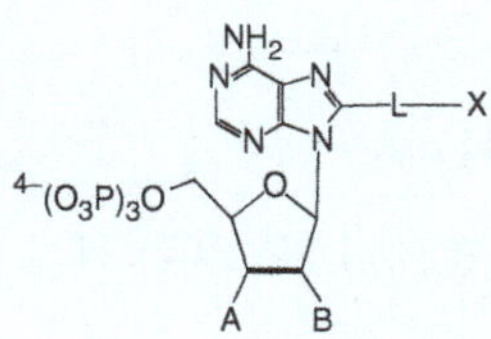

A	B	L	X	Ref.
OH	OH	-NH$(CH_2)_6$NH-	H	121
OH	OH	-NH$(CH_2)_6$NH-	DNP, biotinyl	119, 120
OH	OH	-NH-	Nitroxyl spin label	327
OH	H, OH	-NH$(CH_2)_6$NH-	FAM; fluorescamine	122
OH	OH	-NH$(CH_2)_6$NH[CO$(CH_2)_5$NHCO$(CH_2)_3$]$_{0.1}$	7-Aminocoumarins	293
OH	H	-NH$(CH_2)_6$NH-	Carbocyanine	123
OH	H	-[NHCOCH_2]$_n$NH-	H (mass tag)	323, 324

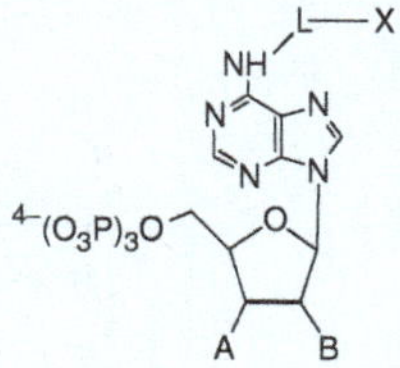

A	B	L	X	Ref.
OH	H	-CH_2CONH$(CH_2)_6$[NHCO$(CH_2)_5$]$_{0,1}$NH-	biotin	126, 350)
OH	OH	-CH_2CONH$(CH_2)_6$NHC(=NH)$(CH_2)_4$S-	H	219
OH	H	-$(CH_2)_{2,6}$[NHCO$(CH_2)_5$]$_{0,1}$NH-	TFA, biotin	126, 350

A	B	L	X	Ref.
OH	OH	$-(CH_2)_4NHCOCH_2-$	ONH_2	222
OH	H	$-(CH_2)_6[NHCO(CH_2)_5]_{0,1}NH-$	FAM, TMR	170
H	H	$-(CH_2)_6NHCO(CH_2)_2C(O)O-$	Carbocyanine dye	343
OH	H	$-(CH_2)_6NHCOOCH(CH_3)[(6-NO_2)C_6H_3-$ 3-]$-CH_2NHC(O)(CH_2)_6NH-$	Biotin (photocleavable)	249, 250
OH	H	$-CONH(CH_2)_6NH-$	TFA, FAM	361
OH	H	$-CONH(CH_2)_6NH-$	Cy5	294
OH	H	$-(CH_2)_2NH-$	4-Azidobenzoyl,	306
OH	OH	$-(CH_2)_4-$	$COCH_3$	223

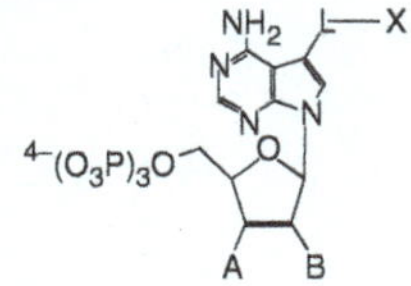

A	B	L	X	Ref.
H	H	$-C\equiv CCH_2NH-$	H, xanthene dyes	5, 132, 311
H	H	$-C\equiv CCH_2NH-$	Rhodamine dyes	233
H	H	$-C\equiv CCH_2[O(CH_2)_2]_{0,1}NH-$	Cl_2rhodamine, FRET dyes	235
H	H	$-C\equiv CCH_2NH-$	BODIPY dyes	312
H	H	$-C\equiv CCH_2[O(CH_2)_2]_{0,1}NH-$	Carbocyanine dyes	313
H	H	$-C\equiv CHCH_2[O(CH_2)_2]_{0,1}NH-$	Diarylrhodamine dyes	314
H	H	$-C\equiv CCH_2[NHCO(CH_2)_5]_{1\text{-}2}-[NHCOCH-(SO_3^{(-)})CH_2]_{1\text{-}3}NH-$	Fluor, rhod, dyes	236, 315
H	H	$-C\equiv CCH_2[NHCOCH((CH_2)_5N^{(+)}-(CH_3)_2)]_{1\text{-}6}NH-$	Fluor, rhod, FRET dyes	316
H	H	$-C\equiv CCH_2NHCOCH_2OPO_2^{(-)}O(CH_2)_2NH-$	FRET dyes	335
H	H	$-C\equiv CCH_2OPO_2^{(-)}O(CH_2)_2NH-$	FRET dyes	335
- acyclo -		$-C\equiv CCH_2NH-$	H, TFA, var. fluors	239
- acyclo -		$-C\equiv CCH_2NH-$	"IRD800" cyanine dye	317
H	H	$-C\equiv CCH_2NH-$	Tricarbocyanine dye	352
OH	H	$-C\equiv CCH_2(OCH_2CH_2)_6[NHCO(CH_2)_5]_2NH-$	Biotin	198, 319
OH	H	$-C\equiv C(CH_2)_2NHCO(CH_2)_5NH-$	FAM, ROX, TMR	353
OH	OH	$-C\equiv CCH_2NH-$; $-CCCH_2O(CH_2)_2NH-$	H, TFA, rhodamine dyes	248
OH	H	$-C\equiv CCH_2NHCO(CH_2)_2SS(CH_2)_3-$	Cy5	92
H	OH	$-C\equiv C(CH_2)_{1,4}NH$ -	H, TFA, rhodamine dye	245

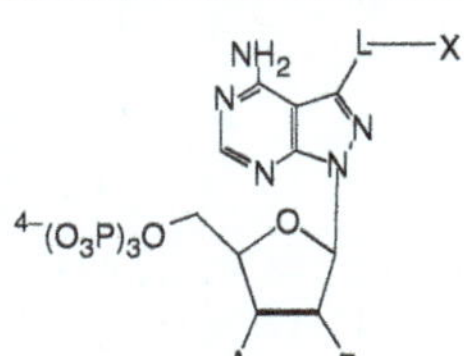

A	B	L	X	Ref.
OH	H	$-(CH_2)_3NH-$	H	144
OH	H	$-(CH_2)_5NHCO(CH_2)_5NH-$	biotin	145
OH	OH	$-C{\equiv}CCH_2NH\text{-}CO(CH_2)_5NH-$	FAM, biotin	136, 346
OH	H	$-C{\equiv}CCH_2NH\text{-}CO(CH_2)_5NH-$	H	110

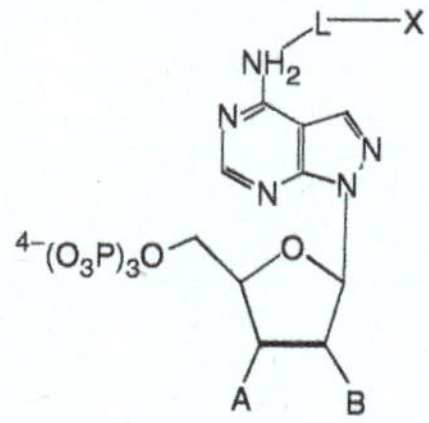

A	B	L	X	Ref.
H	H	$-(CH_2)_4NH(CO(CH_2)_5NH)_{0\text{-}2}-$	FAM, biotin	136, 346

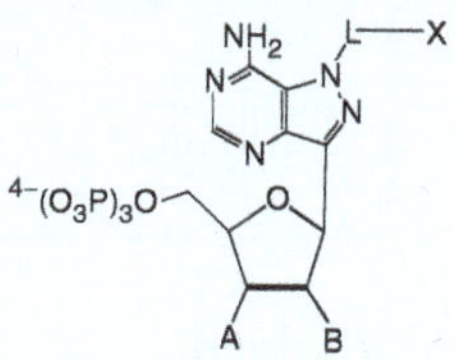

A	B	L	X	Ref.
OH	H	$-CH_2CONH(CH_2CH_2O)_2CH_2CH_2NHCO-(CH_2)_5NH-$	DIG	135, 328
OH	OH	$-CH_2CONH(CH_2CH_2O)_2CH_2CH_2NHCO-(CH_2)_5NH-$	biotin	136

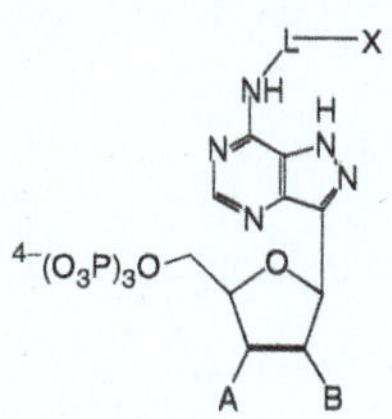

A	B	L	X	Ref.
OH	H	$-(CH_2)_6[NHCO(CH_2)_5]_{0,1}NH-$	FAM (n = 0), DIG (n = 1)	135, 328

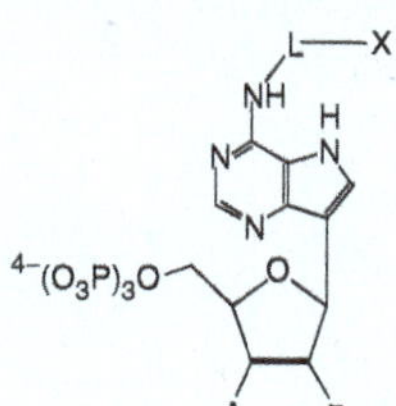

A	B	L	X	Ref.
OH	H	$-(CH_2)_6NH-$	FAM	135, 328

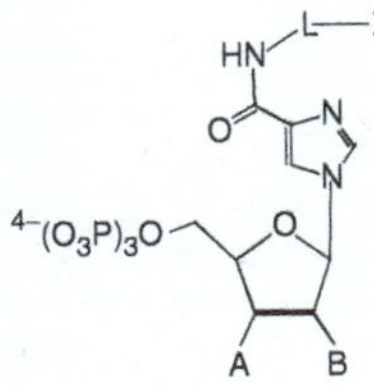

A	B	L	X	Ref.
H	H	-$(CH_2)_2$NH-CO$(CH_2)_5$NH-	FAM, biotin	136, 346
OH	H	- -	NH_2, N = CHAr	354, 355

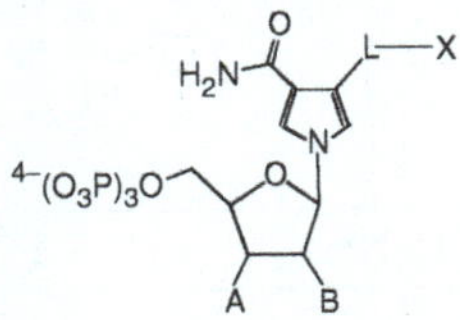

A	B	L	X	Ref.
OH	H	-CONH$(CH_2)_3$NH-; -CH_2NHCO$(CH_2)_5$-NH-	FAM	149, 375

A.10.4 Guanine nucleotide analogs

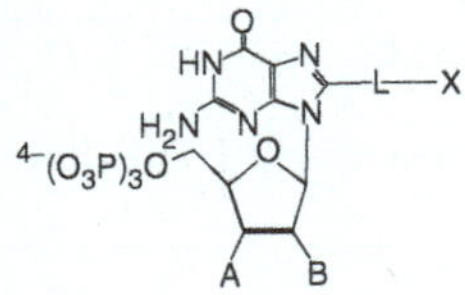

A	B	L	X	Ref.
OH	H, OH	-NH$(CH_2)_6$NH-	H, TFA	356

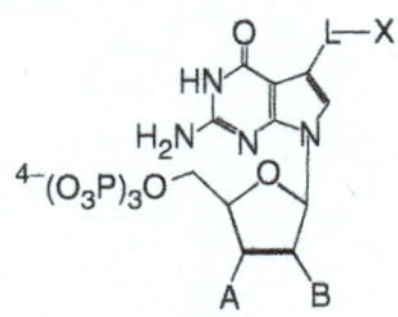

A	B	L	X	Ref.
H	H	-C≡CCH_2NH-	H, xanthene dyes	5, 132, 311
H	H	-C≡CCH_2NH-	Rhodamine dyes	223
H	H	-C≡CCH_2[O$(CH_2)_2$]$_{0,1}$NH-	Cl_2rhodamine, FRET dyes (235)	
H	H	-C≡CCH_2NH-	BODIPY dyes	312
H	H	-C≡CCH_2[O$(CH_2)_2$]$_{0,1}$NH-	Carbocyanine dyes	313
H	H	-C≡CCH_2[O$(CH_2)_2$]$_{0,1}$NH-	Diarylrhodamine dyes	314

A	B	L	X	Ref.
H	H	$-C{\equiv}CCH_2[NHCO(CH_2)_5]_{1-2}-[NHCOCH(SO_3^{(-)})CH_2]_{1-3}NH-$	Fluor, rhod, dyes	236, 315
H	H	$-C{\equiv}CCH_2[NHCOCH((CH_2)_5N^{(+)}(CH_3)_2)]_{1-6}NH-$	Fluor, rhod, FRET dyes	316
H	H	$-C{\equiv}CCH_2O(CH_2)_2NHCO(C_6H_3-3-(SO_3^{(-)})-4-)-CH_2NHCO(CH_2)_5NH-$	FRET dye	335/ khan
H	H	$-C{\equiv}CCH_2O(CH_2)_2NHCO(C_6H_4-4-)-CH_2OPO_2^{(-)}O(CH_2)_2NH-$	FRET dye	335
H	H	$-C{\equiv}C(C_6H_4-4-)C\ CCH_2OPO_2^{(-)}O(CH_2)_2NH-$	FRET dye	335
- acyclo -		$-C{\equiv}CCH_2NH-$	H, TFA, var. fluors	239
- acyclo -		$-C{\equiv}CCH_2NH-$	"IRD800" cyanine dye	317
OH	H	$-C{\equiv}CCH_2NHCO(CH_2)_2SS(CH_2)_3-$	Cy5	92
OH	OH	$-C{\equiv}CCH_2NH-$; $-C\ CCH_2O(CH_2)_2NH-$	H, TFA, rhodamine dyes	248
OH	H	$-C{\equiv}CCH_2NHCO_2CH(CH_3)[(6-NO_2)C_6H_3-3-]-CH_2[NHCO(CH_2)_5]_{0,1}NH-$	H, biotin, Bodipy, TMR (Photocleavable)	92
H	OH	$-C{\equiv}C(CH_2)_{1,4}NH-$	H, TFA, rhodamine dye	245
OH	H	$-C{\equiv}C(CH_2)_2NHCO(CH_2)_5NH-$	FAM, ROX, TMR	353

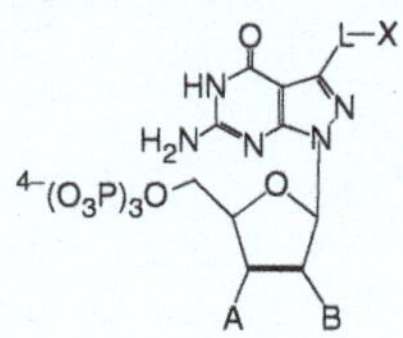

A	B	L	X	Ref.
OH	H	$-C{\equiv}C(CH_2)_3NH-CO(CH_2)_5NH-$	H, rhodamine dye	146

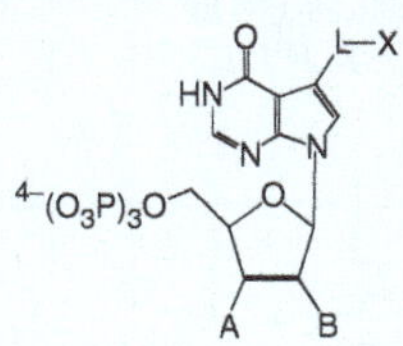

A	B	L	X	Ref.
H	H	-C≡CCH2NH-	H	132, 311

A.10.5 Sugar-modified nucleotide analogs

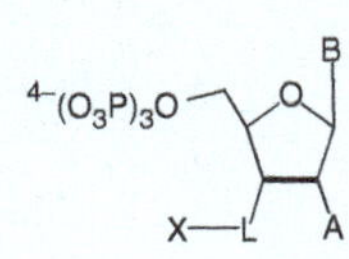

A	B	L	X	Ref.
OH	A,G	-O-	ANT	357

A	B	L	X	Ref.
H	A	-O-	ANT	358
H	T	$-OCO(CH_2)_2CONH(CH_2)_4NH-$	Dansyl	359
H	A,G,C, T	$-OCO(CH_2)_5NH-$	FAM, ANT	193, 360
H	A, T	$-O(CO_2)_{0,1}CH_2(3\text{-}OCH_3\text{-}6\text{-}NO_2\text{-}C_6H_2\text{-}4\text{-})(CH_2)_5NH-$	DAN	88, 361
H	A,G,C, T	$-OC(CH_3)(OCH_3)(CH_2)_{2\text{-}5}CONH(CH_2)_3NH-$	H, FAM	362
H	T	$-O(CH_2)_2SS(CH_2)_2NH-$	DAN	363
H	A,G,C, T	-NH-	H	364, 365
H	T	-NH-	H, biotin, fluorescamine	188
OH	A	-NH-	H, biotin, fluorescamine	190
H	T	-NH-	ANT	193
H	A,G,C, T	$-(NHCOCH_2)_{0,1,2,4}NH-$	Var. fluors	191, 367, 368
H	A,G,C, T	$-NHCONH(CH_2)_6NH-$	H, fluors	342, 369
H	T	$-NHCO(CH_2)_5NH-$	FAM	137
H	T	$-(NHCOCH_2)_nNH-$; $-NHCO(CH_2)_2NH-$	H	323, 324
H	T	$-S(CH_2)_2NH-$	H, tBoc, Cy5	370

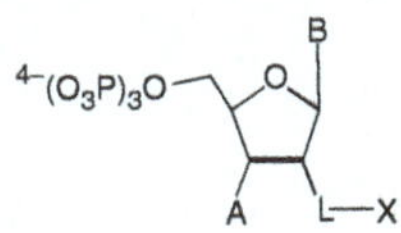

A	B	L	X	Ref.
OH	U	$-NHCO(CH_2)_5NH-$	FAM	137
OH	U	$-O(CH_2)_6[NHCO(CH_2)_5]_{0,1}NH-$	FAM, biotin	137
OH	U	$-(NHCOCH_2)_nNH-$; $-NHCO(CH_2)_2NH-$	H	323, 324
OH	C	-S-	H, 2-nitrophenyl-thio	371

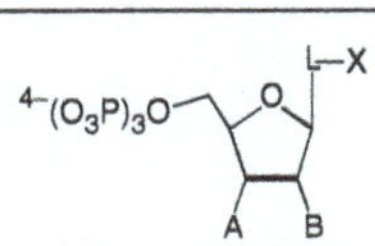

A	B	L	X	Ref.
OH	H	$-CONH(CH_2)_2NHCO(CH_2)_5NH-$	FAM, biotin	137, 139

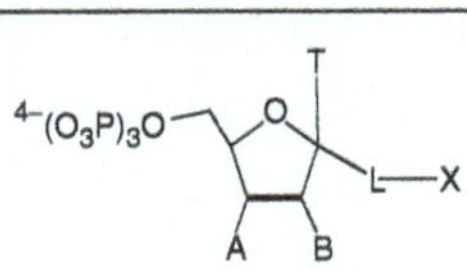

A	B	L	X	Ref.
H	H	$-CH_2NHCO(CH_2)_5NH-$	H, TFA, biotin	372

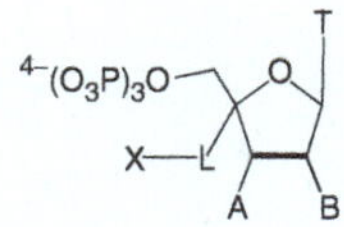

A	B	L	X	Ref.
OH	H	$-CH_2NH-$	H, TFA	89

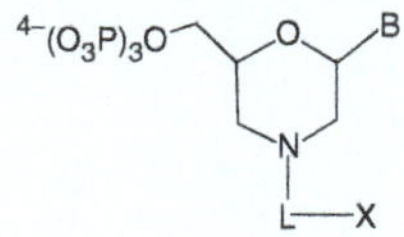

B	L	X	Ref.
A	$-(CH_2)_2NH-$	FAM	373
A,G,C,T	- -	H, FAM	374
U	$-CO(CH_2)_5NH-$	FAM	137

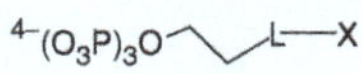

L	X	Ref.
$-(CH_2)_{0\text{-}4}[NHCO(CH_2)_5]_{0,1}NH-$	FMOC	140

A.10.6 Phosphate-modified nucleotide analogs

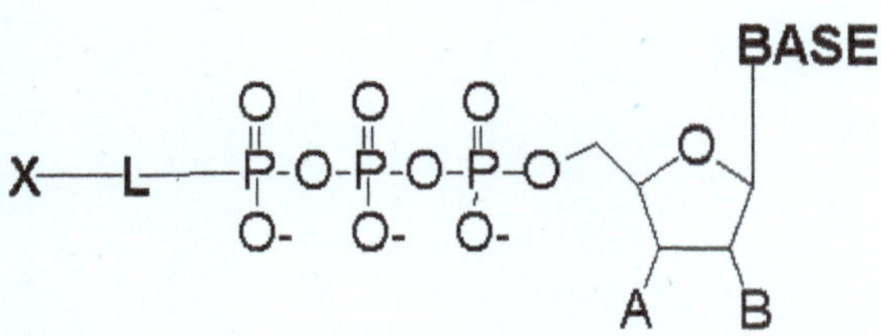

A	B	BASE	L	X	Ref.
OH	OH	A, U	-NH-	Aminonaphthyl- enesulfonyl	379, 380
OH	OH	G	-NH-	7-Coumarinyl	381
OH	OH	A, U	$-NH(CH_2)_2NH-$	Aminonaphthyl- enesulfonyl382	
OH	OH	A, G	$-NHCH_2NHCOCH_2S-$	Var. BODIPY dyes	383, 384
OH	H	A, G, C, T	$-NH(CH_2)_3N^+(CH_3)_2(CH_2)_4N^+(CH_3)_2(CH_2)_3O-$	Var. xanthene dyes	100
OH	H	A, G, C, T	-NH-	Aminonaphthyl- enesulfonyl	101
OH	H	U*	$-NH(CH_2)_4NH-$(*5-CH = $CHCH_2NHCO(CH_2)_5NH$-FAM)	DABCYL quencher	385
H	H	A, G, C, T	-O-; $-O(PO_2^-)O-$	7-Coumarinyl, DDAO dye	386, 387

chapter eleven

Antiviral drugs: triphosphates of nucleoside analogues active as antiviral drugs

Erik De Clercq
Rega Institute for Medical Research, K.U. Leuven, Leuven, Belgium

Contents

11.1 Introduction

Most of the nucleoside analogues currently used as antiviral agents act through their 5′-triphosphate as the active metabolite. Their conversion to the 5′-triphosphate metabolite requires, in principle, three (intracellular) phosphorylation reactions carried out subsequently by (1) a nucleoside kinase; (2) a nucleoside 5′-monophosphate kinase; and (3) nucleoside 5′-diphosphate (NDP) kinase. These phosphorylating enzymes are generally of cellular origin, but in some cases (particularly nucleoside kinases), they may be encoded by the viral genome. In their 5′-triphosphate form, the compounds may interact with the viral polymerase as competitive inhibitors/alternate substrates with respect to the natural substrates (dNTPs: dATP, dGTP, dCTP, dTTP). If acting as alternate substrates — thus being incorporated into the DNA chain — the compounds may, depending on whether they contain a 3′-hydroxyl group, allow or stop further chain elongation, with consequent disruption of viral DNA functioning or synthesis.

The 5′-triphosphate-based modus operandi is followed by virtually all nucleoside/nucleotide analogues active against DNA (herpes, adeno, papova, pox, hepadna) viruses and retroviruses. They can be schematically divided into the following classes:

- 5-Substituted 2′-deoxyuridines
- Acyclic guanosine analogues
- β-D-2′,3′-Dideoxynucleoside analogues
- Acyclic nucleoside phosphonates
- β-L-nucleoside analogues

A sixth class, the ribavirin type of compounds, primarily acts in the 5′-monophosphate form; however, because these compounds can be converted to their 5′-triphosphate form, they will also be briefly discussed.

Principal indications for the clinical use of nucleoside 5′-triphosphate analogues (the first five classes) include herpesvirus infections (herpes simplex virus [HSV-1, HSV-2], varicella-zoster virus [VZV], cytomegalovirus [CMV]); retrovirus infections (human immunodeficiency virus [HIV-1, HIV-2]); and hepadnavirus infections (human hepatitis B virus [HBV)]).[1]

11.2 5-Substituted 2′-deoxyuridines

Currently available 5-substituted 2′-deoxyuridines include idoxuridine (IDU), trifluridine (TFT), and brivudin (BVDU) (Figure 11.1); they are used for the topical treatment of HSV-1 infections of the eye (i.e., herpetic keratitis; IDU as 0.1%, and TFT as 1% eye drops) and systemic treatment of VZV infections (i.e., herpes zoster; BVDU at 125 mg orally, once daily). In fact, brivudin has been registered (as Zostex® or Brivirac®) for the treatment of herpes zoster in several European countries.

Whereas IDU and TFT are indiscriminately phosphorylated by cellular and HSV-induced kinases, BVDU is preferentially phosphorylated by HSV- and VZV-encoded thymidine kinase (TK). HSV-1 TK and VZV TK, but not HSV-2 TK, are able to further phosphorylate BVDU-MP onto BVDU-DP. For TFT, the active metabolite is believed to be the 5′-monophosphate metabolite, TFT-MP, which principally acts as an inhibitor of dTMP synthase, a key enzyme in the *de novo* biosynthesis of dTTP.

For IDU and BVDU, conversion to the 5′-triphosphate metabolites is required for these compounds to exert their antiviral effects (Figure 11.2).

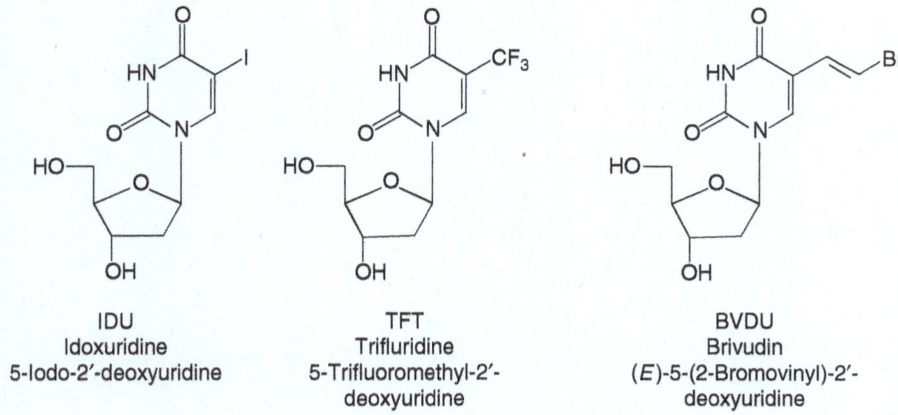

Figure 11.1 Structural formulae of IDU, TFT, and BVDU.

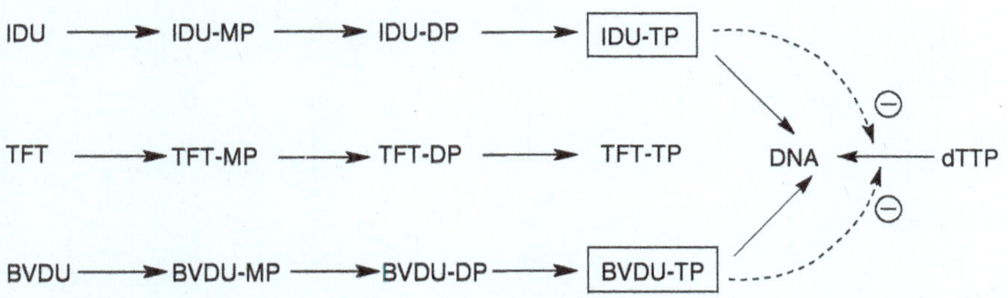

Figure 11.2 Intracellular metabolism and mechanism of action of IDU, TFT, and BVDU.

IDU-TP and BVDU-TP will then interact with the DNA polymerase as competitive inhibitors or alternate substrates (with respect to dTTP). If incorporated, they permit further chain elongation; however, the resulting DNA products containing the fraudulent bases 5-iodouracil or 5-(2-bromovinyl)uracil are likely to be hampered in their functioning and/or stability. It should be remembered that, because the first step in the phosphorylation cascade of BVDU is confined essentially to the virus-infected cell,[2] the ultimate effect on DNA synthesis is expected to be specific for the virus-infected cell. On the other hand, IDU in its effect on DNA synthesis does not discriminate between virus-infected and uninfected cells.

11.3 Acyclic guanosine analogues

Three acyclic guanosine analogues are routinely used in the treatment of viral infections: acyclovir (ACV); ganciclovir (GCV); and penciclovir (PCV), and their oral prodrugs, valaciclovir; valganciclovir; and famciclovir (Figure 11.3):

- Intravenous acyclovir (Zovirax®) at 30 mg/kg/day for treatment of severe mucocutaneous HSV and VZV infections

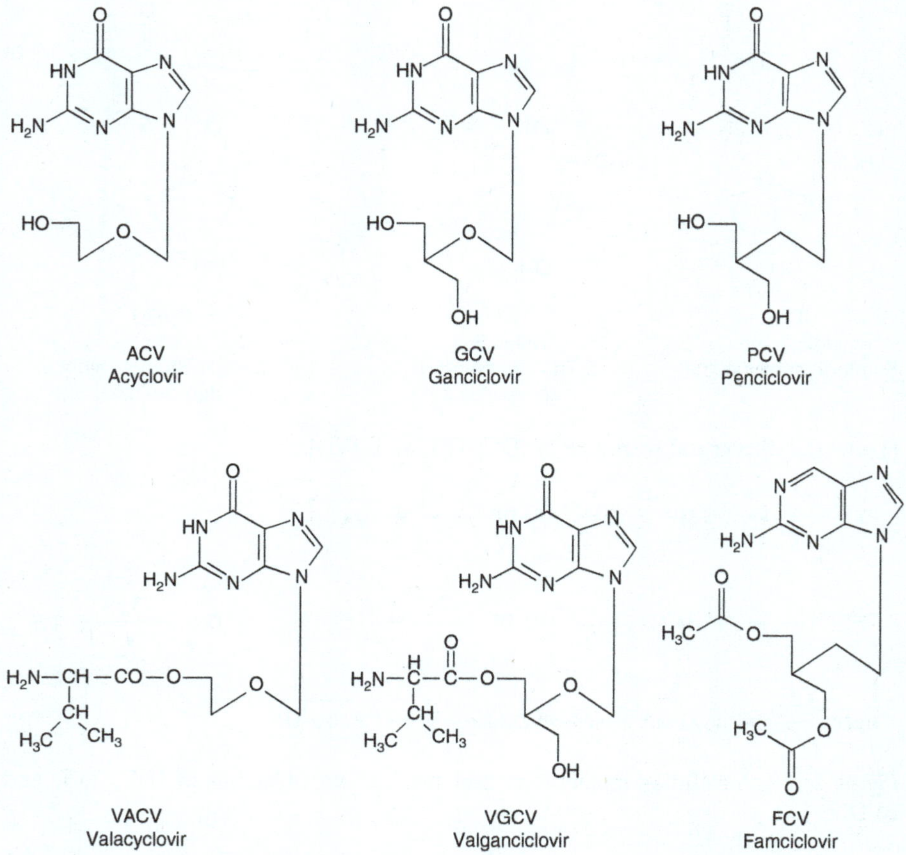

Figure 11.3 Structural formulae of ACV, GCV, and PCV and of their oral prodrugs VACV, VGCV, and FCV.

- Topical acyclovir (5%) for treatment of herpes labialis
- Intravenous ganciclovir (Cymevene®) at 10 mg/kg/day for treatment of CMV infections (i.e., CMV retinitis) in AIDS patients
- Oral valaciclovir (Valtrex®, Zelitrex®) (3 × 1000 mg, daily) and oral famciclovir (Famvir®) (3 × 500 mg, daily) for treatment of VZV infections (i.e., herpes zoster)
- Oral valganciclovir (Valcyte®) (2 × 450 mg, twice daily) for maintenance therapy and prophylaxis of CMV infections

ACV, GCV, and PCV are specifically phosphorylated by the HSV-encoded TK[3] and, to a lesser extent, by the VZV-encoded TK, which preferentially limit their antiviral activity to the virus-infected cell. In addition, ganciclovir can be phosphorylated by the CMV UL97 gene product (a protein kinase held responsible for the egress of the CMV nucleocapsids

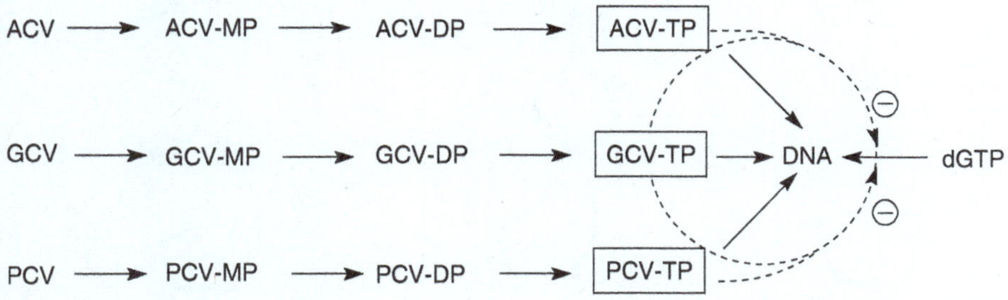

Figure 11.4 Intracellular metabolism and mechanism of action of ACV, GCV, and PCV.

from the nucleus), which explains the selective antiviral activity of ganciclovir against CMV.

Following phosphorylation of ACV, GCV, and PCV to the corresponding 5′-monophosphate, the latter are further phosphorylated by the cellular enzyme, GMP kinase, to the 5′-diphosphates, which are then converted to the 5′-triphosphates by NDP kinase (Figure 11.4). The 5′-triphosphates will then interact with the DNA polymerase as competitive inhibitor/substrate and, if incorporated, terminate chain elongation. Lacking an hydroxyl group equivalent to the 3′-hydroxyl group of the regular nucleosides, acyclovir must function as an obligate chain terminator, whereas ganciclovir and penciclovir may at least theoretically allow limited chain elongation.

11.4 β-D-2′,3′-Dideoxynucleoside analogues

A variety of β-D-2′,3′-dideoxynucleoside analogues[4] that are currently licensed for the treatment of HIV infections have been described: AZT, ddI, ddC, d4T, and ABC (Figure 11.5); a sixth compound (DAPD) is still in the developmental stage. Standard daily dosage for these compounds, all by the oral route, is 600 mg for AZT (Retrovir®); 400 mg for ddI (Videx®); 2.25 mg for ddC (Hivid®); 80 mg for d4T (Zerit®); and 600 mg for ABC (Ziagen®). Moreover, AZT can be combined with 3TC (lamivudine) as capsules (Combivir®) containing 300 mg of AZT and 150 mg of 3TC (to be administered twice daily). A third drug (ABC) could be added to this combination (Trizivir®) so as to contain 300 mg of AZT, 150 mg of 3TC, and 300 mg of ABC (to be administered twice daily).

In order to achieve their antiviral effect, all β-D-2′,3′-dideoxynucleoside analogues need to be phosphorylated successively to the 5′-mono-, 5′-di-, and 5′-triphosphate.[5] All three phosphorylation steps depend on cellular enzymes, including the first phosphorylation step carried out by a nucleoside kinase that, for most of the dideoxynucleoside analogues, represents the bottle neck in the phosphorylation pathway (Figure 11.6). In addition to these three phosphorylation reactions (Figure 11.6), the ddI, ABC, and DAPD pathways require some compound-specific metabolic conversions,[6] i.e.:

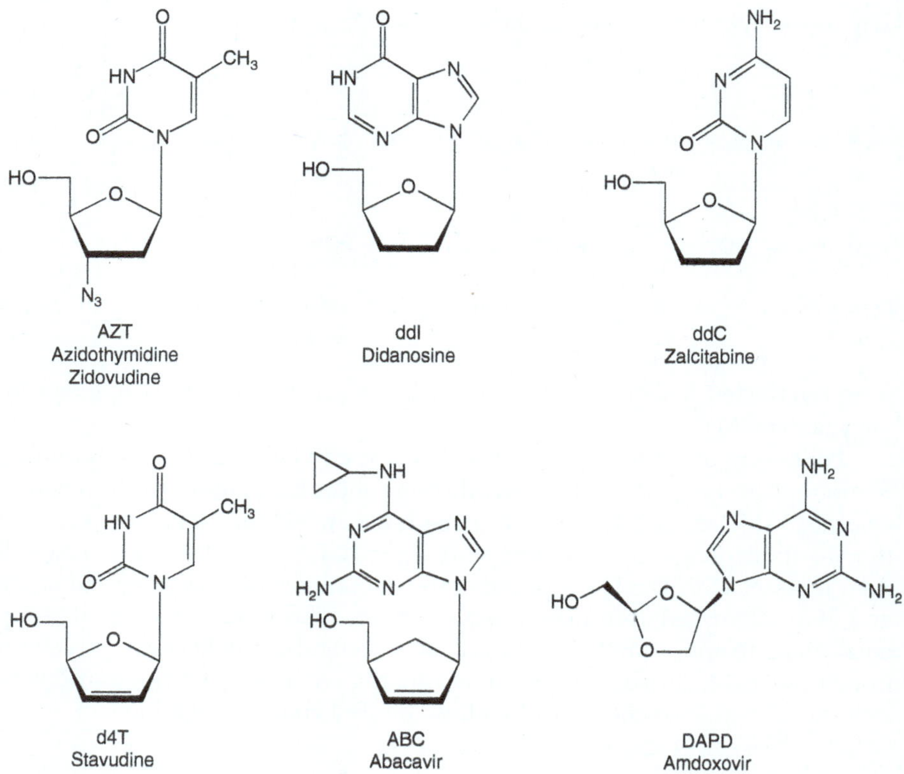

Figure 11.5 Structural formulae of AZT, ddI, ddC, d4T, ABC, and DAPD.

- ddIMP → ddAMP by adenylosuccinaat synthetase/lyase
- ABC → ABC-MP → CBV-MP by a specific phosphotransferase followed by a (deoxy)adenylate deaminase (leading to the formation of carbovir monophosphate)
- DAPD → DXG by a (deoxy)adenosine deamine (leading to the formation of dioxolane guanine)

In all cases, 5′-triphosphate metabolite generated after the three phosphorylation steps will interact with the target enzyme, the HIV reverse transcriptase, as a competitive inhibitor/alternate substrate with respect to dTTP (AZT, d4T), dCTP (ddC), dATP (ddI), or dGTP (ABC, DAPD). Because all the 5′-triphosphate metabolites derived from the 2′,3′-dideoxynucleosides lack the 3′-OH functionality, they will be obliged to act as chain terminators after their incorporation into the DNA chain.

11.5 *Acyclic nucleoside phosphonates*

The acyclic nucleoside phosphonates[7] licensed for clinical use are cidofovir, adefovir, and tenofovir (Figure 11.7). Adefovir and tenofovir are actually

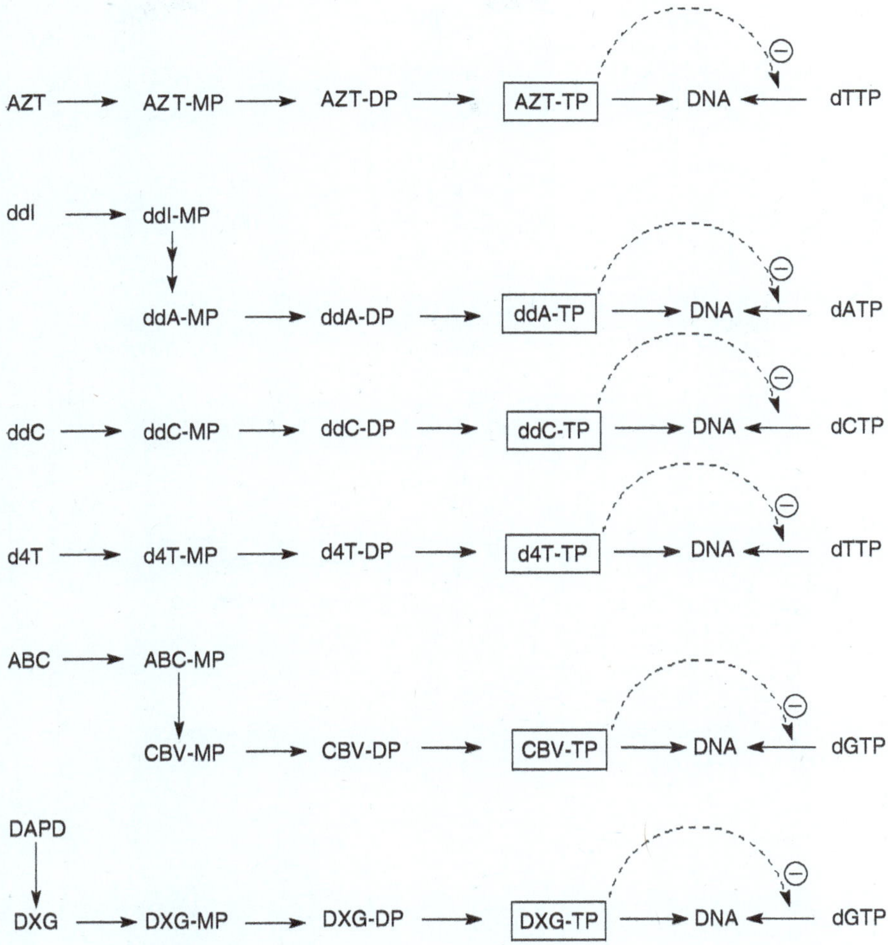

Figure 11.6 Intracellular metabolism and mechanism of action of AZT, ddI, ddC, d4T, ABC, and DAPD.

available in their oral prodrug forms: adefovir as adefovir dipivoxil (Hepsera®) for the treatment of HBV infections (daily single dose: 10 mg) and tenofovir as tenofovir disoproxil fumarate (Viread®) for the treatment of HIV infections (daily single dose: 300 mg). Cidofovir (Vistide®) is made available as solution (containing 375 mg per 5 ml) for intravenous use in the treatment of CMV retinitis in AIDS patients (5 mg/kg/week for the first 2 weeks, then 5 mg/kg every other week, under hydration and probenecid cover).

Cidofovir can also be applied topically (as a 1% cream or gel [Forvade™]), intranasally (as aerosol), intralesionally, etc., but not orally because of poor oral bioavailability. An oral prodrug form, termed HDP (hexadecyloxypropyl)-cidofovir, has been developed that yields high therapeutic promise as an oral formulation for treatment of all infections for which cidofovir is indicated (i.e., herpesvirus infections [HSV, VZV, CMV, etc.];

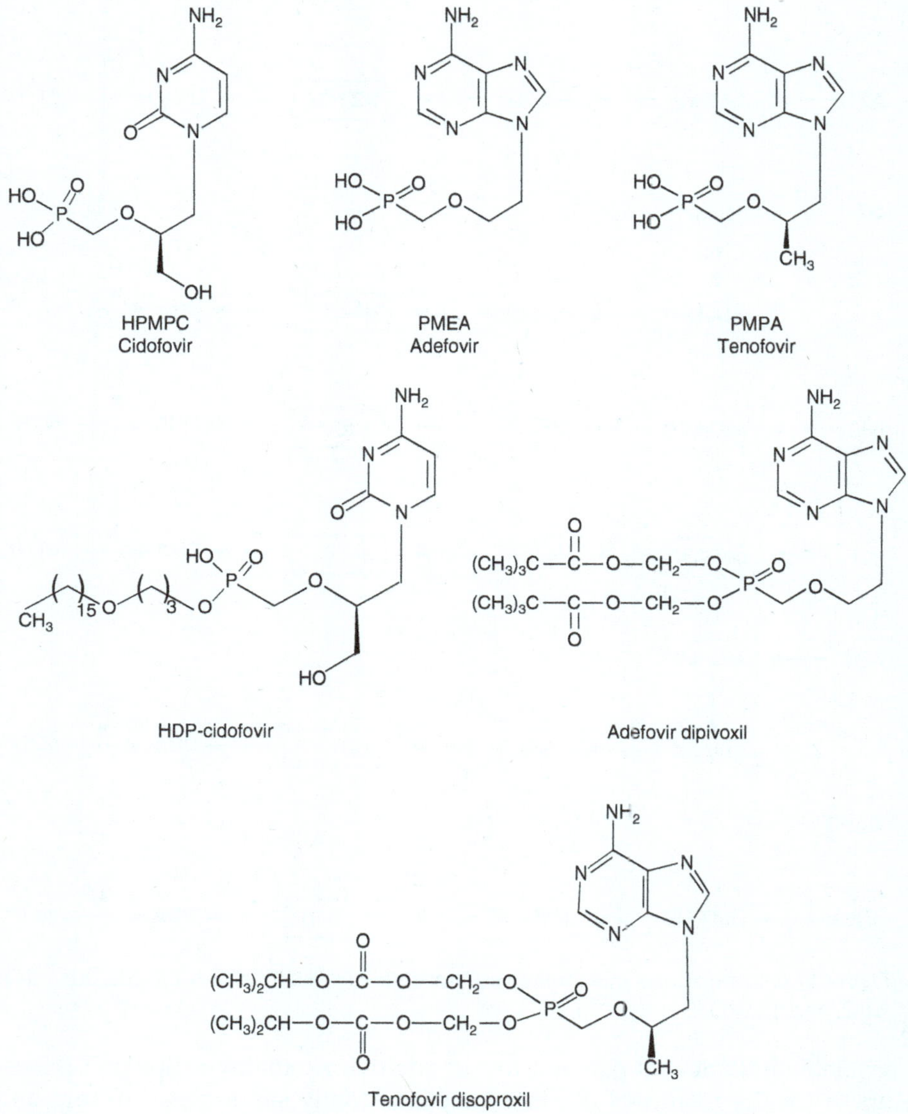

Figure 11.7 Structural formulae of cidofovir, adefovir, and tenofovir and of their oral prodrugs HDP-cidofovir, adefovir dipivoxil, and tenofovir disoproxil.

adeno-, polyoma-, and papillomavirus infections; and poxvirus [smallpox, vaccinia, monkeypox, molluscum contagiosum, orf] infections).

In contrast with the nucleoside analogues, the acyclic nucleoside phosphonates only need two phosphorylation steps to be converted to their active forms because they are nucleoside 5′-monophosphate analogues (Figure 11.8).[8] These diphosphoryl derivatives can be considered directly analogous to the nucleoside triphosphates and will, therefore, act as competitive inhib-

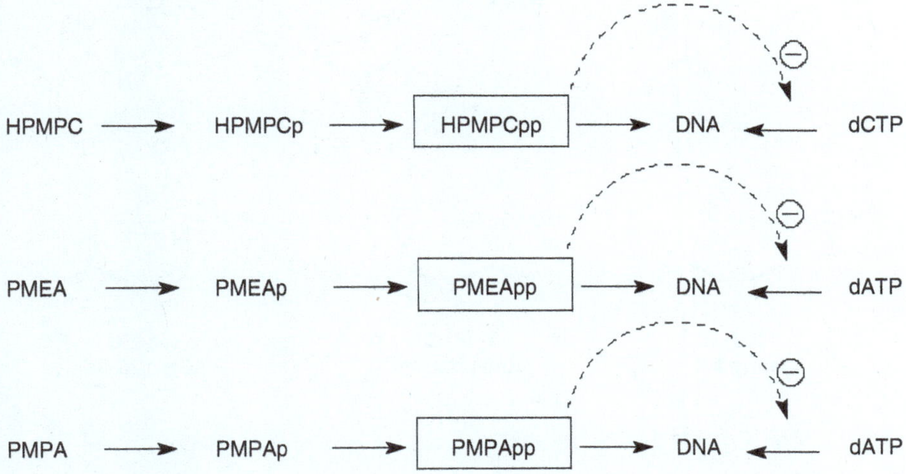

Figure 11.8 Intracellular metabolism and mechanism of action of HPMPC, PMEA, and PMPA.

itors/alternate substrates with respect to the natural substrates, i.e., dATP (for adefovir and tenofovir) or dCTP (for cidofovir).

If incorporated, HPMPC (cidofovir), PMEA (adefovir), and PMPA (tenofovir) will act as chain terminators; however, because they will be linked to the penultimate nucleotide through a phosphonate rather than phosphate group, they may not be readily excised through the action of a 3′ → 5′ exonuclease, pyrophosphorylase, or ATP phosphorylase. This resilience to excision from the DNA chain in which they have been incorporated might explain why it has proven so difficult for the different viruses (herpes, pox, adeno, papilloma, hepadna, retro) to develop resistance to any of the acyclic nucleoside phosphonates (cidofovir, adefovir, tenofovir). On the other hand, acyclic nucleoside phosphonates such as cidofovir do not depend for their phosphorylation on the virus-induced thymidine kinase (TK). Thus, cidofovir may be expected, and has proven, to be effective against those herpes virus mutants that, because the are TK-deficient (TK^-), have acquired resistance to viral TK-dependent compounds such as brivudin and acyclovir.

11.6 L-nucleoside analogues

The L-nucleoside analogues are represented by 3TC, (–)FTC, L-d4FC, L-dC, L-dT, and L-FMAU (Figure 11.9). At present, only one of these compounds, namely, 3TC (lamivudine), has been licensed (Epivir®) for clinical use for the treatment of HIV (AIDS) and HBV (chronic hepatitis B) infections (dosage: 300 mg, once daily for AIDS, and 100 mg, once daily for chronic hepatitis B). Emtricitabine (Coactinon™) should be licensed soon for the treatment of HIV infections and L-d4FC is being developed for the treatment of HIV and HBV infections; L-dC (as its oral prodrug form, val-L-dC), L-dT, and

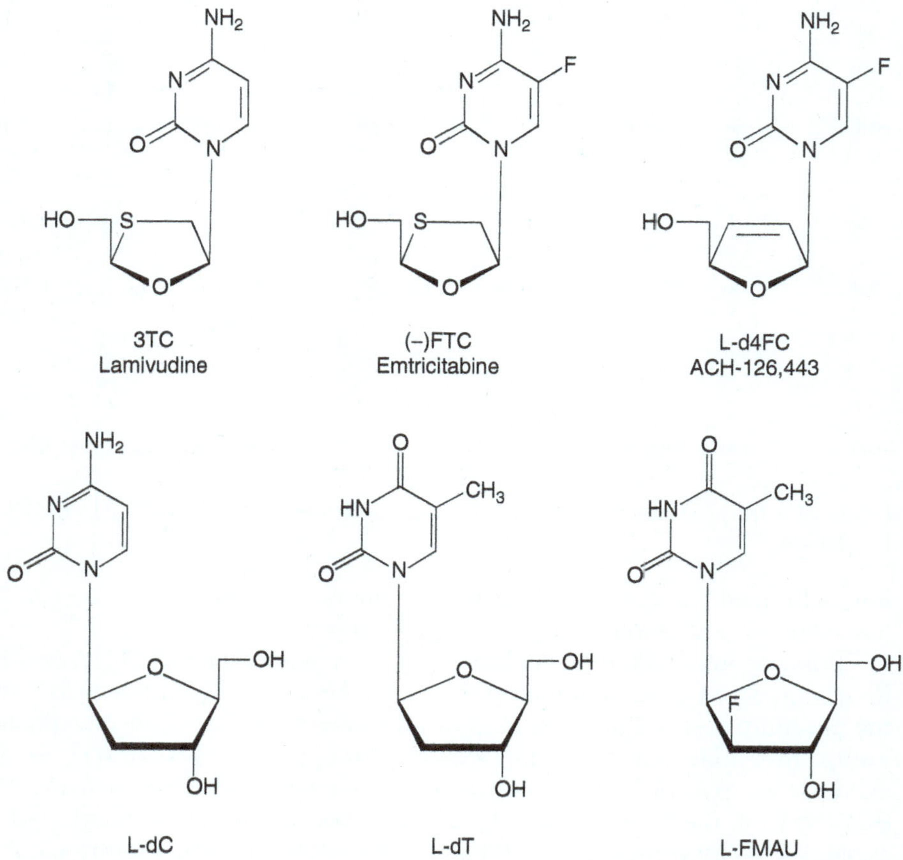

Figure 11.9 Structural formulae of 3TC, (–)FTC, L-d4FC, L-dC, L-dT, and L-FMAU.

L-FMAU are under development for the treatment of HBV infections. In this sense, the discovery of 3TC (lamivudine) as an L-nucleoside antiretroviral agent has heralded the development of a number of other L-nucleoside analogues[9] for treatment of HIV and HBV infections.

Akin to their D-counterparts, 3TC, (–)FTC, L-d4FC, and L-dC need three phosphorylation steps to be converted to their active 5′-triphosphate metabolites, which will then compete with the natural substrate, dCTP, for incorporation into the viral DNA (Figure 11.10). Such incorporation should obviously lead to chain termination — at least as far as 3TC, (–)FTC, and L-d4FC are concerned. Whether the ultimate phosphorylation step (from the di- to the triphosphate) is carried out by NDP kinase or other cellular enzymes has not been unequivocally established. The thymidine analogues, L-dT and L-FMAU, follow a similar pathway to their 5′-triphosphates and, again, the latter correspond to the active metabolites of L-dT and L-FMAU. However, the way in which they precisely interact at the viral DNA polymerase level remains to be clarified further.

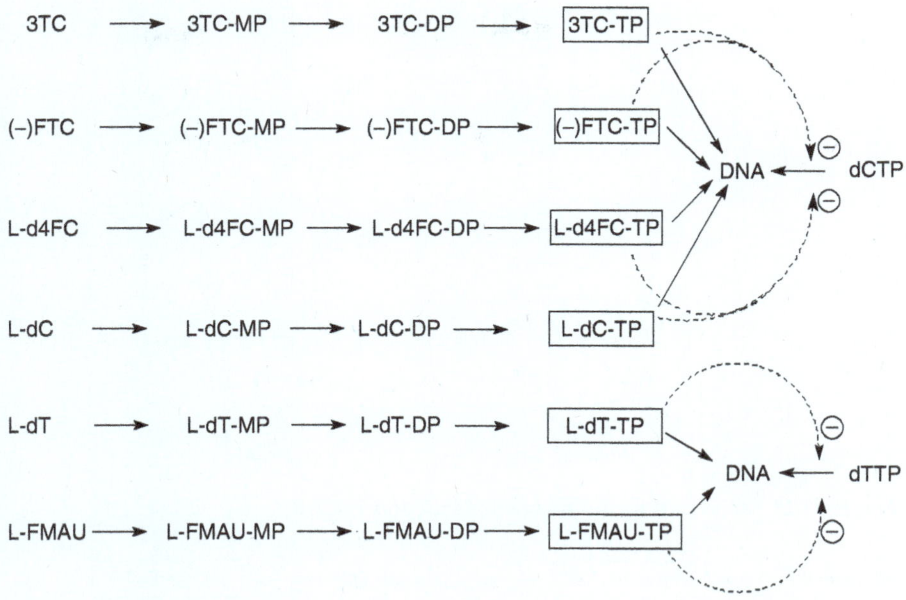

Figure 11.10 Intracellular metabolism and mechanism of action of 3TC, (-)FTC, L-d4FC, L-dC, L-dT, and L-FMAU.

11.7 AICAR analogues

Ribavirin and EICAR (5-ethynylimidazole-4-carboxamide riboside) (Figure 11.11) can be considered structurally related to AICAR (5-aminoimidazole 4-carboxamide ribonucleotide), the immediate precursor of IMP in the *de novo* biosynthetic pathway leading to GTP and ATP and their 2′-deoxynucleotide counterparts (dGTP and dATP).[10] Ribavirin (Virazole®) has been formally approved, as an aerosol (20 mg/ml), for the treatment of respiratory syncytial virus (RSV) infections in high-risk infants. Ribavirin is also administered orally (at doses of 800 up to 1200 mg daily) in combination with interferon α-2a (replaced recently by pegylated interferon α-2a) in the treatment of chronic hepatitis C virus (HCV) infections. The combination of pegulated interferon with ribavirin is commercially available as Rebetron® and PEG INTRON plus ribavirin, respectively.

Ribavirin and EICAR can, in principle, be metabolized intracellularly to their 5′-mono-, 5′-di-, and 5′-triphosphate forms. The latter may then interfere with the incorporation of GTP into RNA. Ribavirin 5′-triphosphate can also be incorporated (as its 5′-monophosphate) into the viral RNA and act as a viral mutagen, at least for poliovirus, when ribavirin is used at sufficiently high concentrations. The main target of action of ribavirin (and EICAR), however, is thought to be IMP dehydrogenase, a key enzyme in the *de novo* biosynthesis of GTP (Figure 11.12). To inhibit the IMP dehydrogenase, ribavirin and EICAR only need to be converted to their 5′-monophosphate. Inhibition of IMP dehydrogenase leads to a shut-off or reduction in the

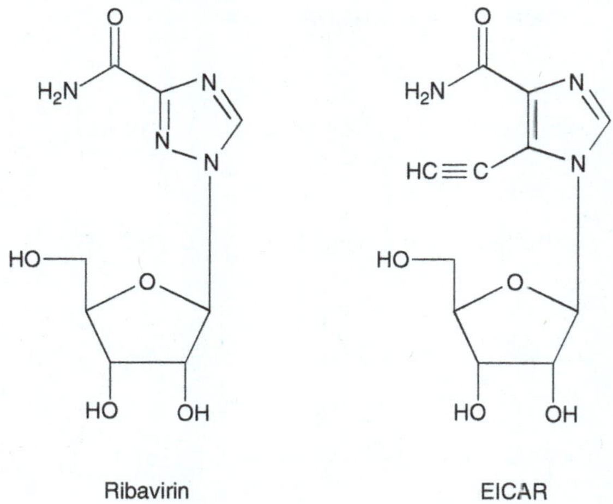

Figure 11.11 Structural formulae of ribavirin and EICAR.

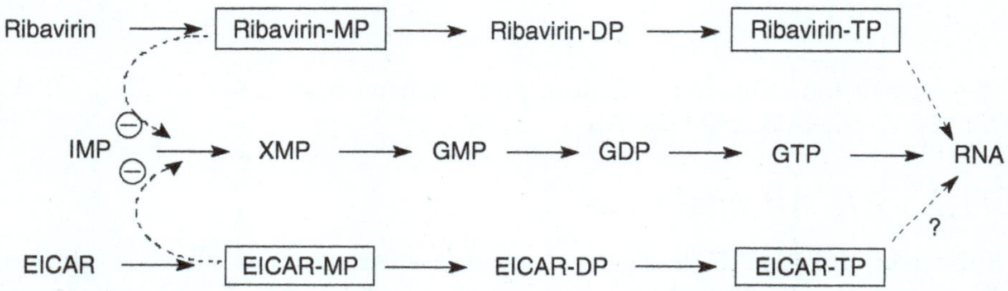

Figure 11.12 Intracellular metabolism and mechanism of action of ribavirin and EICAR.

supply of GTP and, consequently, decreased viral RNA synthesis, thus explaining the broad-spectrum antiviral activity of ribavirin. EICAR is assumed to adhere to a similar modus operandi as ribavirin, but is about 10- to 100-fold more potent an antiviral agent than ribavirin.

11.8 Conclusion

A large variety of antiviral agents currently on the market (or possibly to be marketed soon) for the treatment of virus infections — particularly DNA virus (i.e., herpes and hepatitis B) and retrovirus (i.e., HIV) infections — owe their antiviral action to their triphosphate derivatives or, in case of the nucleotide analogues (acyclic nucleoside phosphonates), their diphosphate derivatives. Representative examples are

- 5′-Triphosphate of brivudin, used in the treatment of herpes zoster
- Triphosphates of acyclovir and penciclovir, used in the treatment of HSV and VZV infections

- Ganciclovir, used in the treatment of CMV infections
- 5′-Triphosphates of azidothymidine, didanosine, zalcitabine, stavudine, and abacavir, used in the treatment of HIV infections
- Diphosphates of cidofovir, adefovir, and tenofovir, licensed for the treatment of CMV retinitis (in AIDS patients), chronic hepatitis B, and HIV infections (AIDS), respectively
- 5′-Triphosphates of lamivudine and other L-nucleosides, approved or are under development for the treatment of HIV and/or HBV infections

References

1. De Clercq E. Antiviral drugs: current state of the art. *J Clin Virol* 2001; 22:73–89.
2. De Clercq E. Antiviral activity of nucleoside analogues: the BVDU connection. In: Chu D, Ed. *Recent Advances in Nucleosides: Chemistry and Chemotherapy.* New York: Elsevier Science B.V., 2002:433–454.
3. De Clercq E. Trends in the development of new antiviral agents for the chemotherapy of infections caused by herpesviruses and retroviruses. *Rev Med Virol* 1995; 5:149–164.
4. De Clercq E. Toward improved anti-HIV chemotherapy: therapeutic strategies for intervention with HIV infections. *J Med Chem* 1995; 38:2491–2517.
5. De Clercq E. Novel compounds in preclinical/early clinical development for the treatment of HIV infections. *Rev Med Virol* 2000; 10:255–277.
6. De Clercq E. New developments in anti-HIV chemotherapy. *Curr Med Chem* 2001; 8:1543–1572.
7. De Clercq E. Clinical potential of acyclic nucleoside phosphonates cidofovir, adefovir, and tenofovir in the treatment of DNA virus and retrovirus infections. *Clin Microbiol Rev* 2003, 16:569–596.
8. De Clercq E. Potential of acyclic nucleoside phosphonates in the treatment of DNA virus and retrovirus infections. *Expert Rev Anti-Infect Ther* 2003; in press.
9. Ying C, De Clercq E, Neyts, J. Selective inhibitors of hepatitis B virus replication. *Curr Med Chem Anti-Infect Agents* 2003, 1:21–43.
10. De Clercq E. Strategies in the design of antiviral drugs. *Nature Rev Drug Discovery* 2002; 1:13–25.

chapter twelve

Other drug applications

Joy Y. Feng*
Gilead Sciences, Durham, NC

Phillip A. Furman
Pharmasset, Inc., Tucker, GA

Contents

* Corresponding author.

12.1 Introduction

This chapter will focus on nucleoside analogs approved for therapeutic use in the treatment of diseases that are nonviral in origin. We will pay particular attention to drugs with proven clinical usage, but also include some experimental drugs currently under development. The chemical structures of the compounds are shown in Figure 12.1 in the order of their appearance in the text.

12.2 Anticancer agents

Nucleoside analogs are widely used for the treatment of hematological malignancies, including acute and chronic leukemia; hairy cell leukemia; indolent non-Hodgkin's lymphomas; and some solid tumors.[1,2] Nucleosides are readily transported into dividing cells through specific nucleoside transport systems, which occur by two mechanistically distinct processes: facilitated diffusion and Na(+)-nucleoside cotransport.[3] Once inside the cell, nucleoside analogs are converted to their active 5′-monophosphate (MP), 5′-diphosphate (DP), and 5′-triphosphate (TP) forms by cellular kinases. The nucleotide form interferes with cellular targets involved in nucleotide *de novo* biosynthesis and DNA synthesis.

The important principle of using antimetabolites in chemotherapy has led to the development of a series of anticancer drugs. A list of Food and Drug Administration (FDA)-approved anticancer nucleoside analogs and their clinical indications is summarized in Table 12.1. We will focus on the development of some of the more important drugs and their mechanism of action.

12.2.1 6-Mercaptopurine and thioguanine

One of the first nucleoside analogs approved for use as an anticancer agent was 6-mercaptopurine (6-MP). This compound was synthesized and developed by George Hitching, Gertrude Elion, and associates at the Wellcome Research Laboratories, in collaboration with Sloan–Kettering Institute in 1951. Only 2 years after its synthesis and preclinical and clinical assessment,

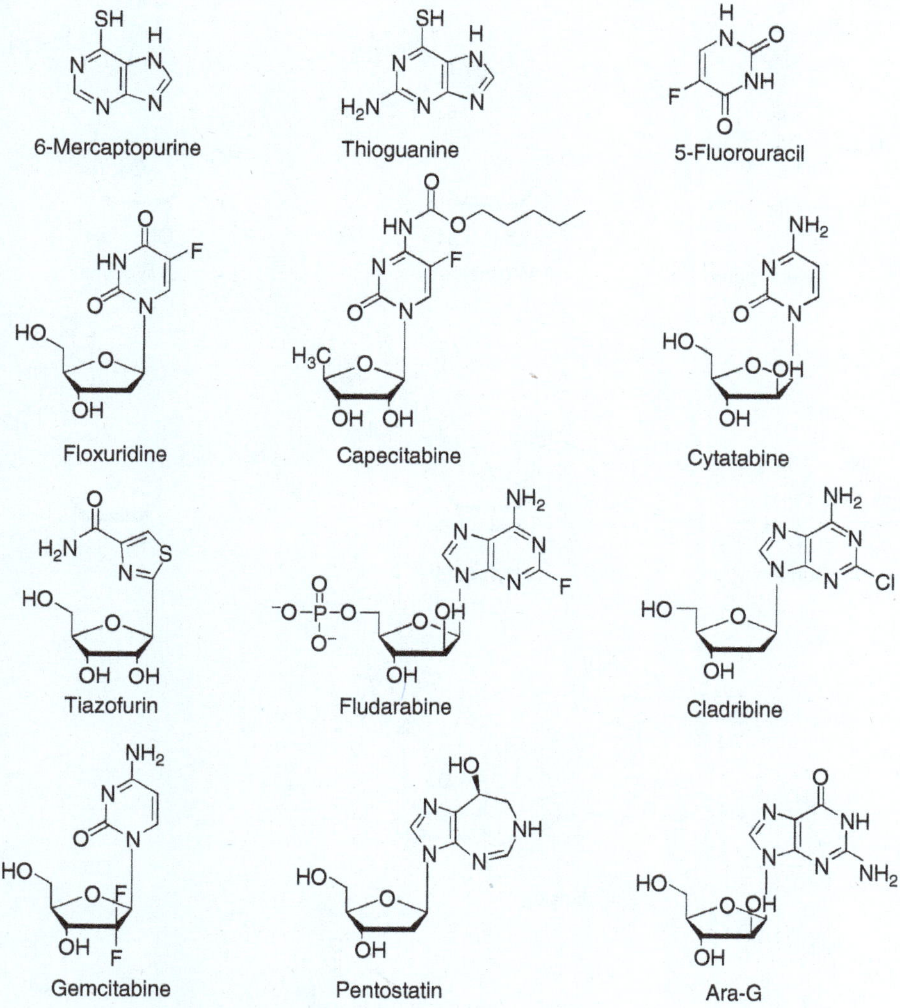

Figure 12.1 Chemical structures of nucleoside analogs.

Continued.

6-MP was approved by the FDA for the treatment of acute leukemia in children. The addition of 6-MP to the antileukemia regimen increased the median survival time from 3 to 4 months to 12 months. Today, children with acute leukemia treated with 6-MP in combination with other drugs have an almost 80% cure rate.

In contrast to its rapid acceptance by physicians, it took scientists nearly 2 decades to unravel the mechanism of action of 6-MP. As illustrated in Figure 12.2, 6-MP is initially activated by hypoxanthine-guanine phosphoribosyltransferase (HGPRT) to 6-thioinosine 5′-monophosphate (t-IMP). This is further converted to 6-thioxanthosine monophosphate (t-XMP) by IMP dehydrogenase (IMPDH); t-IMP is then further metabolized into its 5′-triph-

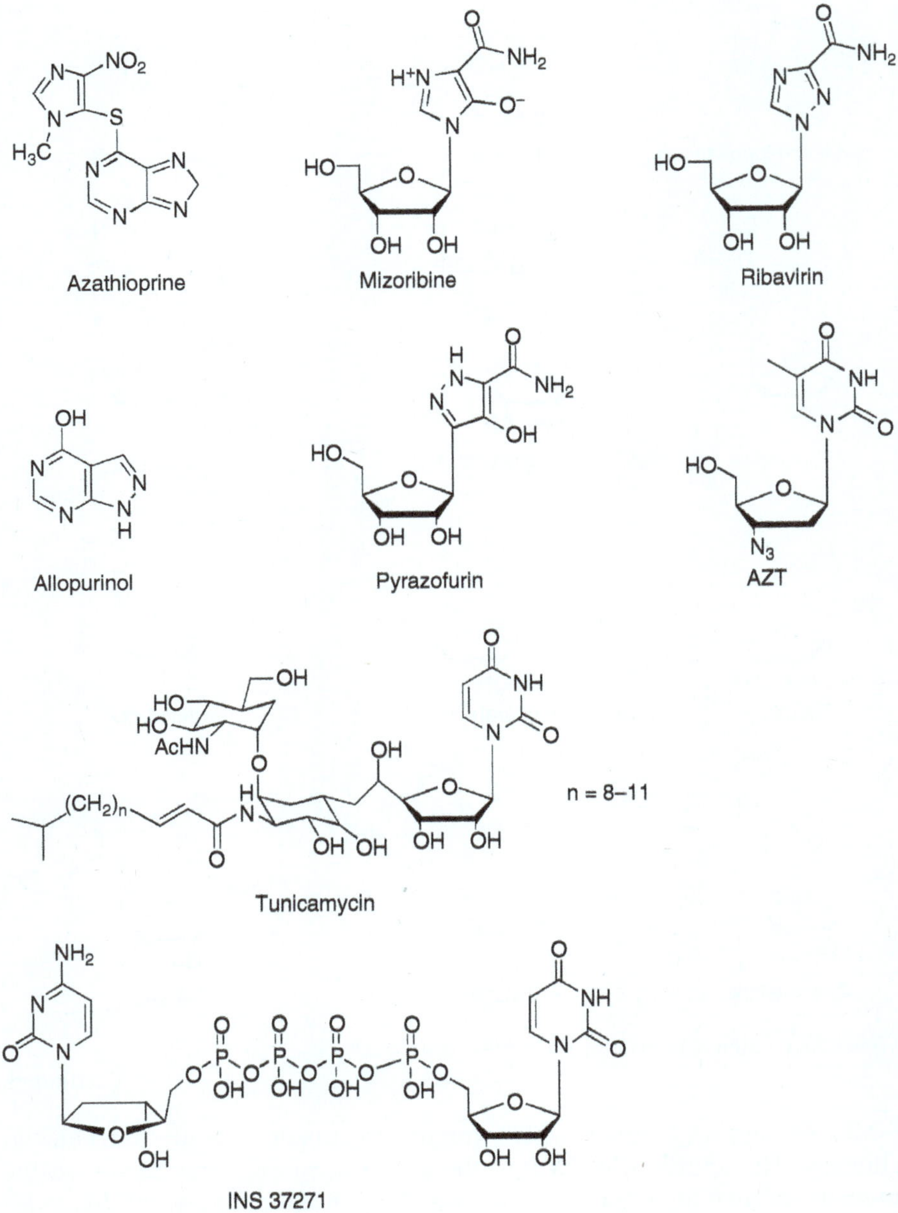

Figure 12.1 Continued.

osphate form, t-ITP. GMP synthase can convert t-IMP into t-GMP by and GMP and GDP kinase and ribonucleotide reductase can further converted into form t-GTP and t-dGTP. Methylation of t-IMP by thiopurine methyltransferase using S-adenosyl-L-methionine forms 6-methythioinosine 5′-monophsophate (me-t-IMP). 6-MP was found to interfere with a variety

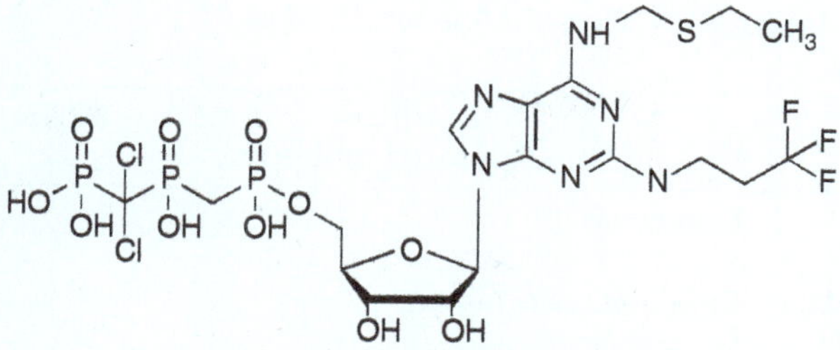

AR-C69931MX

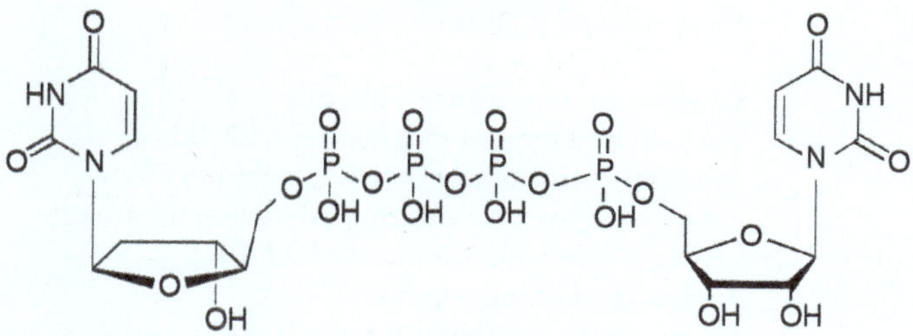

INS-365

Figure 12.1 Continued.

of vital biochemical pathways. However, the principle sites of action in humans can be summarized into the following mechanisms (Figure 12.3)[4,5]:

- Feedback of *de novo* purine synthesis can be inhibited. Me-t-IMP is a strong inhibitor of the enzyme 5-phosphoribosyl-α-pyrophosphate (PRPP) amidotransferase, which catalyzes the conversion of PRPP into phosphoribosylamine, an important precursor during IMP synthesis. t-IMP is also an inhibitor of PRPP amidotransferase, but less potent than me-t-IMP.
- Inhibition of IMP synthesis by me-t-IMP and t-IMP leads to decrease in the formation of XMP, GMP, and AMP.
- t-dGTP is incorporated into DNA and leads to DNA damage, such as single-strand breaks, DNA–protein cross-links, interstrand cross-links, and sister chromatid exchanges.
- RNA transcription can be inhibited through t-ITP-mediated inhibition of RNA polymerase.
- Synthesis of nicotinamide adenosine dinucleotide (NAD^+) from ATP can be inhibited.

Selectivity for neoplastic cells most likely depends on the levels of individual anabolic and catabolic enzymes. Because the catabolic enzyme levels are generally much lower in tumor cells than in normal cells, it is expected that the anticancer agents have a longer half-life in the tumor cells. In addition, mitotic rate, drug transport, and metabolite pool sizes can contribute to the selective cytotoxicity.

Table 12.1 FDA-Approved Nucleoside Analogs Used as Anticancer Drugs

Drug name	Approved indication	Approval date
Mercaptopurine (6-MP)	Leukemias	1953
Fluorouracil (5-FU)	Common solid tumors	1962
Thioguanine (6-TG)	Leukemias	1966
Cytarabine (ara C)	Leukemias	1969
Floxuridine (intra-arterial)	Stomach and gastrointestinal cancer	1970
Tiazofurin	Chronic myelogenous leukemia	1991
Fludarabine	Palliative treatment of patients with B-cell lymphocytic leukemia (CLL) who have not responded or have progressed during treatment with at least one standard alkylating agent-containing regimen	1991
Pentostatin	Single-agent treatment for adult patients with alpha interferon refractory hairy cell leukemia	1991
Cladribine	Active hairy cell leukemia	1993
Pentostatin	Single-agent treatment for untreated hairy cell leukemia patients with active disease as defined by clinically significant anemia, neutropenia, thrombocytopenia, or disease-related symptoms	1993
Allopurinol	Patients with leukemia, lymphoma, and solid tumor malignancies receiving cancer therapy that causes elevations of serum and urinary uric acid levels who cannot tolerate oral therapy.	1996
Gemcitabine	Treatment of patients with locally advanced (nonresectable stage II or III) or metastatic (stage IV) adenocarcinoma of the pancreas; indicated for first-line treatment and for patients previously treated with a 5-FU-containing regimen	1996
Capecitabine	Accelerated approval (clinical benefit subsequently established); treatment of metastatic breast cancer resistant to paclitaxel and an anthracycline-containing chemotherapy regimen or resistant to paclitaxel and for whom further anthracycycline therapy may be contraindicated, e.g., patients who have received cumulative doses of 400 mg/m^2 doxorubicin or doxorubicin equivalents	1998
Gemcitabine	For use in combination with cisplatin for the first-line treatment of patients with inoperable, locally advanced (stage IIIA or IIIB) or metastatic (stage IV) non-small cell lung cancer	1998

Table 12.1 FDA-Approved Nucleoside Analogs Used as Anticancer Drugs (Continued)

Drug name	Approved indication	Approval date
Cytarabine liposomal	Accelerated approval (clinical benefit not established) for intrathecal therapy of lymphomatous meningitis	1999
Capecitabine	Initial therapy of patients with metastatic colorectal carcinoma when treatment with fluoropyrimidine therapy alone is preferred; combination chemotherapy has shown a survival benefit compared to 5-FU/LV alone; a survival benefit over 5-FU/LV has not been demonstrated with this drug as monotherapy	2001
Capecitabine	Treatment in combination with docetaxel of patients with metastatic breast cancer after failure of prior anthracycline-containing chemotherapy	2001

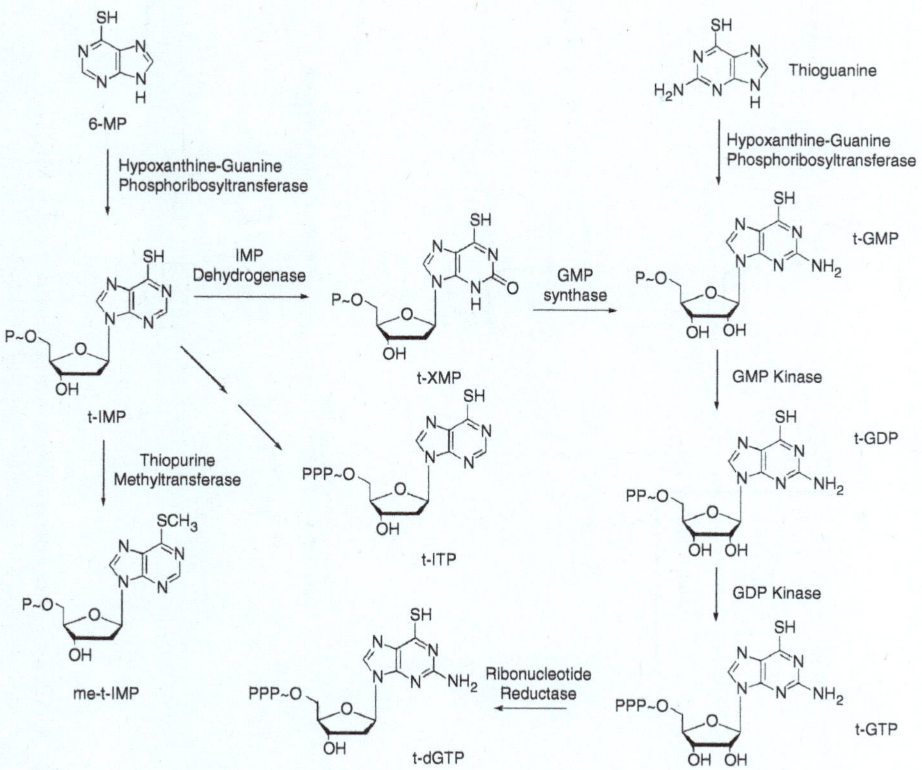

Figure 12.2 Metabolism of 6-MP in humans.

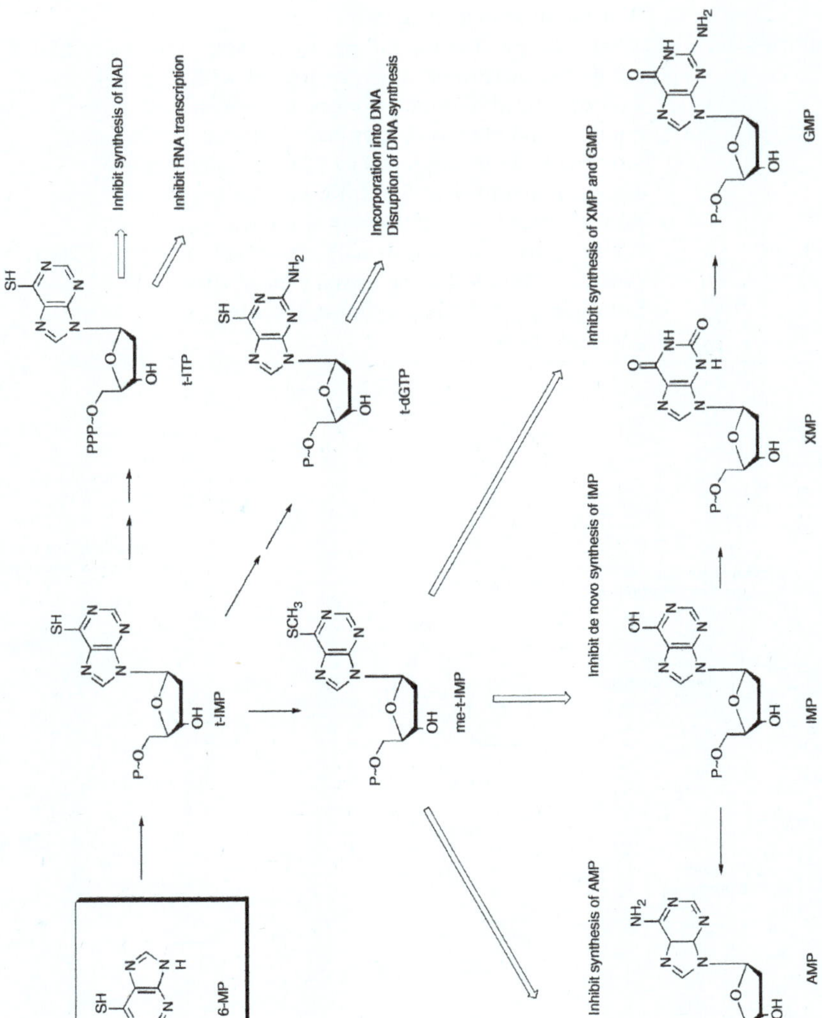

Figure 12.3 Mechanism of action for the anticancer activity of 6-MP.

Thioguanine (6-thiol-2-aminopurine) is another thiopurine analog found to be active against leukemia cell lines. Synthesized by Elion and Hitchings in 1955, it was found to be more active but also more toxic than 6-MP.[4] Thioguanine was approved by the FDA in 1966 and is mainly used in the treatment of acute myelocytic leukemia in adults, in combination with cytarabine. Thiogunine is initially activated by HGPRT to t-GMP, and further to t-GDP, t-GTP, and t-dGTP, as described previously for 6-MP metabolism[5] (Figure 12.2). The active metabolite of thioguanine is believed to be t-dGTP because it is known to be a good substrate for human DNA polymerases α, δ, and γ, with K_m values similar to those of natural substrates.[6] Incorporation of t-dGMP into DNA leads to notable alterations in DNA properties[7] and affects interaction with several DNA-binding proteins, including DNA polymerases, DNA ligases, endonucleases, and topoisomerases.[8,9]

12.2.2 5-Fluorouracil (5-FU), floxuridine, and capecitabine

In 1957, Heidelberger[10] synthesized 5-fluorouracil, which is primarily used for the treatment of common solid tumors, particularly breast, ovarian, and gastrointestinal carcinomas.[1,11] In tumor cells and in sensitive normal tissue, 5-FU is converted to the active form 5-fluorodeoxyuridylate (FdUMP) by thymidine kinase.[11] FdUMP is an irreversible inhibitor of thymidylate synthase, an enzyme essential for *de novo* synthesis of dTMP from dUMP using N^5,N^{10}-methylenetetrahydrofolate (N^5,N^{10}-methylene-THF) as the phosphate donor.[12] Like dUMP, FdUMP binds to the enzyme and undergoes the first two steps of the normal enzymatic reaction. However, in the third step, in which the enzyme normally extracts the 5-hydrogen from dUMP, it encounters the irremovable 5-fluoro of 5-FU. As a result, the enzyme is all but permanently immobilized as an enzyme-FdUMP-THF ternary complex. Consequently, the conversion of uridilate to thymidilate is halted, thereby depriving cells of dTTP needed for DNA synthesis.

Additionally, 5-FU is metabolized to ribo- and deoxyribonucleotides, which act as false bases for incorporation into RNA and DNA.[11] This metabolic error can interfere with DNA replication, RNA processing, and protein synthesis. The FDA approved 5-FU in 1962, and it is currently used to treat colon, rectal, breast, stomach, and pancreatic cancer.

Floxuridine (5-fluorodeoxyuridine, 5-DFUR), the nucleoside analog of 5-FU was developed to gain a better therapeutic index in the treatment of stomach and gastrointestinal cancer. It was approved in 1970; however, contrary to what was anticipated, this compound never underwent wide clinical application because it did not demonstrate toxicological or therapeutic superiority over 5-FU.[1] Recent studies have suggested that in human colorectal cancer cell lines and various animal tumor model systems, 5-DFUR is more effective than 5-FU.[13]

Capecitabine, a fluoropyrimidine carbamate, is an orally administered systemic prodrug of 5′-DFUR that is converted to 5-FU. As shown in Figure 12.4, capecitabine is metabolized to 5-FU by carboxylesterase, cytidine deam-

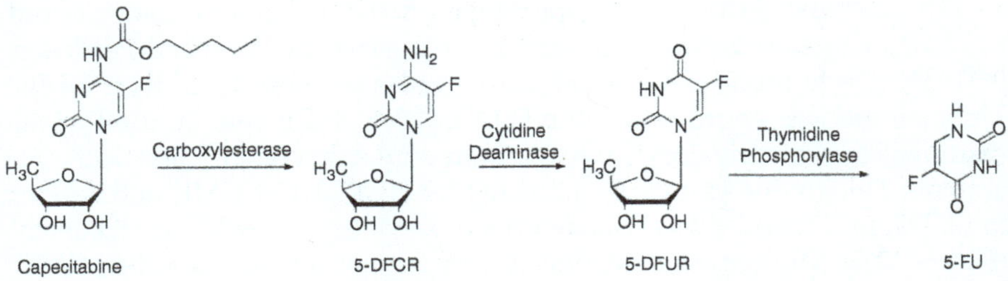

Figure 12.4 Activation pathway of capecitabine.

inase, and thymidine phosphorylase. In 1998, capecitabine was approved by the FDA and is used for the treatment of breast, colon, and rectal cancer. Combination chemotherapy has shown a survival benefit compared to 5-fluorouracil (5-FU)/oral leucovorin (LV) alone; however, a survival benefit over 5-FU/LV has not been demonstrated with capecitabine monotherapy.

12.2.3 *Cytarabin (cytosine arabinoside, ara-C) and 5-azacytidine (5-AzCR)*

Cytarabine (1-β-D-arabinosylcytosine, ara C) is one of the most thoroughly studied drugs in cancer chemotherapy.[2] The drug was approved in 1969 and is used predominantly to treat adult acute myelogenous leukemia. The active form of cytarabine is the 5′-triphosphate form, with the arabinosyl carbohydrate moiety to deliver the cytotoxic properties. The rate-limiting step for the activation of cytarabine is the conversion to cytarabine-MP by deoxycytidine kinase — the same enzyme responsible for the phosphorylation of deoxycytidine to dCMP. A strong relationship has been observed between the amount of cytarabine incorporated into cellular DNA and the extent of subsequent decrease in clonogenicity,[14] indicating that incorporation is related to cytotoxicity.

Studies have demonstrated that once cytarabine becomes incorporated into DNA, multiple mechanisms might affect DNA synthesis.[2] First, cytarabine-TP is a strong alternative substrate inhibitor of DNA polymerase through competition with the natural nucleotide and serves as a nonobligating chain terminator. Incorporation of cytarabin-MP reduces the ability of the DNA polymerase to use the 3′-terminal nucleotide analog as a substrate for addition of a subsequent nucleotide; this causes a decrease in the rate of DNA synthesis. Second, incorporated cytarabine-MP was less readily removed from the DNA by the 3′ → 5′-exonuclease activity of DNA polymerase ε. Third, when cytarabine-MP was present at the 3′-terminus of an oligodeoxynucleotide, ligation to a 5′-end of an adjacent oligodeoxynucleotide by DNA ligase was inhibited.[15]

12.2.4 *Tiazofurin*

Tiazofurin is a C-nucleoside designed and synthesized by Robins in 1982 to target the elevated IMPDH activity in tumors.[16] Tiazofurin is phosphorylated

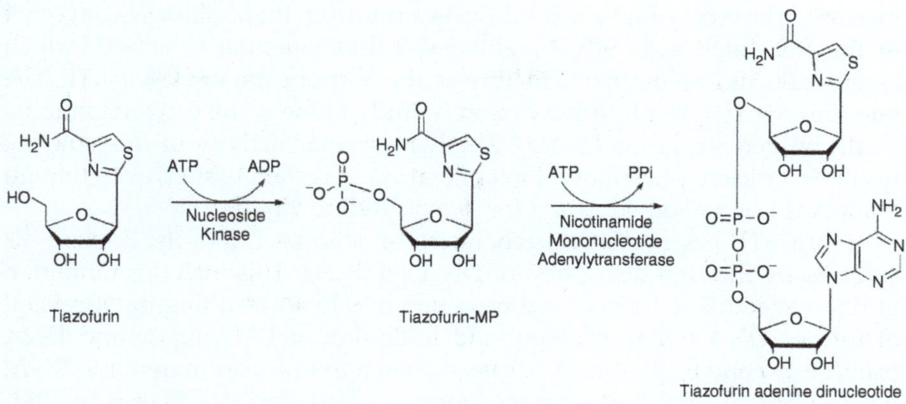

Figure 12.5 Activation pathway of tiazofurin.

to its MP derivative by adenosine kinase, 5-nucleotidase, and nicotinamide ribonucleoside kinase.[17,18] Tiazofurin-MP is then converted to the active-form tiazofurin adenine dinucleotide (TAD, Figure 12.5), as an analog of NAD, by human nicotinamide mononucleotide adenylytransferase (NMNAT). The conversion of tiazofurin to TAD in leukemia cells was significantly (20-fold) higher than the synthesis of TAD that occurred in normal leukocytes.[19] Low levels of NMNAT activity in tumor cells have been associated with the development of tiazofurin resistance.[20,21]

TAD serves as a potent inhibitor of IMPDH, an enzyme that catalyzes the rate-limiting step in the *de novo* synthesis of the guanine nucleotides, which catalyzes the oxidation of IMP to XMP using NAD.[12,22] TAD binds to the NAD/NADH site of IMPDH with $K_i = 0.13$ μ*M*.[23,24] This inhibition of IMPDH leads to significantly decreased GTP concentration and is believed to be the main mechanism of action for tiazofurin. It is worth mentioning that IMPDH activities are markedly elevated in various human and animal cells, particularly in the rapidly growing neoplastic cells.[25] In addition, tiazofurin also leads to down-regulation of ras and myc oncogenes and induced maturation of blast cells.[25] In 1991, the FDA granted orphan-drug status designation to tiazofurin for the treatment of chronic myelogenous leukemia (CML) in accelerated phase or blast crisis.

12.2.5 Fludarabine (2-fluorovidarabine, 2-F-araA)

The success of cytarabine inspired the synthesis and evaluation of other arabinosyl analogs. Arabinosyladenine (vidarabine, ara A) was found to be ineffective as an intravenous drug, mainly due to its susceptibility to rapid metabolic clearance by deamination by the ubiquitous adenosine deaminase. Fludarabine emerged as a second generation of arabinosyladenosine compounds.[2] The substitution of a fluorine atom at the 2-carbon of the adenosine ring made the amino group resistant to deamination, and the addition of the phosphate group at the 5-carbon of the arabinose ring drastically increased the compound's solubility. As a prodrug, fludarabine is converted to the free nucleoside 9-β-D-arabinosyl-2-fluoroadenine (F-ara-A), which enters cells and accumulates mainly as the 5′-triphosphate, F-ara-ATP. The rate-limiting step in phosphorylation of fludarabine is the conversion to its 5′-MP by deoxycytidine kinase.[2] The high specific activity of this enzyme results in efficient phosphorylation of F-ara-A in certain tissues, even though F-ara-A is not a good substrate for deoxycytidine kinase.

F-ara-ATP has multiple mechanisms of action.[26] Primarily, it exerts its cytotoxicity through disruption of DNA synthesis. This includes inhibition of ribonucleotide reductase; incorporation into DNA resulting in repression of further DNA polymerization; and inhibition of DNA ligase and DNA primase. Secondarily, F-ara-A has been shown to be incorporated into RNA, which leads to inhibition of transcription in leukemia cell lines.[26] In 1991, fludarabine was approved by the FDA for the treatment of B-cell lymphocytic leukemia. In the past decade, as a single agent or in combination with DNA damaging agents, fludarabine has had a major impact on increasing the effectiveness of treatment of patients with indolent B-cell malignancies.[26]

12.2.6 *Cladribine (2-chloro-2′-deoxyadenosine, 2-CdA, leustatin)*

Cladribine is a deoxyadenosine analog with a chlorine atom at the 2-carbon of the adenosine ring. It was first synthesized as an intermediate during synthesis of 2-deoxynucleoside analogs by Venner in 1960.[27] Unlike ara A, cladribine is resistant to deamination by adenosine deaminase.[28] Its specific toxicity to lymphocytes was reported by Carson in 1980.[29] Subsequently, cladribine was shown to have activity against hairy cell leukemia[30] and many indolent B-cell diseases.[31,32]

Cladribine is unique compared with traditional antimetabolite drugs in that it is equally active against dividing and resting lymphocytes; this makes the drug especially important in treating indolent lymphoid malignancies, such as chronic lymphocytic leukemia, because most cells in these disorders are in the resting phase. Similar to many antitumor nucleoside analogs, the active metabolite of cladribine is its 5′-triphosphate form; however, the rate-limiting step for its activation is believed to be the phosphorylation of cladribine-MP to DP by a monophosphate kinase, probably AMP kinase.[33] Cladribine has little or no effect on DNA synthesis and only a moderate effect on DNA repair.[2] The main mechanism of action for this drug is believed to be inhibition of ribonucleotide reductase by the cladribine-TP.[2] Cladribine was approved by the FDA in 1993 to treat active hairy cell leukemia.

12.2.7 *Gemcitabine (dFdC, dFdCyd, Gemzar™)*

Gemcitabine, 2′,2′-difluorodeoxycytidine, is an analog of deoxycytidine with two fluorines placed in the geminal configuration on the 2-carbon. It was first synthesized in 1988[34] and was soon found to be active against human leukemic cell lines and murine solid tumors.[35,36] The FDA approved it in 1996 for the treatment of advanced pancreatic cancer and it remains the treatment of choice for this disease.

Gemcitabine is readily taken up by tumor cells and is phosphorylated to dFdCMP, dFdCDP, and dFdCTP, with the phosphorylation of gemcitabine to dFdCMP as the rate-limiting step.[2] Interestingly, all of the gemcitabine nucleotide metabolites facilitate the cytotoxic effects of gemcitabine through different mechanisms.[2,37] dFdCMP is an inhibitor of dCMP deaminase, the enzyme that inactivates gemcitabine to 2′,2′-difluorodeoxyuridine. dFdCDP is a competitive inhibitor of ribonucleotide reductase, which is required for the *de novo* synthesis of dCDP from CDP. Inhibition of ribonucleotide reductase by dFdCDP and dCMP deaminase by dFdCMP resulted in decrease in cellular dCDP polls, elevation of the intracellular dFdCTP/dCTP ratio and, ultimately, in enhanced incorporation of dFdCTP into DNA. Even though dFdCTP was a poor substrate for DNA polymerase with an efficiency of incorporation only 3% that of dCTP, the elongation of the 3′-dFdCMP DNA was found to be efficient.[2]

In whole cells, gemcitabine nucleotides were also found predominantly in internal positions in DNA.[38] In addition, the incorporated dFdCMP showed significant resistance to the 3′ → 5′ excision.[2] The cytotoxicity of gemcitabine has long been strongly associated with the amount of its incorporation into cellular DNA[38,39]; however, it is still unclear by which mechanism(s) the incorporation leads to cell death. Recent studies showed that gemcitabine-substituted DNA poisons topoisomerase I by selective stabilization of the cleavage complexes, consequently leading to DNA-strand breaks.[40] The mechanisms associated with gemcitabine resistance include increased cytidine deaminase activity,[41] a cellular deficiency in deoxycytidine kinase,[42] and reduced uptake by nucleoside transporter.[43]

12.2.8 *Pentostatin (2′-deoxycoformycin, Nipent™)*

Pentostatin, (8*R*)-3-(2-deoxy-β-D-erythro-pentofuranosyl)-3,4,7,8-tetrahydroimidazole[4,5,-*d*]-[1,3]diazepin-8-ol, an analog of 2-deoxyadenosine, was first isolated in 1974 from the soil organism *Streptomyces antibioticus*.[44] Pentostatin is a potent inhibitor of adenosine deaminase, which is found in high concentrations in lymphatic tissue.[45] During preclinical screening, it did not demonstrate antitumor activity; however, because it was a potent inhibitor of adenosine deaminase, it continued to be studied in combination with vidarabine (ara-A), which is rapidly deactivated by adenosine deaminase.[46] The unexpected effect of tumor shrinkage in patients with lymphoid malignancies led to its intensive development.[47]

The exact mechanism of action of this drug is unknown.[48] Inhibition of adenosine deaminase mainly results in disruption of the deamination of adenosine and deoxyadenosine to inosine and deoxyinosine, causing an accumulation of dATP and ATP, which are cytotoxic, especially to lymphocytes. This is partly by inhibition of an enzyme responsible for cellular methylation reactions, S-adenosyl-L-homocysteine hydrolase (AdoHcyase),[49] through functional changes of adenosine receptors.[50] Also, a buildup of intracellular levels of dATP results in the inhibition of ribonucleotide reductase, which ultimately leads to an inhibition of DNA synthesis.[51] Taken together, this results in the arrest of cells in the G1 or S phase of the cell cycle and subsequent cell dysfunction or death. Pentostatin was approved by the FDA in 1991 for treatment of hairy cell leukemia.

12.2.9 Ara-G (9-β-D-arabinosylguanine)

Ara-G, a sugar-modified deoxyguanosine analog, was first synthesized in 1964 by Reist and Goodman.[52] Like deoxyguanosine, it is selectively cytotoxic to leukemia cells from patients with T-acute lymphoblastic leukemia.[53] However, ara-G was resistant to degradation by purine nucleoside phosphorylase activity in T-lymphoblasts.[54] Ara-G was phosphorylated to ara-GTP by cytoplastic deoxycytidine kinase and mitochondrial guanosine kinase,[55] and the phosphorylation occurred only in T-cells.[53] Studies showed that a low level of mitochondrial deoxyguanosine kinase is the dominant factor in acquired resistance to ara-G cytotoxicity.[56] Although ara-G was incorporated into nuclear DNA and mitochondrial DNA, the former is believed to account for its cytotoxicity.[54,57–59] Ara-GTP was a potent inhibitor of DNA polymerase α, β, and γ[60] and telomerse.[61] A water-soluble prodrug of Ara-G, 2 amino-6-methyoxypurine arabinoside (nalarabine) is currently being evaluated in clinical trials for therapy of T-cell lymphoid malignancies.[62]

12.3 Immunosuppressants

During the investigation of nucleoside analogs to treat malignant diseases, it was found that one of the major side effects was their profound immunosuppressive property. This has been exploited for preventing rejection following organ transplant and for the treatment of nonmalignant disorders, including immunological and rheumatological diseases and multiple sclerosis.

12.3.1 Antirejection agent

12.3.1.1 Azathioprine

Azathioprine, a 1-methyl-4-nitro-5-imidazolyl derivative of 6-MP, was developed as a prodrug of 6-MP in the early1960s.[4] Based on the rationale that the immunoblastic lymphocyte formed during an immune response closely resembled leukemic lymphocytes, the effects of 6-MP on the immune response of mammals were first reported by Schwartz[63] and Calne.[64] Pharmacokinetic studies of 6-MP revealed that it undergoes extensive metabolic transformations *in vivo*. The major metabolites include 6-thiouric acid formed by the action of xanthine oxidase on 6-MP; products

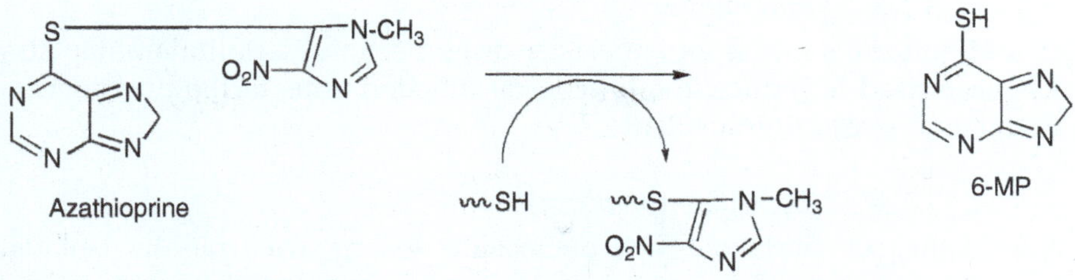

Figure 12.6 Reaction of sulfhydryl ion on azathioprine to release 6-MP.

formed from sulfur-methylation; and further oxidation of the methylthio derivative on the sulfur or on the purine ring. In an attempt to modify the metabolism of 6-MP by protecting the sulfur from oxidation and hydrolysis by adding blocking groups, which could be removed intracellularly to release 6-MP, a prodrug for 6-MP, 1-metyl-4-nitro-5-imidazolyl derivative, was developed.[4]

The compound, now known as azathioprine, is subject to attack by sulfhydryl groups and other nucleophiles once the drug is transported into the cell. In particular, the glutathione present in red blood cells reacts with azathioprine to give 6-MP (Figure 12.6). Studies showed that azathioprine improved the oral bioavailability of 6-MP and also was superior to 6-MP for preventing rejection of canine kidney homografts.[65,66] In 1962, azathioprine was used successfully for transplantation of kidney to unrelated recipients; by 1964, the combination of steroids and azathioprine had become the standard of care.[67] This compound was approved by the FDA in 1968 as an immunosuppressant. For a long time, azathioprine was the only drug available to prevent rejection of transplanted organs, and today it remains a mainstay in kidney transplantation.

The immunosuppressive effects of azathioprine have been studied in a wide variety of immunological systems.[68] It is well known that T-lymphocytes play a primary role in many autoimmune disorders and in allograft rejection; however, the mechanism of action for azathioprine has not been fully understood. It has been shown that, once metabolized to 6-methyl-MP and 6-thioguanine, azathioprine blocks the *de novo* pathway of purine synthesis, which is critical in lymphocytes due to their lack of a salvage pathway.[69,70] In addition, various metabolites of azathioprine can be incorporated into replicating DNA.[69,70] However, none of these effects can fully explain all of the laboratory and clinical findings of azathioprine-induced immunosuppression.[71] Most recently, a report by Tiede et al.[72] linked the immunosuppressive effects of azathioprine and 6-MP to their role in the control of T-cell apoptosis. It has been well known that optimal activation of T-lymphocytes requires a cascade of signal transduction that, when blocked, can lead to apoptosis.[71] Tiede's study suggested that azathioprine and its metabolite 6-MP caused T-cell death through modulation of Rac1 activation upon CD28 constimulation.[72]

12.3.1.2 *Fludarabine*

In addition to its use as an anticancer drug (Section 12.2), fludarabine also has been used to reduce immunological function, thus facilitating nonmyeloablative stem cell transplants.[73,74]

12.3.1.3 *Mizoribine (bredinin)*

Mizoribine, an imidazole-type nucleoside analog, was initially isolated from the culture medium of the mold *Eupenicillium brefedianum* in 1974.[75] Mizoribine is phosphorylated by adenosine kinase to its active form, mizoribin-MP, which inhibits GMP synthesis by blocking two enzymes in the *de novo* synthesis pathway of GMP, IMPDH, and GMP-synthetase.[76,77] Drug resistance is associated with defective adenosine kinase activity.[76] The drug was developed and registered in Japan for the prevention of rejection in renal transplantation[78] and for several other autoimmune diseases, including idiopathic glomerulonephritis, lupus nephritis, and rheumatoid arthritis.[79]

12.3.1.4 *Ribavirin*

Ribavirin, a purine analog, is phosphorylated to its active form, ribavirin-MP, which blocks the synthesis of GMP through inhibition of IMDPH. The high-resolution (1.85 Å) crystal structure of ribavirin-MP bound to human IMPDH II showed that the phosphate and ribose portions of the molecule interact with the enzyme exactly as IMP, indicating that ribavirin-MP is an excellent substrate mimic of IMP.[80] Ribavirin is used widely as an IMPDH inhibitor in clinical and basic research[24,79] and is also an FDA-approved agent for the treatment of HCV infection.

12.3.2 *Autoimmune disease and anti-inflammatory agent*

Azathioprine is an approved drug for the treatment of severe rheumatoid arthritis,[81,82] a chronic systemic inflammatory disease that mainly affects the synovial membranes of multiple joints in the body. The cause of rheumatoid arthritis is not yet known; however, it is considered an autoimmune disease.

Azathioprine is also used to treat vasculitis and lupus. Vasculitis is a group of diseases that involves inflammation in blood vessels due to tissue damage caused by blood cells entering the tissues. Vasculitis can be caused by bacterial, viral, or fungal infection of the blood vessel walls or by an immune reaction to the vessel walls. The latter is the more common cause of vasculitis. Used in combination with cortisone-type medications, azathioprine is effective in treating severe vasculitis or vasculitis that does not respond well to cortisone-type drugs.[83] Systemic lupus erythematosus, or lupus, is a chronic autoimmune disease that causes inflammation in many parts of the body, including the joints, skin, kidneys, heart, lungs, blood vessels, and brain. Azathioprine is one of the most commonly used cytotoxic agents for treating this disease.[84]

Another use of azathioprine is to treat multiple sclerosis.[85,86] Multiple sclerosis is a chronic disease of the central nervous system predominantly affecting young adults during their most productive years. It is caused by a combination of factors, including autoimmune response. However, the specific cause of the disease is not known. Azathioprine is the most widely used immunosuppressive drug for treating multiple sclerosis.

Claribine is a nucleoside analog used "off-label" for the treatment of multiple sclerosis.[87–89] Between 1993 and 1997, claribine was developed to treat chronic progressive multiple sclerosis; however, the studies were subsequently terminated due to a less favorable efficacy profile in phase III studies.* Its mechanism of action is probably through the inhibition of ribonucleotide reductase by claribine-TP in T-lymphocytes. The FDA-approved indication of claribine is the treatment of lymphoid malignances.

Ulcerative colitis and Crohn's disease, collectively know as inflammatory bowel disease, afflict an estimated 1 million Americans and produce symptoms that impair quality of life and ability to function. Azathioprine has been effectively treating the disease, especially in maintenance of the disease in remission.[90,91]

12.4 Metabolic disease — gout and hyperuricemia

Hyperuricemia is defined as a serum urate concentration greater than 70 mg per liter (416 μmol per liter) — the approximate level at which urate is supersaturated in plasma.[92] The hyperuricemia of gout produces deposits of small crystals of uric acid in joints; this results in extreme pain or large deposits, tophi, that result in gout arthritis and restricted joint movement.[4] In patients who excrete excessive amounts of uric acid in the urine, urate stones often form in the kidney.

The most prevalent cause of gout is impaired uric acid excretion. It may also result from a number of metabolic insufficiencies.[12] Gout can be caused by stimulated purine biosynthesis associated with HGPRT and glucose-6-phosphatase deficiency. As a result, elevated levels of hypoxanthine and xanthine are further metabolized to high levels of uric acid by xanthione oxidase. Allopurinol, 4-hydroxypyrazolo (3,4-d) pyrimidine, a hypoxanthine analog that interchanges the N(7) and C(8) positions, is a potent substrate inhibitor of xanthione oxidase in humans (Figure 12.7, lower section). In addition, oxypurinol, the product formed from allopurinol, remains tightly bound to the reduced form of the enzyme, thereby inactivating it.[4] Allopurinol helps to alleviate several of the clinical problems associated with gout and also to reduce the secondary hyperuricemia associated with the therapy of malignancy.[4] Allopurinol was approved by the FDA in 1966 for the treatment of hyperuricemia and chronic gout.

* Medical information was provided by Ortho Biotech.

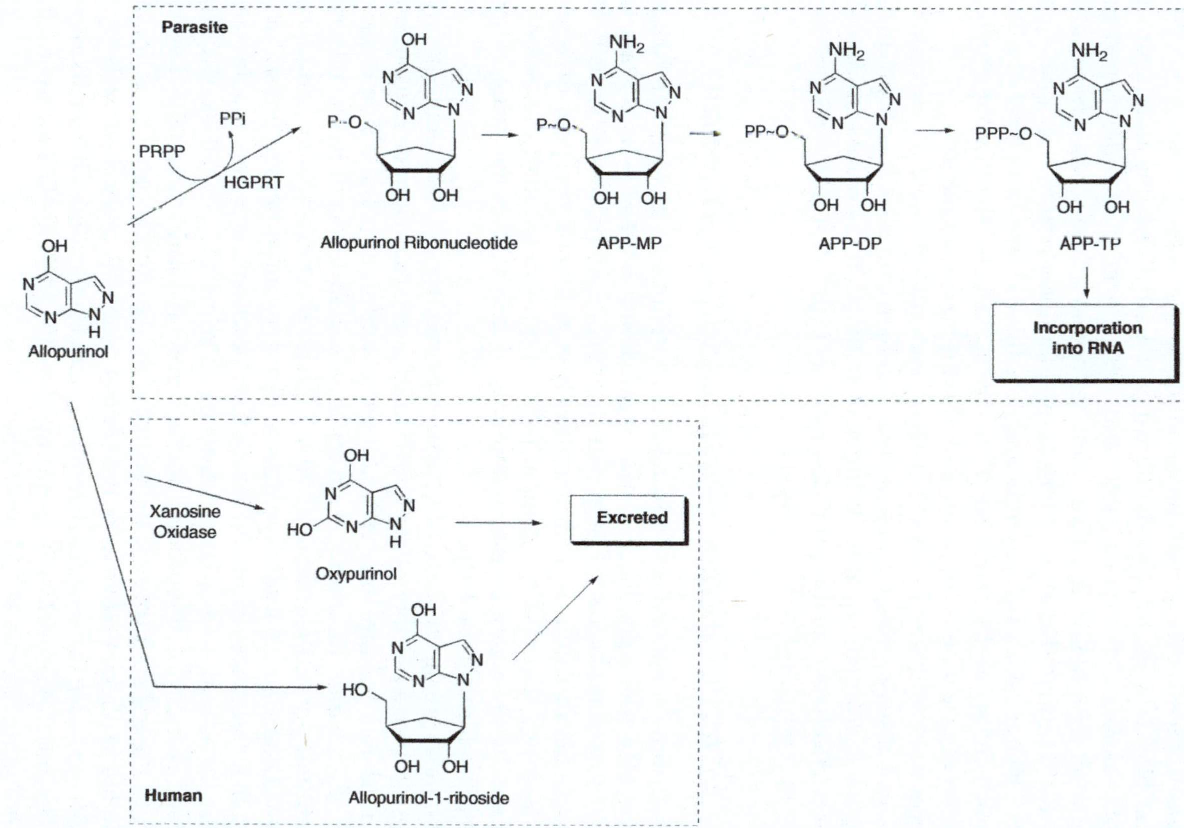

Figure 12.7 Different metabolic pathways of allopurinol in parasite and in human.

12.5 Cardiovascular drugs

12.5.1 Antiarrhythmic agent

Adenosine, a natural purine nucleoside, serves important autocrine and paracrine roles in numerous tissues. Its action on the mammalian heart was first described in 1929.[93] In 1930, Honey et al. reported its AV nodal blocking properties in the human heart.[94] In 1989, it was introduced into the American clinical setting as an antiarrhythmic drug for the acute management of re-entrant supraventricular tachycardia involving the atrioventricular node. In addition, adenosine has been used "off label" for several other cardiovascular conditions.[95] The related purine nucleotide, adenosine 5′-triphosphate (ATP), has been used in Europe for identical indications for more than five decades.[96] The mechanism of action for adenosine involves direct and indirect, anti-β-adrenergic, eletrophysiologic effects.[97] The direct effect includes the activation of a potassium outward current and modulation of the kinetics of the inward Ca^{2+} current (I_{Ca}). The indirect effect is manifested by the diminution of the catecholamine-induced enhancement of I_{Ca} and the hyperpolarization-activated inward current.

12.5.2 Cardioprotection agent

Prevention and attenuation of ischemia and reperfusion injury in patients with acute coronary syndrome are critically important to cardiologists.[98] When cells are under conditions of stress, adenosine is released and provides negative feedback regulation to maintain cellular preservation. For this reason, adenosine is also called a "retaliatory metabolite."[99] Exogenous adenosine is known to reduce the severity of ischemia and reperfusion injury because of its cytoprotection effect.[100]

Several mechanisms of cytoprotection have been proposed.[99] First, adenosine reduces heart rate and force of contractivity, thereby reducing the oxygen demand of the heart. At the cellular level, this action occurs through adenosine activation of the acetylcholine-sensitive K^+ channels in the atrium; inhibition of catecholamine-stimulated cyclic AMP generation; and Ca^{2+} channels in the ventricle. Second, adenosine reduces the local buildup of reactive oxygen species through inhibition of superoxide anion generation by neutrophils and increases myocardial glucose uptake, which could result from the increase in coronary blood flow. Third, adenosine serves as substrate for the regeneration of ATP. Adenosine therapy is currently being pursued as a new approach to chronic heart failure, which can be caused by events such as myocardial ischemia and reperfusion injury.[100,101]

12.6 Antiprotozoal agents

Protozoa are unicellular eukaryotic microbes existing as free-living organisms or parasites. Parasite-related diseases pose major health threats in many parts of the world. Differences in the purine and pyrimidine biosynthetic pathways of parasites and humans, along with the differences in the specificities between the parasite and human enzymes, are used to target the parasites selectively.

12.6.1 Agents to treat leishmaniasis and Chaga's disease

Leishmaniasis is a major tropical and subtropical parasitic disease caused by *Leishmania donovani*. Chaga's disease is a parasitic disease only found in Latin America and is caused by the parasite *Tryeanosoma cruzi*. Leishmaniae and trypanosomes, like many other unicelluar parasites, lack the ability to synthesize purines *de novo*. However, they do have a large repertoire of salvage enzymes, which enable them to utilize preformed purines and nucleosides present in the human blood stream.

In 1974, allopurinol, a hypoxanthine analog, was found to inhibit replication of *Leishmania donovani*.[102] The mechanism of action for allopurinol was discovered in the next decade.[4] As shown in Figure 12.7, allopurinol is metabolized differently in the parasite compared to in humans. In the parasite, allopurinol is converted to allopurinol ribonucleotide by HGPRT using PRPP. Leishmaniae and trypanosomes have large amounts of HGPRT with significantly higher substrate specificity for allopurinol than does the human enzyme. The protozoal enzymes consequently convert allopurinol ribonucleotide to the adenylate analog, 4-aminopyrazolopyrimidine (APP-MP), which is then converted to APP-DP, and APP-TP and incorporated into the parasite's RNA. Because allopurinol contains a modified purine ring, its presence in the RNA could lead to the disruption in RNA translation and parasite replication. Allopurinol is used for the treatment of leishmaniasis[103,104] and Chaga's diseases.[105,106]

12.6.2 Antimalarial

Malaria is a tropical disease caused by the malarial parasite, *Plasmodium falciparum*, which can synthesize pyrimidine nucleotides only via the *de novo* pathway and is unable to salvage preformed uridine.[107] Pyrazofurin (pyrazomicin), a C-nucleoside analog, is an effective antimalarial.[107] Discovered in 1968 from *Aspergillus fumigatus*,[108] pyrazofurin was extensively investigated as an anticancer agent in the 1970s.[109–112] It also has long been known to inhibit the replication of a large variety of viruses.[113] Pyrazofurin is readily taken up by the parasite and phosphorylated to the pyrazofurin-MP — presumably by adenosine kinase — as observed in human cancer cells.[114] Pyrazofurin inhibits malarial orotate phosphoribosyl transferase (OPRTase), while pyrazofurin-MP strongly inhibits orotidine 5′-monophosphate decarboxylase (ODCase).[115] OPRTase and ODCase are two metabolic enzymes involved in the pathway of pyrimidine nucleotide *de novo* synthesis.

12.7 Antibacterial agents

Nucleoside analogs have long been found to have antibacterial activivy. One example, the thymidine analog 3′-azido-3′-deoxythimidine (AZT), is known today for its use in treating HIV infection and has been reported to have potent antibacterial activity. In 1987, scientists at Burroughs Wellcome reported that AZT had antibacterial activities against many members of the family Enterobacteriaceae, including strains of *Escherichia coli*, *Salmonella typhimurium*, *Klebsiella pneumoniae*, *Shigella flexneri*, and *Enterrobater aerogenes*.[116] Evidence suggested that AZT was phosphorylated to AZT-TP by bacterial kinases, including thymidine kinase. AZT-TP was incorporated into bacterial DNA, where it then served as a chain terminator and inhibited bacterial DNA synthesis. Resistance to AZT developed rapidly as the result of mutations in the bacterial thymidine kinase. Currently, no FDA-approved nucleoside analogs are used as antibacterial agents.

It is worth mentioning that a group of antibiotics called nucleoside antibiotics have a nucleoside moiety as part of their structure. One example is tunicamycin, a family of compounds produced by *Streptomyces iysosuperificus*.[117] They contain uracil, N-acetylglycoamine, an 11-carbon aminodialdose called tunicamine, and a fatty acid linked to the amino group of tunicamine. Tunicamycin is an inhibitor of bacterial translocase, which catalyzes the first reaction in bacterial peptidoglycan biosynthesis. In this reaction, phosphor-MurNac-L-Ala-γ-D-Glu-*m*-DAP-D-Ala-D-Ala was transferred from UMP to a membrane-bound carrier, undecaprenyl phosphate.[118] Tunicamycin is widely used as a tool in studying glycoprotein in biological systems.

12.8 Drugs targeting P1 and P2 receptors

Adenosine and ATP are endogenous modulators of many physiological responses throughout the body, mediating their effects preferentially at the P1 and P2 receptors. P1 receptors, also called adenosine receptors, have been classified into four subtypes (A_1, A_{2A}, A_{2B}, and A_{3A}) — all of which are G protein-coupled receptors generally coupled to adenylate cyclase. P2 receptors, also called extracellular nucleotide receptors, are activated mainly by ATP, ADP, UTP, and UDP and contain two families of membrane-bound P2 receptors: P2X and P2Y subtypes. Virtually all cells possess endogenous P2 receptors. By the year 2002, seven P2X receptors ($P2X_{1-7}$) had been cloned and expressed, and seven P2Y receptors ($P2Y_{1-4,\ 6,\ 11-13}$) had been cloned in humans.[119] Agonists of P1 and P2 receptors are almost exclusively nucleosides and nucleotides, respectively; antagonists of these receptors are structurally more diverse.[120]

At the time at which this chapter was prepared, no nucleoside or nucleotide analog in this category had been approved. However, intensive research is taking place and several drug candidates are worth mentioning.[119,120] They are listed next according to their clinical targets.

- *Sputum expectoration.* INS-316 (uridine 5′-triphosphate, UTP) is a naturally occurring agonist for $P2Y_2$ receptors on the apical surface of ciliated respiratory epithelium. It stimulates salt and water transport and cilia beat frequency in human airway epithelium *in vitro*.[122] In pulmonology and oncology clinics, physicians rely on a deep-lung specimen collected from patients to diagnose lung cancer and lung infections, including pneumonia and tuberculosis. As an alternative to an invasive medical procedure called bronchoscopy, a well tolerated and effective inhaled solution is needed that will induce adequate production of the deep-lung specimen. INS 316 is being developed for the cytologic diagnosis of lung cancer and is currently in phase III clinical trials.[123]
- *Respiratory for the treatment of cystic fibrosis.* INS-37217 (dCP_4U) is a deoxycytidine-uridine dinucleotide with agonist activity at the $P2Y_2$ receptor. Like the natural nucleotide UTP, it acts as a homeostatic mediator modulating normal lung function, leading to increased chloride and water secretion, increased cilia beat frequency, and decreased mucin release. INS-37217 is being developed for the treatment of cystic fibrosis and is currently in Phase II clinical trials.
- *Hemostasis.* $P2Y_{11}$ and $P2Y_{12}$ receptors appear to play a role in hematopoesis. AR-C69331MX is one of a series of systemically active, synthetic $P2Y_{12}$ antagonists that has a safer side effect profile than asprine and has superior antithrombotic properties compared to other modulators of platelet activity.[119] This compound is now in phase III clinical trials as an antithrombotic agent. An orally active form of this drug is currently in phase I trials.
- *Visual function.* In the ocular mucosa, $P2Y_2$ receptor activation increases salt, water, and mucus secretion. INS-365 (diquafosol tetrasodium), a uridine–uridine dinucleotide analog of UTP, activates $P2Y_2$ and is at the late stage of development for the treatment of dry eye disease.[119,121]

Acknowledgments

We would like to thank Dr. Janet L. Rideout, Dr. Thomas A. Krenitsky, and Dr. Sammy R. Shaver for their advice during the preparation of this chapter. We are also grateful for valuable input given by Katyna Borroto–Esoda, Nikki McMillian, Dr. Michael Miller, and Dr. Mick Hitchcock.

References

1. Rossi, A. The clinical uses of nucleoside analogues in malignant disease. In: Walker, R.T., De Clercq, E., and Eckstein, F., Eds. *Nucleoside Analogues: Chemistry, Biology, and Medical Applications.* (Part of NATO Advanced Study Institutes Series). New York: Plenum Publishing Co., 1979, 409–436.

2. Plunkett, W. and Gandhi, V. Nucleoside analogs: cellular pharmacology, mechanism of action, and strategies for combination therapy. In: Cheson, B.D., Keating, M.J., and Plunkett, W., Eds. *Nucleoside Analogs in Cancer Therapy.* New York: Marcel Dekker, Inc., 1997, 1–36.
3. Cass, C.E., Young, J.D., Baldwin, S.A., Cabrita, M.A., Graham, K.A., Griffiths, M., Jennings, L.L., Mackey, J.R., Ng, A.M., Ritzel, M.W., Vickers, M.F., and Yao, S.Y. Nucleoside transporters of mammalian cells. *Pharm. Biotechnol.*, 1999, 12, 313–352.
4. Elion, G.B. The purine path to chemotherapy. In: Frangsmyr, T. and Lindsten J., Eds. *Nobel Lectures, Physiology or Medicine 1981–1990.* Singapore: World Scientific Publishing Co., 1988.
5. De Abreu, R.A. Nucleotide metabolism: mode of action of thiopurines in leukemia. In: Sahota, A. and Taylor, M., Eds. *Purine and Pyrimidine Metabolism in Man* VIII. New York: Plenum Press, 1995.
6. Ling, Y.H., Nelson, J.A., Cheng, Y.C., Anderson, R.S., and Beattie, K.L. 2′-Deoxy-6-thioguanosine 5-triphosphate as a substrate for purified human DNA polymerase *in vitro*. *Mol. Pharmacol.*, 1991, 40, 508–514.
7. Maybaum, J. and Mandel, H.G. Unilateral chromatid damage: a new basis for 6-thioguanine cytotoxicity. *Cancer Res.*, 1983, 43(8), 3852–3856.
8. Ling, Y.H., Chan, J.Y., Beattie, K.L., and Nelson, J.A. Consequences of 6-thioguanine incorporation into DNA on polymerase, ligase, and endonuclease reactions. *Mol. Pharmacol.*, 1992, 42(5), 802–807.
9. Krynetskaia, N.F., Cai, X., Nitiss, J.L., Krynetski, E.Y., and Relling, M.V. Thioguanine substitution alters DNA cleavage mediated by topoisomerase II. *FASEB J.*, 2000, 14(14), 2339–2344.
10. Heidelberger, C., Chaudhuri, N.K., Danneberg, P., Mooren, D., Griesbach, L., Duschinsky, R., Pleven, E., and Scheiner, J. Fluorinated pyrimidines, a new class of tumour-inhibitory compounds. *Nature*, 1957, 179, 663–666.
11. Thomas, D.M. and Zalcberg, J.R. 5-Fluorouracil: a pharmacological paradigm in the use of cytotoxics. *Clin. Exp. Pharmacol. Physiol.*, 1998, 25(11), 887–895.
12. Voet, D. and Voet, J., Eds. Nucleotide metabolism. In: *Biochemistry.* New York: John Wiley & Sons, 1990, chap. 26, 757–768.
13. van Laar, J.A., Rustum, Y.M., Ackland, S.P., van Groeningen, C.J., and Peters, G.J. Comparison of 5-fluoro-2′-deoxyuridine with 5-fluorouracil and their role in the treatment of colorectal cancer. *Eur. J. Cancer*, 1998, 34(3), 296–306.
14. Kufe, D.W., Major, P.P., Egan, E.M., and Beardsley, J.P. Correlation of cytotoxicity with incorporation of ara-C into DNA. *J. Biol. Chem.*, 1980, 255, 8997–9000.
15. Mikita, T. and Beardsley, G.P. Functional consequences of the arabinosylcytosine structural lesion in DNA. *Biochemistry*, 1988, 27, 4698–4705.
16. Robins, R.K. Nucleoside and nucleotide inhibitors of inosine monophosphate (IMP) dehydrogenase as potential antitumor inhibitors. *Nucleosides Nucleotides*, 1982, 1, 35–44.
17. Fridland, A., Connelly, M.C., and Robbins, T.J. Tiazofurin metabolism in human lymphoblast cells: evidence for phosphorylation by adenosine kinase and 5′-nucleotidase. *Cancer Res.*, 1986, 46(2), 532–537.
18. Sauders, P.P., Spindler, C.D., Tan, M.T., Alvarez, E., and Robins, R.K. Tiazofurin is phosphorylated by three enzymes from Chinese hamster ovary cells. *Cancer Res.*, 1990, 50(17), 5269–5274.

19. Jayaram, H.N., Pillwein, K., Nichols, C.R., Hoffman, R., and Weber, G. Selective sensitivity to tiazofurin of human leukemic cells. *Biochem. Pharmacol.*, 1986, 35, 2029–2032.
20. Boulton, S., Kyle, S., and Durkacz, B.W. Low nicotinamide mononucleotide adenylyltransferase activity in a tiazofurin-resistant cell line: effects on NAD metabolism and DNA repair. *Br. J. Cancer*, 1997, 76(7), 845–851.
21. Zhou, T., Kurnasov, O., Tomchick, D.R., Binns, D.D., Grishin, N.V., Marquez, V.E., Osterman, A.L., and Zhang, H. Structure of human nicotinamide/nicotinic acid mononucleotide adenylyltransferase. Basis for the dual substrate specificity and activation of the oncolytic agent tiazofurin. *J. Biol. Chem.*, 2002, 277(15), 13148–13154.
22. Weber, G., Prajda, N., and Jackson, R.C. Key enzymes of IMP metabolism: transformation and proliferation-linked alterations in gene expression. *Adv. Enzyme Regul.*, 1976, 14, 3–24.
23. Cooney, D.A., Jayaram, H.N., Glazer, R.I., Kelley, J.A., Marquez, V.E., Gebeyehu, G., Van Cott, A.C., Zwelling, L.A., and Johns, D.G. Studies on the mechanism of action of tiazofurin metabolism to an analog of NAD with potent IMP dehydrogenase-inhibitory activity. *Adv. Enzyme Regul.*, 1983, 21, 271–303.
24. Yamada, Y., Natsumeda, Y., and Weber, G. Action of the active metabolites of tiazofurin and ribavirin on purified IMP dehydrogenasechemistry. *Biochemistry*, 1988, 27, 2193–2196.
25. Weber, G., Prajda, N., Abonyi, M., Look, K.Y., and Tricot, G. Tiazofurin: molecular and clinical action. *Anticancer Res.*, 1996, 16(6A), 3313–3322.
26. Gandhi, V.W.P. Cellular and clinical phamramcology of fludarabine. *Clin. Pharmacokinet.*, 2002, 41(2), 93–103.
27. Venner, H. and Zinner, H. Preparation of 1,3,5-tri-O-acetyl-2-deoxy-b-D-ribofuranose. *Chemische Berichte*, 1960, 93, 137–140.
28. Simon, L.N., Bauer, R.J., Tolman, R.L., and Robins, R.K. Calf intestine adenosine deaminase: substrate specificity. *Biochemistry*, 1970, 9, 573–577.
29. Carson, D.A., Wasson, D.B., Kaye, J., Ullman, B., Martin, D.W.J., Robins, R.K., and Montgomery, J.A. Deoxycytidine kinase-mediated toxicity of deoxyadenosine analogs toward malignant human lymphoblasts *in vitro* and toward murine L1210 leukemia *in vivo*. *Proc. Natl. Acad. Sci. USA*, 1980, 77(11), 6865–6869.
30. Piro, L.D., Carrera, C.J., Carson, D.A., and Beuter, E. Lasting remission in hairy-cell leukemia induced by a single infusion of 2-chlorodeoxyadenosine. *N. Engl. J. Med.*, 1990, 322, 1069–1073.
31. Juliusson, G. and Liliemark, J. High complete remission rate from 2-chloro-2′-deoxyadenosine in previously treated patients with B-cell chronic lymphocytic leukemia: response predicted by rapid decrease of blood lymphocyte count. *J. Clin. Oncol.*, 1993, 11(4), 679–689.
32. Saven, A. and Piro, L.D. 2-Chlorodeoxyadenosine: a newer purine analog active in the treatment of indolent lymphoid malignancies. *Ann. Intern. Med.*, 1994, 120(9), 784–791.
33. Avery, T.L., Rehg, J.E., Lumm, W.C., Harwood, F.C., Santana, V.M., and Blakley, R.L. Biochemical pharmacology of 2-chlorodeoxyadenosine in malignant human hematopoietic cell lines and therapeutic effects of 2-bromodeoxyadenosine in drug combinations in mice. *Cancer Res.*, 1989, 49(18), 4972–4978.

34. Hertel, L.W., Kroin, J.S., Misner, J.W., and Tustin, J.M. Synthesis of 2-deoxy-2,2-difluoro-D-ribose and 2-deoxy-2,2′-difluoro-D-ribofuranosyl nucleosides. *J. Org. Chem.*, 1988, 53, 2406.
35. Heinemann, V., Hertel, L.W., Grindey, G.B., and Plunkett, W. Comparison of the cellular pharmacokinetics and toxicity of 2′,2′-difluorodeoxycytidine and 1-beta-D-arabinofuranosylcytosine. *Cancer Res.*, 1988, 48(14), 4024–4031.
36. Hertel, L.W., Boder, G.B., Kroin, J.S., Rinzel, S.M., Poore, G.A., Todd, G.C., and Grindey, G.B. Evaluation of the antitumor activity of gemcitabine (2′,2′-difluoro-2′-deoxycytidine). *Cancer Res.*, 1990, 50(14), 4417–4422.
37. Gmeiner, W.H. Antimetabolite incorporation into DNA: structural and thermodynamic basis for anticancer activity. *Biopolymers*, 2002, 65, 180–189.
38. Huang, P., Chubb, S., Hertel, L.W., Grindey, G.B., and Plunkett, W. Action of 2′,2′-difluorodeoxycytidine on DNA synthesis. *Cancer Res.*, 1991, 51, 6110–6117.
39. Jiang, H.Y., Hickey, R.J., Abdel–Aziz, W., and Malkas, L.H. Effects of gemcitabine and ara-C on *in vitro* DNA synthesis mediated by the human breast cell DNA synthesome. *Cancer Chemother. Pharmacol.*, 2000, 45, 320–328.
40. Pourquier, P., Gioffre, C., Kohlhagen, G., Urasaki, Y., Goldwasser, F., Hertel, L.W., Yu, S., Pon, R.T., Gmeiner, W.H., and Pommier, Y. Gemcitabine (2′,2′-difluoro-2′-deoxycytidine), an antimetabolite that poisons topoisomerase I. *Clin Cancer Res.*, 2002, 8(8), 2499–2504.
41. Neff, T. and Blau, C.A. Forced expression of cytidine deaminase confers resistance to cytosine arabinoside and gemcitabine. *Exp. Hematol.*, 1996, 24(11), 1340–1346.
42. Ruiz van Haperen, V.W., Veerman, G., Eriksson, S., Stegmann, A.P., and Peters, G.J. Induction of resistance to 2′,2′-difluorodeoxycytidine in the human ovarian cancer cell line A2780. *Semin. Oncol.*, 1995, 22, 35–41.
43. Mackey, J.R., Mani, R.S., Selner, M., Mowles, D., Young, J.D., Belt, J.A., Crawford, C.R., and Cass, C.E. Functional nucleoside transporters are required for gemcitabine influx and manifestation of toxicity in cancer cell lines. *Cancer Res.*, 1998, 58(19), 4349–4357.
44. Woo, P.W.K., Dion, H.W., Lange, S.M., Dahl, L.F., and Durham, L.J. A novel adenosine and ara-A deaminase inhibitor, (R)-3-(2-deoxy-b-D-erythro-pentofuranosyl)-3,6,7,8-tetrahydroimidazo[4,5-d][1,3]diazepin-8-ol. *J. Heterocycl. Chem.*, 1974, 11, 641–643.
45. Rogler–Brown, T., Agarwal, R.P., and Parks, R.E.J. Tight binding inhibitors — VI. Interactions of deoxycoformycin and adenosine deaminase in intact human erythrocytes and sarcoma 180 cells. *Biochem. Pharmacol.*, 1978, 27(19), 2289–2296.
46. Ho, D.H., Pincus, C., Carter, C.J., Benjamin, R.S., Freireich, E.J., and Bodey, G.P.S. Distribution and inhibition of adenosine deaminase in tissues of man, rat, and mouse. *Cancer Treat Rep.*, 1980, 64(4–5), 629–633.
47. Kane, B.J., Kuhn, J.G., and Roush, M.K. Pentostatin: an adenosine deaminase inhibitor for the treatment of hairy cell leukemia. *Ann. Pharmacother.*, 1992, 26, 939–947.
48. Saven, A. and Piro, L.D. Treatment of hariy cell leukemia with nucleoside analogs. In: Cheson, B.D., Keating, M.J., and Plunkett, W., Eds. *Nucleoside Analogs in Cancer Therapy.* New York: Marcel Dekker, Inc., 1997.
49. de la Haba, G., Agostini, S., Bozzi, A., Merta, A., Unson, C., and Cantoni, G.L. S-adenosylhomocysteinase: mechanism of reversible and irreversible inactivation by ATP, cAMP, and 2′-deoxyadenosine. *Biochemistry,* 1986, 25(25), 8337–8342.

50. Daly, W. Adenosine receptors. *J. Med. Chem.*, 1982, 25, 197–207.
51. Lee, N., Russell, N., Ganeshaguru, K., Jackson, B.F., Piga, A., Prentice, H.G., Foa, R., and Hoffbrand, A.V. Mechanisms of deoxyadenosine toxicity in human lymphoid cells *in vitro*: relevance to the therapeutic use of inhibitors of adenosine deaminase. *Br. J. Haematol.*, 1984, 56(1), 107–119.
52. Reist, E.J. and Goodman, L. Synthesis of 9-beta-D-arabinofuranosylguanine. *Biochemistry*, 1964, 127, 15–18.
53. Cohen, A., Lee, J.W., and Gelfand, E.W. Selective toxicity of deoxyguanine and arabinosyl guanine for T-leukemic cells. *Blood*, 1983, 61(4), 660–666.
54. Shewach, D.S., Daddona, P.E., Ashcraft, E., and Mitchell, B.S. Metabolism and selective cytotoxicity of 9-beta-D-arabinofuranosylguanine in human lymphoblasts. *Cancer Res.*, 1985, 45(3), 1008–1014.
55. Rodriguez, C.O.J., Mitchell, B.S., Ayres, M., Eriksson, S., and Gandhi, V. Arabinosylguanine is phosphorylated by both cytoplasmic deoxycytidine kinase and mitochondrial deoxyguanosine kinase. *Cancer Res.*, 2002, 62(11), 3100–3105.
56. Lotfi, K., Mansson, E., Peterson, C., Eriksson, S., and Albertioni, F. Low level of mitochondrial deoxyguanosine kinase is the dominant factor in acquired resistance to 9-beta-D-arabinofuranosylguanine cytotoxicity. *Biochem. Biophys. Res. Commun.*, 2002, 293(5), 1489–1496.
57. Muller, W.E., Geurtsen, W., Beyer, R., Schuster, D.K., Zahn, R.K., and Arendes, J. Arabinosyl nucleosides. XXXII. mechanism of inhibition of L5178y mouse lymphoma cells by 9-beta-D-arabinofuranosylguanine. *Chem. Biol. Interact.*, 1981, 34(3), 301–313.
58. Rodriguez, C.O.J. and Gandhi, V. Arabinosylguanine-induced apoptosis of T-lymphoblastic cells: incorporation into DNA is a necessary step. *Cancer Res.*, 1999, 59, 4937–4943.
59. Curbo, S., Zhivotovsky, B., Jonansson, M., and Karlsson, A. Effects of 9-arabinofuranoylguanine on mitochondria in CEM T-lymphoblast leukemia cells. *Biochem. Biophys. Res. Commun.*, 2003, 307(4), 942–947.
60. Ono, K., Ohashi, A., Yamamoto, A., Matsukage, A., Takahasi, T., Saneyoshi, M., and Ueda, T. Inhibitory effects of 9-beta-D-arabinofuranosylguanine 5′-triphosphate and 9-beta-D-arabinofuranosyladenine 5′-triphosphate on DNA polymerases from murine cells and oncornavirus. *Cancer Res.*, 1979, 39(11), 4673–4680.
61. Jinmei, H., Takahashi, H., Amano, R., Suzuki, K., Saneyoshi, M., and Yamaguchi, T. Telomerase-inhibitory effects of sugar-modified nucleotide analogs. *Nucleic Acids Res. Suppl.*, 2002, 2, 221–222.
62. Gandhi, V., Plunkett, W., Weller, S., Du, M., Ayres, M., Rodriguez, C.O.J, Ramakrishna, P., Rosner, G.L., Hodge, J.P., O'Brien, S., and Keating, M.J. Evaluation of the combination of nelarabine and fludarabine in leukemias: clinical response, pharmacokinetics, and pharmacodynamics in leukemia cells. *J. Clin. Oncol.*, 2001, 19(8), 2142–2152.
63. Schwartz, J., Stack, J., and Dameshek, W. Effect of 6-mercaptopurine on antibody production. *Proc. Soc. Exp. Biol. Med.*, 1958, 99(1), 164–167.
64. Calne, R.Y. The rejection of renal homografts. Inhibition in dogs by 6-mercaptopurine. *Lancet*, 1960, 1, 417–418.
65. Calne, R.Y. Inhibition of the rejection of renal homografts in dogs by purine analogues. *Transplant Bull.*, 1961, 28, 65–81.

66. Calne, R.Y., Alexandre, G.P.J., and Murray, J.E. A study of the effects of drugs in prolonging survival of homologous renal transplants in dogs. *Ann. N.Y. Acad. Sci.*, 1962, 99, 743–761.
67. Murray, J.E., Merrill, J.P., Harrison, J.H., Wilson, R.E., and Dammin, G.J. Prolonged survival of human-kidney homografts by immunosuppressive drug therapy. *N. Engl. J. Med.*, 1963, 268, 1315–1323.
68. Wolberg, G. *Pharmacology of Lymphocytes*. Vol. Berlin: Springer–Verlag, 1988.
69. Elion, G.B. The George Hitchings and Gertrude Elion lecture. The pharmacology of azathioprine. *Ann. N.Y. Acad. Sci.*, 1993, 685, 400–407.
70. Aarbakke, J., Janka–Schaub, G., and Elion, G.B. Thiopurine biology and pharmacology. *Trends Pharmacol. Sci.*, 1997, 18(1), 3–7.
71. Maltzman, J.S. and Koretzky, G.A. Azathioprine: old drug, new actions. *J. Clin. Invest.*, 2003, 111(8), 1122–1124.
72. Tiede, I., Fritz, G., Strand, S., Poppe, D., Dvorsky, R., Strand, D., Lehr, H.A., Wirtz, S., Becker, C., Atreya, R., Mudter, J., Hildner, K., Bartsch, B., Holtmann, M., Blumberg, R., Walczak, H., Iven, H., Galle, P.R., Ahmadian, M.R., and Neurath, M.F. CD28-dependent Rac1 activation is the molecular target of azathioprine in primary human CD4+ T lymphocytes. *J. Clin. Invest.*, 2003, 111(8), 1133–1145.
73. Maris, M. and Storb, R. The transplantation of hematopoietic stem cells after non-myeloablative conditioning: a cellular therapeutic approach to hematologic and genetic diseases. *Immunol. Res.*, 2003, 28(1), 13–24.
74. Bashey, A. Immunosuppression with limited toxicity: the characteristics of nucleoside analogs and anti-lymphocyte antibodies used in non-myeloablative hematopoietic cell transplantation. *Cancer Treat. Res.*, 2002, 110, 39–49.
75. Mizuno, K., Tsujino, M., Takada, M., Hayashi, M., and Atsumi, K. Studies on bredinin. I. Isolation, characterization and biological properties. *J. Antibiot.* (Tokyo), 1974, 27(10), 775–782.
76. Koyama, H. and Tsuji, M. Genetic and biochemical studies on the activation and cytotoxic mechanism of bredinin, a potent inhibitor of purine biosynthesis in mammalian cells. *Biochem. Pharmacol.*, 1983, 32(23), 3547–3553.
77. Mitchell, B.S., Dayton, J.S., Turka, L.A., and Thompson, C.B. IMP dehydrogenase inhibitors as immunomodulators. *Ann. N.Y. Acad. Sci.*, 1993, 685, 217–224.
78. Inou, T., Kusaba, R., Takahashi, I., Sugimoto, H., Kuzuhara, K., Yamada, Y., Yamauchi, J., and Otsubo, O. Clinical trial of bredinin in renal transplantation. *Transplant Proc.*, 1981, 13, 315–318.
79. Sintchak, M.D. and Nimmesgern, E. The structure of inosine 5′-monophosphate dehydrogenase and the design of novel inhibitors. *Immunopharmacology*, 2000, 47, 163–184.
80. Nimmesgern, E., Black, J., Futer, O., Fulghum, J.R., Chambers, S.P., Brummel, C..L, Raybuck, S.A., and Sintchak, M.D. Biochemical analysis of the modular enzyme inosine 5′-monophosphate dehydrogenase. *Protein Exp. Purif.*, 1999, 17(2), 282–289.
81. Hunter, T., Urowitz, M.B., Gordon, D.A., Smythe, H.A., and Ogryzlo, M.A. Azathioprine in rheumatoid arthritis: a long-term follow-up study. *Arthritis Rheum.*, 1975, 18(1), 15–20.
82. Willkens, R.F. and Stablein, D. Combination treatment of rheumatoid arthritis using azathioprine and methotrexate: a 48-week controlled clinical trial. *J. Rheumatol. Suppl.*, 1996, 44, 64–68.

83. Hahn, B.H. Lupus and vasculitis. 2001, http://www.hamline/edu/.
84. Ioannou, Y. and Isenberg, D.A. Current concepts for the management of systemic lupus erythematosus in adults: a therapeutic challenge. *Postgrad. Med. J.*, 2002, 78(924), 599–606.
85. Ellison, G.W., Myers, L.W., Mickey, M.R., Graves, M.C., Tourtellotte, W.W, and Nuwer, M.R. Clinical experience with azathioprine: the pros. *Neurology*, 1988, 38(7 suppl 2), 20–23.
86. Confavreux, C. and Moreau, T. Emerging treatments in multiple sclerosis: azathioprine and mofetil. *Multiple Sclerosis*, 1996, 1(6), 379–384.
87. Janiec, K., Wajgt, A., and Kondera–Anasz, Z. Effect of immunosuppressive cladribine treatment on serum leucocytes system in two-year clinical trial in patients with chronic progressive multiple sclerosis. *Med. Sci. Monit.*, 2001, 7(1), 93–98.
88. Beutler, E., Sipe, J.C., and Romine, J.S. The treatment of chronic progressive multiple sclerosis with cladribine. *Proc. Natl. Acad. Sci. USA*, 1996, 93, 1716–1720.
89. Tortorella, C., Rovaris, M., and Filippi, M. Cladribine. Ortho Biotech Inc. *Curr. Opin. Invest. Drugs*, 2001, 2(12), 1751–1756.
90. Sandborn, W.J. Azathioprine: state of the art in inflammatory bowel disease. *Scand. J. Gastroenterol. Suppl.*, 1998, 225, 92–99.
91. Hanauer, S.B. and Present, D.H. The state of the art in management of inflammatory bowel disease. *Rev. Gastroenterol. Disord.*, 2003, 3(2), 81–92.
92. Loeb, J.N. The influence of temperature on the solubility of monosodium urate. *Arthritis Rheum.*, 1972, 15, 189–192.
93. Drury, A.N. and Szent–Gyorgyi, A. The physiologic actvity of adenosine componds with special reference to their action upon the mammalian heart. *J. Physiol.*, 1929, 68, 213–237.
94. Honey, R.M., Ritchie, W.T., and Thomson, W.A.R. The action of adenosine upon the human heart. *Q. J. Med.*, 1930, 23, 485–489.
95. Pelleg, A., Pennock, R.S., and Kutalek, S.P. Proarrhythmic effects of adenosine: one decade of clinical data. *Am. J. Ther.*, 2002, 9, 141–147.
96. Belhassen, B. and Pelleg, A. Electrophysiologic effects of adenosine triphosphate and adenosine on the mammalian heart: clinical and experimental aspects. J. *Am. Coll. Cardiol.*, 1984, 4, 414–424.
97. Wilbur, S.L. and Marchlinski, F. Adenosine as an antiarrhytymic agent. *Am. J. Cardiol.*, 1997, 79, 30–37.
98. Kitakaze, M. and Hori, M. It is time to ask what adenosine can do for cardioprotection. *Heart Vessels*, 1998, 13, 211–228.
99. Ramkumar, V., Hallam, D.M., and Nie, Z. Adenosine, oxidative stress and cytoprotection. *Jpn. J. Pharmacol.*, 2001, 86, 265–274.
100. Sommerschild, H.T. and Kirkeboen, K.A. Adenosine and cardioprotection during ischaemia and reperfusion, an overview. *Acta Anaesthesiol. Scand.*, 2000, 44, 1038–1055.
101. Kitakaze, M. and Hori, M. Adenosine therapy: a new approach to chronic heart failure. *Expert Opin. Invest. Drugs*, 2000, 9, 2519–2535.
102. Pfaller, M.A. and Marr, J.J. Antileishmanial effect of allopurinol. *Antimicrob. Agents Chemother.*, 1974, 5(5), 469–472.
103. Kager, P.A., Rees, P.H., Wellde, B.T., Hockmeyer, W.T., and Lyerly, W.H. Allopurinol in the treatment of visceral leishmaniasis. *Trans. R. Soc. Trop. Med. Hyg.*, 1981, 75(4), 556–559.

104. Moss, J.T. and Wilson, J.P. Current treatment recommendations for leishmaniasis. *Ann. Pharmacother.*, 1992, 26(11), 1452–1455.
105. Meirovich, C.I., Montrull, H.L., Gallerano, R.H., and Sosa, R.R. Allopurinol in the treatment of chronic Chaga's disease (paper in Spanish). *Arq. Bras. Cardiol.*, 1985, 45, 217–223.
106. Apt, W., Aguilera, X., Arribada, A., Perez, C., Miranda, C., Sanchez, G., Zulantay, I., Cortes, P., Rodriguez, J., and Juri, D. Treatment of chronic Chagas disease with itraconazole and allopurinol. *Am. J. Trop. Med. Hyg.*, 1998, 59, 133–138.
107. Scheibel, L.W. and Sherman, I.W. *Malaria: Principles and Practice of Malariology.* Vol. 1. Melbourne: Churchill Livingston, 1988.
108. Plotnikov, N.P., Silaev, A.B., Maksimova, R.A., and Semenov, M.N. Antibiotic from *Aspergillus fumigatus* with cytotoxic and antiviral action (article in Russian). *Antibiotiki*, 1968, 13(4), 316–319.
109. Ohmana, T., Roboz, J., Shapiro, M.L., and Holland, J.F. Pharmacological and biochemical effects of pyrazofurin in humans. *Cancer Res.*, 1977, 37, 2043–2049.
110. Plagemann, P.G. and Behrens, M. Inhibition of de novo pyrimidine nucleotide and DNA synthesis and growth of cultured Novikoff rat hepatoma cells and other cell lines by pyrazofurin (NSC 143095). *Cancer Res.*, 1976, 36(10), 3807–3812.
111. Nichols, W.C., Kvols, L.K., Ingle, J.N., Edmonson, J.H., Ahmann, D.L., Rubin, J., and O'Connell, M.J. Phase II study of triazinate and pyrazofurin in patients with advanced breast cancer previously exposed to cytotoxic chemotherapy. *Cancer Treat. Rep.*, 1978, 62(5), 837–839.
112. Weber, G. Recent advances in the design of anticancer chemotherapy. *Oncology,* 1980, 37, 19–24.
113. Descamps, J. and De Clercq, E. Broad-spectrum antiviral activity of pyrazofurin (pyrazomycin). Current chemotherapy. *Proc. 10th Int. Congr. Chemother.* Zurich, Switzerland, 1978.
114. Cadman, E.C., Dix, D.E., Hill, S.C., and Handschumacher, R.E. Pyrazofurin: biochemical and clinicopharmacological studies. *Proc. Am. Assoc. Cancer Res.*, 1976, 17, 208.
115. Christopherson, R.I., Lyons, S.D., and Wilson, P.K. Inhibitors of de novo nucleotide biosynthesis as drugs. *Acc. Chem. Res.*, 2002, 35, 961–971.
116. Elwell, L.P., Ferone, R., Freeman, G.A., Fyfe, J.A., Hill, J.A., Ray, P.H., Richards, C.A., Singer, S.C., Knick, V.B., and Rideout, J.L. Antibacterial activity and mechanism of action of 3′-azido-3′-deoxythymidine (BW A509U). *Antimicrob. Agents Chemother.*, 1987, 31(2), 274–280.
117. Kimura, K. and Bugg, T.D. Recent advances in antimicrobial nucleoside antibiotics targeting cell wall biosynthesis. *Nat. Prod. Rep.*, 2003, 20(2), 252–273.
118. Brandish, P.E., Kimura, K.I., Inukai, M., Southgate, R., Lonsdale, J.T., and Bugg, T.D. Modes of action of tunicamycin, liposidomycin B, and mureidomycin A: inhibition of phospho-N-acetylmuramyl-pentapeptide translocase from *Escherichia coli*. *Antimicrob. Agents Chemother.*, 1996, 40(7), 1640–1644.
119. Jacobson, K.A., Jarvis, M.F., and Williams, M. Purines and pyrimidines (P2) receptors as drug targets. *J. Med. Chem.*, 2002, 45(19), 4057–4093.
120. Fredholm, B.B., Abbracchio, M.P., Burnstock, G., Dubyak, G.R., Harden, T.K., Jacobson, K.A., Schwabe, U., and Williams, M. Towards a revised nomenclature for P1 and P2 receptors. *Trends Pharmacol. Sci.*, 1997, 18(3), 79–82.
121. Shaver, S.R. P2Y receptors: biological advances and therapeutic opportunities. *Curr. Opin. Drug Discovery Dev.*, 2001, 4(5), 665–670.

122. Brown, H.A., Lazarowski, E.R., Boucher, R.C., and Harden, T.K. Evidence that UTP and ATP regulate phospholipase C through a common extracellular 5′-nucleotide receptor in human airway epithelial cells. *Mol. Pharmacol.*, 1991, 40, 648–655.
123. Johnson, F.L., Donohue, J.F., and Shaffer, C.L. Improved sputum expectoration following a single dose of INS 316 in patients with chronic bronchitis. *Chest*, 2002, 122, 2021–2029.

appendix one

Phosphorus nuclear magnetic resonance chemical shifts of nucleotide derivatives and some related organophosphorus compounds

Morteza M. Vaghefi
ChemCyte Corporation, San Diego, CA

Phosphorous Nuclear Magnetic Resonance spectroscopy (^{31}P NMR) has been used as a valuable tool to study the structure of nucleosides mono- and polyphosphates, as well as oligonucleotides and their derivatives. Proton (^{1}H) and ^{31}P are the two major nuclei in the nucleotides that have a nuclear spin quantum value of 1/2. Their natural form is 100% measurable by NMR; however, the resonance frequency of the ^{1}H nucleus is much higher than ^{31}P. Most other nuclei have a spin value of 0 (e.g., carbon, oxygen, and sulfur), or 1 (e.g., nitrogen) and contain traces of their corresponding isotopes with spin values of 1/2 or 3/2 (e.g., ^{13}C and ^{15}N). NMR detection of these nuclei requires more material.

The theory of ^{31}P chemical shifts is described in many detailed reviews and textbooks.[1,7] The differences in resonance frequency between the nuclei in different chemical sites are caused by variations in shielding by the electronic environment of the nuclei. A nuclei exhibits similar chemical shifts when it is located in a similar chemical environment, independent of the chemical structure of the molecule. Electronegativity and back donation of electrons by pi bonding are the two major contributors to the chemical shifts of the ^{31}P nuclei. Generally, for ^{31}P NMR spectra, 85% orthophosphoric acid (0 ppm) is used as an external reference.

Typical chemical shift spreads for trivalent ^{31}P nuclei are within 500 ppm, while the more shielded pentavalent ^{31}P nuclei display signals within 100 ppm. Similar to ^{1}H NMR, the intensities of the ^{31}P signals are directly proportional to the number of nuclei that produces them. In general, the ^{31}P NMR signals of naturally occurring phosphates appear in distinct regions depending on the connectivity of the P-bound oxygen atoms.

Alkylation of oxygen has a shielding effect relative to the alkyl groups and results in moving the phosphorus signal to a lower frequency. Phosphorylation of the oxygen has a greater influence, which moves the signal to an even lower frequency. The effect is greatest when two oxygen atoms are attached to another phosphate, such as the middle phosphorus in the triphosphates (e.g., β-phosphate in ATP). In this situation, the signal appears in the region of –21 to –25 ppm. Most pyrophosphate esters (e.g., α-phosphphate of ADP or ATP) appear in the range of –10 to –14 ppm. The chemical shifts of the terminal phosphate groups (e.g., γ-phosphate of ATP) are highly pH dependent. At a high pH (7–10), they appear at a low frequency range of –5.5 to –6 ppm and at a low pH, signals move to –10 to –11 ppm. The ortho-phosphate esters are routinely observed in the range of 5 to –3 ppm. Monoesters appear in the range of 4.8 to –2.2 ppm, while most phospho-diesters demonstrate chemical shifts at 1.1 to –3 ppm. To reduce the overlap of the mono- and diesters, it is necessary to run the NMR at a higher pH. At pH 8 to 12, chemical shifts of most monoesters are between 5 to 2 ppm, whereas diesters appear at 1 to 0 ppm (except diesters of phospholipids, which appear at 3 ppm at pH 10). Depending on the pH, inorganic phosphates also appear in the range of 2.6 to 0 ppm.

The signals for cyclic diesters appear at higher frequencies relative to the phosphodiester. A relatively small change (–2.6 ppm) is observed for the 6-membered ring (e.g., uridine-3′,5′-cyclic monophosphate), but for the strained 5-membered ring (uridine 2′,3′-cyclic monophosphate), the change is greater and signals appear between 10 and 21 ppm. Primary amides and phosphonates usually resonate at 6–12 ppm. The phosphonic acid anhydride chemical shifts are detected around 10 ppm and the phosphonate anions resonate between 16 to 22 ppm.

The following tables are a compilation of ^{31}P NMR chemical shifts of nucleotides, as well as related organophosphorus compounds. The data have been organized into groups according to the chemical environment of the phosphorous atom. Only ^{31}P NMR spectra recorded with spin-spin decoupling $\{^{1}H\}$-^{31}P have been reported.

References

1. Lebedev, A.V. and Rezvukhin, A.I. Tendencies of ^{31}P chemical shifts changes in NMR spectra of nucleotide derivatives, *Nucl. Acids. Res.*, 12:5547–5566, 1984.

2. ^{31}P Nuclear Magnetic Resonance in *Topics in Phosphorous Chemistry*, Vol. 5, M.M. Crutchfield, Ed., New York, Interscience Publishers, 1967.
3. TriLink Biotechnologies, Inc. data.
4. Ludwig, J. and Eckstein, F. Rapid and efficient synthesis of nucleoside 5′-0-(1-thiotriphosphates), 5′-triphosphates and 2′,3′-cyclophosphorothioates using 2-chloro-4H-1,3,2-benzodioxaphosphorin-4-one, *J. Org. Chem.*, 54:631–635, 1989.
5. Higuchi, H., Endo, T., and Kaji, A. Enzymic synthesis of oligonucleotides containing methylphosphonate internucleotide linkages, *Biochemistry*, 29:8747–9753, 1990.
6. Victorova, L.S., Dyatkina, N.B., Mozzherin, D.J., Atrazhev, A.M., Krayevsky, A., and Kuhanova, M.K. Formation of phosphonester bonds catalyzed by DNA polymerase, *Nucleic Acids Res.*, 20:783–789, 1992.
7. *CRC Handbook of Phosphorus-31 Nuclear Magnetic Resonance Data*, John Debby, Ed., CRC Press, Boca Raton, FL, 1991.

A.1 ^{31}P Chemical shifts in NMR spectra of nucleotide derivatives and some related organophosphorous compounds

A.1.1 Pentavalent phosphorous

A.1.1.1 Phosphate mono-, di-, triesters and their mono- and diamides

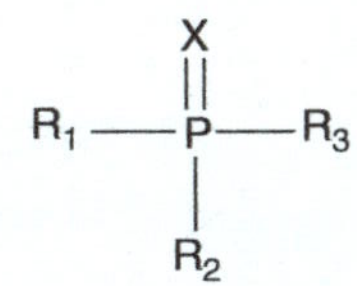

X	R_1	R_2	R_3	δ, ppm	Solvent	Ref.
S	5′-O-Nucleoside	-OAlkyl(-CNEt)	3′-O-Nucleoside	69.0*	H_2O	3
S	5′-O-Nucleoside	-OH	5′-or 3′-O-Nucleoside	55.0–57.0*	H_2O	1, 3
S	-OH	-OH	5′-or 3′-O-Nucleoside	43.5*	H_2O	1
O	5′-or 3′- O-Nucleoside	$-NH_2$	5′-or 3′- O-Nucleoside	10.8–12.1	D_2O, Pyridine	1
O	$-NH_2$	-OH	5′- or 3′- O-Nucleoside	8.8–9.3	D_2O	1
O	5′-or 3′- O-Nucleoside	-NHAlkyl	5′-or 3′- O-Nucleoside	8.4–9.2	Pyridine	1
O	-NH Alkyl	-OH	5′- or 3′- O-Nucleoside	5.0–6.5	Pyridine	1
O	5′-or 3′- O-Nucleoside	-NHAryl	-NHAryl	2.9	Pyridine	1
O	-OH	-OH	5′-O-Nucleoside	2.8–3.6	H2O	2
O	5′-or 3′- O-Nucleoside	-NHAryl	5′-or 3′-Nuc (or Alkyl)	2.6–2.8	Pyridine	1
O	-OH	-OH	5′-O-Nucleoside	1.2	H_2O, pH 7.0	3
O	-OH	-OH	3′-O-Nucleoside	1.0	H_2O, pH 7.0	3
O	-OH	-OH	5′-O-Nucleoside	0.4–1.2	Pyridine	1
O	-OAlkyl	-OAlkyl	5′- or 3′- O-Nucleoside	–(0.7–3.0)	Pyridine	1
O	3′-O-Nucleoside(or -OAlkyl)	-OH	5′-O-Nucleoside	–2.0–1.0	Pyridine	1

O	-NH Aryl	-OH	5′- or 3′- O-Nucleoside	–1.0	Pyridine	1
O	5′-or 3′- O-Nucleoside	-NHAryl	-OAryl	–(2.2–2.8)	Pyridine	1
O	-OAryl	-OH	5′- or 3′-O-Nucleoside	–(5.0–6.0)	1,4dioxane; Pyridine	1
O	-OAryl	-OAlkyl	5′- or 3′-O-Nuc (or-Alkyl)	–7.7	Pyridine	1
O	-OAryl	-OAryl	5′- or 3′- O-Nucleoside	–(12.0–13.0)	Pyridine	1

* Indicates range of ^{31}P chemical shifts for two diastereomers

A.1.2 *Pyrophosphate derivatives of nucleotides*

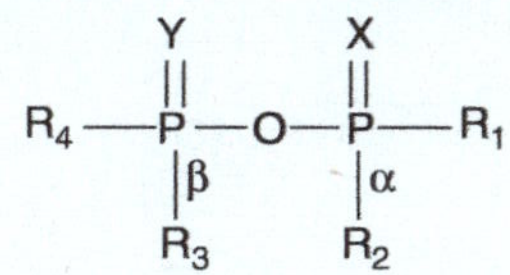

X	Y	R_1	R_2	R_3	R_4	δα, ppm	δβ, ppm	Jαβ, Hz	Solvent	Ref.
S	O	5′-O-Nucleoside	-OH	-OH	-OH	42.1*	–6.3*	30.8	H_2O	1
O	S	5′-O- Nucleoside	-OH	-OH	-OH	–11.7*	34.0*	31.2	H_2O	1
O	O	5′-O- Nucleoside	-OH	-OH	-OH	–10.0	–6.0	20.7	H_2O	3
O	O	3′-O- Nucleoside	-OH	-OH	-OH	–(10.0–11.0)	–(5.0–6.0)	21.4	H_2O	3
O	O	5′-O- Nucleoside	-OH	-OH	5′- or 3′-O-Nucleoside	–(10.0–11.0)	–10.0	n/a	Pyridine	1
O	O	5′-O- Nucleoside	-OH	-OAryl	-OH	–11.2	–16.6	17.0	Pyridine	1
O	O	5′-O- Nucleoside	-OAryl	-OAryl	5′-O- Nucleoside	–19.0	–19.0	n/a	Pyridine	1
O	O	5′-O- Nucleoside	-OH	-OH	-NH2	–11.3	–0.8	20.0	D_2O, pH10	1
O	O	5′-O- Nucleoside	-OH	-NHAryl	5′-O- Nucleoside	–11.2	–6.8	17.5	Pyridine	1
O	O	5′-O- Nucleoside	-NHAryl	-NHAryl	5′-O- Nucleoside	–(7.0–7.4)	–(7.0–7.4)	n/a	Pyridine	1

* Indicates range of 31P chemical shifts for two diastereomers

A.1.3 Triphosphate derivatives of nucleosides

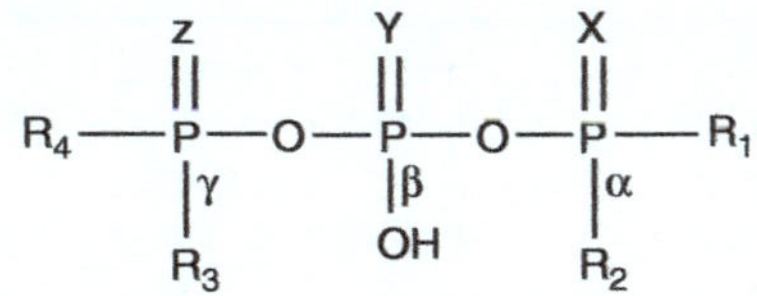

X	Y	Z	R_1	R_2	R_3	R_4	δα, ppm	δβ, ppm	δγ, ppm	Jαβ, Hz	Jβγ, Hz	Solvent	Ref
S	O	O	5'-O-Nucleoside	-OH	-OH	-OH	42.0–44.0*	–(21.1–22.0)	–(4.7–9.6)	23.0–31.8	19.3–20.5	H_2O, pH 7.0	3, 4
O	S	O	5'-O-Nucleoside	-OH	-OH	-OH	–11.4	30.0*	–6.0	6.5	27.3	H_2O, pH 7.0	1
O	O	S	5'-O-Nucleoside	-OH	-OH	-OH	–10.6	–22.0	35.0	19.6	29.0	H_2O, pH 7.0	1
O	O	O	-OH	-OH	-OH	-OH	–5.3(d)	–18.6(t)	–5.3(d)			H_2O	3
O	O	O	5'-O- Nucleoside	-OH	-OH	-OH	–10.0–11.0(d)	–(20.0–23.0)(t)	–(5.0–7.7)(d)	20.0	20.0	H_2O,pH 7.0	1, 2, 3
O	O	O	5'-O-Nucleoside	-OH	-OH	5'-O-Nucleoside	–11.0(d)	–22.6(t)	–11.0(d)	n/a	n/a	H_2O, pH7.0	3
O	O	O	5'-O-Nucleoside	-OH	-OH	-OAryl	–14.0	–24.5	–19.7	n/a	n/a	DMF:Pyridine=1:1	1
O	O	O	5'-O-Nucleoside	-Alkyl	-OH	-OH	25.0(d)*	–(23.5–24.0)*	–9.7(d)*	21.8	19.5	D_2O	3, 5, 6
O	O	O	5'-O-Nucleoside	-OH	-OH	-NHAlkyl	–11.8	–22.0	–3.6	22.5	22.5	DMSO:Pyridine=1:1	1
O	O	O	5'-O-Nucleoside	-OH	-OH	-NHAryl	–11.4	–23.2	–10.2	n/a	n/a	H_2O, pH11	1

* Indicates range of ^{31}P Chemical shifts for two diastereomers

A.1.4 Polyphosphate derivatives of Nucleosides

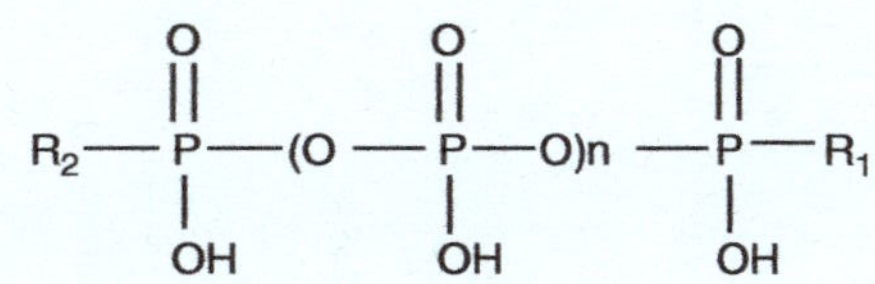

n=2, 3,..

N	R_1	R_2	$\delta\rho_1$, ppm	$\delta\rho_n$, ppm	δ n+1, ppm	$J_{1,2}$, Hz	J(n+1)n, Hz	Solvent	Ref.
2	5′-O-Nucleoside	-OH	–11.0 (d)	–22.7(d)	–10.3	19.0–20.0	19.0–20.0	H_2O	3
2, 3	5′-O-Nucleoside	5′-O-Nucleoside	–(10.8–11.2)(d)	–22.7(d)	–(10.8–11.2)	19.0–20.0	19.0–20.0	H_2O, pH 7.0	1, 3

A.1.5 Nucleoside 3′,5′-cyclopyrophosphate

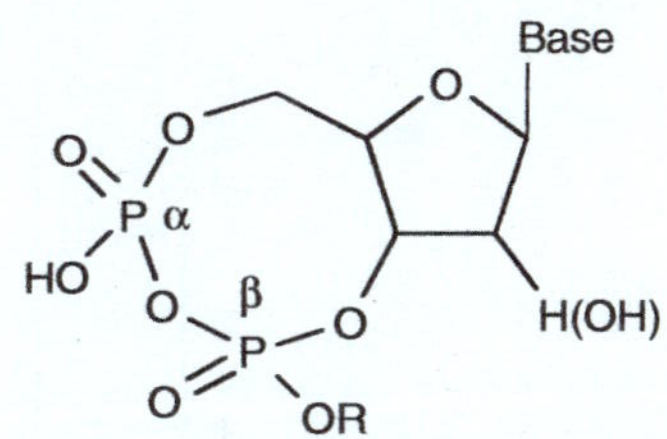

R	$\delta\rho\alpha$, ppm	$\delta\rho\beta$, ppm	$J\alpha\beta$, Hz	Solvent	Ref.
H	–13.7	–14.4	27.6	DMF	1
3′-O-Nucleoside	–13.0	–15.0	22.0	H_2O, pH 9	1

A.1.6 Nucleoside 5′-O-trimetaphosphate

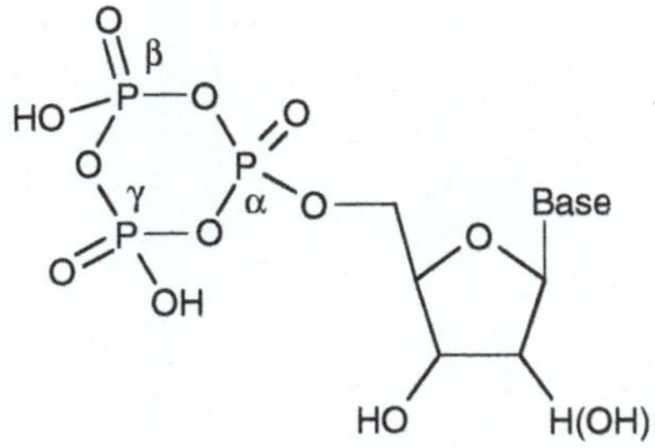

δα	δβ	δγ	Jαβ, Hz	Jβγ, Hz	Jαγ, Hz	Solvent	Ref.
–23.0	–24.0	–24.6	23.5	23.0	23.7	Pyridine	1

A.1.7 Nucleoside 3′,5′-cyclophosphate

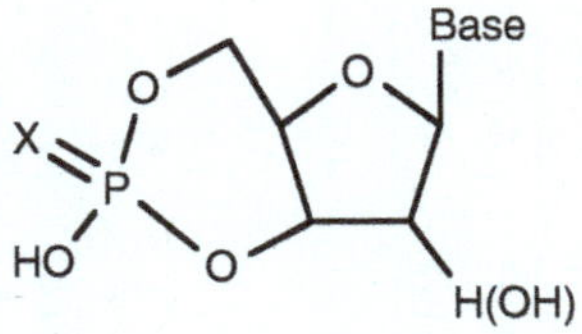

X	δ, ppm	Solvent	Ref.
S	52.0–55.0*	Pyridine	1
O	–(1.5–2.0)	D_2O, pH7	1, 7

* Indicates range of ^{31}P Chemical shifts for two diastereomers.

A.1.8 Nucleoside 2′,3′-cyclophosphate

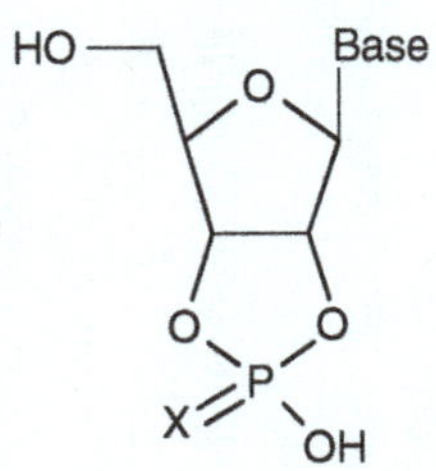

X	δ, ppm	Solvent	Ref.
S	75.0–76.0*	H_2O	1
O	17.6–20.3	DMF, Pyridine, MeOH	1, 7

* Indicates range of ^{31}P chemical shifts for two diastereomers

A.1.2 Trivalent phosphorous compounds

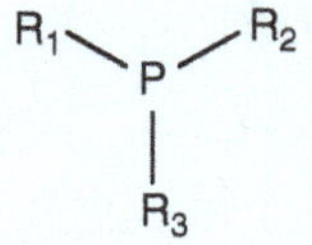

R_1	R_2	R_3	δ, ppm	Solvent	Ref.
-Cl	-Cl	-Cl	215.0–220.0	**	2
-OAlkyl	-Cl	-Cl	177.0–180.0	**	2
-OAryl	-OAryl	-Cl	153.0–157.0	**	2
-OAlkyl	-N(Alkyl)$_2$	-N(Alkyl)$_2$	131.0–138.0	**	2
-OAryl	-Cl	-Cl	176.0–184.0	**	2
-OAryl	-N(Alkyl)$_2$	-N(Alkyl)$_2$	131.0	**	2
-OAryl	-OAryl	-N(Alkyl)$_2$	141.0	**	2
3′-O-Nucleoside	-Cl	-OAlkyl	165.0–170.0	**	2
3′-or 5′- O-Nucleoside	-N(Alkyl)$_2$	-OAlkyl	149.0–150.0*	CD_3CN, CDCl3	3
3′-or 5′-Nucleoside	3′- or 5′-Nucleoside	-OAlkyl	140.0–141.0*	$CDCl_3$	1

* Indicates range of [31]P Chemical shifts for two diastereomers.

** See Solvents information in Ref. 2.

Index

E

L

M

N

O

P

T

U

V

W

X

Y

Z